THE

BOTANICAL TEXT-BOOK.

THE

BOTANICAL TEXT-BOOK,

AN

INTRODUCTION TO SCIENTIFIC BOTANY,

BOTH STRUCTURAL AND SYSTEMATIC.

FOR COLLEGES, SCHOOLS, AND PRIVATE STUDENTS.

FOURTH EDITION.

ILLUSTRATED WITH TWELVE HUNDRED ENGRAVINGS ON WOOD.

By ASA GRAY, M. D.

FISHER PROFESSOR OF NATURAL HISTORY IN HARVARD UNIVERSITY.

NEW YORK:

GEORGE P. PUTNAM & CO.

1853.

CAMBRIDGE: METCALF AND COMPANY,
PRINTERS TO THE UNIVERSITY.

PREFACE

TO THE THIRD EDITION.

THIS compendious treatise is designed to furnish classes in our schools and colleges with a suitable text-book of Structural and Physiological Botany, as well as private students with a convenient introductory manual, adapted to the present condition of the science. The favor with which the former editions have been received, while it has satisfied the author that the plan of the work is well adapted to the end in view, has made him the more desirous to improve its execution, and to render it a better exponent of the present state of Physiological Botany. To this end the structural and physiological part of the work has been again almost entirely rewritten for this Third Edition, and much enlarged. The chapter on the Elementary Structure of Plants, or Vegetable Anatomy, the sections on the Internal Structure of the Stem, on Phyllotaxis and its relations to floral structure, and on the Symmetry and Morphology of the Flower, may be particularly adverted to, as having been altogether recast and greatly extended. The want of space and time has prevented a similar extension of the systematic part of the work, especially of the

Illustrations of the Natural Orders. This portion, however amplified, could never take the place of a Flora, or System of Plants, but is designed merely to give a general idea of the distribution of the vegetable kingdom into families, &c., with a cursory notice of their structure, distribution, properties, and principal useful products. The student who desires to become acquainted, as he should, with the plants that grow spontaneously around him, will necessarily use some local Flora, such, for example, as the author's *Manual of the Botany of the Northern United States.* For particular illustrations the botanist may advantageously consult the *Genera of the Plants of the United States, illustrated by Figures and Analyses from Nature*, of which two volumes have been published.

By permission of the Secretary of the Smithsonian Institution, the figures No. 20–22, 33, 37, 105–110, 130–133, 135, 136, 159, 160, and 161–164, are copied from original sketches made for the Introduction to a *Report on the Trees of the United States*, now in preparation by the author, for that Institution.

The changes in this Fourth Edition are comparatively small; consisting of corrections and minor alterations, especially in the parts which relate to Vegetable Anatomy and Physiology, and in the addition of a short chapter on the Fecundation of Cryptogamous or Flowerless Plants.

HARVARD UNIVERSITY, CAMBRIDGE, March, 1853.

CONTENTS.

PART I.

STRUCTURAL AND PHYSIOLOGICAL BOTANY.

PART II.

SYSTEMATIC BOTANY.

THE

BOTANICAL TEXT-BOOK.

INTRODUCTION.

GENERAL SURVEY OF THE SCIENCE.

1. BOTANY is the Natural History of the Vegetable Kingdom. The vegetable kingdom consists of those beings (called plants) which derive their sustenance from the mineral kingdom, that is from the earth and air, and create the food upon which animals live. The proof of this proposition will be hereafter afforded, in the chapter upon the food and nutrition of plants. The vegetable kingdom, therefore, occupies a position between the mineral and the animal kingdoms. Comprehensively considered, Botany accordingly embraces every scientific inquiry that can be made respecting plants, — their nature, their kinds, the laws which govern them, and the part they play in the general economy of the world, — their relations both to the lifeless mineral kingdom below them, from which they draw their sustenance, and to the animal kingdom above them, endowed with higher vitality, to which in turn they render what they have thus derived.

2. There are three aspects under which the vegetable world may be contemplated, and from which the various departments of the science naturally arise. Plants may be considered either as individual beings; or in their relations to each other, as collectively constituting a systematic unity, that is, a *vegetable kingdom;* or in their relations to other parts of the creation, — to the earth, to animals, to man.

3. Under the first aspect, namely, when our attention is directed to the plant as an individual, we study its nature and structure,

the kind of life with which it is endowed, the organization through which its life is manifested ; — in other words, how the plant lives and grows, and fulfils its destined offices. This is the province of PHYSIOLOGICAL BOTANY. It comprises a knowledge, 1st, of the intimate structure of the plant, the minute machinery through which its forces operate ; — this is the special field of VEGETABLE ANATOMY ; — and, 2d, of the plant's external conformation, the forms and arrangement of the several organs of which it is composed, the laws of symmetry which fix their position, and the modifications they respectively undergo, whether in different species, under different conditions, or in a single individual during the successive stages of its development. This branch of the science is variously called ORGANOGRAPHY (the study of the organs), or MORPHOLOGY (the study of their various modifications in form, according to the office they are destined to subserve), or STRUCTURAL BOTANY ; and nearly corresponds with what is termed Comparative Anatomy in the animal kingdom. Under both these aspects, (whether we study their *interior structure*, or their *external conformation*,) the plant is viewed as a piece of machinery, adapted to effect certain ends. The study of this apparatus in action, endowed with life, and fulfilling the purposes for which it was constructed, is the province of VEGETABLE PHYSIOLOGY, strictly so called.

4. The subjects which Physiological Botany embraces, namely, Vegetable Anatomy, Organography, and Physiology, therefore, spring naturally from the study of vegetables as individuals, — from the contemplation of an isolated plant throughout the course of its existence, from germination to the flowering state, and the production of a seed like that from which the parent stock originated. These branches would equally exist, and would form a highly interesting study, (analogous to human anatomy and physiology,) even if the vegetable kingdom were restricted to a single species.

5. But the science assumes an immeasurably broader interest and more diversified attractions, when we look upon the vegetable creation as consisting, not of wearisome repetitions of one particular form, in itself however perfect or beautiful, but as composed of thousands of species, all constructed upon one general plan, indeed, but this plan modified in each according to the rank it holds, and the circumstances in which it is placed. This leads to the

second great department of the science, namely, SYSTEMATIC BOTANY, or the study of plants in their relations to one another; as forming a *vegetable kingdom*, which embraces an immense number of species, more or less like each other, and therefore capable of being grouped into *kinds* or *genera*, into *orders*, *classes*, &c.

6. Thus arises CLASSIFICATION, or the arrangement of plants in systematic order, so as to show their relationships; also SPECIAL DESCRIPTIVE BOTANY, embracing a scientific account of all known plants, designated by proper names, and distinguished by clear and exact descriptions. Necessarily connected with these departments is TERMINOLOGY or GLOSSOLOGY, which relates to the application of distinctive names or terms to the several organs of plants, and to their numberless modifications of form, &c. The accomplishment of this object renders necessary a copious vocabulary of technical terms; for the current words of ordinary language are not sufficiently numerous or precise for this purpose. New terms are therefore introduced, for accurately expressing the great variety of new ideas to which the exact comparison of plants gives rise; and thus a technical language has gradually been formed, in this as in every other science, by which the botanist is able to describe the objects of his study with a perspicuity and brevity not otherwise attainable.

7. These several departments include the whole natural history of the vegetable kingdom, considered independently. But, under a third point of view, plants may be contemplated in respect to their relations to other parts of the creation; whence arises a series of interesting inquiries, which variously connect the science of Botany with Chemistry, Geology, Physical Geography, &c. Thus, the relations of vegetables with the mineral kingdom, considered as to their influence upon the soil and the air, — as to what vegetation draws from the soil and what it imparts to it, what it takes from and what it renders to the air we breathe; and, again, the relations of the vegetable to the animal kingdom, considered as furnishing sustenance to the latter, and the mutual subservience of plants and animals in the general economy of the world, — all these inquiries belong partly to Chemistry, and partly to Vegetable Physiology; while the practical deductions from them lay the foundation of scientific Agriculture, &c. The relations of plants to the earth, considered in reference to their natural distribution

over its surface and the laws that regulate it, especially as connected with the actual distribution of those natural agents which chiefly influence vegetation, such as heat, light, water, &c., (in other words, with climate,) give rise to GEOGRAPHICAL BOTANY, a subject which connects Botany with Physical Geography. Under the same general department naturally falls the consideration of the changes which the vegetable kingdom has undergone in times anterior to the present state of things, as studied in their fossil remains, (a contribution which Botany offers to Geology,) as well as of those changes which man has effected in the natural distribution of plants, and the alterations in their properties or products which have been developed by culture.

8. Of these three great departments of the science, that of Physiological Botany, forming as it does the basis of all the rest, first demands the student's attention.

PART I.

STRUCTURAL AND PHYSIOLOGICAL BOTANY.

---

9. THE principal subjects which belong to this department of Botany may be considered in the most simple and natural order by tracing, as it were, the biography of the vegetable through the successive stages of its existence, — the development of its essential organs, *root*, *stem*, and *foliage*, the various forms they assume, the offices they severally perform, and their combined action in carrying on the processes of vegetable life and growth. Then the ultimate development of the plant in flowering and fructification may be contemplated, — the structure and office of the flower, of the fruit, the seed, and the embryo plant it contains, which, after remaining dormant for a time, is at length aroused by the influence of common physical agents, (warmth, air, and moisture conjoined,) and in germination developes into a plant like the parent; thus completing the cycle of vegetable life. A preliminary question, however, presents itself. To understand how the plant grows and forms its various parts, we must first ascertain *what plants are made of*.

CHAPTER I.

OF THE ELEMENTARY STRUCTURE OF PLANTS.

SECT. I. OF ORGANIZATION IN GENERAL.

10. **The Elementary Constitution of Plants.** In considering the materials of which vegetables are made, it is not necessary at the outset to inquire particularly into their *chemical* or *ultimate* composition, that which they have in common with the mineral world.

2*

The chemistry of vegetation may be more advantageously treated of hereafter. As they derive all the materials of their fabric from the earth and air, plants can possess no simple element which these do not supply. They may take in, to some extent, almost every element which is thus supplied. Suffice it for the present to say, however, that, of the about sixty simple substances now recognized by chemists, only four are essential to vegetation and are necessary constituents of the vegetable structure. These are *Carbon*, *Hydrogen*, *Oxygen*, and *Nitrogen.* Besides these, a few *earthy* bodies are regularly found in plants, in small and varying proportions. The most important of them are *Sulphur* and *Phosphorus*, which are thought to take an essential part in the formation of certain vegetable products, *Potassium* and *Sodium*, *Calcium* and *Magnesium*, *Silicon* and *Aluminum*, *Iron* and *Manganese*, *Chlorine*, *Iodine*, and *Bromine.* None of these elements, however, are of universal occurrence, or are actual components of any vegetable tissue; they occur either among the materials which are deposited on the walls of the cells or collected within them.

11. **Their Organic Constitution.** Although plants and animals have no peculiar elements, though the materials from which their bodies spring, and to which they return, are common earth and air, yet in them these elements are wrought into something widely different from any form of lifeless mineral matter. Under the influence of *the principle of life*, in connection with which alone such phenomena are manifested, the three or four simple constituents effect peculiar combinations, giving rise to a few *organizable elements* (27), as they may be termed; because of them the *organized* fabric of the vegetable or animal is directly built up. This fabric is in a good degree similar in all living bodies; the solid parts or *tissues* in all assuming the form of thin membranes or filaments, arranged so as to surround cavities, or form the walls of tubes, in which the fluids are contained. It is called *organized structure*, and the bodies so composed are called *organized* bodies, because such fabric consists of parts coöperating with each other as instruments or *organs* adapted to certain ends, and through which alone the living principle, under whose influence the structure itself was built up, is manifested in phenomena which the plant and animal exhibit. There is in every organic fabric a necessary connection between its conformation and the actions it is destined to perform. This is equally true of

the minute structure, or tissues, themselves, as revealed by the microscope, and of the larger *organs* which the tissues form in all plants and animals of the higher grades, such as a leaf, a petal, or a tendril, a hand, an eye, or a muscle. The term *organization* formerly referred to the possession of organs in this larger sense. It is now recognized to apply quite as well to the intimate structure of these larger parts, themselves made up of smaller organs through which the vital forces directly act.

12. **Distinctions between Minerals and Organized Beings.** In no sense can mineral bodies be said to have organs, or parts subordinate to a whole, and together making up an individual, or an organized structure in any respect like that which has just been spoken of, and is soon (in respect to plants) to be particularly described. Without attempting to contrast mineral or unorganized with organized bodies in all respects, we may briefly state that the latter are distinguished from the former, — 1. By *parentage:* plants and animals are always produced under the influence of a living body similar to themselves, or to what they will become, in whose life the offspring for a time participates; while in minerals there is no relation like that of parent and offspring, but they are formed directly, either by the aggregation of similar particles, or by the union of unlike elements combined by chemical affinity, independent of the influence, and utterly irrespective of the previous existence, of a similar thing. 2. By their *development:* plants and animals develope from a germ or rudiment, and run through a course of changes to a state of maturity; the mineral exhibits no phases in its existence answering to the states of germ, adolescence, and maturity, — has no course to run. 3. By their *mode of growth:* the former increasing by processes through which foreign materials are taken in, made to permeate their interior, and deposited *interstitially* among the particles of the previously existing substance; that is, they are nourished by food; — while the latter are not nourished, nor can they properly be said to grow in any way; if they increase at all, it is merely by *juxtaposition*, and because fresh matter happens to be deposited on their external surface. 4. By the power of *assimilation*, or the faculty that plants and animals alone possess of converting the proper foreign materials they receive into their own peculiar substance. 5. Connected with assimilation, as a part of the function of nutrition, which can in no sense be predicated of minerals, is the state of

internal activity and unceasing change in living bodies; these constantly undergoing decomposition and recomposition, particles which have served their turn being continually thrown out of the system as new ones are brought in. This is true both of plants and animals, but more fully of the latter. The mineral, on the contrary, is in a state of permanent internal repose: whatever changes it undergoes are owing to the action of some extraneous force, not to any inherent power. This holds true even in respect to the chemical combinations which occur in the mineral and in the organic kingdoms. In the former they are stable; in the latter they are less so in proportion as they are the more under the influence of the vital principle; as if in the state of unstable equilibrium, a comparatively slight force induces retrograde changes, through which they tend to reassume the permanent mineral state. 6. Consequently the *duration* of living beings is limited. They are developed, they reach maturity, they support themselves for a time, and then perish by death sooner or later. Mineral bodies have no life to lose, and contain no internal principle of destruction. Once formed, they exist until destroyed by some external power; they lie passive under the control of physical forces. As they were formed irrespective of the existence of a similar body, and have no self-determining power while they exist, so they have no power to determine the production of like bodies in turn. The organized being perishes, indeed, from inherent causes; but not until it has produced new individuals like itself, to take its place. The faculty of *reproduction* is, therefore, an essential characteristic of organized beings.

13. **Individuals.** The mass of a mineral body has no necessary limits; a piece of marble, or even a crystal of calcareous spar, may be mechanically divided into an indefinite number of parts, each one of which exhibits all the properties of the mass. It is only figuratively that we speak of a mineral individual. Plants and animals, on the contrary, exist only as *individuals;* that is, as beings composed of parts together constituting an independent whole, which can be divided only by mutilation. Each may have the faculty of *self-division*, or of making offshoots, which become new and complete individuals. It is in this faculty, indeed, comprehensively considered, that reproduction consists. The individuality is no less real in those animals of lower grades, and in plants, where successive generations of individuals remain more

or less united with the parent, instead of separating while the offspring is in the embryo or infantile state.

14. **Species.** This succession of individuals, each deriving its existence with all its peculiarities from a similar antecedent living body, and transmitting it with its peculiarities essentially unchanged from generation to generation, gives the idea of *species ;* a term which essentially belongs to organic nature, and which is applicable only by a figure of speech to inorganic things. By species we mean, abstractly, the *type* or original of each sort of plant, or animal, thus represented in time by a perennial succession of like individuals : or, concretely, the species is the sum of such individuals.

15. **Life.** All these peculiarities of organized, as contrasted with inorganic bodies, will be seen to depend upon this ; that the former are living beings or their products. The great characteristic of plants and animals is *life*, which these beings enjoy, but minerals do not. Of the essential nature of the vitality which so controls the matter it becomes connected with, and of the nature of the connection between *the living principle* and *the organized structure*, we are wholly ignorant. We know nothing of life except by the phenomena it manifests in organized structures. We have adverted only to some of the most universal of these phenomena, those which are common to every kind of organized being. But these are so essentially different from the manifestations of any recognized physical force, that we are compelled to attribute them to a special, superphysical principle. As we rise in the scale of organized structure through the different grades of the animal creation, the superadded vital manifestations become more and more striking and peculiar. But the fundamental characteristics of living beings, those which all enjoy in common, and which necessarily give rise to all the peculiarities above enumerated (12), are reducible to two ; namely, — 1. the power of *self-support*, or *assimilation*, that of nourishing themselves by involving surrounding mineral matter and converting it into their own proper substance ; by which individuals increase in bulk, or grow, and maintain their life : 2. the power of *self-division* or *reproduction*, by which they increase in numbers and perpetuate the species.*

---

* A single striking illustration may set both points in a strong light. The larva of the flesh-fly possesses such power of assimilation, that it will increase

16. **Difference between Vegetables and Animals.** The distinction between vegetables and minerals is therefore well defined. But the line of demarcation between plants and animals — the two kingdoms of organized beings subject to the same general laws — is by no means so readily drawn. Ordinarily, there can be no difficulty in distinguishing a vegetable from an animal. But the questionable cases occur on the lower confines of the two kingdoms, which descend to forms of the greatest possible simplicity of structure, and to a minuteness of size that baffles observation. Even here the uncertainty is probably attributable rather to the imperfection of our knowledge, than to any confusion of the *essential* characteristics of the two kinds of beings. It may therefore be less difficult to define them, than to apply the definitions to the actual discrimination of the lowest plants from the lowest animals. The essential characteristics of vegetables are doubtless to be sought in the position which the vegetable kingdom occupies between the mineral and the animal, and in the general office it fulfils. Plants, according to the definition given at the outset (1), are those organized beings that live directly upon the mineral kingdom, that grow at the immediate expense of the surrounding earth and air. They alone convert inorganic, or mineral, into organic matter; while animals produce none, but draw their whole sustenance from the organized matter which plants have thus elaborated. Plants, having the most intimate relations with the mineral world, are generally fixed to the earth, or other substance upon which they grow, and the mineral matter on which they feed is taken directly into their system by absorption from without, and assimilated under the influence of light in organs exposed to the air; while animals, endowed with volition and capable of receiving external impressions, have the power of selecting the food ready prepared for their nourishment, which is received into an internal reservoir or stomach.* The proper tissue

---

its own weight two hundred times in twenty-four hours; and such consequent power of reproduction, that Linnæus perhaps did not exaggerate, when he affirmed that "three flesh-flies would devour the carcass of a horse as quickly as would a lion."

* The faculty of locomotion, and even that of "making movements tending to a determinate end," cannot be denied to many plants. Doubtless the sensibility to external impressions, which some plants so strikingly manifest, does not amount to perception: on the other hand, that the lowest animals possess

of plants, moreover, is composed of three elements only; namely, Carbon, Hydrogen, and Oxygen. The tissue of animals comprises a fourth element, Nitrogen. Plants, as a necessary result of assimilating their inorganic food, decompose carbonic acid and restore its oxygen to the atmosphere. Animals in respiration continually recompose carbonic acid, at the expense of the oxygen of the atmosphere and the carbon of plants. These definitions will be verified, extended, and illustrated in the progress of this work.

SECT. II. OF THE CELLS AND CELLULAR TISSUE OF PLANTS IN GENERAL.

17. THE question recurs, What is the organized fabric or tissue of plants, and how is vegetable growth effected? The stem, leaves, and fruit appear to ordinary inspection to be formed of smaller parts, which are themselves capable of division into still smaller portions. Of what are these composed?

18. **Cellular Structure.** To obtain an answer to this question, we examine, by the aid of a microscope, thin slices or sections of any of these parts, such, for example, as the young rootlet of a seedling plant. A magnified view of such a rootlet, as in Fig. 1, presents on the cross-section the appearance of a network, the meshes of which divide the whole space into more or less regular cavities. A part of the transverse slice more highly magnified (Fig. 2) shows the structure with greater distinctness. A perpendicular slice (Fig. 3) exhibits somewhat similar meshes, showing that the cavities do not run lengthwise through the whole root without interruption. In whatever direction the sections are made, the cavities are seen to be equally circumscribed, although the outlines may vary in shape. Hence, we arrive at the conclusion, that the fabric, or tissue, consists of a multitude of separate cavities, with

---

consciousness is not certainly made out. But it is becoming more and more apparent, that the absolute distinctions between plants and animals are not to be drawn from this class of characters. Dr. Lindley's definition, that "a plant is a cellular body, possessing vitality, living by absorption through its outer surface, and *secreting starch*," is so far good, that it indirectly recognizes the essential function of vegetation, starch being one of its organic products; yet it is only one special form under which the nutritive matter created by the plant occurs. It is much as if animals were characterized by the faculty of secreting fat.

closed partitions; forming a structure not unlike a honeycomb. This is also shown by the fact, that the liquid contained in a juicy fruit, such as a grape or currant, does not escape when it is cut in two. The cavities being called CELLS, the tissue thus constructed is termed CELLULAR TISSUE. When the body is sufficiently translucent to be examined under the microscope by transmitted light, this structure may usually be discerned without making a section. We may often look directly upon a delicate rootlet (as in Fig. 1), or the petal of a flower, or a piece of thin and transparent sea-weed, and observe the closed cavities, entirely circumscribed by nearly transparent membranous walls.

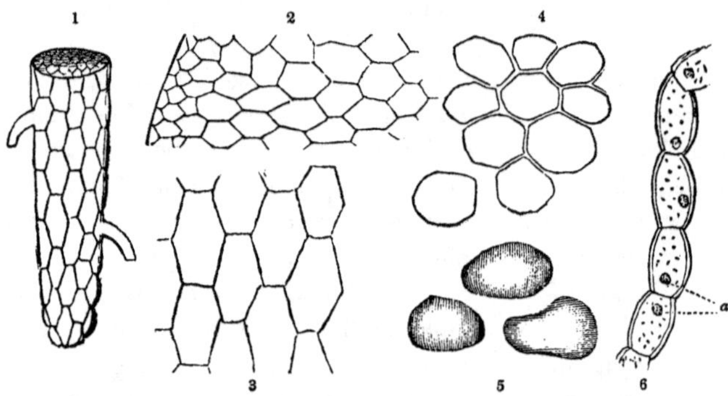

19. Does this *cellular tissue* consist of an originally homogeneous mass, filled in some way with innumerable cavities? Or is it composed of an aggregation of little bladders, or sacs, which, by their accumulation and mutual cohesion, make up the root or other organ? Several circumstances prove that the latter is the correct view. 1. The partition between two adjacent cells is often seen to be double; showing that each cavity is bounded by its own special walls. 2. There are vacant spaces often to be seen between contiguous cells, where the walls do not entirely fit together. These *intercellular spaces* are sometimes so large and numerous, that many of the cells touch each other at a few points only; as in the lower stratum of the green pulp of leaves (Fig. 7).

FIG. 1. Portion of a young root, magnified. 2. A transverse slice of the same, more magnified. 3. A smaller vertical slice, magnified.

FIG. 4. Cellular tissue from the apple, as seen in a section. 5. Some of the detached cells from the ripe fruit; magnified.

FIG. 6. Portion of a hair from the filament of the Spider Lily (Tradescantia), magnified: *a*, vestige of the nucleus or cytoblast.

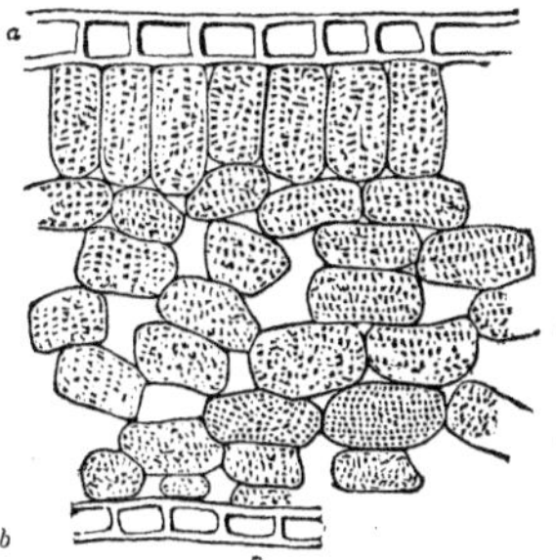

3. When a portion of any young and tender vegetable tissue, such as an Asparagus shoot, is boiled, the elementary cells separate, or may readily be separated by the aid of fine needles, and examined by the microscope. 4. In pulpy fruits, as in the Apple, the walls of the cells, which at first cohere together, spontaneously separate as the fruit ripens (Fig. 4, 5).

20. The vegetable, then, is constructed of these cells or vesicles, much as a wall is built up of bricks. When the cells are separate, or do not impress each other, they are generally rounded or spherical. By mutual compression they become polyhedral. As in a mass of spheres each one is touched by twelve others, if equally impressed in every direction, the yielding cells become twelve-sided ; and in a section, whether transverse (as in Fig. 2) or longitudinal (as in Fig. 3), the meshes consequently appear six-sided. If the organ is growing in one direction more than another, the cells commonly lengthen more or less in that direction, and thus become oblong, cylindrical, or tubular when nearly free, or prismatic when laterally impressed. If the force of extension, compression, or nutrition be greater in one direction than another, or unequal on corresponding sides, a corresponding variety of form is produced. It is not necessary to detach a cell in order to ascertain its shape ; that may usually be inferred from the outlines of their section in two or three directions. Nor have the forms precise geometrical regularity ; they merely approach more or less closely the figures to which they are likened.

21. The walls of the cells are transparent, at least in their early state, and almost always colorless. In a few cases the membrane itself is said to have a tinge of green, and in the stems of Ferns it is often brown. The various colors which the parts of the plant present, the green of the foliage, or the vivid hues of the corolla,

FIG. 7. A magnified section through the thickness of a leaf of Illicium Floridanum, showing the irregular spaces or passages between the cells, which are small in the upper layer of the green pulp, the cells of which (placed vertically) are well compacted, so as to leave only minute vacuities at their rounded ends; but they are large and copious in the rest of the leaf, where the cells are very loosely arranged. *a*, The epidermis or skin of the upper, *b*, of the lower surface of the leaf, composed of perfectly combined thick-walled cells.

do not belong to the tissues themselves, but to the matters of different colors which the cells contain (87).

22. The cells vary greatly in size, not only in different plants, but in different parts of the same plant. The largest are found in aquatics, and in such plants as the Gourd, where some of them are as much as one thirtieth of an inch in diameter. Their ordinary diameter is about $\frac{1}{400}$ or $\frac{1}{500}$ of an inch. In the common Pink, it has been computed that more than 5,000 cells are contained in the space of half a cubic line, which is equivalent to almost 3,000,000 in a cubic inch.

23. Cells are sometimes drawn out into tubes of a considerable length, as in hairs, and the fibres of cotton, which are long and attenuated cells. The hairs, or hair-like prolongations from the surface of rootlets, are good examples of the kind. Two short ones are seen in Fig. 1. In Fig. 13, 14, they are more fully illustrated.

24. Some idea may be formed respecting the rate of their production, by comparing their average size in a given case with the known amount of growth. Upon a fine day in the spring, many stems shoot up at the rate of three or four inches in twenty-four hours. When the Agave or Century-plant blooms in our conservatories, its flower-stalk often grows at the rate of a foot a day; it is even said to grow with twice that rapidity in the sultry climes to which it is indigenous. In such cases, new cells must be formed at the rate of several millions a day. The rapid growth of Mushrooms has become proverbial. A gigantic Puff-ball has been known to grow from an insignificant size to that of a large gourd during a single night; when the cells of which it is entirely composed are computed to have been developed at the rate of three or four hundred millions per hour. But this rapid increase in size is owing, in great part, to the expansion of cells already formed.

25. **Development of Cells.** The whole potentiality of the plant exists in the individual cells of which it is made up. In them its products are elaborated, and all the vital operations carried on. Growth consists in their production, multiplication, and enlargement. A knowledge of these processes is therefore requisite in almost every inquiry that arises in physiological botany. Systematic botany and zoölogy, moreover, as well as anatomy and physiology, both animal and vegetable, have advanced to the point at which investigations into the development of organs are of the

utmost consequence. The formation, propagation, and growth of cells, forming, as they do, the groundwork of anatomy and physiology, are subjects which for the last few years have tasked the powers of the ablest investigators. Such, however, are the intrinsic difficulties of these investigations, that the subject is still involved in much obscurity, especially in regard to the formation of cells; and great differences of opinion prevail upon many other essential points. At present, it is hardly possible to separate what is known or reasonably well settled from what is conjectural, unproved, or untrue; nor can the more or less conflicting views of the most experienced observers be presented and explained in such an elementary treatise as this.* In respect to cellular development in plants, however, now that Schleiden has greatly modified his views,† the highest authorities, namely, Mohl, Schleiden, and Nägeli, have arrived at substantially similar conclusions. These, in their general outlines, may be here presented.

26. We must distinguish between the original formation of cells and their multiplication. We must also distinguish between the young, vitally active cell, and the completed cell, no longer capable of multiplication or of having new cells formed within it.

27. **Formation of Cells.** Cells originate within other cells, or at least within living tissues.‡ They are formed from organizable

* The best authorities for the student to consult upon the subject are,— 1. The memoirs of Mohl in the *Linnæa*, the *Botanische Zeitung*, &c., the most important of which are translated in the *Annales des Sciences Naturelles*, the *Annals and Magazine of Natural History*, and in Taylor's *Scientific Memoirs*. 2. Those of Nägeli in the *Zeitschrift für Wissensch. Botanik*, whose principal memoir has been translated by Henfrey for the Ray Society. 3. Schleiden's *Principles of Scientific Botany*, translated into English by Dr. Lankester. 4. Lindley's *Introduction to Botany*, 4th edition. 5. Henfrey's *Outlines of Structural and Physiological Botany;* a compendious work, of which the chapters on elementary structure, and all of this author's writings upon the subject, are especially excellent.

† *Grundüze der Wissenschaftl. Botanik*, ed. 3, reproduced in the Appendix to the English translation, cited above.

‡ The Yeast-plant, developed in fermenting fluids, if that be a true vegetation, is an exception to the rule. According to Schleiden, this is a case of "the formation of cells without the influence of another cell previously existing." The material has of course been elaborated in former vegetable cells; and, according to Karsten, the ferment-cells, with which the development commences, already exist in the juice of the fruit, and pass through the filter into the solution; which makes this a case of cell-multiplication, rather than of cell-formation.

matter (11, vegetable mucilage, protoplasm, &c.) assimilated in previously existing cells, and dissolved in the water which the tissue of growing parts contains.* This organizable material always and necessarily consists of a mixture of two classes of assimilated matter, one of which is azotized, the other is not. That is, one is composed of three elements, carbon, hydrogen, and oxygen, and exists in the liquid form in the state of vegetable mucilage, dextrine, sugar, &c., or collects in a peculiar solid form in the cells, as starch, or finally constitutes the proper and permanent wall of the cell, under the name of CELLULOSE. The other is composed of nitrogen in addition to these three elements, and exists in growing parts in solution, as some state of what is called *proteine*, and is known among vegetable products in the forms of diastase, albumen, gluten, fibrine, &c. The latter makes no portion of the permanent fabric, indeed; but it plays an indispensable part in the production of cells, and always exists in young and vitally active cells, as a mucilaginous lining. A weak solution of iodine causes it to turn brown, and detaches it from the proper wall of the cell. According to Mohl, it appears earlier than the proper cell-wall, which is formed under its influence, and is, as it were, moulded upon it. Mohl has therefore given the appropriate name of *protoplasm* to this azotized mucilaginous matter.

28. *From a Nucleus or Cytoblast.* When new cells are produced by original formation within the cavity of a parent cell, the following processes appear to take place. Portions of "the protoplasm collect into a more or less perfectly spherical body, at length sharply defined, the nucleus of the cell (*cytoblast*); upon this is deposited a layer of protoplasm, which expands as a vesicle, and forms the subsequent lining of the cell; at a very early period the whole becomes inclosed by a wall of cellulose, and the cell is completed."† This plan, under a more restricted form, was propounded, and until recently maintained, by Schleiden, as the universal mode of cell-development. It is now maintained as one principal mode only, and in a form essentially agreeing with

---

* "Cells can be formed only in a fluid which contains sugar, dextrine, and proteine compounds." — *Schleiden, l. c.*

† *Schleiden, l. c.*, ed. 3; from the Appendix to the English translation. "This appears to occur especially in the embryo-sac and the embryonal vesicle."

Mohl's view.* The gelatinous nucleus of the cell often remains adherent to some part of the wall, where its vestiges frequently appear as a dark spot after the cell is full grown. Otherwise it lies free in the cavity, the forming cell-wall being disengaged from it on every side; and sooner or later it is dissolved or absorbed.

29. *Without an antecedent Nucleus.* Some observers do not admit that the nucleus plays an essential part in cell-formation, or that it exists in the first instance. Nor does it have a place in Schleiden's account of the formation of free cells in fermenting fluids, viz.: † "A globule of nitrogeneous substance originates; in this a cavity is formed, it grows, and the complete cell has a delicate coat of cellulose, without our being able to determine the epoch of its production." ‡

30. **Multiplication of Cells.** It is not by original cell-formation, however, but by the multiplication of cells already existing, that the fabric of the vegetable is built up. A cell once originated, in

---

* In *Botanische Zeitung*, Vol. 2, 1844. The abstract of Mohl's view is thus rendered, in the Appendix, *l. c.* p. 571, translated from Schleiden's 3d ed.: — "In all vitally active cells a living membrane occurs, consisting of a nitrogeneous layer; this membrane exists earlier than the cell-wall formed of cellulose, and therefore Mohl calls it the 'primordial utricle.' The new cells probably originate by the solution of the old primordial utricle, and the formation of several new ones effected through a nucleus, which always precedes the cell-formation."

† Schleiden, in *App'x to Engl. Trans.*, *l. c.* And Nägeli, as rendered in an abstract by Schleiden, *l. c.* p. 572. "1. There is a free cell-formation without a nucleus in certain of the lower Algæ, and in the formation of the spores of Lichens and Fungi. Sometimes a nucleus is subsequently produced in the completed cell. 2. Perfectly homogeneous globules of mucilage are formed, the nucleoli; around these a perfectly homogeneous nucleus, on which a proper membrane is soon to be distinguished. A homogeneous layer of mucilage is deposited around the nucleus; this gradually becomes thick, especially at one side; then granular in the interior; next it is enveloped by a membrane, and the cell with a parietal nucleus is complete." On the other hand, "Hoffmeister holds that, in the formation of a nucleus, a spherical drop of mucilaginous fluid becomes coated by a membrane, and thus individualized, without the presence of a corpuscle of denser substance (a nucleolus) inside the spherical mass of mucilage either being essential or contributing to the process." Henfrey, *Bot. Gazette*, 1. p. 128.

‡ There seems to be little real discrepancy between this view and those of Grew, Bauer, Mirbel, Unger, and Endlicher, which agree in this: that cells originate as cavities in a mucilaginous matrix, and at length acquire independent walls.

whatever manner, has the power of propagating itself by division into parts, each of which forms a new cell. The modes by which cells are thus multiplied, diverse as they appear to be in the various processes of vegetable growth, are evidently reducible to two; and even these, if they are now rightly understood, are only two modifications of one and the same process of *division*, or *merismatic multiplication.* Taking the most distinct cases for examples, we may say that, in the first mode,

31. *The cell is propagated by the division of its living contents into two, four, or sometimes a greater number of free new cells;* the wall of the original cell perishing or losing its vitality in the process. This can occur only in cells whose walls have not been thickened by internal deposition (39), and while yet lined with the vitally active layer of protoplasm * (26, 27). This mucilaginous lining becomes constricted or infolded around the middle, and the fold extends inward until it is divided, with the whole contents, into two parts (Fig. 64); at the same time, or immediately following the division, a wall of cellulose is deposited around each portion. The two new cells thus produced may at once divide again in the same way, giving rise to four cells in a parent cell (as in Fig. 65); or the division may be again and again repeated. The delicate wall of the parent cell is either absorbed or obliterated as the new ones it incloses enlarge, or it remains, for a while at least, although no longer in a living state. By this method the cells of pollen formed in the anther of all Flowering plants (110), and

* This layer, according to Mohl, is a delicate and soft membrane of protoplasm (called by him the *primordial utricle*), formed earlier than the cellulose cell-wall which is soon deposited around it. Schleiden has not been able to satisfy himself that this matter is organized into a *membrane*, or that it *precedes* the proper wall of cellulose. By terming it, without reference to these points, the mucilaginous lining, or vitally active layer of protoplasm, interposed between the proper wall of the cell and its contents (*nucleus, gelatinous mass, endochrome*, or whatever they may be called), their views are brought into agreement with each other. Those of Mr. Thwaites do not essentially differ, except in his pushing too far, as I should suppose, the inference, "that cell-membrane is quite a subordinate part of living structure; that its functions are of a purely physical character; that its principal office is to protect, locate, or isolate the matter it contains, and that any vitality it possesses is derived from the presence within it of its endochrome." *Ann. & Mag. Nat. Hist.*, Vol. 18. — The movement of the cilia on the surface of the cell-wall, seen in so many spores, surely shows that this possesses for a time a vitality of its own.

the *spores* of most Flowerless plants (101, 109), originate.* It is subservient to reproduction, as these examples show, rather than to vegetation. On the one hand, it might be ranked as a mode of original cell-formation; on the other, it passes by insensible gradations into the next mode, — where

32. *The cell is multiplied by the formation of a partition which divides its cavity into two;* the original wall remaining. In this way, a single cell gives rise to a row of connected cells, when the division takes place in one direction only, or a plane or solid mass of such cells, when it takes place in two or more directions; thus producing a tissue. It is in this way that all ordinary vegetating or growing parts are produced and increased. The division is effected, as before, by the annular constriction and infolding of the mucilaginous lining of the cell (the primordial utricle of Mohl); the circular fold meeting at the centre divides the contents into two portions, and a layer of permanent cell-membrane, which is somewhat later deposited upon each lamella of the fold, forms a complete double partition; thus converting one cell into two, and so on.†

33. Although connected in their origin, such cells may break

---

* Some spores are produced by the condensation of the whole contents of the parent cell and the acquisition of an investing cell-membrane, without any division, as in Conferva glomerata, &c., or of the undivided contents of one end of a cell, as in Vaucheria, Fig. 71.

† This mode of cell-multiplication was first shown and most ably maintained by Mohl, as the universal mode of increase in growing parts. It has been illustrated from independent observations by Henfrey, in a paper read before the British Association at Cambridge, in 1846; and has recently received new confirmation from Mitscherlich's researches upon the development of *Conferva glomerata*, the plant upon which Mohl's observations upon cell-divisions were principally made. Henfrey has given an abstract of Mitscherlich's paper in *Ann. & Mag. Nat. Hist.*, Vol. 1, new ser., 1848, p. 436. Schleiden's statement of the process, as rendered by his English translator (p. 572), is: "This fold of the primordial utricle is followed somewhat later by a fold of the cell-membrane itself, which, finally arriving at the axis of the cell, blends, and from the nature of its origin forms a complete double septum." But Mohl, Henfrey, and Mitscherlich appear to agree that the proper wall of the parent cell is not constricted, only its lining or primordial utricle; and that "the septum is certainly a new structure, a double layer of membrane formed in the fold," yet deposited, according to Mohl and Henfrey, "gradually from the circumference to the centre." "The layers of the partition are therefore continuous with the layers of thickening in the interior of the lateral walls," as Henfrey states.

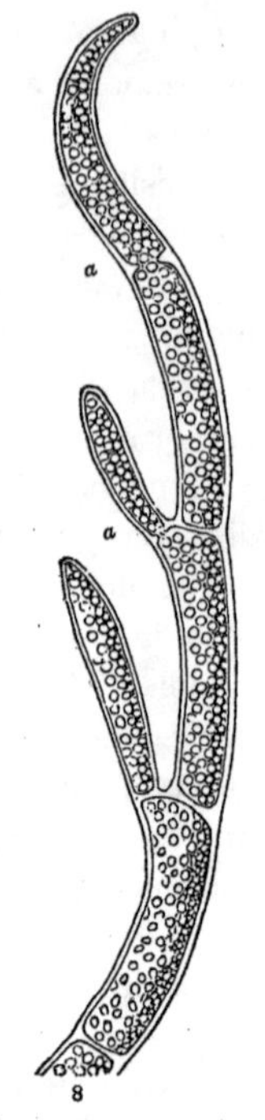

apart at an early period into separate individuals. In that case, the result is the same as in the preceding (31); especially when the cells are globular and divide first in one direction and next in the opposite direction; except that here the parent cell is, as it were, broken up into two or four, each invested with its portion of the original cell-membrane. While in the former, the old cell-wall is destroyed or remains distinct, and the new cells formed in it acquire each its own independent coating of cell-membrane. This is the more apparent where the cell is elongated and goes on to form a chain of cells, as in the green Confervas of streams and pools. Fig. 8 represents a portion of a Conferva, magnified, so as plainly to exhibit the formation of the partitions. Here the process of division goes on *pari passu* with that of

34. Gemmation or Budding; namely, with continuous growth from their free extremity, or the shooting forth of a protrusion or branch from some part of the surface of a cell, which grows onward from its apex in the same way. A cell thus prolonged into a tube is divided by a transverse partition; the upper joint, after elongating from its apex, has its cavity likewise divided into two by a transverse partition; the lowest of these remaining stationary, the upper elongates and continues the same process; which may thus go on indefinitely. Fig. 9 – 12 show modifications of this *gemmiparous* (or budding) mode of growth, as seen in some of the microscopic plants of doubtful nature which develope in fermenting infusions.

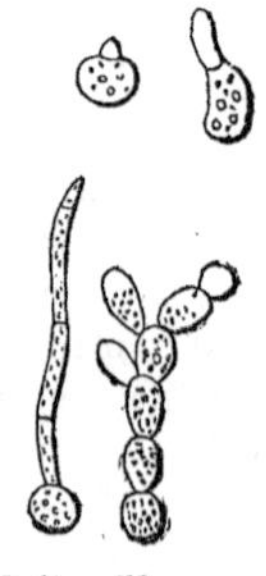

FIG. 8. Branching summit of a plantlet of Conferva glomerata, magnified; after Mohl showing, at *a*, *a*, the partitions forming by the infolding process.

FIG. 9 – 12. The minute infusory plant which developes in yeast and fluids which are in vinous fermentation. 9. The original vesicle or cell, which is forming a second by a kind of budding. 10. The same, farther advanced. 12. The plant fully developed by the successive production of new cells in this manner. 11. The same, or a similar plant, developing in a slightly different mode, nearly as in Fig. 8. All the figures are magnified.

35. Elongating and Ramifying Cells. This onward growth may take place, moreover, without the formation of partitions at all; when elongated, vegetating cells are produced, whether simple or branched. The hair-like bodies that copiously appear on the surface of young rootlets furnish examples of the kind, as is shown in Fig. 13, 14. More conspicuous examples are furnished by certain Algæ of the simplest structure, where the cell grows out into a tube of uninterrupted calibre, or branches as it grows into a series of such tubes with the cavity perfectly continuous throughout; as in Botrydium (Fig. 67 – 70), where an originally spherical cell is extended and ramified below in the fashion of a root; in Vaucheria (Fig 71), where a slender tube forks or branches sparingly; and in Bryopsis (Fig. 73), where numerous branches are very regularly produced. In these cases, the fully developed plant, with all its branches, is only one proliferous cell, extended from various points by this faculty of continuous budding growth. The mycelium or spawn of Mushrooms, and the intricate threads of Moulds (Fig. 74 – 76) are formed of very attenuated branching cells. And in Lichens, cells of the same kind are densely interwoven into a filamentous tissue (Fig. 15).

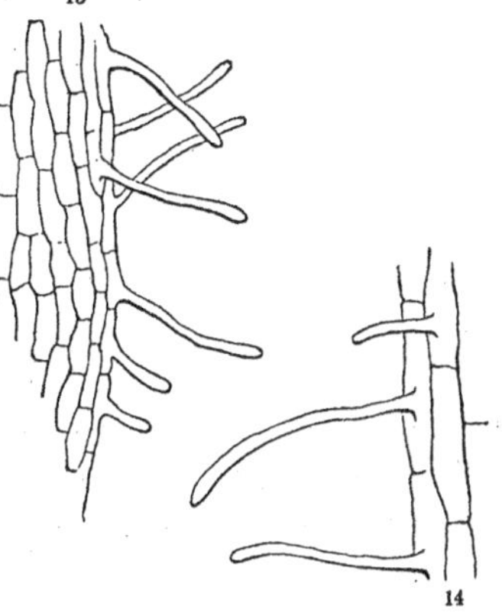

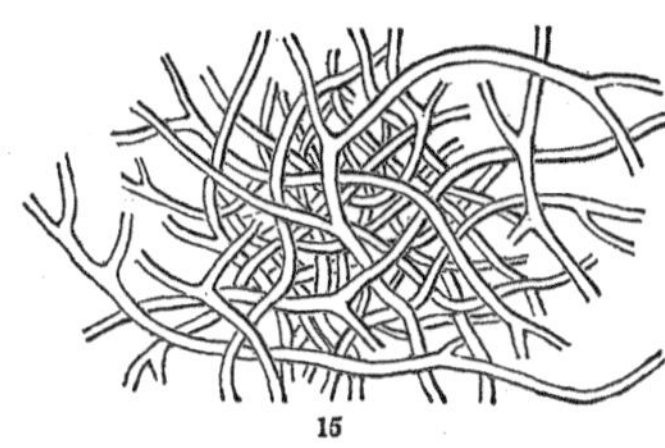

36. Circulation in young Cells. A kind of circulation or movement of rotation has been observed in numerous cells, particularly in those that form the hairs of many plants, which are well situated for observation; and it probably takes place in most cells at an early period, while yet filled with fluid. The string of bead-like

FIG 13. Magnified cellular tissue from the rootlet of a seedling Maple; some of the external cells growing out into root-hairs. 14. A few of the cells more highly magnified.

FIG. 15. Entangled, filamentous, branching cells from the fibrous tissue of the Reindeer Lichen (Cladonia rangiferina), magnified.

cells which compose the jointed hairs of the common Spider Lily (Tradescantia, Fig. 6) show this circulation well, under a magnifying power of about four hundred diameters. With this power, a network of anastomosing currents, rendered visible by the little globules they carry with them, will be seen to move between the transparent and glassy cell-membrane and the inclosed colored contents, traversing the cell in various directions, without much regularity, except that the streamlets appear to radiate from, and return to, the parietal cytoblast (28). In this instance, it is easy to see that the currents belong to the layer of mucilaginous fluid, or protoplasm, interposed between the cell-membrane and the colored aqueous contents. The same is the case, according to Mohl's thorough observations, in the tubular cells of Chara, where they may be observed with an ordinary lens; and in our Vallisneria, where a moderate magnifying power shows, in the cells of the leaves, a continuous rotation round the whole wall of the cell, the stream rising on one side and descending on the other. The current is powerful enough to carry along, not only minute granules, but small grains of chlorophyll or green coloring matter (87), which renders it abundantly visible; and sometimes, where the green granular contents cohere in a mass filling the centre of the cell, it throws this whole mass into slow revolution on its axis. In these instances, the whole layer of mucilaginous fluid takes part in the movement. The cause of this motion is wholly unknown, as also the office it subserves. We shall have occasion to refer to it in another chapter, in connection with other vegetable movements. At present, we may merely remark that it is not like a true circulation, through vessels, which is characteristic of animals.

37. **Permeability and Imbibition.** The wall of the cells, at least in their living or vitally active state, is a perfectly closed sac, destitute of openings or visible pores (although perforations sometimes appear in old or effete cells, as in those of Peat-Moss); but, like all organic membranes, it is permeable to fluids. The cell constantly contains a fluid thicker than water, and therefore tends to imbibe water by *endosmosis*,* as well as to yield by *exosmosis* * a

---

* Endosmosis and exosmosis are names given by Dutrochet (who first illustrated them in liquids) to a physical process of permeation and interchange which takes place in fluids, according to the following law, briefly stated. When two liquids of unequal density are separated by a permeable membrane, the lighter liquid or the weaker solution will flow into the denser or

portion of its liquid contents to a contiguous cell, which may be charged with contents of greater density than its own. From the nature of the process of assimilation and other operations carried on in the interior of cells, they must always contain a denser fluid than the water in which aquatic plants live, or which is presented to the roots or other parts of the surface of terrestrial plants. This, with the gaseous and other matters it holds in solution, the vegetable must constantly tend to imbibe by endosmosis. In virtue of the same law, as will hereafter be explained, not only is the crude food imbibed by the roots, but transferred from cell to cell to the place where assimilation is principally effected or growth is going on. In addition to the simpler process, animals, even of the lowest grades, have a proper circulation through vessels. There is no such circulation in plants.

38. **Growth of Cell-Membrane interstitially.** By appropriating the assimilated matter it contains or imbibes, the young cell increases rapidly in size ; its wall is extended equally on every side (unless something interferes with its expansion in particular directions), so that a larger space is surrounded. Meanwhile, instead of becom-

---

stronger, with a force proportioned to the difference in density (*endosmosis*); but at the same time, a smaller portion of the denser liquid will flow out into the weaker (*exosmosis*). Thus, if the lower end of an open tube, closed with a thin membrane, such as a piece of moistened bladder, be introduced into a vessel of pure water, and a solution of sugar in water be poured into the tube, the water from the vessel will shortly be found to pass into the tube, so that the column of liquid it contains will increase in height to an extent proportionate to the strength of the solution. At the same time, the water in the vessel will become slightly sweet; showing that a small quantity of syrup has passed through the pores of the membrane into the water without, while a much larger portion of water has entered the tube. The water will continue to enter the tube, and a small portion of syrup to leave it, until the solution is reduced to the same strength as the liquid without. If a solution of gum, salt, or any other substance, be employed instead of sugar, the same result will take place. If the same solution be employed both in the vessel and the tube, no transference or change will be observed. But if either be rendered stronger than the other, a circulation will be established, and the stronger solution will increase in quantity until the two attain the same density. If two different solutions be employed, as, for instance, sugar or gum within the tube, and potash or soda without, a circulation will in like manner take place, the preponderance being *towards* the denser fluid, and in a degree exactly proportionate to the difference in density. Instead of animal membrane, any vegetable matter with fine pores, such as a thin piece of wood, or even a porous mineral substance, may be substituted with the same result.

ing thinner as it expands, it grows thicker; although the increase of surface at this time is much greater than that of thickness. Therefore it not merely enlarges, but grows. That is, it incorporates new assimilated matter, which penetrates the membrane and is deposited in it, not as a new layer, lining and strengthening the old, but *interstitially;* so that the enlarging cell-wall is still as homogeneous and simple as before. After attaining, for the most part rapidly, a definite size, the cell ceases to enlarge, and its wall no longer incorporates new materials. Some cells remain in this condition, with walls of great tenuity, as do the parent cells in which grains of pollen or other new cells are produced (31); in which case they seldom endure, but are soon destroyed or absorbed. The assimilated matters they contained were wholly diverted to the new product to which they give rise.

39. **Thickening by Deposition.** In most cells that make part of a permanent structure, however, the membrane continues to thicken after it has ceased, or nearly ceased, to enlarge, no longer interstitially, but by a deposit on its inner surface. The nature of the contained assimilated matter is such, that, by the mere abstraction of water, it readily passes into a solid state (81). As it organizes (doubtless under the influence of the living lining of protoplasm), it solidifies on the surrounding cell-wall, which is thus strengthened by a new layer of cellulose, or by a succession of such layers. Every degree of this secondary deposition occurs, from a slight increase in the thickness of the membrane to the filling up of the greater part of the cavity of the cell. The older wood-cells of any hard wood furnish good illustrations of such solidification. Indeed, the difference between *sap-wood* and *heart-wood* of trees is principally owing to the increase of this secondary deposit, which converts the former into the latter; as may be seen by comparing, under the microscope, the tissue of the older with that of the newest rings of wood, taken from the same tree. In an ensuing chapter (on the internal structure of the stem), this is shown in a piece of oak wood. Fig. 18 represents a highly magnified cross-section of some wood-cells from the bark of a Birch, with their calibre almost obliterated in this way. It is by the same process that the tissue of the stone of the peach, cherry, and other stone-fruits acquires its extreme hardness. Indurated cells of the same kind are met with even in the pulp of some fruits, as in the gritty grains, which every one has noticed, scattered through the flesh of many pears, espe-

cially of the poorer sorts. A section of a few cells of the kind is represented in Fig. 16, with their cavity much reduced and rendered very irregular in outline by such incrustation. Similar cells are readily seen, with a moderate magnifying power, to form a part of the tissue even of such juicy fruits as the cranberry and the blueberry (Fig. 17).

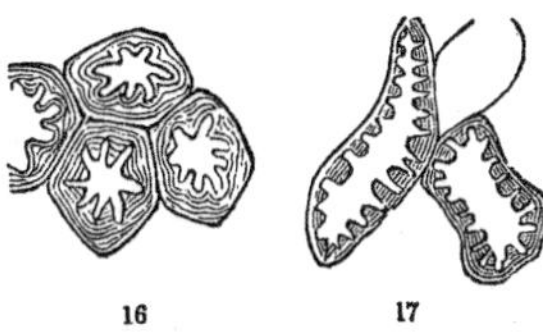

40. This deposited matter can rarely consist of pure cellulose, but may include all the various matters capable of solidification, of whatever nature, which are introduced into the cells from without, or are elaborated there. As, for example, mineral matters, small quantities of which must needs be dissolved in the water which the plant imbibes by its roots, and be deposited in the cells of the wood through which it passes, and especially in those of the leaves where it is concentrated by evaporation (**311 – 313**); also, coloring matters, such as give the different tints to heart-wood, and other special solidified products formed in the cells themselves. The cells fill up, therefore, partly by organic deposition, and partly by incrustation.

41. Even when purified as much as possible from all admixture of foreign materials, the secondary deposit is found to differ a little from cellulose, or original cell-membrane, in chemical composition. It contains a somewhat larger proportion of carbon and hydrogen, and is therefore richer in combustible matter. Forming as it does the principal part of the weight of wood (*lignum*), it has received the name of LIGNINE (also that of *Sclerogen*); but it is probably only cellulose a little modified or altered. This difference in chemical composition, however, shows why the hard woods, such as hickory and oak-wood, which abound in this lignified deposit, should be more valuable for fuel, weight for weight, than the soft woods, which have little of it (such as basswood, &c.); at least, when the latter are not charged with resinous matter.

42. The secondary deposit often forms an even and continuous increase in the thickness of the walls (as is shown in the White Oak, in the section on the Internal Structure of the Stem): but it is

FIG. 16. Magnified section of the gritty cells of the pear; the cavity almost filled with incrusting matter. 17. Similar cells from the pulp of the blueberry (Vaccinium corymbosum).

not unfrequently distinguishable, when highly magnified, into more or less defined concentric layers; as shown in Fig. 18, from the inner bark of the Birch, and in Fig. 19, in some cells of proper wood. Whether the thickening deposit is distinguishable into layers or not, it is more commonly interrupted at certain points and in a definite way, so as to give the diminished cavity very irregular outlines; as we see in Fig. 16 and Fig. 17. This occurs in wood-cells as well as in ordinary rounded cells, and is partly shown in Fig. 19. The earliest layers of thickening fail to be deposited at certain points, consequently leaving thinner spots; the succeeding layers are exactly applied to the next preceding, and leave precisely the same intervals: consequently, these unthickened spots become grooves or canals running from the cavity of the cell to the original wall, or in that direction. And it is noticeable that the pits or canals of contiguous cells usually correspond: an obvious effect or use of this adaptation is to maintain a lateral communication between contiguous cells of the kind, notwithstanding the thickening of their walls. No tissue which we have seen shows these lateral passages and their nature more clearly than the wood of the American Plane-tree, or Buttonwood (Fig. 22), which at the same time demonstrates the true character of one large class of the

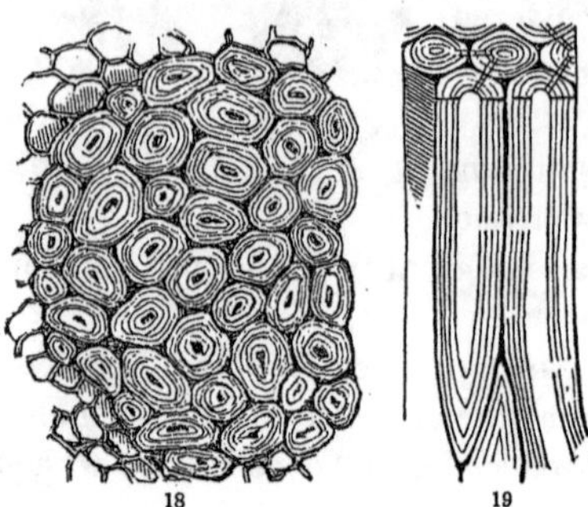

43. Markings of the Walls of Cells. These, whether in the form of bands, spiral lines, dots, or apparent pores, all arise from the unequal distribution of the secondary deposit. They are portions of the walls which are either thinner or thicker than the rest. These markings display the greatest variety of forms, many of them of surpassing elegance. The principal kinds occur with perfect uniformity in each species or family, and in definite parts of the plant; so that, in a multitude of cases, a given species or genus

FIG. 18. Highly magnified cross-section of a bit of the old liber of the bark of the Birch; the tubes nearly filled with a deposit of solid matter in concentric layers. (From Link.)

FIG. 19. Highly magnified wood-cells (seen in transverse and longitudinal section), from the root of the Date Palm; showing the internal deposit in layers, and some connecting canals or pits. (From Jussieu, after Mirbel.)

may be as certainly identified by the minute sculpture of its cells alone, as by more conspicuous external characters. They are preserved even when the tissue is fossilized, and the external form, with every outward appearance of organization, is obliterated. Through thin slices and other contrivances, the hidden structure is revealed under the microscope, and thus the true nature of our earth's earliest vegetation may be often satisfactorily made out.* The simplest cases of these markings are those of

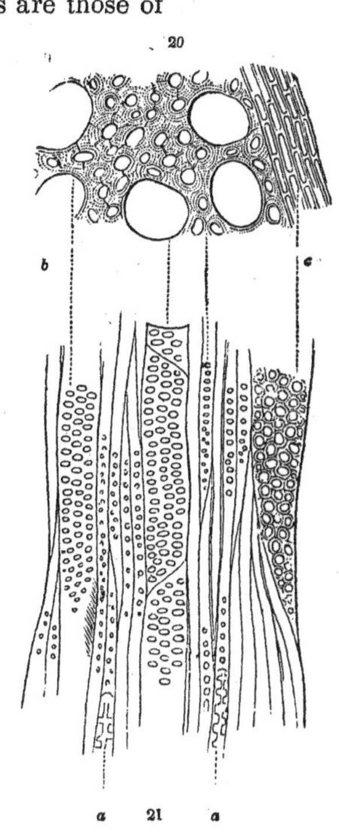

44. **Dots or Pits**, often taken for *pores*, such as those on the cells of the pith of Elder (Fig. 25), and upon those that are called *dotted ducts;* as in Fig. 39, and Fig. 21, *b*. All markings of this kind are thin spots, which, for some reason, have not partaken in the general thickening of the wall. It has been explained by supposing that a slight enlargement of the original wall takes place, which stretches the nascent lining, so as to break or fray it into slits or holes here and there. But their remarkable regularity, and the uniformity with which each successive layer is moulded on the preceding, with exactly corresponding interruptions (42), forbid our adopt-

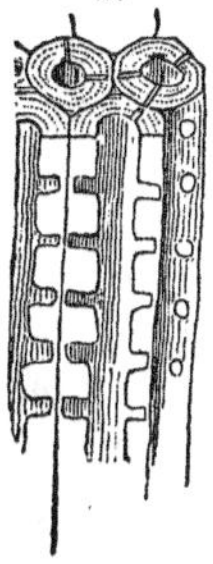

* In this way, and by taking advantage of the fact, that the secondary deposits in the cells consist in part of mineral matter, which is left behind in the

FIG. 20. Magnified cross-section of a small portion of heart-wood of the Plane-tree or Buttonwood (Platanus occidentalis). 21. A corresponding longitudinal section, parallel with the circumference. *a*, The dotted woody tissue; the lower ends of the two cells to which the letters are appended are divided lengthwise, so as to show the irregularly thickened calibre; the others are mostly entire, showing the dots: in the cross-section the secondary deposit is seen to form indistinct layers, and some of the dots to form canals of lateral communication. *b*, Dotted ducts: the middle one in the longitudinal section is obliquely jointed. *c*, Medullary ray.

FIG. 22. Portion of four cells of the woody tissue, with both transverse and longitudinal section, highly magnified, showing the canals or deep pits in the thickened walls, and their apposition in adjoining cells: on the cross-section the layers of deposit are more plainly visible.

ing this mechanical explanation. Although they are not primarily pores or real perforations of the wall, as has been thought by some, yet they often become so with age, by the breaking away of the thin primary membrane, after the cell has lost its vitality. The subjoined dissections of the wood of the American Plane-tree, already referred to, clearly show the true nature of these dots, which here abound on the proper wood-cells as well as the larger ducts. Except in their lesser size and greater depth, arising from the more extensive thickening of the tubes, they do not essentially differ from the well-known

45. **Discs**, or *large circular dots*, which mark nearly all the wood-cells of the Pine Family (Fig. 23, 24). These are thinner spaces, which consequently appear more transparent than the rest of the tube (except when filled with a film of air), when viewed by transmitted light. The discs of contiguous tubes are applied directly to each other, face to face (just as the canals or thin places of other cells thickened by secondary deposits correspond, 42), and each is a little depressed, so that a lenticular space is left between them, as between two watch-glasses put together by their circumferences. They are seldom found on the sides of the wood-cells that look towards the bark or towards the pith; while they abound in a section made in the direction of the lines of silver-grain. The dots on the wood-cells of the Plane-tree, on the contrary, are most abundant on the sides that look towards the centre and the circumference of the trunk. Although of universal occurrence in the Pine Family and the related order Cycadaceæ, these discs are not restricted to them, as was once supposed. Mr. Brown long since showed that the wood of the Winter's-bark tree was similarly marked; and our Fig. 33 represents them as they appear in the Star-Anise of Florida, which belongs to the same natural group of plants. They are said to be

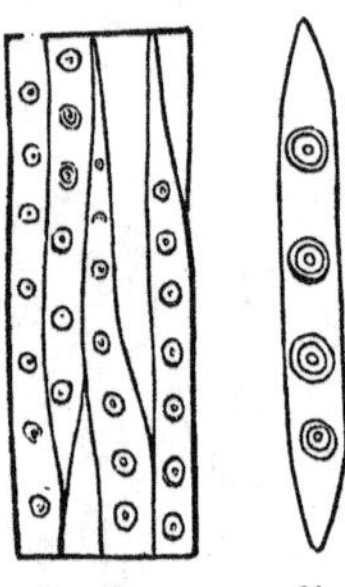

---

ashes, Prof. Bailey, of West Point, has enabled us to detect and distinguish vegetable tissues in anthracite coal. See *Silliman's Journal*, Vol. I., New Series.

FIG 23. Piece of a Pine shaving, magnified, to show the dots or discs which appear on the cells of all Coniferous wood. 24. A separate cell of the above, more strongly magnified.

"common in aromatic trees"; probably under forms scarcely if at all distinct from ordinary dotted wood-cells.

46. **Bands, Rings, or Spiral Markings.** These are, in most cases at least, definite portions of the wall more thickened than the rest; as is shown by the spiral vessel, where the secondary formation is restricted to a delicate thread, capable of being unwound (60); and particularly by the thick plate which winds around in the cells of certain Cacti, like a spiral staircase (Fig. 29). Markings of this kind (which are rarely thick and projecting as in the last example) occur as rings (Fig. 43), or fragments of rings (Fig. 44), but more frequently as spiral threads or bands (Fig. 26), sometimes as branching threads (Fig. 27); all of which, however, exhibit a spiral tendency. The elongated cells which form the hairs on the seeds of many Acanthaceous plants exhibit these markings in great variety. Two such cells from the same seed, one with a series of rings, the other with a continuous spiral thread, are represented in Fig. 31. Sometimes a band of fibres appears to ascend in the same direction: occasionally two spiral threads seem to wind in opposite directions; and sometimes branching threads inosculate and form a kind of network on the membrane, as in Fig.

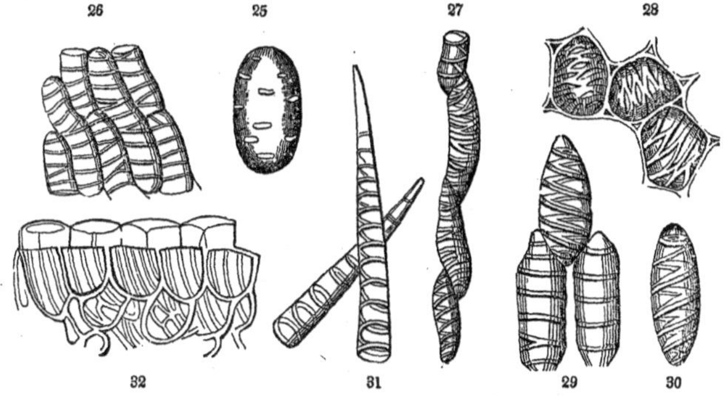

28. Often the rings or turns of the spiral thread are nearly in contact (Fig. 45); while as frequently they are separated more or

FIG. 25. Cell of the pith of Elder, marked with oblong dots.

FIG. 26. Cells of the leaf of Sphagnum, or Peat Moss, marked with a spiral fibre.

FIG. 27-30. Spirally banded cells from species of Cactus, after Schleiden.

FIG. 31. Hairs from the seed-coat of Ruellia strepens; one with a spiral band, the other with a set of rings developed on the inner surface of the tube.

FIG. 32. Tissue from the lining of the anther of Cobæa scandens; where, the delicate walls of the cells being soon obliterated, the fibrous bands with which they were marked remain.

less; as if the cell-membrane had extended after the thread was deposited, which is probably the case.

47. The delicate walls of some such cells are torn or obliterated at maturity, while the firmer bands or fibrous markings remain in the form of separate threads; as in the tissue that lines the walls of the anther (Fig. 32). In a similar manner the spirally-marked tubes that are mingled with the seeds of the Hepatic Mosses are converted into elastic spiral threads (Fig. 85). So, also, the delicate cells or hairs that invest the coat of some seeds, which contain a spirally-coiled thread, give way when moistened, or are torn asunder by the force with which the thread uncoils.

48. **Free Gelatinous Coils in Cells.** In many cases, however, the spiral deposit in the cells which form the hairs on the surface of seeds, and of some seed-like fruits, remains of a gelatinous consistence, and lies loose in the cell. When moistened, water is absorbed by endosmosis, the gelatinous contents swell, burst the cell-membrane (at the same time frequently forcing it away from its attachment), and gush out in the form of uncoiling mucilaginous threads. Examples of the kind are furnished by the seeds of Collomia and Gilia; and very striking ones by hairs or papillæ on the seed-like fruits of numerous species of Senecio and the allied genera. Those of Crocidium project a thick, mucilaginous twisted band, in place of a thread. This structure is known to be common on the surface of seeds and seed-like fruits; one purpose which it subserves will be pointed out in a future chapter.

49. Similar in their formation, probably, are the loose, spirally-coiled bodies, which are found in the antheridia of Mosses and Liverworts, in seedling Ferns, &c.; which, on account of their exhibiting a vermicular motion in water when first extricated from the cell, were denominated *Phytozoa* by Grisebach. The functions which these bodies are supposed to perform, in reproduction, will be explained hereafter.

SECT. III. OF THE KINDS OR TRANSFORMATIONS OF CELLULAR TISSUE; VIZ. WOODY TISSUE, DUCTS, ETC.

50. THE statements of the preceding section apply in general to the cells of which all plants are composed, irrespective of the manifold forms they may assume, and of some peculiar transformations they may undergo. Some of these should now be speci-

fied ; as they give rise to kinds of tissue so unlike the ordinary cellular, in outward appearance at least, that they have always been distinguished by special names. We allude particularly to what is called *Woody Tissue* or *Woody Fibre*, and *Vascular Tissue* or *Vessels*, of various forms. These have been considered as essentially different kinds of tissue, of independent origin. But it is now known that they are modifications of one common type, the cell, and are produced in the same mode as ordinary cells ; so all the statements of the foregoing section, in respect to the formation, multiplication, and growth of cells, are equally applicable to these also. Some kinds differ from ordinary cells in shape alone ; others result from their combination or confluence. This is shown in two ways: first, by noting the intermediate gradations which may be found between every particular sort; and second, by watching their development and tracing them directly from their earliest condition, as ordinary cells, to the peculiar forms they soon assume. The first of the kinds enumerated below is typical cellular tissue ; the second, through a slight change in the development, introduces the special forms.

51. **Parenchyma** is the substantive name applied to ordinary membranous cellular tissue in general, such as that which forms the pith of stems, the outer bark, &c. In the most restricted application, it belongs to such tissue when composed of angular or polyhedral cells (as in Fig. 1 – 3, 13, &c.) ; the distinctive name of *Merenchyma* having been proposed for the looser tissues (as in Fig. 7, and in the pulp of leaves and fruits generally), formed of rounded or ellipsoidal cells, that is, where they do not mutually impress each other into plane faces. But this distinction vanishes in the numberless intermediate states; and the name of *Parenchyma* is applied to both. That in which the walls barely touch each other, more or less extensively, and leave intervening spaces where the ends or sides are rounded off, is termed by Schleiden *incomplete parenchyma*. The principal forms of *complete parenchyma*, where the cells are in perfect contact on every side, and the sections are consequently several-sided, are designated by adjective terms; as the *regular*, when the cells are dodecahedral or cubical; the *elongated* or *prismatic*, when extended longitudinally; and the *tabular*, when cubical cells are much flattened ; one kind of which, called the *muriform*, because the laterally compressed cells appear in the magnified section like

courses of bricks in a wall, is seen in the silver-grain of wood (Fig. 20, *c*).

52. **Prosenchyma** is the general name to designate tissues formed of elongated cells, with pointed or conical extremities; their narrowed ends overlapping and thus filling up the intervening spaces which must otherwise exist. Every gradation may be traced between this and incomplete parenchyma. As to length, such cells vary from *fusiform*, or spindle-shaped, only three or four times longer than broad, to *tubular*, and to tubes so long and narrow that they are commonly called fibres. As to their extremities, they are often so blunt, and applied to each other with such moderate obliquity, that they are more properly said to be placed end to end than side by side; while, again, precisely similar cells, sometimes even in the same bundle, exhibit flattened ends resting directly one over the other.* Nor can we draw any fixed line of distinction from the thickness of the walls. Indeed, no one can diligently examine the tissues of two or three of the commonest plants, without perceiving that there is no essential difference between cellular and

53. **Woody Tissue.** (*Pleurenchyma* of Meyer and Lindley. *Woody Fibre* of the older authors.) Wood, which makes up so large a part of trees and shrubs, and a distinguishable portion in all Phænogamous (110) herbaceous plants, is wanting in Mosses and plants of still lower grades, such as Lichens, Sea-weeds, and Fungi. That is, in the latter there is no formation corresponding to the wood of higher plants, although many of them exhibit, at least in certain parts, prosenchymatous cells, and others drawn out into tubes or hollow fibres of greater length and tenuity than are those of ordinary wood; such, for instance, as the interlaced fibrous tissue of Lichens (Fig. 15). Nor, on the other hand, does the proper woody system of trees (except in the Pine Family) consist entirely of that form which has received the special name of woody tissue, but three or four other sorts are variously intermingled with it. Indeed, there are some trees whose wood is almost entirely composed of true parenchyma, or of large dotted (58) cells; while in stone-fruits, and many like cases, common parenchymatous

* The forming woody tissue, as seen in a germinating plant or young rootlet, consists of prismatic cells, with square ends; as these lengthen, their ends push by each other, and so become oblique and wedged together, or converted into prosenchyma.

cells acquire by incrustation a ligneous consistence and even greater density than wood (39). Nevertheless, the principal and characteristic component of wood in general is thick-walled prosenchyma. So that this takes the name of woody tissue even in the bark, leaves, &c. Fig. 21 represents some pleurenchyma along with the other usual elements of the wood, and shows the manner in which these woody tubes are spliced together, as it were, by their overlapping, pointed ends. Their diameter, in this instance, is about $\frac{1}{2000}$ of an inch. Those of our Linden or Bass-wood (a few of which are shown in Fig. 36, 37) are rather larger, but not more than $\frac{1}{1500}$ of an inch in diameter.* Their size varies in different plants almost as much as ordinary cells do, but they are usually much smaller than parenchyma, especially in herbaceous plants. Perhaps the largest are found in the Pine Family, where they are of a peculiar sort, and are often as much as $\frac{1}{300}$ or $\frac{1}{200}$ of an inch in diameter. The density or closeness of grain in wood, however, does not depend so much on the fineness of the wood-cells as upon the intermixture of other kinds of tissue, and the thickness of their walls. This is much greater in proportion to their diameter than in ordinary parenchyma, and, with their slenderness and their very compact arrangement into threads or masses which run lengthwise through the stem, conspires to give the toughness and strength which characterize those parts in which this tissue abounds. A transverse section under the microscope shows that woody tissue is composed of lengthened cells, that is, of hollow tubes and not of solid fibres (Fig. 20, 36, &c.). But as their walls thicken by the secondary or incrusting deposit to which they are especially liable (39 – 41), the calibre diminishes, and in old wood sometimes becomes nearly obliterated. This thickening usually occurs evenly in woody tissue; at least, bands or spiral lines are seldom seen in it; but small dots or pores, the nature of which has already been explained (44), are not uncommon. They are well shown in the wood of the Plane-tree (Fig. 20 – 22). Of similar character, only more conspicuously marked, is the

54. *Disc-bearing Woody Tissue* (*Glandular Woody Tissue* of Lindley), which forms the wood in the Pine Family. The nature

* Lindley states that the woody tubes of the Linden are as much as $\frac{1}{150}$ of an inch in diameter; but I find none of any thing like this size.

of the discs, or thin spots, has just been explained (45). On account of their markings and unusually large size, and because in the Pine Family they make up the wood without any admixture of ducts, these peculiar wood-cells have been thought to be rather a form of vascular tissue. But in the Star-Anise the same kind of markings is found on undoubtedly genuine woody tissue (Fig. 33). In the Yew, on the other hand, where the discs are few, delicate spiral markings appear (Fig. 34), showing a perfect transition between the proper woody and the vascular tissues; as is seen by comparing the figure with that of a spirally marked duct of Bass-wood, Fig. 36, *a*.

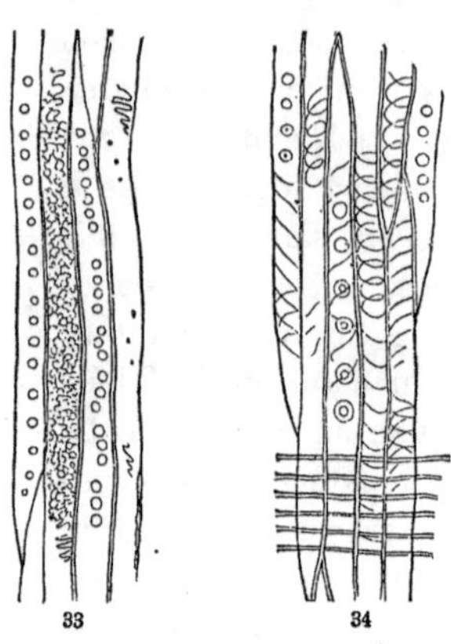

55. **Bast Tissue**, or *Woody Tissue of the Liber*. The *bast* or *bass*, fibrous inner bark, or liber, as it is variously termed, of those plants that have a true bark separable from the wood of the stem, is principally pleurenchyma, consisting of much longer, very thick-sided, and usually tougher, but more soft and flexible cells, than those of the wood itself. These properties are "probably given them that they may possess the strength, combined with flexibility, which their position near the circumference of a branch renders necessary." These especially adapt them to the useful purposes they so largely subserve for clothing and cordage. The textile fibres of flax, hemp, &c. are all derived from this woody tissue of the bark, separated from the brittle cells of the wood itself, and freed from the surrounding thin-sided parenchyma by maceration (which soon decomposes the latter) and mechanical means. *Cotton* differs from linen in many respects; it consisting of hairs, or long tubular cells, growing on the seeds, with very thin walls, that collapse so that they twist variously, which gives them a peculiar adaptation to be spun, or drawn out together by torsion into a thread. But the walls have none of the

FIG. 33. Magnified woody tissue of Illicium Floridanum (longitudinal view), marked with large dots, like the discs on the wood-cells of the Pine Family.

FIG. 34. Magnified woody tissue from the American Yew (longitudinal view), showing delicate spiral markings; some of the cells also showing the disc-like markings or dots of ordinary Coniferæ. Across the base is seen a portion of a medullary ray.

thickness and toughness which characterize the liber-cells. Fig. 35 represents one of the bast-cells of our Bass-wood or Linden, with a portion of another; while Fig. 36, 37, represent a few of the cells of the wood from the same stem, and equally magnified; showing the great difference in the length of the fibre-shaped cells. Being a soft wood, the cells of the latter have thin walls, as is seen on the cross-section of two of them at the top; while the section of one of the bast-cells shows a thick wall and very small calibre. The disproportion in length is still greater in our Leather-wood, which has a bark of extraordinary toughness, used for thongs, while the wood itself is very brittle and tender. Its capillary bast-cells measure from an eighth to a sixth of an inch in length, with an average diameter of $\frac{1}{2000}$ of an inch (so that, if the whole length of a cell, magnified as in Fig. 38, were given, the figure would be from a foot to a foot and a half in length), while those of the wood itself are only the $\frac{1}{100}$ of an inch long. Among the bast-cells are found the longest cells which occur in any tissue. Schleiden says that he has measured those which were four or five inches long. They are of great length in the Milkweed Family, and in the Dogbane, or Indian Hemp, the tough bark of which accordingly furnishes the aborigines a sort of ready-made cordage. In these families they are said by Schleiden frequently to exhibit "very delicate spiral fibres, crossing each other. In some spots their cavity becomes entirely obliterated; whilst in others they are swollen and vesicular, and contain a true milky juice." So that they are the milk-vessels in these plants; at least in part. The ribs, with the veins and veinlets, that form the fibrous framework of leaves, giving to them the requisite firmness, are chiefly of the same kind of woody tissue as those of the bark.

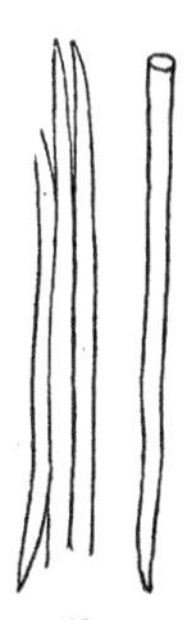

FIG. 35. Two bast-cells from the bark of the American Bass-wood, magnified.

FIG. 36. Some woody tissue from the wood of the same; *a*, upper end of a spirally-marked duct. 37. A separate cell from the wood. All magnified to the same degree as Fig. 35.

FIG. 38. Ends of some bast-cells from the bark of the Leather-wood (Dirca palustris), highly magnified.

56. The woody tissue runs lengthwise through the stem, root, or other organ (except in reticulated leaves, and there its ramifications all spread in one plane); for this reason, it is sometimes designated as *Longitudinal Tissue*, the *Vertical* or *Longitudinal System* of the stem, &c. It shares this name, however, with some other forms of tissue which accompany it, particularly in the wood. These all agree in exhibiting markings of some kind on their walls, and in being larger than woody tissue: they are all more or less tubular, or conspire to form tubes of considerable length, and hence they have all been combined, in a general way, under the name of

57. **Vascular Tissue or Vessels.** This is an unfortunate name, however, and apt to mislead, like most of those in botany that are based on loose analogies with the animal kingdom. To avoid the erroneous impressions that are so prevalent, it should be remembered that these so-called *vessels* are mere modifications of cellular tissue, and are wholly unlike the veins and arteries of animals. It is much better to call them *ducts*, a name appropriate to their nature and office, and leading to no false inferences. Their true nature is most readily shown in the largest and most conspicuous form, which often exhibits unequivocal indications of its cellular origin, namely,

58. **Dotted Ducts,** called also *Pitted* or *Vasiform Tissue*, *Bothrenchyma*, &c. (Fig. 39, 40). They have likewise been termed *Porous Cells* or *Porous Vessels;* but the round or oblong dots that characterize them are thin places where the wall has not been thickened by an internal incrusting deposit, as has already been explained (44), and not perforations, except in old cells where the primary membrane is obliterated at these points. Sometimes they are continuous tubes of considerable length (Fig. 40); but commonly, the circular lines which they exhibit at short intervals (as in Fig. 39), and the imperfect transverse partition which is often found at these points, plainly indicate their composition; showing that they are made up of a row of cells, with the intervening partitions more or less obliterated. In Fig. 21, some

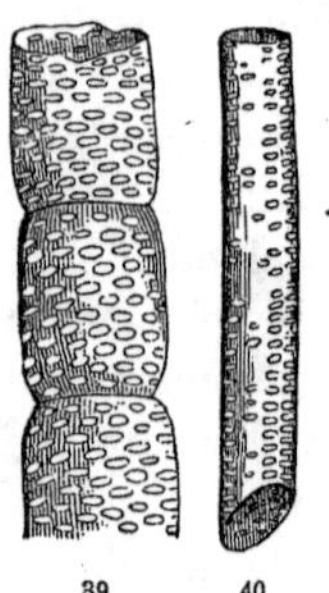

FIG. 39 Portion of a dotted duct from the Vine, evidently made up of a series of short cells.
FIG. 40. Part of a smaller dotted duct, showing no appearance of such composition.

of these ducts, shown in place among the woody tissue, are seen to have oblique partitions of the same kind. An examination in the forming state confirms this view; and, in the young stems of herbaceous plants, they may often be separated artificially into their primitive elements. These jointed ducts are occasionally branched, giving further proof that they are aggregations of confluent cells. Dotted ducts are usually met with in the wood alone, where they commonly abound. Being of greater calibre than any other cells or vessels found there, they form the pores so conspicuous to the naked eye on the cross-section of many kinds of wood, such as of Oak, Chestnut, and Mahogany, as well as the lines or channels seen on the longitudinal section. Their size, compared with that of the wood-cells in the wood of the Plane-tree, is shown both in longitudinal and transverse section, in Fig. 20, 21.

59. **Reticulated, Banded, and Scalariform Ducts** are the modifications of what is more strictly called vascular tissue (*Trachenchyma* of Morren and Lindley) which most resemble dotted ducts; and which usually take their place, or occur with them, in the stems of herbaceous and small woody plants. There is no essential difference between them: indeed, they are often distinguishable

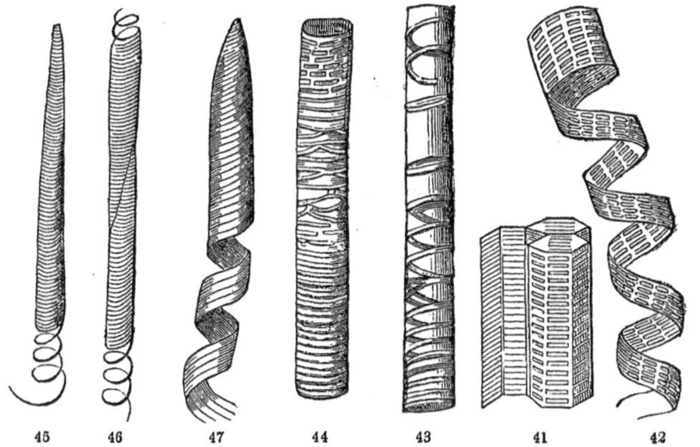

FIG. 41. Scalariform ducts of a Fern, rendered prismatic by mutual pressure.
FIG. 42. Similar duct of a Fern, torn into a spiral band.
FIG. 43. Duct from the Wild Balsam or Jewel-weed; the coils of the thread distant; a portion forming separate rings.
FIG. 44. A portion of a duct from the leafstalk of Celery; the lower part *annular;* the middle *reticulated*, and the thread at the upper part broken up into short pieces.
FIG. 45. A simple spiral vessel, torn across, with the thread uncoiling. 46. Two such vessels joined at their pointed extremities.
FIG. 47. A compound spiral vessel, partly uncoiled, from the Banana.

only by the form of the markings; and these vary so greatly in the same tissue, and even in the very same duct (Fig. 44), that it would be an endless and useless task to describe all their varieties. A continuous dotted duct with oblong spots is nearly the same as the large ducts with rather larger markings, disposed so as to form a series of regular bands, which abound in Ferns (Fig. 42). When the markings are a little longer, and the walls are rendered prismatic by mutual pressure (as in parenchyma) we have the *Scalariform Ducts* of Ferns (Fig. 41), so named because the lines (or slits as they become in old tissue) form transverse bars resembling the rounds of a ladder. In many cases, it is uncertain whether the lines or narrow bands are spots thinner than the rest of the wall, as they certainly are in dotted ducts, and probably in the scalariform vessels; or whether they are places where the secondary deposit is thickened. Probably there are *Reticulated Ducts* (those where the lines branch and run together here and there, forming a network) of both sorts; — certainly of the latter; for we occasionally meet with such markings (as in the middle of Fig. 44) on a part of the walls of true

60. **Annular and Spiral Ducts** (*Tracheæ*). The nature of their markings is explained in Paragr. 46. They are elongated cells (or ducts formed by the confluence of several cells), with their delicate membranous walls strengthened by the deposition of fibres within. Sometimes the fibre is deposited in unbroken rings (as in the middle of Fig. 43, and in Fig. 48, *d*), which forms the *Annular Duct*. More commonly it is deposited as a continuous spiral coil, producing the *Spiral Duct* or *Spiral Vessel* (Fig. 45 – 47); which is taken as the typical or pattern form of vascular tissue, because of its universal occurrence in Flowering Plants, and because of the general tendency of such definite secondary deposits to assume a spiral form. That these markings are thickened, and not thinner lines, is well shown in those remarkable cells from Cacti, already described (Fig. 29, 30), in which the fibre thickens into a band, with its edge, as it were, applied to the wall: also in those cells which have a loose spiral fibre generated within (48). Moreover, in what is called the true *Spiral Vessel* (Fig. 45 – 47), the fibre is so strong and tough, in comparison with the delicate cell-wall on which it is deposited, that it may be torn out and uncoiled when the vessel is pulled asunder, the membrane being destroyed in the operation. This is seen by breaking almost any young shoot or

leafstalk, or the leaf of an Amaryllis, and gently separating the broken ends; when the uncoiled threads appear to the naked eye like a fine cobweb. In stems furnished with pith, the spiral vessels usually occupy a circle immediately around it. They occur also in the veins of the leaves, and in all parts which are modifications of leaves. More commonly the spire is formed of a single fibre, as in Fig. 45, 46: it rarely consists of two fibres; but not uncommonly of a considerable number, forming a band, as in Fig. 47. Such *Compound Spiral Vessels* are to be found in an Asparagus shoot; and are finely seen in the stems of the Banana, from which the fibres may be extracted in large quantities. From the Musa textilis of Manilla, of the same genus as the Banana, these cobwebby fibres are procured and used in the production of the most delicate of textile fabrics. By comparing Fig. 47 with Fig. 42, we may readily perceive that the wall of those ducts in Ferns which tear into a band when pulled asunder, may have an indistinct spiral deposit, composed as it were of a band of fibres that are confluent into a lining, but are individually separated at certain points, so as to leave interstices in the form of bars, &c.

61. These ducts or vessels usually have tapering extremities (Fig. 45 – 47), as in prosenchyma. Like prosenchyma, they vary greatly in length; some of them are barely oblong or cylindrical, and are manifestly only simple cells, of the same character as the fibrous-walled cells formerly mentioned (46, Fig. 26, 29), which no one would think of calling vessels. Others, though still nothing but single cells, are more prolonged. But those which form tubes of much greater length usually consist (as their development shows), like dotted ducts, of a row of cells formed by multiplication (32 – 34, and therefore produced from one cell), with the intervening walls obliterated, so as to give a continuous calibre. This origin is well shown in some of the spiral ducts in Fig. 48 (*a*, *b*, *c*), which are conspicuously jointed, or composed of a series of cells directly confluent by their

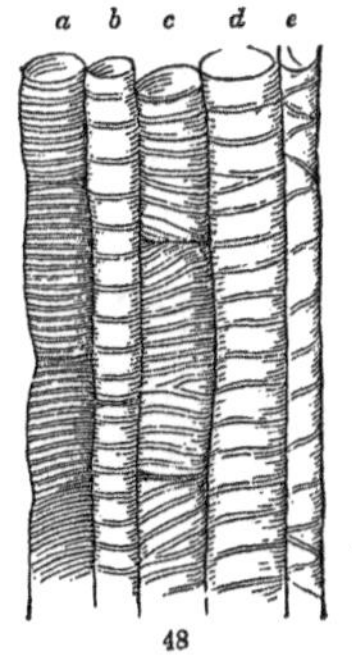

FIG. 48. A bundle of spiral ducts from the stem of Polygonum orientale, magnified: *a*, one composed of short cells and with the fibre closely coiled: the next, *b*, is composed of much longer joints and has a very loose coil: *c* is short-jointed, and the fibre of the loose coil is occasionally forked: *d* and *e* show no appearance of joints or partitions, and the turns of the spiral fibre are still more remote.

abrupt extremities. Even the pointed overlapping ends of two contiguous ducts frequently communicate at maturity, by the obliteration of the membrane between the coils of the fibre. The turns of the spiral fibre are more commonly close, as in Fig. 48, *a*; but they are often separated, even widely, as if the thread had been extended by the elongation of the cell after the spiral deposition had been formed. Fig. 48 exhibits several degrees of this, in different vessels of the very same bundle.

62. **Interlaced Fibrilliform Tissue.** This is quite as distinct from ordinary cellular tissue, and as worthy of a special name, as is any sort of the so-called vascular tissue of plants. It is the more worthy of notice from its near resemblance to ordinary forms of animal tissue. It consists of very long and much attenuated, simple or branching, fibre-like cells, or strings of cells, inextricably entangled or interwoven without order, so as to make up a loose, fibrous tissue. It is principally met with in Fungi, Moulds, &c., where the cells are extremely soft and destructible; and in Lichens (Fig. 15), where it is dry and much firmer.

63. **Laticiferous Tissue.** (*Vessels of the Latex or Milky Juice. Cinenchyma* of Morren and Lindley.) This, the only remaining kind of vegetable tissue, is of an ambiguous character. It consists of long and irregularly branching tubes or passages, lying in no definite position with respect to other tissue, and when young of such extreme tenuity (their average diameter being less than the fourteen-hundredth of an inch) and transparency that they are hardly visible, even under powerful microscopes, except by particular manipulation. But their older trunks are much larger than this, when gorged with the milky or other special juices which it is their office to contain, and when their sides are thickened by the deposition of such matters. Another peculiarity is, that they *anastomose* or inosculate, forming a sort of network by the union of their branches, so that there is a free

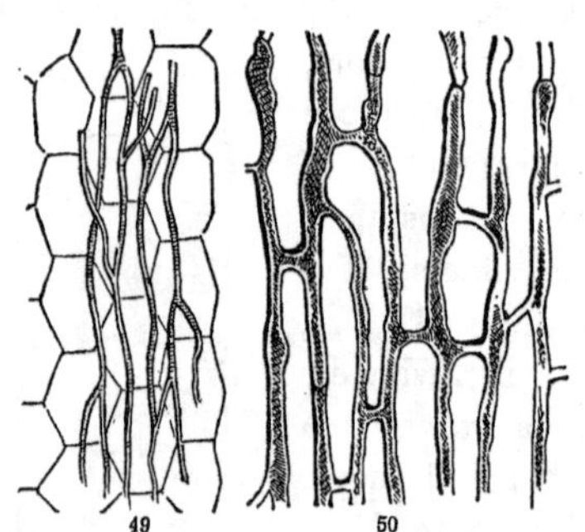

FIG. 49. Vessels of the latex, ramifying among cellular tissue, in the Dandelion; and 50, older and larger vessels from the same plant; all highly magnified.

communication throughout the whole system. In this respect, as well probably as in the mode of their formation, they resemble the veins of animals. But their branches do not proceed from larger trunks, and in turn divide into smaller branchlets. They merely fork and inosculate here and there, the branches being commonly as large as the trunk before division. The articulations which they often present (as in the upper part of Fig. 50) would seem to prove that they are formed by the confluence of cylindrical cells. It appears altogether most probable, however, that they are not composed of cells at all; but are, at first, mere passages in the intercellular spaces, which in time acquire a proper membrane by deposition from the contained fluid. The regular circulation which Schultz, the discoverer of this tissue, so elaborately described, has been shown to have no real existence. There is merely a mechanical flow from a part subject to pressure, or towards a place where the latex is escaping, as from a wound. These vessels are found in the bark, especially in the liber, in the leafstalks, and in the leaves. They are most numerous or conspicuous in those plants in which the fluid they contain becomes white or colored, that is, in those which have a milky juice.

64. All the different kinds of tissue that enter into the composition of the plant have now been described, and referred to the cell as their original. Every plant or each organ consists at first of one or more cells of proper cellular tissue; each doubtless commencing with a single *specialized* cell. In many of the simpler vegetables, the cells multiply in this primitive form solely; and the fully developed plant consists of *parenchyma* alone. But in all plants of the higher grades, some of them early assume the forms, or undergo the transformations, by which they give rise to woody tissue, ducts, or vessels. All these various sorts of modified cells lie vertically in, or conspire to form bundles or cords that run lengthwise through, the stem or other organ they occur in; so that they may be collectively called the *Vertical System* or *Longitudinal System* (56). They accompany each other, and together make up the woody parts, as in the wood proper, in the liber or inner bark, and in the fibrous framework of the leaves. Although the various kinds run into each other through every manner of intermediate forms (as in the wood of the Yew, for instance, 54), the whole, taken together, compose tissues which are almost always manifestly different from the parenchyma in which they are imbedded. It

5 *

is convenient, therefore, to give to these the collective name of *Fibro-vascular Tissues*, or the FIBRO-VASCULAR SYSTEM, as distinguished from the *Horizontal*, PARENCHYMATOUS, or common CELLULAR SYSTEM of the plant.

65. **Intercellular System.** The only exception, if such it be, to the statement that all the vegetable tissues are formed of cells, is that of the so-called vessels of the latex, which, according to the view now best supported (63), do not so originate, but are a secondary formation, resulting from the transudation of peculiar assimilated matters into the interspaces between the cells; and are therefore rather to be classed with other receptacles, canals, or intervals that are found among or between the cells. Some of these are due to imperfect contact or cohesion, and are in some sort accidental, or at least are irregular and indefinite: such are the INTERCELLULAR SPACES or PASSAGES, left when the angles in parenchyma do not accurately fit throughout. Such are the larger and irregular winding passages in the looser tissues called merenchyma (51), as in the lower stratum of the leaf (Fig. 7), or those formed by the lobed or branching shape of the cells themselves, so disposed as to join each other only by their extremities, as is seen in many water-plants. These spaces are soon filled with air. There are besides, in the stems and foliage of aquatic and marsh plants, an abundance of much larger AIR-CELLS or AIR-PASSAGES, usually of many times greater diameter than the cells of the tissue, and produced by their particular arrangement. These are as elaborately built up as any proper organ can be, are constructed upon a uniform plan in each species, and are evidently essential to its existence, such plants requiring a full supply of air in their interior. Other air-spaces or empty intervals, apparently less essential to the life of the plant, arise from the destruction of a part of the parenchyma, either by absorption, or by distention, from the more rapid enlargement of the outer part. In this way, the stem or the pith of many plants becomes hollow.

66. **Receptacles of Special Secretions.** These arise from the exudation of the proper juices of the cells into the intercellular passages, which are distended by the accumulation; or from the obliteration of contiguous cells, so as to form cavities of considerable size. Such are the turpentine canals of the Pines, &c.; the oil-cells of the fruit of the Umbelliferæ, and in the rind of the orange and lemon; the latex-canals in Sumach, &c.

67. **Internal Glands,** such as those which form the translucent dots in the leaves of the Orange and Myrtle, are compact little clusters of cells filled with essential oil.

68. **Epidermal System.** In most plants, except of the lowest grades, the superficial layer or layers of cells are different from those they envelope. Also certain appendages grow from the surface, which may be briefly noticed here.

69. The **Epidermis,** or skin of the plant, is formed of one or more layers of empty cells, with thick walls, cohering so as to form a firm and close membrane, which may be torn off from the subjacent tissue. It covers all parts of the plant that are directly exposed to the air, except the stigma. Its structure and office will be more particularly described, (and the nature of what has been specially termed the *Cuticle* explained,) in the chapter on the Leaves.

70. **Stomates** (Stomata), or *Breathing-pores*, are orifices connected with a peculiar structure in the epidermis of leaves and other green parts: their structure and office will likewise be described in the chapter on the Leaves, to which organ they more particularly belong.

71. **Hairs** are exterior prolongations of cells of the epidermis, consisting either of single elongated cells, or of several cells placed end to end, or of various combinations of such cells. They are simple or branched, single or clustered (stellate, &c.), and exhibit the greatest variety of forms. They are called *Glandular Hairs*, or *Stalked Glands*, when the upper cell or cluster of cells elaborates peculiar (usually odorous) products, such as the fragrant volatile oil of the Sweet Brier.

72. **Glands.** This name is applied to any secreting apparatus, and also to superficial appendages of diverse kinds.

73. **Bristles** (*Setæ*) are rigid, thick-walled hairs, usually of a single cell. But the name is likewise given to any setiform body, of whatever nature.

74. **Prickles** are larger and indurated, sharp-pointed processes of the epidermis or bark; such as those of the Rose and Blackberry.

75. **Stings,** or *Stinging Hairs*, such as those of the Nettle, generally consist of a rigid and pointed cell, terminating in an expanded, globular base, which secretes an irritating fluid.

76. **Scurf,** or *Lepidote, Scale-like Hairs*, are flattened, star-like clusters of cells, united more or less into a flat scale, which is fixed

by its centre to the epidermis. They are well shown in the Oleaster, Shepherdia, and most silvery leaves like theirs. Our species of Vesicaria exhibit beautiful gradations between these and stellate hairs.

SECT. IV. OF THE CONTENTS OF THE TISSUES.

77. THESE comprise all the products of plants, and the materials they take in from which these products are elaborated. To treat of them fully would anticipate the topics which belong to the chapter on Nutrition. Some of the contents of cells, however, have already been mentioned, in the account of their production and growth (27 – 39): others require a brief notice here, especially two solid products which are of nearly universal occurrence and great importance in the vegetable economy, namely, *Chlorophyll* and *Starch;* and a third, which, however constant, may be regarded as a kind of accidental deposit, namely, *Raphides* or *Crystals.*

78. The same cells contain liquids, solids, and air, at different ages. Growing and vitally active cells are filled with liquid (at least while vital operations are carried on), namely, with water charged more or less with nutritive assimilated matters, the prepared materials of growth (11, 27). Any gaseous matter they may contain at this period is, for the most part, held in solution. Completed cells may still be filled with liquid, or with air, or with solid matter only. The liquid contents of the vegetable tissues, of whatever nature or complexity, are often spoken of under the common name of

79. **Sap.** In employing this name we must distinguish, first, CRUDE SAP; the liquid which is imbibed by the roots and carried upwards through the stem. This is water, impregnated with certain gaseous matters derived from the air, and with a minute portion of earthy matter dissolved from the soil. It is therefore inorganic (12). But as it enters the roots and traverses the cells in its ascent, it mingles and necessarily becomes impregnated with the liquid or soluble assimilated matters which these contain (37). On reaching the leaves, the inorganic materials are transformed, under the influence of light, into organizable or assimilated matter; and the liquid, thus charged with the ready prepared materials of growth, is now ELABORATED SAP. The two classes of

nutritive matter thus produced, and which all forming and vitally active cells necessarily contain, namely the ternary (of which *sugar* and *dextrine* are representatives), and the quaternary (*proteine*, *protoplasm*, &c.), have already been mentioned (27).

80. **Proper Juices, Caoutchouc, Essential Oils, Turpentines, &c.** Of the *peculiar products* of plants, which occur under an infinite variety of forms in different species, it is only needful to say here, that they doubtless arise from one or the other of the two classes of assimilated matter just mentioned, by chemical transformations which throw them out of the ranks of nutritive bodies. They seem to be turned to no account in vegetable growth; they undergo changes on exposure to the air, by which they become resins, gums, wax, &c.; they incline to extravasate into intercellular spaces or into cavities of dead or effete tissues, or to be directly excreted from the surface. So that we may regard them all, perhaps, as of the nature of excretions, even where they are stored up in the interior of the plant. For we must remember that the vegetable has no organs or apparatus for eliminating and casting out excreted matters, except to a very limited extent by a few superficial glands, which are found in some plants and in some organs only. *Caoutchouc* exists in the form of minute globules, diffused as an emulsion in the milky juice of plants, most abundantly in Urticaceæ, Euphorbiaceæ, and Apocynaceæ. *Gutta percha* is a similar product of the milky juice of a Sapotaceous plant.

81. **Starch** (*Farina*, *Fecula*) is one of the most important and universal of the contents of cells, in which it is often accumulated in great quantity, so as to fill them completely (Fig. 52), as in farinaceous roots, seeds, &c. It occurs in the parenchyma of almost every part of the plant, excepting the epidermis: but while chlorophyll is nearly restricted to the superficial parts, directly exposed to the light, starch is most abundant

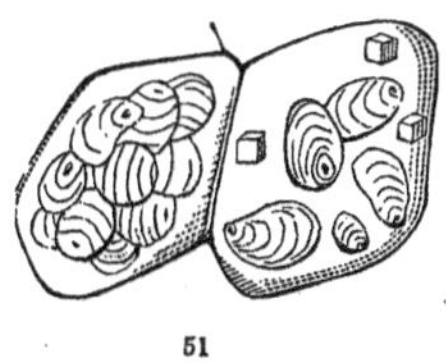

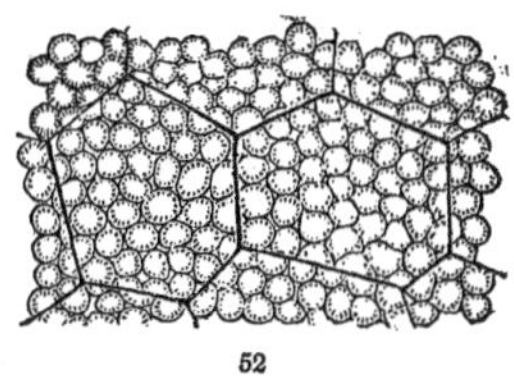

FIG. 51. Two cells of a potato, with some contained starch-grains, highly magnified; one of the cells contains a few cubical crystals also.

FIG. 52. A minute portion of Indian meal, strongly magnified; the cells absolutely filled with grains of starch.

in internal or subterranean parts, most concealed from the light, as in roots and tubers, the pith of stems, and seeds. Starch consists of oval or rounded grains, usually somewhat irregular in outline, and sometimes becoming polyhedral by mutual pressure, as in rice. The size of the grains varies extremely in different plants, and even in the same cell; as in the potato, where the larger grains measure from $\frac{1}{300}$ to $\frac{1}{500}$ of an inch in their larger diameter, but the smallest only $\frac{1}{4400}$ of an inch. In wheat-flour the larger grains are $\frac{1}{800}$ to $\frac{1}{900}$ of an inch in diameter. And the largest starch-grains known are $\frac{1}{240}$ of an inch long. Indeed, from their manner of growth, we might expect that their bulk would be somewhat indefinite. The mode of their formation is indicated by the peculiar markings, by which starch-grains may almost always be recognized; namely, by the dot or darker point which is seen commonly at one end of the grain, and the fine concentric lines drawn around it, which present the appearance of a succession of irregular circles over the whole surface, in whatever direction the grain is turned. These appearances are best seen in starch from the potato, one of the most characteristic forms and easiest to be examined, under a magnifying power of from 250 to 500 diameters (Fig. 51). The chemical composition of starch is exactly the same as that of cellulose (27); and the grains are solid throughout, but their interior usually softer or more gelatinous. The lines, therefore, it is evident, mark the concentric layers, or hollowed scales, of different density, which are successively deposited on an original nucleus. The dot (or hilum, as it has been called) that indicates the position of the nucleus, becomes a concavity, from its not receiving a part of the successive deposits, which are greatest on the opposite side, or very eccentric. The grains lie loose in the cell, and are probably formed so; although it is thought by some that the nucleus or hilum was in contact with the cell-wall, so that the increase by deposition must necessarily have taken place on the other sides. On the whole, there is reason to conclude that starch-grains are formed on nuclei or cytoblasts, that is, on minute solidified portions of protoplasm, like those from which cells primarily originate, by the deposition of layer over layer of ternary assimilated matter (dextrine, &c.), essentially like that which constitutes the secondary deposit that thickens the cell-membrane (39). Their origin, therefore, would be closely analogous to that of cells formed directly from a cytoblast in the manner propounded by Schleiden; only that the deposit in the case of starch is *exogenous*, by layer over layer upon a solid nucleus; while in the cell it is *endogenous*, or by layer within layer, lining the walls. In both, the solidified matter is insoluble in cold water; but in starch it dissolves (or rather swells up into a jelly) and is diffused in boiling water. The deposit on the walls of the cell is of various degrees of density, and sometimes exhibits the chemical peculiarity of starch. Though usually permanent, probably it is sometimes redissolved, to be appropriated elsewhere. But starch is a temporary formation, for future use; in which respect it may be compared with the fat of animals. When required for nutrition, the grains are restored to a liquid state in the plant, at the natural temperature; that is, they are reconverted into *Dextrine*,—a modification of the same substance which is soluble in cold water,—and this passes, in part, at least, into *Sugar*, which is still more soluble; and thus a syrup is formed, which the sap dilutes and conveys to the adjacent parts wherever the process of growth is going on. Physiologically considered, therefore, starch is unappropriated cellulose, stored up in a particular form, as the ready-prepared material of new tissues: while dextrine and sugar are forms in which the same unazotized assimilated matters are prepared for the immediate purposes of nutrition. The part which these substances play in the vegetable economy will be more fully explained elsewhere.

82. A distinguishing character of starch is that it is turned blue or deep violet by iodine, even in the most dilute solution. Starch-grains are usually simple and separate; but occasionally two or more young grains join, and are enwrapped by new layers into one. In some plants the grains regularly cohere in united clusters. Compound grains of the kind are seen in West Indian Arrow-root, the corms of Colchicum, Arum,* &c. The starch-grains are nearly uniform in the same plant or organ, and of very different appearance in different plants: so that the smallest quantity of starch from the potato, wheat, rice, maize, &c. may at once be distinguished under the microscope.

83. **Vegetable Jelly** (*Bassorin*, *Salep*, *Pectine*, *Vegetable Mucilage*

---

* The rootstocks of Brasenia and Nymphæa exhibit oblong or club-shaped compound starch-grains of great size, very much like those from Arum, represented by Schleiden, on page 17, Engl. translation.

in part) has the chemical composition and nearly the properties of starch after it has been diffused in hot water. It is not only one of the contents of cells, as in the tubers of Orchises, in many fruits, &c., and largely in those of Algæ, but it also forms in great part the cell-wall of Algæ, as in the Carragheen Moss (Chondrus crispus), from which vegetable jelly is obtained for culinary purposes. When dry, it is horny or cartilaginous; when moist, it swells up, becomes gelatinous, and is capable of being diffused perfectly through cold water. It passes by various modifications, on the one hand into cellulose, and on the other into starch and dextrine. We have it as an excretion in Gum Tragacanth. True gums, such as Gum Arabic, &c., are altered states of the same substance, or of dextrine, and are likewise formed only as excretions.

84. **Sugar** (of which there are two distinct kinds, *Cane* and *Grape Sugar*) is the most soluble of the many forms of ternary organizable matter, as already stated. Though sometimes crystallized as an excretion in the nectaries of flowers, yet in the plant it exists only in solution. It abounds in growing parts, in many stems just before flowering, as those of the Sugar-cane, Maize, Maple, &c. and in pulpy fruits.

85. **Fixed Oils** belong to the class of ternary assimilated products, but they contain little or no oxygen. The fatty oils take the place of starch (from which they are probably formed) in the seeds of many plants (as in flax-seed, walnuts, &c.), and of sugar in some fruits, such as the olive. They also exist in the herbage, and in some smaller proportion in the cells, perhaps, of almost all plants.

86. **Wax** is a product of nearly the same nature as the fixed oils (only it is solid at the ordinary temperature), which is extensively found in plants as an excretion, particularly on the surface of leaves and fruits, forming the bloom or *glaucous* surface which repels water, and so prevents such surfaces from being wetted. It forms a thick coating on some fruits, as the bayberry. As bees convert sugar into wax, and as the sugar-cane yields a kind of wax which "sometimes passes into sugar," we may infer that wax in the vegetable is formed of sugar or its kindred products. Wax also exists as one of the contents of cells, of leaves especially; where a substance allied to it in composition abounds, namely,

87. **Chlorophyll**, the substance which gives the universal green color to the leaves and herbage. It is formed principally in parts exposed to the light, such as the green bark, and especially the

leaves; not, however, in the external layer of cells, or epidermis (69), but in the parenchyma, especially in the superficial strata. It consists of minute soft granules, of no particular form, either separate or in clusters, forming grains of considerable size, which lie free in the cells (Fig. 53, 180), or loosely adhere to their sides. In some Confervæ they collect in the form of spiral lines or bands (as in Fig. 81, the lower part). They often adhere to the surface of starch-grains. Indeed, Mr. Henfrey plausibly considers chlorophyll to arise from altered starch (with the evolution of oxygen); which is the more likely, as it is said to appear in the cells later than starch.* It belongs to the class of waxy bodies; and is soluble in alcohol or ether, but not in water. Chlorophyll undergoes certain changes, in autumn foliage especially, by which it turns to red or yellow. CHROMULE is a name applied to coloring matters not green, and mostly in a liquid form, as in the cells of petals, giving to them their peculiar tints. These coloring matters are probably a mixture of very various products.

88. **Alkaloids** (such as *Morphine*, *Strychnine*, and *Quinine*) are quaternary products of plants, principally formed in the cells or interspaces of the bark. Unlike the proteine compounds (27, 79, gluten, fibrine, &c.), they appear to bear no part in vegetation, but to be completed results of vegetation, and of excretory nature. In these substances reside the most energetic properties of the vegetable, considered as to its action on the animal economy, the most powerful medicines, and the most virulent poisons. That they are of the nature of excretions may be inferred from the fact, that a plant may be poisoned by its own products.

89. **Tannin or Tannic Acid**, which most abounds in older bark, is probably a product of the oxidation or commencing decomposition of the tissues. So, also, *Humus*, *Humic Acid*, *Ulmine*, *Ulmic Acid*, and the numerous related substances distinguished by the chemists, are products of further decomposition of vegetable tissue, and not products of vegetation.

---

* In that case, the nitrogen obtained in Mulder's incomplete analysis (which gave C^{18}, H^{18}, N^2, O^8, with some nitrogeneous matter not determined) must belong to the mucous matter, or protoplasm, which invests the green granules. According to M. Verdeil (in *Comptes Rendus*, Dec. 22, 1851), the green grains consist of a mixture of a colorless, fatty matter, and a coloring matter analogous to the red coloring matter of the blood in composition, and like it containing a considerable proportion of iron!

90. **Vegetable Acids.** *Tartaric*, *Citric*, and *Malic* acids are the principal kinds, which occur in leaves and those succulent stems which have a sour juice, and in all acidulated fruits. They are ternary products, with an excess of oxygen. *Oxalic Acid*, which is an almost universal vegetable product, is a binary body, differing from carbonic acid in ultimate composition only in having a small proportion more of oxygen. (*Hydrocyanic* or *Prussic Acid* is one of the special products peculiar to certain plants, and of very different composition, containing a large proportion of nitrogen.) These vegetable acids do not appear to play any leading part in vegetation. They seldom exist in a free state, but are combined with the alkaloids, and with the inorganic or earthy alkalies (Potash, Soda, Lime, and Magnesia) which are introduced into plants from the soil with the water imbibed by the roots. The more soluble salts thus produced are found dissolved in plants ; the more insoluble are frequently deposited in the cells in the form of

91. **Crystals or Raphides.** These exist in more or less abundance in almost every plant, especially in the cells of the bark and leaves, as well as in the wood and pith of herbaceous plants. Far the most common, and the principal kind formed with a vegetable acid, are those of oxalate of lime. In an old stem of the Old-man Cactus (*Cereus senilis*), the enormous quantity of 80 per cent. of the solid matter left after the water was driven off was found to consist of these crystals. In the thin inner layers of the bark of the Locust, for example, each cell contains a single crystal, as is seen in Fig. 57. And Professor Bailey, who has devoted particular attention to this subject, computed that, in a square inch of a piece of Locust-bark, no thicker than ordinary writing-paper, there are more than a million and a half of these crystals. There is frequently a group of separate crystals in the same cell ; or a conglomerate cluster, as in Fig. 58. In the leaves of the Fig, and many other Urticaceous plants, a globular crystalline mass is suspended in the cell by a kind of stalk. Oxalate of lime crystallizes in octahedra (as in Fig. 56, the crystal in the lower right-hand cell), and in right-angled four-sided prisms (as in Fig. 59, 60), with variously modified terminations. The crystals are frequently acicular, or needle-shaped, either scattered or packed in bundles of from twenty to some hundreds (as in Fig. 53 – 55). It is to this form that the name of *Raphides* (which is the Greek

word for needles) was originally applied, and to which it properly belongs; although it has been indiscriminately extended to all kinds of crystals which occur in the cells of plants. In the common Arum or Indian Turnip, as well as in the Calla Æthiopica and other plants of that family, the crystal-bearing cells (Fig. 54) may readily be detached from the rest of the tissue; and when moistened and distended by endosmosis, they forcibly discharge their contents, in a curious manner, from an orifice at each end, as is shown in Fig. 55. These acicular crystals are generally thought

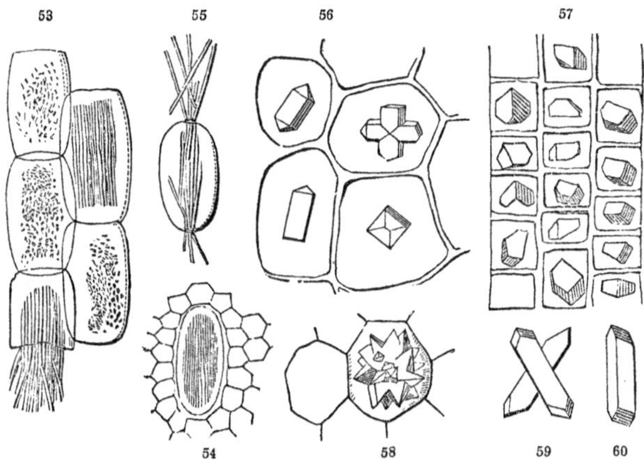

to consist of oxalate of lime; according to Quekett, they are phosphate of lime. Of other crystals composed of inorganic acids and an earthy base, the more usual are rhombic crystals of carbonate of lime, found in Cacti; and tabular, often twin crystals of sulphate of lime, which are "found in the Musaceæ and many Scitamineæ." Such are wholly formed of inorganic materials, derived from the soil.

92. **Silex**, likewise derived from the soil, very generally occurs

FIG. 53. Raphides, or acicular crystals, from the stalk of the Rhubarb: three of the cells contain chlorophyll, and two of them raphides.

FIG. 54. Raphides of an Arum, contained in a large cell; and 55, the same, detached from the surrounding tissue, and discharging its contents upon the application of water.

FIG. 56. Crystals from the Onion; one of them a hemitrope.

FIG. 57. Crystals of the inner bark of the Locust.

FIG. 58. A glomerate mass of crystals from the Beet-root.

FIG. 59, 60. Crystals from the bark of Hickory. Figures 55-60, and also 51, are from sketches kindly supplied by Professor Bailey of West Point.

as a part of the deposit or incrustation on the walls of cells ; * but it is not found in the form of crystals in their interior. In the Diatomaceæ nearly the whole cell-wall is composed of this indestructible material ; consequently, the remains of these minute organisms accumulate at the bottom of the water in which they live, so as to form immense strata in many places.

CHAPTER II.

OF THE GENERAL MORPHOLOGY OF THE PLANT.

93. **The Individual Plant.** The organic elements, or cells in their various forms, which have been treated of in the preceding chapter, make up the individual plant. Looking now upon plants as individual beings, we observe that they present themselves under the greatest variety of forms ; some of them are of the utmost simplicity, and many of these are so minute, that they are individually undistinguishable or invisible to the naked eye, and only become conspicuous by their aggregation in great numbers : others are highly complex in structure, and attain to a vast size, such as gigantic trees, some of which have flourished for a thousand years or more. All the larger vegetables are formed of a countless number of cells ; which, as they increase, arrange themselves so as to shape the fabric into definite parts, such as stem, leaves, and roots, each having distinct offices to fulfil, while all are subservient to the nutrition and perfection of the individual whole. These parts are called the *Organs* of the plant ; or, more technically, the *Compound Organs*, since it is the cells of which they are composed that are the real instruments, and carry on the operations of the vegetable economy. These organs are most distinct, and at the same time most diversified, in the highest grade of plants ; in the lower, they are successively less and less evolved, until all such distinction of parts vanishes, and the plant is reduced to a rounded or flattened mass of cells, to a row of cells strung end

---

* This may be shown by carefully burning off the organized matter of the tissue, and examining the undisturbed ashes by the microscope (311, 312).

to end, or even to a single cell. Since these last are the simplest plants, and the higher acquire their more complex structure (as will hereafter be shown) from an equally simple beginning, the most natural order for exhibiting the principal grades of vegetation is to commence with the lowest and simplest possible kinds, namely with

94. **Plants of a Single Cell.** There are several kinds of such plants among the Algæ (Sea-weeds, &c.), which rank as the lowest order of the vegetable kingdom. They are especially interesting here, because they furnish the readiest illustrations of the various methods of cell-formation which have been described in the preceding chapter (26 – 35). For in them vegetation is reduced to its simplest terms: the *plant* and the *cell* are here identical. The cell constitutes *an entire vegetable without organs*, imbibing its food by endosmosis (37) through its permeable walls, assimilating this food in its interior, and converting the organizable products at first into the materials of its own enlargement or growth, or finally into new cells, which constitute its progeny. Thus we have an epitome of all that is essential in vegetation, even on the largest scale, namely, the *imbibition of inorganic materials ;* their *assimilation ;* their application to the growth of the individual, or *nutrition*, and the formation of new individuals, or *reproduction.* But even while thus organically simple, the plant is not restricted to one monotonous pattern. On the contrary, different species, each in its own uniform manner, develope the cell and give rise to their progeny in all the various ways that have been mentioned when describing the forms and the development of cells. The simplest case is that of

95. 1st, *Plants of a Single Globular Cell ;* that is, of a cell which grows equally in every direction, and therefore is neither elongated nor branched. Of this, the microscopic plant known as giving rise to the phenomenon of *red snow* (but which also occurs on damp earth, &c.) furnishes a good illustration. Each indi-

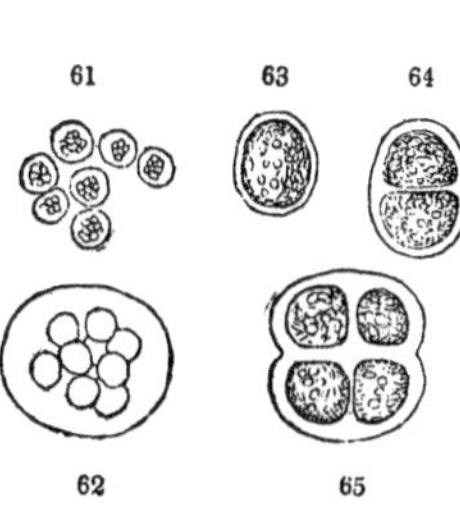

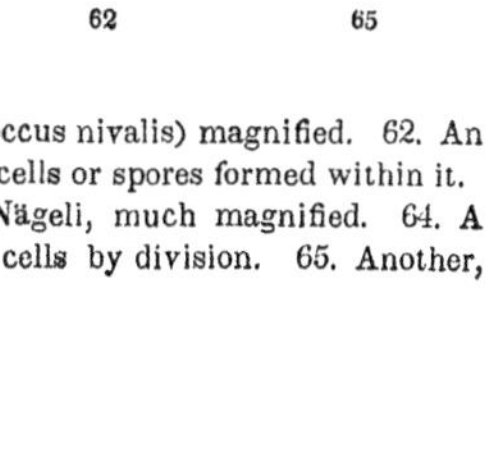

FIG. 61. Several individuals of the Red-snow Plant (Protococcus nivalis) magnified. 62. An individual highly magnified, showing more distinctly the new cells or spores formed within it.

FIG. 63. An individual of Chroococcus rufescens, after Nägeli, much magnified. 64. A more advanced individual, with the contents forming two new cells by division. 65. Another, with the contents divided into four new cells.

vidual is a single cell (Fig. 61), which quickly attains its growth, and produces (by original cell-formation, it is thought) a considerable number of minute free cells in its interior. The mature mother-cell now decays; and the new generation it contained enlarge into similar cells or plants, which give rise to their progeny and perish in their turn. Some other globular one-celled plants (like Chroococcus, Fig. 63), are very similar, except that they propagate by division of the whole contents, and finely illustrate that general process of free cell-multiplication (37). The layer of protoplasm which lines the cell-wall forms a constriction in the middle, and soon separates the whole inclosed contents into two parts; a layer of cellulose is at the same time deposited on the surface, and thus two new cells are produced (Fig. 64), which usually subdivide each into two (Fig. 65). Four new cells are thus formed within a mother-cell; and the latter is destroyed in the process, all its living contents having been employed in the formation of the progeny, and its effete wall is obliterated by softening or decay, or by the enlargement of the contained cells. Thus the simplest vegetation goes on, from generation to generation. The softened remains or products of the older cells often accumulate and form a gelatinous stratum or nidus, in which the succeeding generations are developed, and from which they doubtless derive a part of their sustenance, — just as a tufted Moss is nourished in part from the underlying bed of vegetable mould which is formed of the decayed remains of its earlier growth. One step in advance brings us to

96. 2d, *Plants of a Single Elongated Cell;* that is, of a cell which grows on in one direction, but without branching. Such plants answer to cells of prosenchyma, or to vessels (52, 57). For an example we may take any species of Oscillaria (Fig. 66); a form of aquatic vegetation of microscopic minuteness, considered as to the size of the individuals, but which rapidly multiply in such inconceivable numbers, that, at certain seasons, they sometimes color the surface

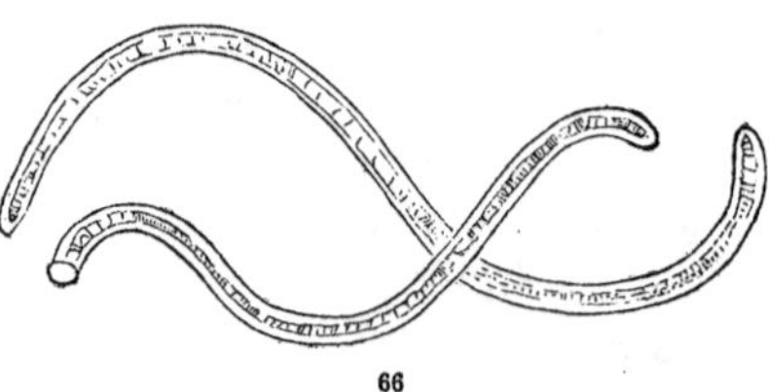

FIG. 66. Two individuals of Oscillaria spiralis, magnified; one of them with one extremity cut off.

of whole lakes of a green hue, as suddenly as broad tracts of alpine or arctic snow are reddened by the Protococcus.*

97. 3d, *Plants of an Elongated and Branching Cell.* Some elongated cells in vegetable tissue fork as they elongate, and become branched; as seen in Fig. 15. Several plants consist of individual cells of this kind; as, for example, the species of Vaucheria, which form one kind of the delicate and flossy green threads which abound in fresh waters, and are known in some places by the name of *Brook-silk.* These, under the magnifying-glass, are seen to be single cells, of unbroken calibre, furnished with branches here and there (Fig. 71). The branches are protrusions, or new growing points, which shoot forth, and have the power of continuous growth from the apex. In Bryopsis (Fig. 73),

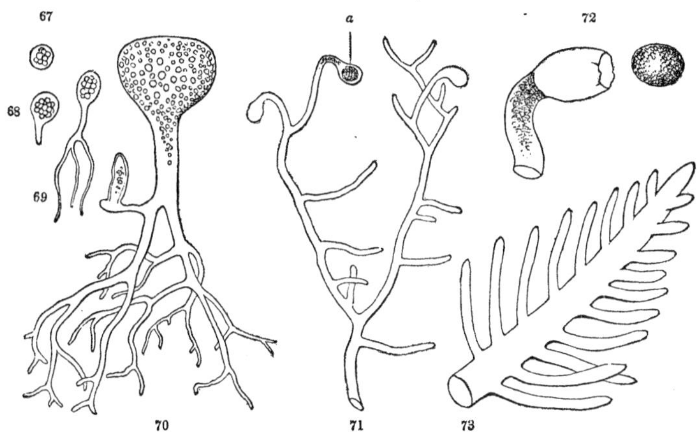

a beautiful small Sea-weed, the branches are much more numerous and regular: they are often constricted where they join the main stem, if we may so call it, but the cavity continues from stem to

* If the transverse markings of Oscillaria arise from imperfect partitions, then the plant corresponds to the duct (58).

FIG. 67-69. Botrydium Wallrothii in its development, and with new cells forming within; after Kützing: 67, the cell still spherical: 68, pointing into a tube below: 69, the tube prolonged and branched: all much magnified.

FIG. 70. Botrydium argillaceum, after Endlicher; the full-grown plant, magnified.

FIG. 71. Vaucheria clavata, enlarged: *a*, a spore formed in the enlarged apex of that branch. 72. End of the branch, more magnified, with the spore escaped from the burst apex.

FIG. 73. Bryopsis plumosa; summit of a stem with its branchlets, much enlarged.

branch; or, in other words, the whole plant consists of a single vegetating cell.

98. While in these cases the ramifications of the cell imitate, or as it were foreshadow, the stem and branches of higher organized plants, we have in Botrydium (Fig. 70) a cell whose ramifications resemble and perform the functions of a root. This is a terrestrial Alga, with a rounded body composed of an enlarged cell, which elongates and ramifies downwards, the slender branches penetrating the loose, damp soil on which the plant grows, exactly in the manner of a subdivided root. Meanwhile, a crop of rudimentary new cells is produced, by original free cell-formation (28), in the liquid which fills the body of the mother-cell: these, escaping when that decays or bursts, grow into similar plants, in the manner shown by Fig. 67 – 69.

99. The new cells by which Vaucheria is propagated are produced in a different way; as is shown in V. clavata (Fig. 71, 72). The apex of a branch enlarges; its green contents thicken, separate from those below, and a membrane of cellulose is formed around it, just as it forms around the contents of the whole cell in the microscopic Chroococcus (Fig. 63), but no further division takes place; the wall of the mother-cell bursts open, and the new-born cell escapes into the water. When it grows, it elongates a little from one end, and by this fastens itself to any solid body it rests on, and then grows from the opposite end into a prolonged tube, with occasional branches, like its parent. In this way, a plant composed of a single cell imitates not obscurely the downward and upward growth (the root and stem) of the more perfect plants. In the foregoing cases we noticed that the production of new cells insured the death of the parent; the whole living contents being appropriated to the new formation. In this case, the progeny originates from the living contents of a part of the cell only, and the walls of that portion alone perish.

100. **Plants of a Single Row of Cells.** To these there is but a single step from plants formed of a single cell (whether branching or unbranched) which has the power of continuous growth from the apex; and that step consists in the formation of transverse partitions. The manner in which these are produced has been already described (Fig. 8), as observed in a species of Conferva. Most of these simple, thread-like Algæ are composed of a single row of cells, produced in this way. The three kinds of Moulds or Mil-

dew Fungi here represented (Fig. 74–76) consist, as to the creeping part at the base (which spreads widely through the substance they live on) of long, thread-like, and usually branching cells (much like those of Fig. 15), for the most part destitute of partitions; while the upright portions are composed of a row of short cells, like those of a Conferva. These are terminated in the Bread-mould (Fig 74) by a much larger cell, which developes numerous and very minute rudimentary ones in its interior. In Fig. 75, we have a different arrangement, namely, a cluster of branches, made up of a series of bead-like, easily separable cells, which are evidently formed by the process of division just illustrated, and which serve as seeds to reproduce the species.

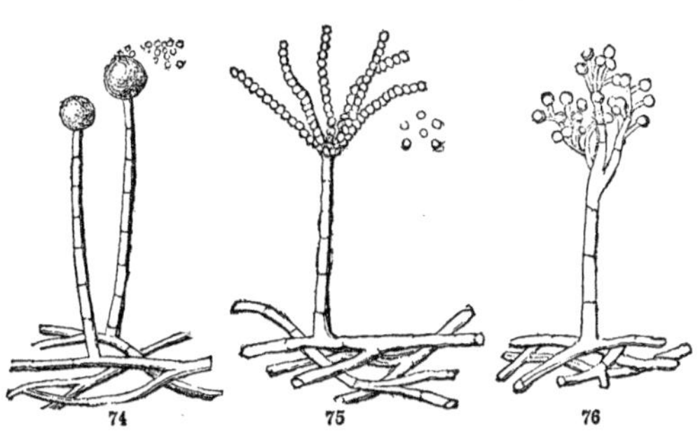

101. Spores. When the cells remain connected as they multiply, they increase the size or complexity of the individual vegetable. When they separate, each becomes the initial cell of a new plant. Any cell is capable of originating a new individual. No sooner, however, does the plant acquire such slight complexity as to consist even of a single series of cells, than a distinction begins to appear between cells adapted for *vegetation*, and those for *reproduction*. Both may propagate the species: the thread-like, vegetating cells which form the base of the Moulds, in Fig. 75, for example, grow with the same readiness as the minute specialized cells which terminate this simple vegetation. But the first appear to do so after the manner in which the higher grades of plants multiply by offshoots or division of the root; while the second are analogous in this respect to the seeds or embryos of such higher plants. These cells specialized for propagation, however they may originate, are accordingly distinguished by a special name, that of SPORES or SPORULES. We have to rise still higher in the scale, however, before a well-marked distinction can be

FIG. 74. The Bread-mould (Mucor) magnified. 75. Another Mould (Penicillum glaucum). 76. Botrytis Bassiana, a parasitic Mould. All magnified.

drawn in all cases between cells for reproduction and cells for vegetation.

102. Conjugation. At this stage of vegetation, however, and even in a large tribe of plants composed of single and simple cells, a process of great physiological importance is first observed,—the evident equivalent of bisexuality in the higher orders,— by which the reproductive cells or *spores* are still further specialized and potentiated. They are formed by *conjugation;* that is, by the mingling of the contents of two cells, both of which take part in the formation of the resulting spore. Fig. 77 – 80 exhibit this conjugation in a minute silicious-coated, one-celled plant, of the family Desmidiaceæ; where the recent discovery of this process, by Mr. Ralfs, has confirmed the vegetable character of these ambiguous microscopic bodies beyond all doubt. Also Figure 81 shows the conjugation of two individuals of Zygnema (Spirogyra), a common plant of our pools, composed of single rows of cells, nearly all of which, in the figure, are represented as taking part in the conjugation.

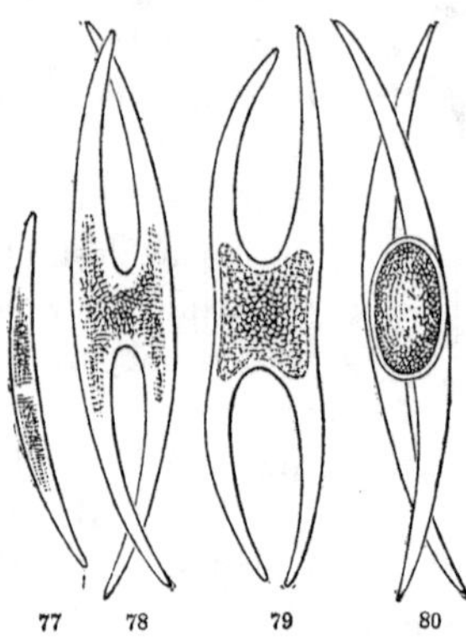

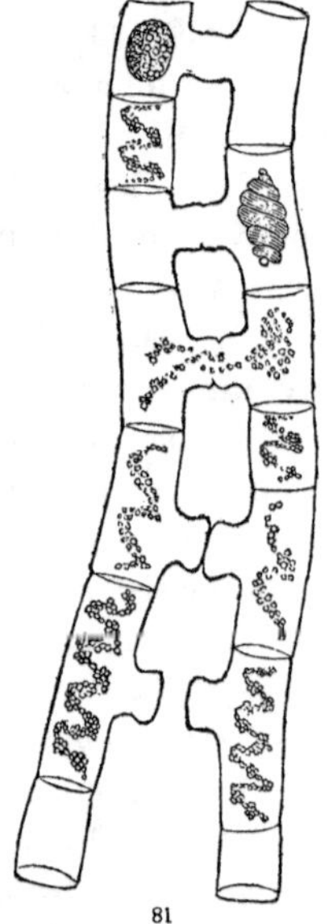

103. Plants of a Tissue of Cells combined in one Plane. The next step in complexity is seen in those Algæ which consist of a few jointed tubes laterally cohering with each other; or of numerous cells united in a single plane, as in the little Sea-weed, Fig. 82.

FIG. 77. Magnified individual of Closterium acutum, after Ralfs. 78. Two individuals more magnified, in conjugation; their cells opening one into the other, and the contents mingled; in 79, condensing; in 80, collected and formed into a spore.

FIG. 81. Magnified view of two conjugating filaments of Zygnema, showing all the stages of the process by which the cells from different approximated filaments form each a corresponding protuberance, these come into contact, the intervening walls are absorbed, and the green contents pass from one cell into the other, condense, acquire an investing membrane, and so form a spore: the several stages are shown from below upwards.

This gives rise to *frondose* or leaf-like forms. The name of FROND is applied to such expanded bodies, which are neither leaf nor stem, but combine the appearance and the office of both. Only the simplest forms, however, consist of a single layer of cells. Most frondose Sea-weeds, as well as Lichens, Liverworts, &c., are made up of several such layers. This is not the place to illustrate the almost endless diversity of forms under which the frond, or, as it is called in Lichens and Fungi, the *Thallus*, appears in these lower grades of plants; nor to notice their particular modes of propagation; except to say, in general, that the spores are still nothing but specialized cells, developed in some one of the ways already explained. But we now begin to meet with special organs or peculiar apparatus in which the reproductive cells are formed, instead of occurring indifferently in any part.

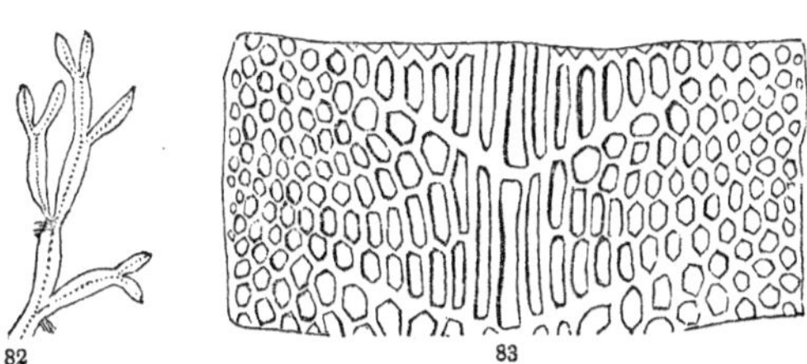

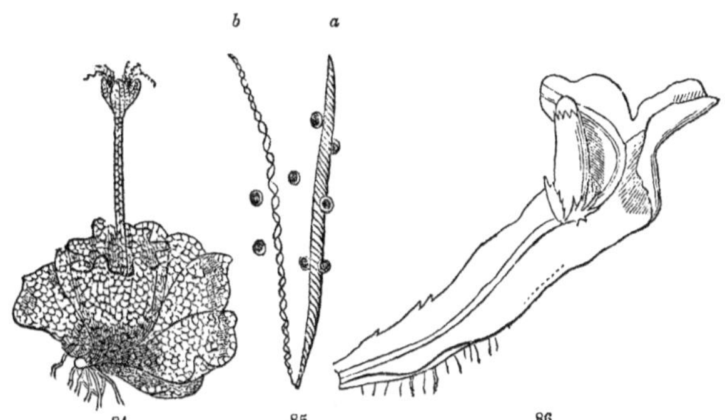

104. Plants of a Tissue of Cells combined into a solid Axis, or with

FIG. 82. A branch of Delesseria? LePrieurei (from the Hudson River), enlarged to twice the natural size. 83. A small portion more magnified, to show the cellular structure. The cells have thick gelatinous walls; those in the middle are elongated, those towards the margins rounded.

FIG. 84. Fruit-stalk, with a portion of the foliage, of a Jungermannia, magnified, to show its entire cellular structure.

FIG. 85. One of the tubular spirally-marked cells from the fruit of a Jungermannia (*a*); and (*b*) the spiral threads which result from its disruption. Some of the spores stick to the tube.

FIG. 86. Jungermannia Lyellii, less than the natural size.

stem and *branches*. Stem-like solid forms occur, perhaps as abundantly as the leaf-like or frondose, in the higher representatives of the lowest orders of plants, in Algæ, Fungi, and Lichens; and occasionally the two are somewhat vaguely presented in the same individual. Thus, many of the larger Sea-weeds display a leaf-like frond on the summit of a solid stalk; this stem, however, has once formed a part of the leaf. But in the Liverwort Family the distinction is first clearly exhibited, and in the true Mosses the higher type of vegetation is fully realized, namely in

105. **Plants with a Distinct Axis and Foliage**; that is, with a stem which shoots upward from the soil, or whatever it is fixed to, or creeps on its surface; which grows onward from its apex, and is symmetrically clothed with distinct leaves as it advances. All these lower vegetables which have now been mentioned, of whatever form, imbibe their food through any or every part of their surface, at least of the freshly formed parts. Their roots, when they have any, are usually intended to fix the plant to the rock or soil, and not to draw nourishment from it. The strong roots of the *Oar-weed*, *Devil's Apron* (Laminaria), and some other large Sea-weeds of our coast, are merely holdfasts, or cords expanding into a disc-like surface at their extremity, which by their adhesion bind these large marine vegetables so firmly to the rock that the force of the waves can seldom carry them away. Mosses also take in their nourishment through their whole expanded surface, principally therefore by their leaves: but the stems also shoot forth from time to time delicate rootlets, composed of slender cells or tubes, which grow in a downward direction and doubtless perform their part in

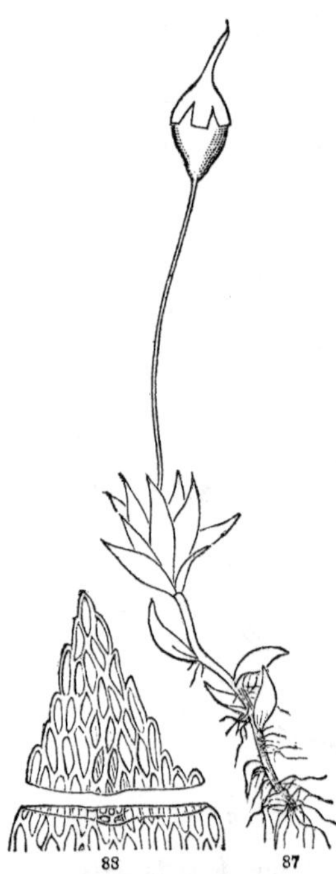

FIG. 87. An individual of a Moss (Physcomitrium pyriforme), enlarged to about 12 times the natural size. 88. Tip of a leaf, cut across, much magnified, to show that it is made up (except the midrib) of a single layer of cells.

absorption. Although sometimes of scarcely higher organization than the root-hairs which grow from the under side of a Liverwort (Fig. 86), yet they distinctly introduce the root. A Moss, therefore, as respects its vegetation, is an ordinary herb in miniature: it presents an epitome of the three universal ORGANS OF VEGETATION, namely *Root*, *Stem*, and *Leaves;* although its roots are of a secondary and subordinate character. In the apparatus of reproduction there is more complexity, but no essential change of plan. The spores of Mosses are formed by division of the contents of mother-cells into fours (31); and are contained in *Spore-cases* (or *Sporangia*) of peculiar structure, which are accompanied with some apparatus too elaborate to be described here, and are commonly elevated, before maturity, on a naked and slender stalk. The reproductive apparatus no longer forms a part of the general tissue, nor is imbedded in it, but special and altogether distinct organs are assigned to this office.

106. **Thallophytes and Cormophytes.** It is convenient to mention here, that these plants of the lower grades, Algæ, Fungi, and Lichens, which exhibit no proper distinction of stem and foliage, are by some botanists collectively called THALLOPHYTES, that is, plants formed of a thallus (103), or bed, as the compound word imports. And the name is appropriate for the greater part of these rootless, stemless, and leafless forms of vegetation, which compose flat crusts or plates, like the common Lichens on rocks, walls, and bark; or spreading Mushrooms; or the broad, membranous Seaweeds, such as the Dulse and Laver: and even the plants of single cells or single rows of cells are more commonly aggregated so as to make up a stratum, or bed of interlaced threads, more or less compact or definite. Such general names are seldom characteristic of every form they are meant to comprise. The contradistinguishing name of CORMOPHYTES (meaning stem-growing plants) is given to the higher forms of vegetation, from Mosses upwards, because they develope a proper stem, usually adorned with distinct foliage.

107. **Cellular and Vascular Plants.** While the Mosses emulate ordinary herbs and trees in vegetation and external appearance, they agree with the lower plants in the simplicity of their internal structure. They are entirely composed of cellular tissue strictly so called, chiefly in the form of parenchyma (51), at least they

have no vessels or ducts* (57) and form no wood. They, with all the plants below them, were therefore denominated CELLULAR PLANTS by De Candolle. Those above, inasmuch as vascular and woody tissues enter into their composition, when they are herbs as well as when they form shrubs or trees, he distinguished by the general name of VASCULAR PLANTS.

108. The strength which these tissues impart — owing to their toughness and the close bundles or masses they form running lengthwise through the stem (53, 56) — enables these vascular and woody plants to attain a great size and height; while Mosses and all other Cellular plants are of humble size, except when they float in water, in which a few of the coarser Sea-weeds do indeed attain a prodigious length and bulk. The lowest forms of Vascular plants, such as the Club-Mosses (Fig. 89), are of humble size, as the name indicates, although the stems are often of a woody texture. Most Ferns, or Brakes, are also herbaceous, or their persistent and more or less woody stems remain underground, in the form of rootstocks, or creep on its surface (as in Fig. 95). A few of them, however, in the warmer parts of the world, rise into trunks, and form palm-like trees (Fig. 91), of graceful port, and sometimes of great altitude. Thus far, the roots are still of a secondary character; that is, they spring from the stem, wherever it is in contact with or covered by the soil. From the mode of development it will hereafter appear that Ferns and Club-Mosses, like true Mosses, can

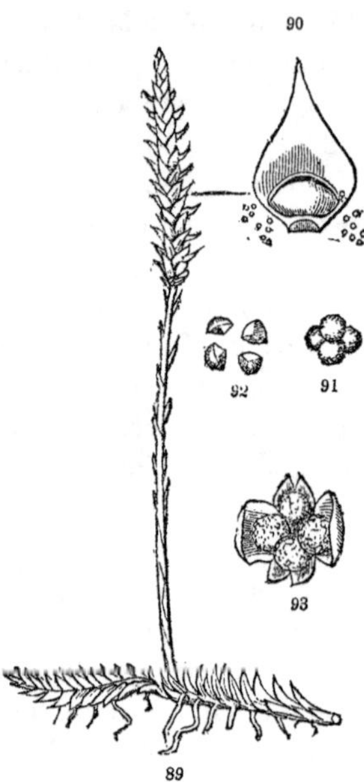

* The spirally marked tubes which are found in the spore-cases of Liverworts (Fig. 85, *a*) offer an exception.

FIG. 89. Lycopodium Carolinianum, of the natural size. 90. A leaf from the spike of fructification, with the spore-case in its axil, and spores falling out. 91. A group of four spores, magnified. 92. The same separated. 93. A burst spore-case of Selaginella apus, with its four large spores.

have no primary root. The axis, therefore, grows from the apex only, and it has no provision for increase in diameter as it increases in age. They have accordingly received the name of ACROGENS or ACROGENOUS PLANTS, — terms of Greek derivation, signifying that they grow from the apex alone. As to their fructification, all these families belong to the great lower series of

109. **Cryptogamous or Flowerless Plants.** Such are all plants which are reproduced by *spores* in place of seeds. *Spores*, as has been already shown, are single specialized cells, which originate in some one of the ordinary modes of cell-production, and without the agency of proper flowers. *Cryptogamous* and *Flowerless* are therefore equivalent terms ; the former denoting, metaphorically, that the flowers or organs of reproduction are concealed or obscure.* The great advance made by Club-Mosses and Ferns in their organs of vegetation is not attended by any corresponding complexity in their mode of reproduction. The spores of Club-Mosses and Ferns are as simple as those of true Mosses themselves, and the apparatus concerned is scarcely more elaborate. Even the tall Tree Ferns spring from spores of the same simple character, and of size so small that they are separately invisible to the naked eye. It is worthy of note, however, that their simple spore-cases are borne on the leaves, either on leaves in their natural state as organs of vegetation, or on those more or

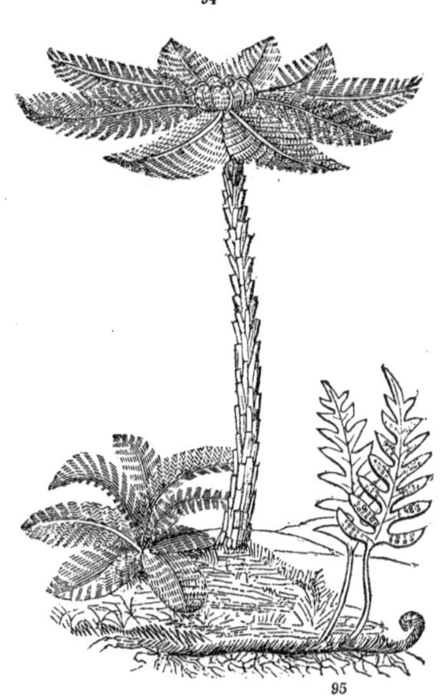

* Most Cryptogamous plants, however, are now known to have organs *analogous* to those of the flower, at least in function. These will be described in another place.

FIG. 94. Sketch of a Tree Fern, Dicksonia arborescens, of St. Helena; after Dr. J. D. Hooker. 95. Polypodium vulgare, with its creeping stem or rootstock.

less altered to subserve the special purpose. For in like manner, on leaves more or less altered or specialized, the seeds are manifestly borne in the simplest form of

110. **Phænogamous* or Flowering Plants.** In these we reach at length the perfected type, the highest grade of vegetation. They are the only flower-bearing plants, as their name indicates. Their reproduction is effected through an apparatus essentially different from that of Cryptogamous plants, namely, by *Stamens* and *Pistils* (the essential organs of the flower); the stamen producing *Pollen*, or free fertilizing cells; the pistil producing bodies to be fertilized, called *Ovules*, and which after fertilization become SEEDS. While Cryptogamous plants are propagated from *spores*, or specialized cells, which in germination multiply into other cells, and at length form a young plant, Phænogamous plants are propagated from *seeds*, which are more complex bodies, essentially characterized by having already formed within them, before they separate from the mother plant, an EMBRYO, that is an organized plantlet, which is only further developed in germination.

111. In the lowest grade of Phænogamous plants (viz. in the Cycadaceæ, and in the Coniferæ or Pine Family), the flowers are of such extreme simplicity that they consist, some of a stamen only, others of one or more naked ovules borne on the margins of an evident leaf, as in Cycas, or on the base or inside of an altered, scale-like leaf, as in the Pine Family. In the former, the ovules answer to the spore-cases of Ferns; in the latter, to the spore-cases of Club-Mosses; thus confirming an analogy which is indicated by general aspect between two of the higher families of Cryptogamous, and the lowest two of Phænogamous plants. These are *Gymnospermous* (that is, naked-seeded) Phænogamous plants. In all the rest, the ovules are perfectly inclosed in the pistil, which forms a pod or closed covering of some sort for the seeds; they are accordingly distinguished by the name of *Angiospermous* (that is, covered-seeded) Phænogamous plants. Their flowers in the simplest cases consist, one sort of a stamen only, the other of a pistil only. But as we rise in the scale, these organs tend to multiply; to be combined so as to have both kinds in the same flower;

* Sometimes written *Phanerogamous*. Both terms are made from the same Greek words, and signify, by a metaphorical expression, the counterpart of Cryptogamous; that is, that the essential organs of the flower are manifest or conspicuous.

to be protected or adorned with a circle of peculiar leaves (the CALYX), or with two such circles (CALYX and COROLLA), of which the inner is commonly more delicate in texture and of brighter color. This, the completed flower, exhibits the ORGANS OF REPRODUCTION in their most perfect form.

112. The Organs of Vegetation also exhibit their most perfect development in Phænogamous plants. The three kinds, *root*, *stem*, and *leaves*, are almost always well defined. In a few exceptional cases, however, we have *frondose* forms; as in the Duck-weed (Fig. 96), where stem and leaf are fused together into a green flat body which floats on the water, emitting roots from the lower surface and exposing the upper like a leaf to the light. So, true leaves seldom appear in the Cactus Family, where the green bark of the whole surface takes their place, although the points from which they should arise are distinctly indicated; nor are they developed at all in the Dodder (**135**, Fig. **122**), and some other parasitic Flowering plants. In all Cryptogamous plants furnished with a distinct axis, or stem, and leaves, this whole structure has to be formed after germination (**110**, in a manner to be hereafter shown); and when formed, the axis grows from its apex only (**108**), so that there is no primary root. Phænogamous plants, on the contrary, are developed directly from an embryo plantlet, namely, from an axis with its appendages, which already exists in the seed, and which grows both ways in germination; from one end to produce the stem, and from the other to form the root, thus exhibiting a regular opposition of growth from the first. To understand this, and to obtain the clearest conception of the plant as a whole and of its mode of growth, we should at the outset attentively consider the

113. **Development of the Embryo.** The Phænogamous plant, then, in the early stage at which we begin its biography, is an EMBRYO (Fig. 100) contained in the seed (Fig. 99). The form of this initial plantlet varies greatly in different species. It is often an oblong or cylindrical body, simple at one extremity, and nicked or lobed at the other, as in the case we have chosen for illustration. The undivided or stem part is called the RADICLE; it is the rudimentary

FIG. 96. A Duck-weed (Lemna minor, the whole plant), in flower; magnified.

axis, the initial stem. The two lobes into which the upper end is split are the COTYLEDONS, or the undeveloped first pair of leaves, often named the *Seed-leaves.* These are often so large as to make up nearly the whole bulk of the seed, as in the pea and bean, or the Apple and Almond (Fig. 97), where the radicle is very short in proportion; and on separating or taking off one of them the minute rudiments of one or more additional leaves may often be detected within (Fig. 98, *a*). The embryo, therefore, consists of a short axis or stem, crowned with two or more undeveloped leaves, or, in other words, with a *Bud.* In germination the axis or radicle elongates throughout, so as usually to elevate the budding apex above the surface of the soil, and its cotyledons expand in the air into the first pair of leaves; and at the same time from the opposite extremity is formed the root, which grows in a downward direction, so as to penetrate more and more into the soil. The two extremities of the embryo are therefore differently affected by the same external in-

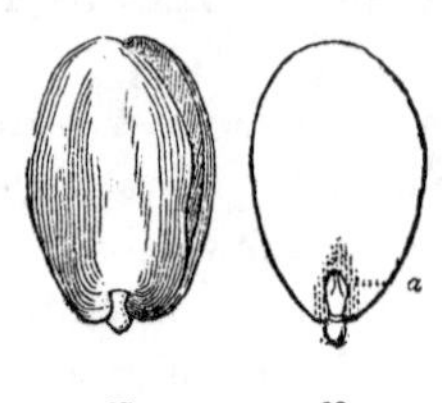

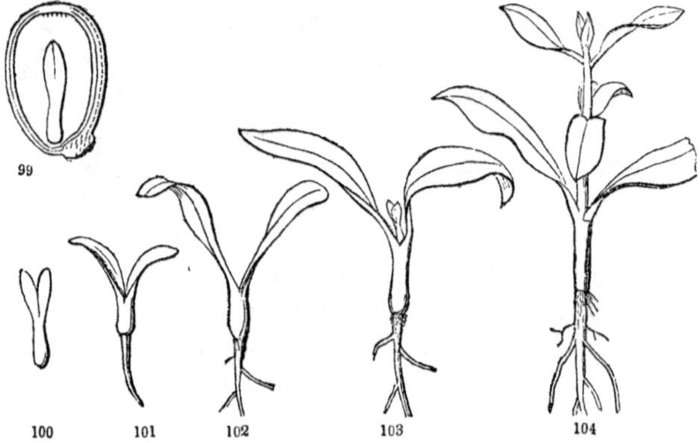

fluences, by light especially, and exhibit exactly opposite tendencies. The budding end invariably rises upwards, as if it sought

FIG. 97. Embryo (the whole kernel) of an Almond, the cotyledons slightly separated. 98. The same, with one cotyledon taken off, to show the *plumule*, or minute undeveloped leaves, *a*, between the two.

FIG. 99. A longitudinal section of a seed, showing the embryo or rudimentary plant it contains. 100. The embryo taken from the seed, and its rudimentary leaves, or cotyledons, a little separated. 101. The same in germination, the cotyledons expanding into the first pair of leaves. 102-104. The seedling plant more advanced. (The radicle, or first joint of stem, should have been drawn more elongated.)

the light and air; the root-end turns constantly from the light, and buries itself in the dark and moist soil. These tendencies are absolute and irreversible. If obstacles intervene, the root will take as nearly a downward, and the stem as nearly an upward direction, as possible. They are only the first manifestation of an inherent property which continues, with only incidental modifications, throughout the whole growth of the plant, although, like instinct in the higher animals, it is strongest at the commencement: and it insures that each part of the plant shall be developed in the medium in which it is designed to live and act, — the root in the earth, and the stem and leaves in the air. The axis, therefore, especially in plants of the highest grade, possesses a kind of polarity; it is composed of two counterpart systems, namely, a *Descending Axis* or root, and an *Ascending Axis* or stem. The point of union or base of the two is termed the *crown*, *neck*, or *collar*. Both the root and stem branch; but the branches are repetitions of the axis from which they spring, and obey its laws. The branches of the root tend to descend; those of the stem tend to ascend.

114. **Organs of Vegetation.** These three organs, *Stem*, *Root*, and *Leaves*, either preëxist rudimentarily in the seed, or appear at the first development of the embryo in germination. Of them, vegetables essentially consist; for they are all that are requisite to, and actually concerned in, their life and growth. Indeed, the whole ulterior evolution of the plant exhibits only repetitions of these essential parts, under more or less varied forms. They are, therefore, properly termed the FUNDAMENTAL ORGANS of plants, or the ORGANS OF VEGETATION. The root absorbs the crude food of the plant from the soil; this is conducted through the stem into the leaves, is in them digested, under the agency of solar light and heat; and the nourishment thus assimilated is returned into the stem and root, to be expended in the formation of new rootlets, new branches, and new leaves. The more the plant grows, therefore, the more it multiplies its instruments and means of growth; and its evolution would seem to be limited only by the failure of food, of a fit temperature, or other external circumstances.

115. Sooner or later, however, the plant changes its mode of development, and bears *Flowers*, or ORGANS OF REPRODUCTION. But even in these, the philosophical botanist recognizes the stem and leaves, under peculiar forms, adapted to special purposes. And the object or consummation of the flower is the production

of seeds, containing an embryo plant which is composed of these same fundamental organs, and which in its development repeats these successive steps, to attain the same ultimate result.

116. Having briefly traced the plan and progress of vegetation from the simplest or lowest through to the highest or most elaborately perfect grade of plants, we may, in the following chapters, leave the Cryptogamous or Flowerless plants entirely out of view (reverting to them only to explain separately their principal peculiarities at the close), and explain the phenomena, first of vegetation, and then of reproduction, as manifested in the higher series of Phænogamous or Flowering plants. The simpler kinds of the lower series doubtless afford peculiar facilities for investigating questions of anatomical structure, and for ascertaining what is really essential to vegetation. But the general scheme of the vegetable kingdom, and the unity of plan which runs through the manifold diversities it displays, enabling us to refer an almost infinite variety of details to a few general laws, must be studied in the higher series of Phænogamous plants, which exhibit, in manifold variety of form, the completed type of vegetation.

CHAPTER III.

OF THE ROOT OR DESCENDING AXIS.

117. THE Organs of Vegetation (114) in Phænogamous plants, namely, the root, stem, and leaves, are to be considered in succession; and it is on some accounts most convenient to begin with the root, charged as it is with the earliest office in the nutrition of the vegetable, that of absorbing its food. According to our view of the matter, however (113), its formation does not precede, but follows, that of the stem.

118. **The Primary Root**, as already defined (112 – 114), is the descending axis, or that portion of the trunk which, avoiding the light, grows downwards, fixing the plant to the soil, and absorbing nourishment from it. The examination of any ordinary embryo during germination, such as that of the Sugar Maple (Fig. 105 – 107), will give a good idea of the formation and entire peculiar-

ities of the root. Its radicle (*a*), or preëxisting axis, first of all grows in such a way as to elongate throughout its whole extent (thus showing that it is not itself *root*, but the first joint of stem); this lengthening, while it thrusts the root-end downwards (**113**) a little deeper into the soil, at the same time raises the cotyledons (*b*) to the surface, and at length elevates them above it, where they expand in the light and air, and begin to perform the office of leaves (Fig. 107). Contemporaneous with this elongation of the radicle, a new and different growth takes place from its lower extremity in a downward direction, which forms the Root (Fig. 107, *r*). The root is therefore a new formation from the root-end of the radicle. It begins by the production of a quantity of new cells (by division) at the extremity of the radicle; not on its surface, however, but beneath its thin epidermis and the superficial cells. The multiplication of cells at this point proceeds from below onwards; those behind quickly expanding to their full size, and then remaining unaltered, while those next the apex continue to multiply by division. In this way the root grows onward by continual additions of new material to its advancing ex-

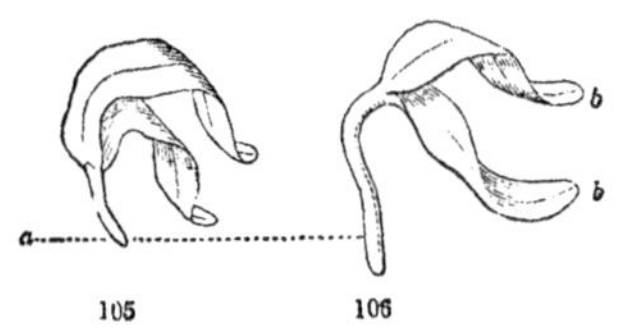

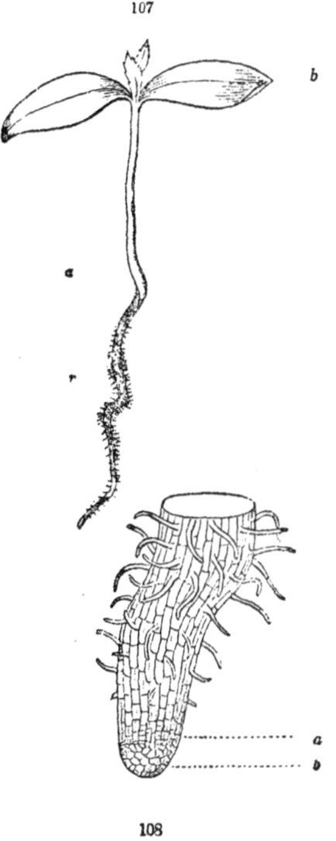

FIG. 105. An embryo of Sugar Maple, just unfolding in germination. 106. Same, a little more advanced; the radicle, *a*, considerably elongated.

FIG. 107. A germinating embryo of Sugar Maple, still more advanced: *a*, the radicle elongated into the first joint of stem, bearing the unfolded cotyledons or seed-leaves, *b*, and between them the rudiments of the next pair of leaves; while from its lower extremity the root, *r*, is formed.

FIG. 108. The lower end of the same root, magnified: *a*, the place where growth, through the multiplication of cells by division, is principally taking place: *b*, the original apex of the radicle, which has been carried onward by the growth that has taken place just behind it.

tremity; lengthening from the lower end entirely or chiefly, so that this part of a growing root always consists of the most newly formed and vitally active tissue.

119. The new cells, however, do not occupy the very point, as is commonly, but incorrectly, stated. This is capped, as it were, by an obtusely conical mass of older cells, consisting of the superficial tissue of the end of the radicle, pushed forward by the cell-multiplication that commenced behind it, as already mentioned (Fig. 108). As the original cells of this apex wear away or perish, they are replaced by the layer beneath; and so the advancing point of the root consists, as inspection plainly shows, of older and denser tissue than that behind it.* The point of every branch of the root is capped in the same way. It follows that the so-called *spongioles* or *spongelets* of the roots have no existence. Not only are there no such special organs as are commonly spoken of, but absorption evidently does not take place, to any considerable extent, through the older tissue of the point itself.

120. As to absorption by roots, the inspection of the root of a germinating plantlet, or of any growing rootlet, even under a low magnifying power, shows that they must imbibe the moisture that bathes them, by endosmosis (37), through the whole recently formed surface, and especially by the hair-like prolongations of the exterior layer of cells, or *fibrils*, as they may be termed, which are copiously borne by all young roots (Fig. 108). Fig. 109, 110, show some of these root-hairs, and the tissue that bears them, more magnified. These capillary tubes, of great tenuity and with extremely delicate walls, immensely increase the surface which the rootlet exposes, and play a more important part in absorption than is generally supposed; for they appear to have attracted little attention. These fibrils perish when the growing season is over, or when the

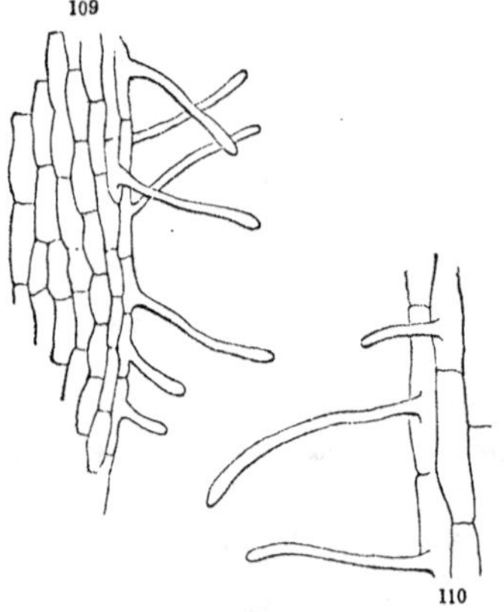

* It is a similar tissue that exfoliates from the point of some aquatic (as in Lemna, Fig. 96), and many aerial roots (as in Pandanus), in the form of a loose cup or sheath.

root gets a little older; at the same time, the external layer of cells that bears them, at first undistinguishable from the parenchyma beneath, except perhaps in the size of the cells, hardens and thickens into a sort of epidermis, or firmer skin, so as to arrest or greatly restrain the imbibition. This epidermis (69) of the root consists of less compressed cells than in parts exposed to the light, and is destitute of stomates or breathing-pores (70).

121. The growth of the root and its branches keeps pace with the development of the stem. As the latter shoots upward and expands its leaves, from which water is copiously exhaled during vigorous vegetation, the former grow onward and continually renew the tender, hygrometric tissue through which the absorption, required to restore what is lost by evaporation or consumed in growth, is principally effected. Hence the danger of disturbing the active roots during the season of growth. In early summer, when the sap is rapidly consumed by the fresh leaves, the rootlets are also in rapid action. The growth of the branches and roots being simultaneous, while new branchlets and leaves are developing, the rootlets are extending at a corresponding rate, and their tender absorbing points are most frequently renewed. They cannot now be removed from the soil without destroying them, at the very time when their action is essential to restore the liquid which is exhaled from the leaves. But towards the close of summer, as the leaves grow languid and the growth of the season is attained, the rootlets also cease to grow, the loose tissue of their extremities, not being renewed, gradually solidifies, and absorption at length ceases. This indicates the proper period for transplanting, namely, in the autumn after vegetation is suspended, or in early spring before it recommences.

122. This elongation of roots by their advancing points alone is admirably adapted to the conditions in which they are placed. Growing as they do in a medium of such unequal resistance as the soil, if roots increased like growing stems, by the elongation of the whole body, they would be thrown, whenever the elongating force was insufficient to overcome the resistance, into knotted or contorted shapes, very ill adapted for the free transmission of fluid. But, lengthening only at their farthest extremity, they insinuate themselves with great facility into the crevices or yielding parts of the soil, and afterwards by their expansion in diameter enlarge the cavity; or, when arrested by insuperable obstacles, their advan-

cing points follow the surface of the opposing body until they reach a softer medium. In this manner, too, they readily extend from place to place, as the nourishment in their immediate vicinity is consumed. Hence, also, may be derived a simple explanation of the fact, that roots extend most rapidly and widely in the direction of the most favorable soil, without supposing any prescience on the part of the vegetable, as some have imagined.

123. The advancing extremity of the root consists of parenchyma alone ; but bundles of vessels and woody tissue appear in the forming root, soon after their appearance in the primordial stem above : these form a central woody or fibrous portion, which continues to descend (by the transformation of a portion of the nascent tissue) as the growing apex advances ; sometimes, although not usually, inclosing a distinct pith, as the wood of the stem does. The surrounding parenchymatous portion becomes the bark of the root. Increase in diameter takes place in the same way as in the stem. (Chap. IV. Sect. IV., V.)

124. We have taken the root of the seedling as an example and epitome of that of the whole herb or tree ; as we rightly may ; for in its whole development the root produces no other parts ; it bears nothing but naked branches, which spring from different portions of the surface of the main root, nearly as this sprung from the radicle, and exactly imitate its growth. They and their ramifications are mere repetitions of the original descending axis, serving to multiply the amount of absorbing surface. The branches of the root, moreover, shoot forth without apparent order ; or at least in no order like that of the branches of the stem, which have a symmetrical arrangement, dependent, as we shall see, upon the arrangement of the leaves.

125. To the general statement that roots give birth to no other organs, there is this abnormal, but by no means unusual exception, that of producing buds and therefore sending up leafy branches. Although not naturally furnished with buds, like the stem, yet, under certain circumstances, the roots of many trees and shrubs, and of some herbs, have the power of producing them abundantly. Thus, when the trunk of a young Apple-tree or Poplar is cut off near the ground, while the roots are vigorous and full of sap, those which spread just beneath the surface produce buds, and give rise to a multitude of young shoots. The roots of the Maclura, or Osage Orange, habitually give rise to buds and branches.

Such buds are said to be irregular, or *adventitious*. This power, however, roots share with every part of the vegetable that abounds with parenchyma: even leaves are known to produce adventitious buds.

126. The root has been illustrated from the highest class of Phænogamous plants; in which the original root, or downward prolongation of the axis, continues to grow, at least for a considerable time, and becomes a *tap-root*, or main trunk, from which branches of larger or smaller size emanate. Often, however, this main root early perishes or ceases to grow, and the branches take its place. In some plants of the highest class (in the Gourd Family, for example), and in nearly the whole great class to which Grasses, Lilies, and Palms belong, there is no one main trunk or primary root from which the rest proceed; but several roots spring forth almost simultaneously from the radicle in germination, and form a cluster of fibres, of nearly equal size (Fig. 111). Such plants scarcely exhibit that distinct opposition of growth in the first instance, already mentioned as one characteristic of Phænogamous vegetation. Most Phænogamous plants likewise shoot forth *secondary roots* from the stem itself, the only kind produced by Cryptogamous plants. To these we must revert, after having considered some diversities connected with the duration and form of roots, and an important subsidiary purpose which they often subserve.

127. **Annual Roots** are those of a plant which springs from the seed, flowers, and dies the same year or season. Such plants always have *fibrous* roots, composed of numerous slender branches, fibres, or rootlets, proceeding laterally from the main or tap-root, which is very little enlarged, as in Mustard, &c.; or else the whole root divides at once into such fibrous branches, as in Barley (Fig. 111) and all annual Grasses. These multiplied rootlets are well adapted for absorption from the soil, but for that alone. The food which the roots of such a plant absorbs, after being digested and elaborated in its leaves, is all expended in the production of new leafy branches, and at length of flowers. The flowering process and the maturing of the fruit exhaust the vegetable greatly (in a manner hereafter to be explained), consuming all the nourishing material which it contains, or storing it up in the fruit or seed for its offspring; and having no stock accumulated in the root or elsewhere to sustain this draught, the plant perishes at the close of the season, or whenever it has fully gone to seed.

128. **Biennial Roots** are those of plants which do not blossom until the second season, after which they perish like annuals. In these the root serves as a reservoir of nourishing, assimilated matter (27, 79); its cells therefore become gorged with starch (81), vegetable jelly (83), sugar (84), &c. Such thickened roots are said to be *fleshy*, and receive different names according to the shapes they assume. When the accumulation takes place in the main trunk or tap-root, it becomes *conical*, as in the Carrot, Fig. 112, when it tapers regularly from the base or crown to the apex; it is *fusiform* or *spindle-shaped* when it tapers upwards as well as downwards, as in the Radish, Fig. 113; or *napiform* or *turnip-shaped*, when much swollen at the base, so as to become broader than long. If some of the branches or fibres are thickened, instead of the main axis, the root is said to be *fasciculated* or *clustered*, as in Fig. 114; or *tuberiferous* or *tuberous*, when they assume the form of rounded knobs, as in Fig. 115; or *palmate*, when the knobs are branched, as in Fig. 116. These must not be confounded with tubers, such as potatoes, which are forms of stems. Most of these are biennial. Such plants (of which the Radish, Carrot, Beet, and Turnip, among our esculents, are familiar exam-

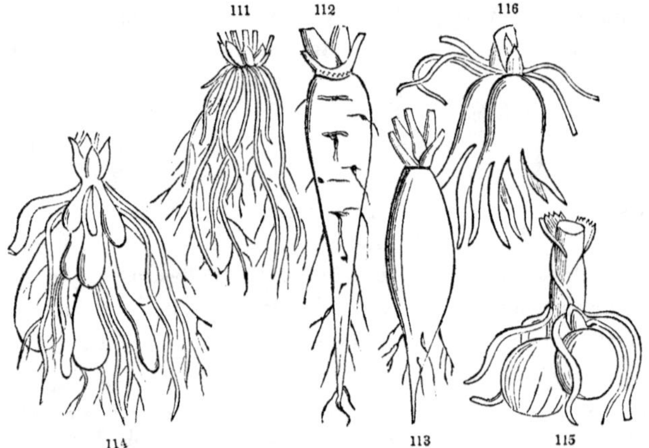

ples) neither flower the first season, nor even expend in the production of stems and branches much of the nourishment they generate; but, forming a large tuft of leaves at the very surface of the ground, they accumulate in the root nearly the whole sum-

FIG. 111-116. Different kinds of roots.

mer's supply of nourishment. When vegetation is resumed, the following spring, they make a strong and rapid growth, shooting forth a large stem, and bearing flowers, fruit, and seed, almost wholly at the expense of the accumulation of the previous year; this store is soon consumed, therefore; and the plant, meanwhile neglecting to form new roots, perishes from exhaustion.

129. **Perennial Roots.** A third class of herbs, and all woody plants, do not so absolutely depend upon the stock of the previous season, but annually produce new roots and form new accumulations; sometimes in separate portions of the root, as in the Dahlia or the Orchis (Fig. 115), where, while one or more of such reservoirs is exhausted each year, others are providently formed for the next year's sustenance; and so on from year to year; a portion annually perishing, but the individual plant surviving indefinitely. More commonly, the whole body and main branches of the root are somewhat thickened; or portions of the stem may subserve this purpose, as in all tuberous herbs; or the nourishing matter may be widely distributed through the trunk, as in shrubs and trees. These are some of the modifications in this respect of *perennial* plants, which survive, or at least their roots, and blossom from year to year indefinitely.

130. **Secondary Roots.** (Also called *Adventitious Roots.*) Thus far, the primitive root, that which originated from the base of the embryo in germination, with its ramifications, has alone been considered. But roots habitually spring from any part of a growing stem that lies on the ground, or is buried beneath its surface, so as to provide the moisture and darkness they require; for such roots obey the ordinary tendency of the organ, avoiding the light, and seeking to bury themselves in the soil. Most creeping plants produce them at every joint; and most branches, when bent to the ground and covered with earth, will strike root. So, often, will separate pieces of young stems, if due care be taken; as when plants are propagated by cuttings. Stems commonly do not strike root, except when in contact with the ground. To this, however, there are various exceptions; as in the case of

131. **Aerial Roots.** Some woody vines climb by such rootlets; as the Ivy, our own Poison Ivy (Rhus Toxicodendron), and the Bignonia or Trumpet-Creeper, which in this way reach the summit of high trees. Such plants derive their nourishment from their ordinary roots imbedded in the soil; their copious aerial rootlets

merely serving for mechanical support. Other plants produce true *aerial roots*, which, emitted from the stem in the open air, descend to the ground and establish themselves in the soil. This may be observed, on a small scale, in the stems of Indian Corn, where the lower joints often produce roots which grow to the length of several inches before they reach the soil. More striking cases of the kind abound in those tropical regions where the sultry air, saturated with moisture for a large part of the year, favors the utmost luxuriance of vegetation. The Pandanus or Screw-Pine (a Palm-like tree, often cultivated in our conservatories) affords a well-known instance. The strong roots, emitted in the open air from the lower part of the trunk, soon reach the soil, as is shown in Fig. 117, giving the tree the appearance of having been partially raised out of the ground. The famous Banyan-tree (Fig. 119) affords a still more striking illustration. Here the aerial rootlets strike from the horizontal branches of the tree, often at a great height, and swing free in the air like pendent cords; but they finally reach and establish themselves in the ground, where they increase in diameter and form numerous accessory trunks, surrounding the original boll and supporting the wide-spread canopy of branches and foliage. Very similar is the economy of the Mangrove (Fig. 118), which inhabits muddy sea-shores throughout the tropics, and even occurs sparingly on the coast of Florida and Louisiana. Its aerial roots spring both from the main trunk, as in the Pandanus,

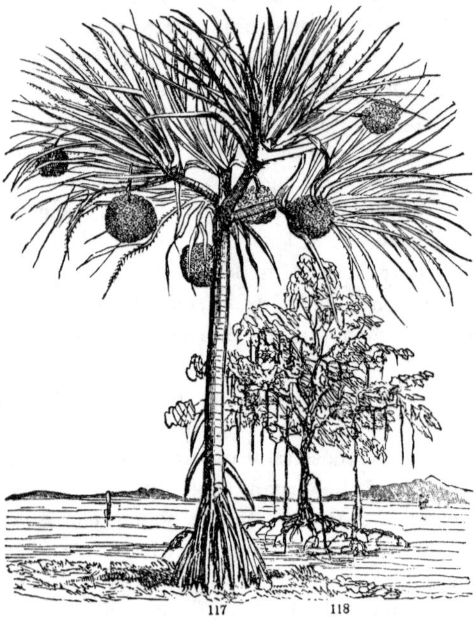

FIG. 117. The Pandanus, or Screw-Pine; with, 118, a Mangrove-tree (Rhizophora Mangle).

and from the branchlets, as in the Banyan. Moreover, this tendency to shoot in the air is shown even in the embryo, which be-

119

gins to germinate while the pod is yet attached to the parent branch; the radicle, or root-end of the embryo, elongating into a slender thread, which may even reach the ground, from the height of many yards, before the pod is detached. In this manner the Mangrove forms those impenetrable maritime thickets which abound on low, muddy shores, within the tropics.

132. **Epiphytes**, or *Air-plants*, exhibit a further peculiarity. Their roots not only strike in the free air, but throughout their life have no connection with the soil. They generally grow upon the trunks and branches of trees; their roots merely adhering to the bark to fix the plant in its position, or else hanging loose in the air, from which such plants draw all their nourishment. Of this kind are a large portion of the gorgeous Orchidaceous plants of very warm and humid climes, which are so much prized in hot-houses, and which, in their flowers as well as their general aspect, exhibit such fantastic and infinitely varied forms. Some of the flowers resemble butterflies, or strange insects, in shape as well as in gaudy coloring; such, for example, as the Oncidium Papilio (Fig. 120), which we have selected for one of our illustrations. To another family of Epiphytic plants belongs the Tillandsia, or Long Moss, which, pendent in long and gray tangled clusters or festoons from the branches of the Live-Oak or Long-leaved Pine, gives such a

FIG. 119. The Banyan-tree, or Indian Fig (Ficus Indica).

peculiar and sombre aspect to the forests of the warmer portions of our Southern States. They are called Air-plants, in allusion to

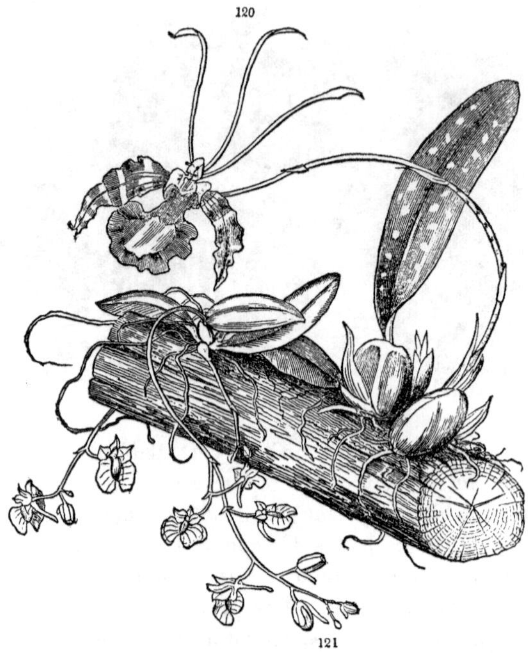

the source of their nourishment; and Epiphytes, from their growing upon other plants, and in contradistinction to

133. **Parasites**, that not only grow upon other vegetables, but live at their expense; which Epiphytes do not. *Parasitic plants* may be divided into two sorts, viz.:—1st, those that have green foliage, and 2d, those that are destitute of green foliage. They may vary also in the degree of parasitism; the greater number being absolutely dependent upon the foster plant for nourishment, while a few, such as the Cursed Fig (Clusia rosea) of tropical America, often take root in the soil, and thence derive a part, or sometimes the whole, of their support. This occurs only in

134. *Green Parasites*, or those furnished with green foliage, or proper digestive organs of their own. These strike their roots through the bark and directly into the new wood of the foster

FIG. 120. Oncidium Papilio, and, 121, Comparettia rosea; two epiphytes of the Orchis Family; showing the mode in which those Air-plants grow.

plant; whence they can draw little except the ascending, mostly crude sap (79), which they have to assimilate in their own green leaves. The Mistletoe is the most familiar example of this class. It is always completely parasitic, being at no period connected with the earth; but the seed germinates upon the trunk or branch of the tree where it happens to fall, and its nascent root, or rather the woody mass that it produces in place of the root, penetrates the bark of the foster stem, and forms as close a junction, apparently, with its young wood as that of a natural branch. Some species of Mistletoe, or of the same family, however, display no proper green foliage, but are of a yellow or brown hue. On the other hand, imperfect root-parasites with green foliage have recently been detected in more than one tribe of plants; * thus exhibiting intermediate states between the Green and the

135. *Pale or Colored Parasites*, that is, of other colors than green; such as Beech-drops, Orobanche, &c. These strike their roots, or sucker-shaped discs, into the bark, mostly that of the root, of other plants, and thence draw their food from the sap already elaborated (79). They have accordingly no occasion for digestive organs of their own, and are in fact always destitute of green foliage. In some cases of the kind, as in the Dodder (Fig. 122 – 124), the seeds germinate in the earth, from which the primitive root derives its nourishment in the ordinary manner; but when the slender twining stem reaches the surrounding herbage, it gives out aerial roots, which attach themselves firmly to the surface of the supporting plant, penetrate its epidermis, and feed upon its juices; while

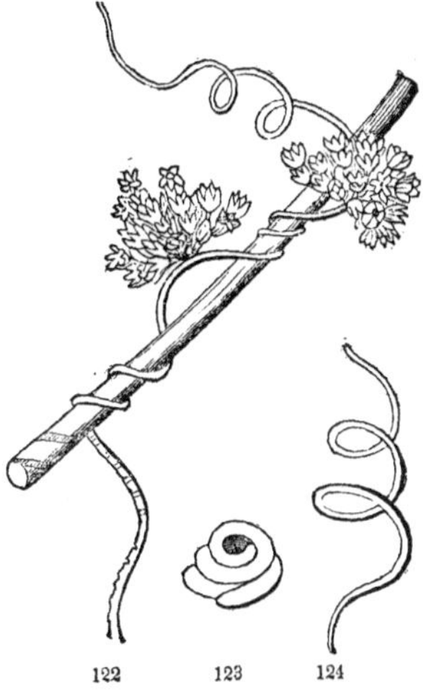

* In England a Thesium was discovered by Mr. Mitten to attach its roots

FIG. 122. The common Dodder of the Northern States (Cuscuta Gronovii), of the natural size, parasitic upon the stem of an herb: the uncoiled portion at the lower end shows the mode of its attachment. 123. The coiled embryo taken from the seed, moderately magnified. 124. The same in germination; the lower end elongating into a root; the upper into a thread-like leafless stem.

the original root and base of the stem perish, and the plant has no longer any connection with the soil. Thus stealing its nourishment ready prepared, it requires no proper digestive organs of its own, and, consequently, does not produce leaves. This economy is, as it were, foreshadowed in the embryo of the Dodder, which is a slender thread spirally coiled in the seed (Fig. 123, 124), and which presents no vestige of cotyledons or seed-leaves. A species of Dodder infests and greatly injures flax in Europe, and sometimes makes its appearance in our own flax-fields, having been introduced with the imported seed. Some species make great havoc in the clover-fields of the Old World.

136. Such parasites do not live upon all plants indiscriminately, but only upon those whose elaborate juices furnish a propitious nourishment. Some of them are restricted, or nearly so, to a particular species; others show little preference, or are found indifferently upon several species of different families. Their seeds, in some cases, it is said, will germinate only when in contact with

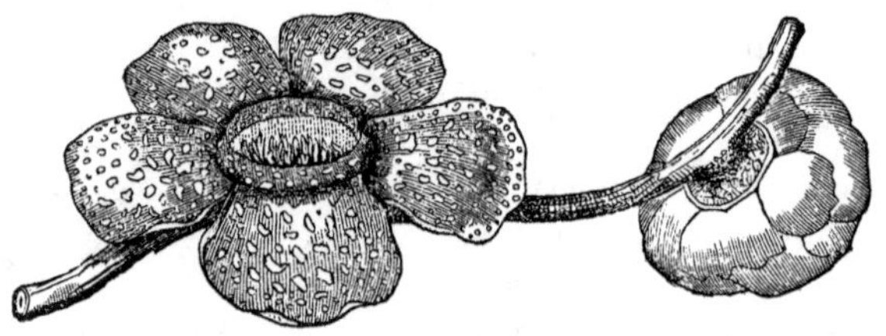

125

the stem or root of the species upon which they are destined to live. · Having no seed or foliage, such plants may be reduced to a

parasitically, by suckers, to the roots of adjacent herbs. (It would be interesting to know if this is the case with our Comandra.) Then Decaisne, recollecting that Rhinanthaceous plants generally, all of which blacken more or less in drying, were known to be uncultivable, and have the reputation, in France and elsewhere, of being injurious to cereal and other plants in their vicinity, was led to the discovery that plants of Rhinanthus, Melampyrum, and of the allied genera, attach themselves by numerous suckers on their roots to the roots of Grasses, shrubby plants, and even of trees, among which they grow. Our handsome species of Gerardia are equally uncultivable, doubtless on account of this partial parasitism.

FIG. 125. Rafflesia Arnoldi; an expanded flower, and a bud, directly parasitic on the stem of a vine: reduced to the scale of half an inch to a foot.

stalk with a single flower or cluster of flowers, as in the different kinds of Beech-drops,* the Cytinus, which is parasitic on the Cistus of the South of Europe, &c. They may even be reduced to a single flower directly parasitic on the bark of the foster plant, without the intervention of any manifest stem. A truly wonderful instance of this kind is furnished by that vegetable Titan, the Rafflesia Arnoldi of Sumatra (Fig. 125). The flower which was first discovered grew upon the stem of a kind of grape-vine; it measured nine feet in circumference, and weighed fifteen pounds! Its color is light orange, mottled with yellowish-white.

137. Among Cryptogamous plants, numerous Fungi are parasitic upon living, especially upon languishing vegetables; others infest living animals; the rest feed on dead or decaying vegetable or animal matters: all are destitute of chlorophyll (87), or any thing like green foliage. It is not improbable that our Monotropa, or Indian Pipe, a pallid and fungus-like Phænogamous plant, draws its nourishment, at least in great part, from the decaying leaves among which it grows.

CHAPTER IV.

OF THE STEM, OR ASCENDING AXIS.

Sect. I. Its General Characteristics and Mode of Growth.

138. Besides the direction of its growth, the descending axis or root we have found to be characterized by producing nothing except naked branches or subdivisions, and these in no definite order; by their continued extension through new formation at the extremity only, and in an uninterrupted manner, so as to give rise to no joints or nodes, and consequently to bear no leaves (141); by the absence of stomates in its epidermis (which, however, is the case in all parts developed under ground); and commonly by having no pith in the centre, or only a minute pith at the base, where it joins the stem. The latter organ differs in nearly all these particulars.

---

* See family *Orobanchaceæ*, in the second part of this work.

139. **The Stem** is the ascending axis, or that portion of the trunk which in the embryo grows in an opposite direction from the root, seeking the light, and exposing itself as much as possible to the air. All Phænogamous plants (110) possess stems. In those which are said to be *acaulescent*, or *stemless*, it is either very short, or concealed beneath the ground. Although the stem always takes an ascending direction at the commencement of its growth, it does not uniformly retain it; but sometimes trails along the surface of the ground, or burrows beneath it, sending up branches, flower-stalks, or leaves into the air. The common idea, therefore, that all the subterranean portion of a plant belongs to the root, is by no means correct.

140. The root gives birth to no other organs, but itself directly performs those functions which pertain to the relations of the vegetable with the soil; — its branches bind the plant to the earth; its newly formed extremities, or fresh rootlets, with the capillary fibrils they bear, imbibe nourishment from it. But the aerial functions of vegetation are chiefly carried on, not so much by the stem itself as by a distinct set of organs which it bears, namely, the leaves. Hence, the production of leaves is one of the characteristics of the stem. These are produced only at certain definite and symmetrically arranged points, called

141. **Nodes**, literally *knots*, so named because the tissues are here condensed, interlaced, or interrupted, more or less, as is conspicuously seen in the Bamboo, in a stalk of Indian Corn, or of any other Grass. Here each node forms a complete indurated ring, because the leaf arises from the whole circumference of the stem at that place. When the base of the leaf or leaf-stalk occupies only a part of the circumference, the nodes are not so distinctly marked, except by the leaves they bear, or by the scars left by their fall (Fig. 127, 130). When distinct they are often called joints, and sometimes, indeed, the stem is actually *jointed*, or *articulated*, at these points; but commonly there is no tendency to separate there. Each node bears either a single leaf, or two placed on opposite sides of the stem (Fig. 104), or three or more, placed in a ring (in botanical language, a *whorl* or *verticil*) around the stem. The naked portions or spaces that intervene between the nodes are termed

142. **Internodes.** The undeveloped stem is, in fact, made up of a certain number of these leaf-bearing points, separated by short

intervals; and its growth consists, primarily, in the elongation of these internodes (much after the mode in which the joints of a pocket-telescope are drawn out, one after the other), so as to separate the nodes to a greater or less distance from each other, and allow the leaves to expand.

143. This brings to view the leading peculiarity of the stem, namely, that it is formed of a succession of similar parts, developed one upon the summit of another, each with its own independent growth: each developing internode, moreover, lengthens throughout its whole body, unlike the root, which elongates continuously from its extremity alone. The nodes or the leaves they bear are first formed, in close contiguity with the preceding; then the internodes appear, and by their elongation separate them, and so carry upward the stem. To have a good idea of this, we have only to observe the gradual evolution of a germinating plant, where each internode developes nearly to its full length, and expands the leaf or pair of leaves it bears, before the elongation of the succeeding one commences. The radicle, or internode which preexists in the embryo (118), elongates, and raises the seed-leaves into the air (Fig. 107); they expand and elaborate the material for the next joint, the leaves of which in turn prepare the material for the third (Fig. 102 – 104), and so on. The internode lengthens principally by the elongation of its already formed cells, particularly in its lower part, which continues to grow after the upper portion has finished.

144. Buds. The apex of the stem, accordingly, at least of every stem capable of further terminal growth, is always crowned with an undeveloped portion, the rudiments of parts similar to those already unfolded, that is, with a BUD (113). The embryo itself may be rightly viewed as the fundamental bud borne on the apex of the radicle or original internode, from which the whole plant is developed; just as an ordinary bud of a tree or shrub developes to form a year's growth. Except that, in the latter case, the different steps follow each other more closely; for the bud usually has a considerable number of parts ready formed in miniature before it begins to grow, and has a full store of assimilated sap accumulated in the parent stem to feed upon. Such buds, which appear at the apex of a stem when it has completed its growth for the season, often exhibit the whole plan and amount of the next year's growth; the nodes, and even the leaves they bear, being already formed, and only requiring the elongation of the internodes for their full expansion. The structure is shown in the annexed diagram (Fig. 126), which represents the vertical section of a bud (like that which crowns the stem in Fig. 127), as it appears in early spring. As the bud is supplied by the stem on which it rests with nourishment sufficient for its whole development, it elongates rapidly ; and although the growth commences with the lowest internode, and follows the same course as in the seedling, yet the second, third, and fourth internodes, &c., have begun to lengthen long before the first has attained its full growth ; as is attempted to be shown by the diagram, Fig. 128. The stem thus continued from a *terminal* bud is, if it survive, again terminated with a similar bud at the close of the season, which in its development repeats the same process.

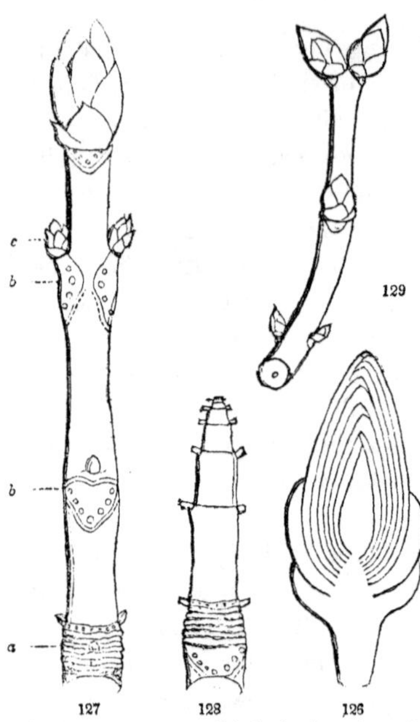

145. These yearly growths, in trees with well-formed *Scaly Buds*, such as the Magnolia (Fig. 130), the Horsechesnut (Fig. 127), &c., are plainly marked by the assemblage of scars or rings on the bark (*a*), which mark the places where the bud-scales were attached. The reason why these, and the leaf-scars, are obliterated after a few years, will appear when the increase of the stem in diameter is considered. The bud-scales themselves, which so closely overlie each other and protect the tender parts within against injury from moisture and sudden changes of temperature during the dormant state,* are only a special modification of

* The more effectually to ward off moisture, they are commonly covered with a waxy, resinous, or balsamic exudation (as in the Poplar especially),

FIG. 126. Diagram of a longitudinal section of a bud, such as that of the Horsechestnut.

FIG. 127. A year's growth of a Horsechestnut branch, crowned with a terminal bud: *a*, scars left by the bud-scales of the previous year: *b*, scars left by the fallen leafstalks: *c*, axillary buds.

FIG. 128. Diagram to illustrate the development of the bud in Fig. 126, 127.

FIG. 129. Branch and buds (all axillary) of the Lilac.

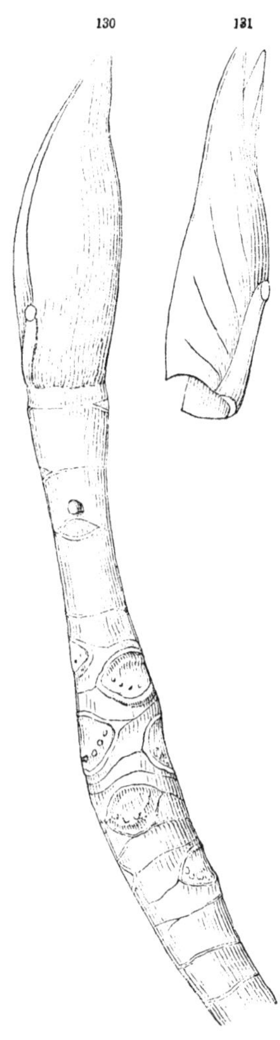

leaves, developed in this shape at a time when the internodes have ceased to elongate; so that the space between each ring in the figure just referred to represents an undeveloped internode. Such a stem displays alternately two modes of growth. First, the internodes elongate and interspace a succession of leaves, making the proper vegetation of the season. Then a series of leaves form as bud-scales, with internodes incapable of extension, and within them the rudiments of the next year's vegetation are prepared, to be developed as before, after a season of repose. As might be expected, therefore, such *scaly* (or *perulate*) *buds* belong to trees and shrubs of countries which have a winter; and are not met with, at least so distinctly, in those of the tropics; where, as there is no danger of injury from cold, the first parts that appear in the bud are ordinary leaves. On the other hand, many trees and shrubs of cold climates bear *naked buds*, as the Locust, Honey Locust, Ailanthus, &c., or buds with little scaly covering, as in the Kentucky Coffee-tree, the Papaw, &c. But in these cases the bud scarcely projects so as to be visible externally until it begins to develope in the spring. In Viburnum, some species, such as V.

---

impervious to rain, but which is melted by the heat of the sun when it stimulates the bud into growth. To guard against sudden changes of temperature, they are often lined, or the rudimentary leaves within are invested, with non-conducting down or wool.

FIG. 130. Branch of Magnolia Umbrella, of the natural size, crowned with the terminal bud; and below exhibiting the large, rounded leaf-scars, and the annular scars left by the fall of the bud-scales, of the previous season. 131. A detached scale from a similar bud; its thickened axis is the base of a leafstalk; the membranous sides consist of the pair of stipules.

Opulus, &c., have proper scaly buds, while in V. lantanoides, V. nudum, &c., they are entirely naked.

146. The bud, it is evident, is nothing more than the first stage in the development of a stem (or branch), the axis still so short that the scales without and the rudimentary leaves within cover or overlap one another. The various ways in which these parts are packed in the bud will be considered under another head (Vernation, 257). That the scales of the bud are of the same general nature as leaves is evident, not only from their position, but from their gradual transition into ordinary leaves in many cases. This is well seen in the expanding buds of the Lilac, Hickory, Horsechestnut, and especially of the Buckeye. The scales represent, sometimes the blade of the leaf, as in the Lilac; but more commonly the dilated base of the leafstalk, as is evident in the Balsam, Poplar, Butternut, and Hickory; or their stipules (259), either combined with this base, as in the Magnolia (Fig. 131), or alone, as in the Tulip-tree. Scales passing into ordinary leaves are abundantly obvious on the *turions*, or subterranean budding shoots, of numerous perennial herbs.

147. By the development of the preëxisting bud in the embryo, the original stem is produced; and it may be continued from year to year by the continued evolution of a *terminal bud*. Growing in this way only, the stem would of course remain simple or unbranched; as is the case with many during the first year, and with others, such as most Palms (Fig. 166) and Reeds, throughout their whole existence. But more commonly branches appear, even during the first year's growth.

SECT. II. RAMIFICATION.

148. Branches spring from *lateral* or *axillary buds*. These are new undeveloped axes or growing points, which habitually appear, or at least may appear, one (or occasionally two or three) in the *axil* of each leaf, that is, in the upper angle which the leaf forms with the stem. (See Fig. 127, *c*, where the point at which the fallen leaves were attached is marked by the broad scar, *b*, just below the bud.) The axillary bud is at first a little cellular nucleus on the surface of the wood, at the end of one of the cellular lines that form the silver-grain (196), and underneath the bark, through which it pushes as it grows, and shapes itself into a rudimentary axis, covered with the little appendages which become scales or leaves. When these buds grow, they give rise to BRANCHES; which are repetitions, as it were, of the main stem, growing just as that did from the seed; excepting merely, that, while that was implanted in the ground, these proceed from the parent stem. The branches thus produced are in turn provided with similar buds in the axils of their leaves, which have the same relation to the primary branch that it has to the main stem, and are capable of developing into branches of a third order, and so on indefinitely, producing the whole ramification of the plant. The whole is merely a series of repetitions, from new starting-points, of what took place in the evolution of the first axis, preëxistent in the seed. In the seed, therefore, or rather in the *embryo* it contains, we have the expression, in a condensed form, of the whole being of the plant. The latest ramifications, or twigs, are termed BRANCHLETS.

149. The arrangement of axillary buds depends upon that of the leaves. When the leaves are *opposite* (that is, two on each node, placed on opposite sides of the stem), the buds in their axils are consequently *opposite;* as in the Maple, Horsechestnut (Fig. 127), Lilac (Fig. 129), &c. When the leaves are *alternate*, or one upon each node, as in the Apple, Poplar, Oak, Magnolia (Fig. 130), &c., the buds implicitly follow the same arrangement. Branches, therefore, being developed buds, their arrangement is not left to chance, but is predetermined, symmetrical, and governed by fixed laws. When the leaves are alternate, the branches will be alternate: when the leaves are opposite, *and the buds develope regularly*, the branches will be opposite. In other words, if a bud in the axil of each leaf is developed into a branch, the relative situation of the branches will be the same as that of the leaves.

150. But the regular symmetry of the ramification is often accidentally interfered with by various causes, especially by the *non-development of a part of the buds.* As the original embryo plant remains for a time latent in the seed, growing only when a conjunction of favorable circumstances calls its life into action, so also many of the buds of a shrub or tree may remain latent for an indefinite time, without losing their power of growth. In our trees, most of the lateral buds generally remain dormant for the first season: they appear in the axils of the leaves early in summer, but do not grow into branches until the following spring; and

even then only a part of them usually grow. Sometimes the non-development or suppression occurs without appreciable order; but it often follows a nearly uniform rule in each species. Thus, when the leaves are opposite, there are usually three buds at the apex of a branch; namely, the terminal, and one in the axil of each leaf; but it seldom happens that all three grow at the same time. Sometimes the terminal bud continues the branch, the two lateral generally remaining latent, as in the Horsechestnut; sometimes the terminal one is regularly suppressed, and the lateral grow, when the stem annually becomes forked, as happens in the Lilac (Fig. 129).

151. The undeveloped buds do not necessarily perish, but are ready to be called into action in case the others are checked. When the terminal buds are destroyed, some of the lateral, that would else remain dormant, develope in their stead, incited by the abundance of nourishment, which the former would have monopolized. In this manner our trees are soon reclothed with verdure, after their tender foliage and branches have been killed by a late vernal frost, or other injury. The buds may remain latent even for years, and become covered with wood. The trunk of a tree, therefore, always contains an immense number; some of which, after a long period, may force their way through the wood to the surface, and break forth into branches; especially when the tree is *pollarded*, or its leading branches injured.

152. **Adventitious Buds.** But many such branches have an *abnormal* origin, from irregular or *adventitious buds*, like those produced by roots under similar circumstances (125). Such buds are still more readily produced on woody stems, when surcharged with sap, as we constantly observe on pollard Willows and Lombardy Poplars. Indeed, in several instances, buds are known to arise even from the surface or margins of leaves, as in Bryophyllum, which derives its name from this unusual circumstance; and the gardener produces them from root-cuttings or leaf-cuttings of certain plants, which he propagates in this way. Adventitious buds originate in the parenchyma, some cells of which are incited to take an independent development. In trees, they form on the surface of the wood, at the ends of the lines of the silver-grain (medullary rays, 191, 196). They are especially liable to spring from the new cellular tissue that forms at the growing season between the wood and bark when the trunk is wounded or cut off.

Thus the predestined symmetry of the branches is obscured or interfered with in two distinct ways; first, by the failure of a part of the regular buds to develope; and secondly, by the irregular or casual development of buds from other parts than the axils of the leaves: to which we may add, that great numbers of branches perish and fall away after they have begun to grow, or have attained considerable size. There is still another source of irregularity, namely, in the production of

153. **Accessory Buds.** These are, as it were, *multiplications* of the regular axillary bud, giving rise to two, three, or more, instead of one; in some cases situated one above another, in others side by side. In the latter case, which occurs occasionally in the Hawthorn, in certain Willows, in the Maples (Fig. 132), &c., the axillary bud seems to divide into three, or itself give rise to a lateral bud on each side, as soon as or before it penetrates the bark. On some shoots of the Tartarean Honeysuckle as many as half a dozen buds are developed independently in each axil, one above another, the lower being successively the stronger and earlier produced, and the one immediately in the axil, therefore, grows in preference; but when two or more grow, superposed accessory branches result. It is much the same in Aristolochia Sipho, except that the uppermost bud is there strongest. So it is in the Butternut (Fig. 133), where the true axillary bud is minute and usually remains latent, while the accessory ones are considerably remote, and the uppermost, which is much the strongest, is far out of the axil; usually this alone developes, and gives rise to an *extra-axillary* branch.

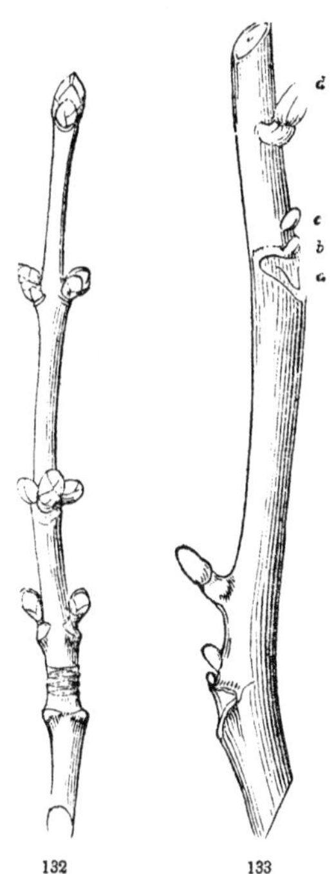

154. The stems of those Cryptogamous plants that possess a

FIG. 132. Branch of Red Maple, with triple axillary buds, placed side by side.

FIG. 133. Piece of a branch of the Butternut, with accessory buds placed one above another: *a*, the leaf-scar: *b*, proper axillary bud: *c*, *d*, accessory buds.

proper trunk (the Horsetails or Scouring Rushes excepted) do not *branch*, by the development of axillary or any kind of lateral buds implanted on its surface; but they often *fork* at the apex, by the division of the terminal bud. Their ramification, like their whole growth, is merely *acrogenous*, or from the apex (108).

155. **Excurrent and Deliquescent Stems.** Sometimes the primary axis is prolonged without interruption, by the continued evolution of the terminal bud, even through the whole life of a tree (unless accidentally destroyed), forming an undivided main trunk, from which lateral branches proceed; as in most Fir-trees. Such a trunk is said to be *excurrent*. In other cases the main stem is arrested, sooner or later, either by flowering, by the failure of the terminal bud, or the more vigorous development of some of the lateral buds, and thus the trunk is lost in the branches, or is *deliquescent*, as in most of our deciduous-leaved trees. The first naturally gives rise to conical or spire-shaped trees; the second, to rounded or spreading forms. As stems extend upward and evolve new branches, those near the base, being overshadowed, are apt to perish, and thus the trunk becomes naked below. This is well seen in the excurrent trunks of Firs and Pines, which, when grown in forest, seem to have been branchless for a great height. But the knots in the centre of the trunk are the bases of branches, which have long since perished, and have been covered with a great number of annual layers of wood, forming the *clear-stuff* of the trunk.

156. **Definite and Indefinite Annual Growth of Branches.** In the larger number of our trees and shrubs, especially those with scaly buds, the whole year's growth is either already laid down rudimentally in the bud (144), or else is early formed; and the development is completed long before the end of summer, and crowned with a vigorous terminal bud (as in the Horsechestnut, Fig. 127, Magnolia, Fig. 130, &c.), or with the uppermost axillary, as in the Lilac (Fig. 129). Such *definite* shoots do not lie down at all the following winter, but grow on directly, the next spring, from the terminal or some of the upper axillary buds, which are generally more vigorous than those lower down. In others, on the contrary, the branches grow onward *indefinitely* through the whole summer, or until arrested by the cold of autumn: they mature no terminal or upper axillary buds; or at least the lower and older axillary buds are more vigorous, and alone develope into

branches the next spring; the later-formed upper portion most commonly perishing from the apex downward for a certain length in the winter. The Rose and Raspberry, and among trees the Sumac and Honey Locust, are good illustrations of this sort; which, however, runs into the other mode through various gradations. Perennial herbs grow after the latter mode, their stems dying down to or beneath the surface of the ground, where the persistent base is charged with vigorous buds, well protected by the ground, for the next year's vegetation.

157. **Propagation from Buds.** Buds, being, as it were, new individuals springing from the original stem, may be removed and attached to other parts of the parent trunk, or to that of another individual of the same, or even of a different, but nearly related species, where they will grow equally well. This is directly accomplished in the operation of *budding*. In *ingrafting*, the bud is transferred, along with a portion of the shoot on which it grew. Moreover, as the cut end of such shoots, when buried in moist and warm soil, will commonly, under due care, send out adventitious roots, they may be made to grow independently, drawing their nourishment immediately from the soil, instead of indirectly through the parent trunk. This is done in the propagation of plants by cuttings. The great importance of these horticultural operations rests chiefly on the well-known fact, that buds propagate *individual peculiarities*, or varieties (668), which are commonly lost in raising plants from the seed.

Sect. III. The Kinds of Stem and Branches.

158. On the size and duration of the stem the oldest and most obvious division of plants is founded, namely, into Herbs, Shrubs, and Trees.

159. **Herbs** are plants in which the stem does not become woody and persistent, but dies annually or after flowering, down to the ground at least. The difference between *annual*, *biennial*, and *perennial* herbs has already been pointed out (127 – 130). The same species is so often either annual or biennial, according to circumstances or the mode of management, that it is convenient to have a common name for plants that flower and fruit but once, at whatever period, and then perish: such De Candolle accordingly designated as Monocarpic plants; while to perennials, whether

herbaceous or *woody*, large or small, he applied the counterpart name of POLYCARPIC plants, signifying that they bear fruit more than once, or an indefinite number of times. Between herbs and shrubs there are the intermediate gradations of

160. **Undershrubs**, or *suffruticose* plants, which are woody plants of humble stature, their stems rising little above the surface. If less decidedly woody, they are *suffrutescent.*

161. **Shrubs** are woody plants, with stems branched from or near the ground, and less than five times the height of a man. Between shrubs and trees there is every intermediate gradation. A shrub which approaches a tree in size, or imitates it in port, is said to be *arborescent.*

162. **Trees** are woody plants with single trunks, which attain at least five times the human stature.

163. **A Culm** is a name applied to the peculiar jointed stem of Grasses and Sedges, whether herbaceous, as in most Grasses, or woody or arborescent, as in the Bamboo.

164. **A Caudex** is a name usually applied to a Palm-stem (Fig. 166), to that of a Tree Fern (Fig. 94), and to any persistent, erect, or ascending, root-like forms of main stems. It is sometimes nearly synonymous with the rhizoma (174).

165. Those stems which are too weak to stand upright, but recline on the ground, rising, however, towards the extremity, are said to be *decumbent :* if they rise obliquely from near the base, they are said to be *ascending.* When they trail flat on the ground, they are *procumbent*, *prostrate*, or *running ;* and when such stems strike root from their lower surface, as they are apt to do, they are said to be *creeping*, or *repent.*

166. They are called *Climbers*, when they cling to neighboring objects for support; whether by tendrils, as the Vine and Passion-flower, by their leafstalks, as the Virgin's Bower (Clematis), or by aerial rootlets, as the Poison Oak (Rhus); and *Twiners*, or twining plants, when they rise, like the Convolvulus, by coiling spirally around stems or other bodies within their reach. Other modifications of the stem or branches have received particular names, some of which merit notice from having undoubtedly suggested several important operations in horticulture.

167. **A Stolon** is a form of branch which curves or falls down to the ground, where, favored by shade and moisture, it strikes root, and then forms an ascending stem, which is thus capable of draw-

ing its nourishment directly from the soil. The portion which connects it with the parent stem at length perishing, the new individual acquires an entirely separate existence. The Currant, Gooseberry, &c., multiply in this way, and doubtless suggested to the gardener the operation of *layering;* in which he not only takes advantage of and accelerates the attempts of nature, but incites their production in species which do not ordinarily multiply in this manner. Plants which spread or multiply by this natural layering are said to be *stoloniferous.*

168. **A Sucker** is a branch of subterranean origin, which, after running horizontally and emitting roots in its course, at length, following its natural tendency, rises out of the ground and forms an erect stem, which soon becomes an independent plant. The Rose, the Raspberry, and the Mint afford familiar illustrations, as well as many other species which shoot up stems "from the root," as is generally thought, but really from subterranean branches. Cutting off the connection with the original root, the gardener propagates such plants *by division.* Plants which produce suckers are said to be *surculose.*

169. **A Runner,** of which the Strawberry furnishes the most familiar example, is a prostrate, slender branch, sent off from the base of the parent stem, which strikes root at its apex, and produces a tuft of leaves; thus giving rise to an independent plant capable of extending itself in the same manner. Branches of this sort are termed *flagelliform.*

170. **An Offset** is a similar, but short, prostrate branch, with a tuft of leaves at the end, which, resting on the ground, there takes root, and at length becomes independent; as in the Houseleek.

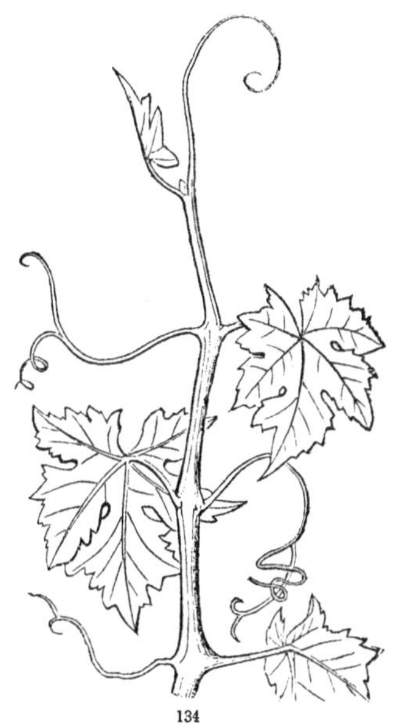

171. **A Tendril** is commonly a thread-like, leafless branch, capa-

FIG. 134. End of a shoot of the Grape-vine, showing the tendrils.

ble of coiling spirally, by which climbing plants attach themselves to surrounding bodies ; as in the Grape-vine (Fig. 134). But sometimes tendrils belong to the leaves, as in the Pea ; when they are slender prolongations of the leafstalk. Stems or stalks which bear tendrils are *cirrhose* or *cirrhiferous.*

172. **A Spine or Thorn** is an imperfectly developed, indurated, leafless branch of a woody plant, attenuated to a point. Their nature is manifest in the Hawthorn (Fig. 136), not only by their position in the axil of a leaf, but often by their bearing imperfect leaves themselves. In the Sloe, Pear, &c., many of the feebler branches become *spinose* or *spinescent* at the apex, tapering off gradually into a rigid, leafless point. These are less liable to appear on the cultivated tree, when duly cared for, such branches being thrown into more vigorous growth. In the Hawthorn, the spines spring from this peculiar growth of the main axillary bud, but it bears an accessory bud (153) on each side, one or the other of which grows into an ordinary branch. In the Honey Locust, it is the uppermost of several accessory buds, placed far above the axil, that developes into the thorn (Fig. 135). In this tree the spine itself branches, and sometimes becomes extremely compound. Sometimes the stipules of the leaves develope into spines, as in the Prickly Ash.

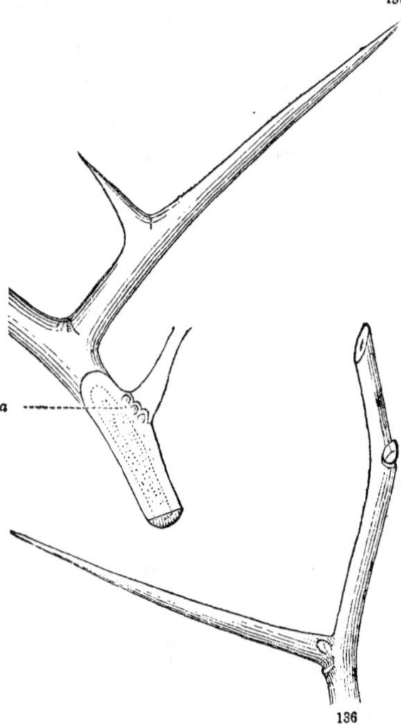

173. **The Subterranean Modifications of the Stem** are scarcely less

FIG. 135. Branching thorn of the Honey Locust (Gleditschia), an indurated branch developed from an accessory bud produced above the axil. *a*, Three buds under the base of the leafstalk, brought to view in a section of the stem and leafstalk below.

FIG. 136. Thorn of the Cockspur Thorn, developed from the central of three axillary buds ; one of the lateral ones is seen at its base.

numerous and diverse than the aerial; but they may all be reduced to a few principal types. They are perfectly distinguishable from roots by producing regular buds; or by being marked with scars, which indicate the former insertion of leaves, or furnished with scales, which are the rudiments or vestiges of leaves. All the *scaly roots* of the older botanists are therefore forms of the stem or branches, with which they accord in every essential respect; they grow, also, in the opposite direction from roots. So, likewise, what were called (as they are still popularly considered) *creeping roots* are really subterranean branches; such as those of the Mint, and of most Sedges and Grasses. Some of these, such as the Carex arenaria (Fig. 137) of Europe, render important service in binding

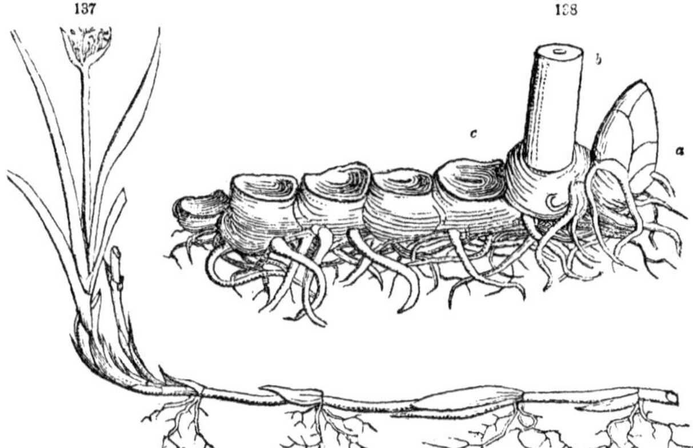

the shifting sands of the sea-shore. Others, like the Couch-Grass, are often very troublesome to the agriculturist, who finds it next to impossible to destroy them by the ordinary operations of husbandry; for, being furnished with buds and roots at every node, which are extremely tenacious of life, when torn in pieces by the plough, each fragment is only placed in the more favorable condition for becoming an independent plant. The Nut-Grass (Cyperus Hydra), an equally troublesome pest to the planters of Carolina and Georgia, is similarly constituted; and besides, the interminable subterranean branches bear tubers, or reservoirs of nutritive matter, in

FIG. 137. Creeping subterranean stem of Carex arenaria.

FIG. 138. Rhizoma of Diphylleia cymosa, showing six years' growth, and a bud for the seventh: *a*, the bud: *b*, base of the stalk of the current year: *c*, scar left by the decay of the annual stalk of the year before; and beyond are the scars of previous years.

their course, which have still greater powers of vitality, as they contain a copious store of food for the development of the buds they bear. The name of

174. **Rhizoma or Rootstock** is applied in a general way to all these perennial, horizontally elongated, and more or less subterranean root-like forms of the stem; and more particularly to those which are thickened by the accumulation of nutritive matter in their tissue (chiefly in the form of starch, 81), such as the so-called roots of Ginger, of the Iris or Flower-de-luce, of the Calamus or Sweet Flag, and of the Blood-root. They grow after the manner of ordinary stems, advancing from year to year by the annual development of a bud at the apex, and emitting roots from the under side of the whole surface; thus established, the most ancient portions die and decay, as corresponding additions are made to the opposite growing extremity. Each year's growth is marked in the rootstock of the Iris, &c., by a set of annular leaf-scars, left by the decay of the foliage of that year. In the Solomon's Seal and the Diphylleia (Fig. 138) it is more indelibly recorded by the series of broad and rounded scars on the upper surface, not unlike the impression of a seal (whence the popular name of Solomon's Seal), which is left by the separation in autumn of the herbaceous stalk of the season. The rootstock of Diphylleia is merely a string of such thickened and extremely abbreviated axes, formed by the annual development of a bud which, without elongation, sends up at once the single herbaceous stalk that bears the foliage and flowers. In our common Dentaria or Toothwort, and in Hydrophyllum, the base of this annual stalk or of the leafstalks partakes in the thickening and persists as a part of the rhizoma, in the form of fleshy scales or tooth-shaped processes. In other scaly rootstocks, these persistent bases of the leaves are thin, and more like bud-scales, and slowly decay after a year or two. All such markings are vestiges of leaves, &c., and indicate the nodes: they show that the body that bears them belongs to the stem; not to the root, which is wholly leafless. Rootstocks branch, like other stems, by the development of lateral buds from the axils of their scales or leaves. Thickened rootstocks serve as a reservoir of nourishing matter, for the maintenance of the annual growth, in the same manner as thickened roots (128). When such subterranean stems are thickened interruptedly, they produce

175. **A Tuber.** This is usually formed by the enlargement of

the apex, or growing bud, of a subterranean branch, the elongation of which is arrested, and the whole excessively thickened, by the deposition of starch, &c. in its tissue. This accumulation serves for the nourishment of the buds (eyes) which it involves, when they develope the following year. The common Potato offers the most familiar example ; and it is very evident on inspection of the growing plant, that the tubers belong to branches, and not to the roots. The nature of the Potato is also well shown by an accidental case (Fig. 140), in which some of the buds or branches above ground showed a strong tendency to develope in the form of tubers. By heaping the soil around the stems, the number of tuberiferous branches is increased. The Jerusalem Artichoke affords a good illustration of the tuber (Fig. 139). A tuber of a rounded form, and with few buds, is nearly the same as

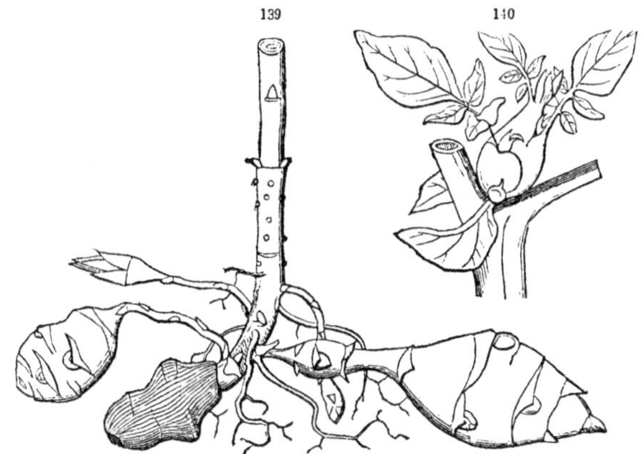

176. **A Corm** (*Cormus*), or *Solid Bulb.* This is a fleshy subterranean stem, of a round or oval figure and a uniform, compact texture ; as in the Arum triphyllum or Indian Turnip (Fig. 144), the Colchicum, the Crocus (Fig. 148), the Cyclamen,* &c.

* The broad and flattened corm of Cyclamen arises from the dilatation of the first internode of the stem, that which preëxists in the embryo below the cotyledons or seed-leaves. In many plants, this internode, or that immediately above the cotyledons, enlarges with the root. This occurs in the Tur-

FIG. 139. Base of the stem of the Jerusalem Artichoke (Helianthus tuberosus), showing the nature of the tubers.

FIG. 140. A monstrous branch or bud of the Potato, showing a transition to the tuber. (From the Gardener's Chronicle.)

It may be compared to the globular stem of a Melon-Cactus, like which it has no power of elongation; or it may be viewed as a tuber or rhizoma reduced to the greatest simplicity, developing one or more buds from its summit, and emitting roots from its base. Corms are often termed *solid bulbs;* and, indeed, they are only a kind of bulb with the axis more enlarged, and the investing scales either wholly wanting, as in the Indian Turnip (Fig. 144), or very few, forming a thin coating, as in the Colchicum and Crocus.

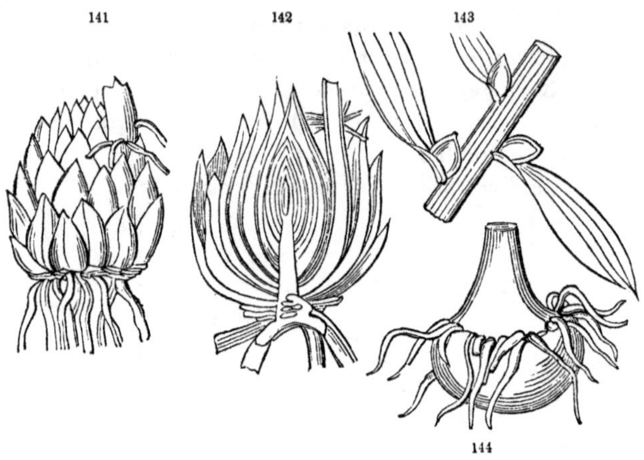

177. **A Bulb** is a permanently abbreviated stem, mostly shorter than broad, and clothed with scales, which are imperfect and altered leaves, or the thickened and persistent bases of ordinary leaves. Or, in other words, it is a scaly and usually subterranean bud, with thickened scales, and a depressed axis which never elongates. Its centre or apex developes above the herbaceous stalk, foliage, and flowers of the season, and beneath it emits roots. In the bulb, the thickening by the deposition of nutritive matter stored for future use takes place in the leaves or scales it bears, instead of the stem itself, as in the preceding forms. The scales are sometimes separate, thick, and in several distinct rows, as in the *scaly bulb* of the Lily (Fig. 141); sometimes broad and encircling each other in concentric layers, as in the *tunicated bulb* of the Onion (Fig. 145).

nip, Radish, Beet, &c.; where the *root* thus produced, or at least the upper part of it, presents the structure of the stem.

FIG. 141. The scaly bulb of a Lily. 142. A vertical section of the same, forming the annual stalk. 143. Axillary bulblets of Lilium bulbiferum. 144. Corm of Arum triphyllum.

178. **Bulblets** are small aerial bulbs, or buds with fleshy scales, which arise in the axils of the leaves of several plants, such as the common Lilium bulbiferum of the gardens (Fig. 143), and at length separate spontaneously, falling to the ground, where they strike root, and grow as independent plants. In the Onion, and other species of Allium, the flower-buds frequently change to bulblets. They plainly show the identity of bulbs with buds.

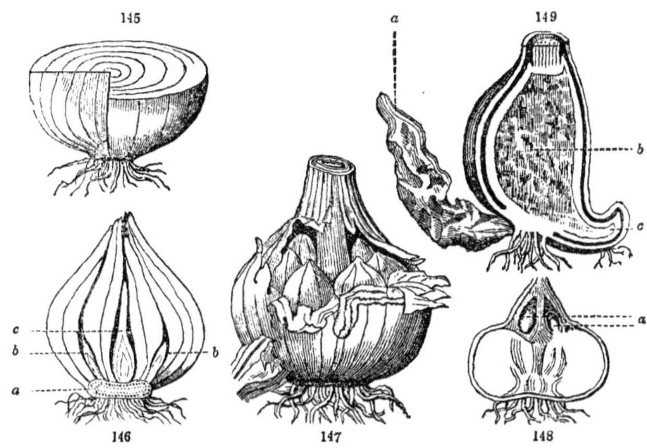

179. The regular plan of increase and ramification already described prevails in these extraordinary, no less than in the ordinary, forms of the stem. They grow and branch, or multiply, by the development of terminal and axillary buds. This is perfectly evident in the rhizoma and tuber, and is equally the case in the corm and bulb. The stem of the bulb is usually reduced to a mere plateau (Fig. 146, *a*), which produces roots from its lower surface, and leaves (the exterior of which are reduced to scales) from the upper surface. Besides the terminal bud (*c*), which usually forms the flower-stem, lateral buds (*b*) may be produced in the axils of the leaves or scales. One or more of these may develope as flowering stems the next season, and thus the same bulb survive and blossom from year to year (as is the case with the Tulip, Hya-

FIG. 145. Section of a tunicated bulb of the Onion.

FIG. 146. Longitudinal section of the bulb of the Tulip, showing its stem (*a*) and buds (*b*, *c*).

FIG. 147. Bulb of the Garlic, with a crop of young bulbs.

FIG. 148. Vertical section of the corm of Crocus: *a*, new buds.

FIG. 149. Vertical section of the corm of Colchicum, with the withered corm of the preceding (*a*), and the forming one (*c*) for the ensuing year.

cinth, &c.); or these axillary buds may themselves become bulbs, feeding on the parent bulb, which in this way is often consumed by its own offspring, as in the Garlic (Fig. 147); or, finally separating from the living parent, just as the bulblets of the Tiger Lily fall from the stem, they may form so many independent individuals. So the old corm of the Crocus (Fig. 148) produces one or two new ones (*a*) near the apex, and gradually dies as they develope. That of the Colchicum produces a new bud near the base of the old, upon which it feeds, and is in turn destroyed by its own progeny the next year; so that we observe (Fig. 149), *a*, the shrivelled corm of the year preceding; *b*, that of the present season (a vertical section); and *c*, the nascent bud for the ensuing season.

180. Many of the forms which the stem assumes when above ground differ as much from the ordinary appearance as do any of these subterranean kinds; as, for example, the globular Melon-Cactus, the columnar Cereus, and the jointed Opuntia or Prickly Pear.

SECT. IV. THE INTERNAL STRUCTURE OF THE STEM IN GENERAL.

181. Having considered the various external forms and appearances which the stem exhibits, and its mode of increase in length, our attention may now be directed to its internal structure, and mode of increase in diameter.

182. The stem embraces in its composition the various forms of elementary tissue that have already been described (Chap. I., Sect. II., III.); namely, ordinary cells, woody fibre, and vessels. At first, indeed, it consists entirely of parenchyma (51), which possesses much less strength and tenacity than woody tissue, and is therefore inadequate to the purposes for which the stem, in all the higher plants, is destined. The stem of a Moss or a Liverwort is, in fact, composed of ordinary cellular tissue alone; and is therefore weak and brittle, well enough adapted to the humble size of that tribe of plants, but incapable of attaining any considerable height. Accordingly, as soon as the stems of all the Phænogamous plants begin to grow, and in proportion as the leaves are developed, woody mingled with vascular tissue is introduced, to afford the requisite toughness and strength, and to facilitate the rise of the ascending sap. If it accumulates only to moderate extent in proportion to the parenchyma, the stem remains *herba-*

ceous (159); if it predominates and continues to accumulate from year to year, the proper woody trunk of a shrub or tree is formed. That the woody and vascular tissues arise from cells, which from an early period take a peculiar development, has already been shown (52-61).

183. The cellular part of the stem grows with equal readiness, in whatever direction the forces of vegetation act. It grows vertically, to increase the stem in length, and horizontally, to increase its diameter. Into this the elongated cells that form the woody tissue and ducts are introduced vertically; they run lengthwise through the stem and branches. Hence, the latter has been called the *longitudinal*, *vertical*, or *perpendicular system* (56, 64); and the cellular part, the *horizontal system* of the stem. Or the stem may be compared to a web of cloth; the cellular system forming the *woof*, and the woody, the *warp*.

184. The diversities in the internal structure of the stem are principally owing to the different modes in which the woody or vertical system is imbedded in the cellular. These diversities are reducible to two general plans; upon one or the other of which the stems of all Flowering Plants are constructed. Not only is the difference in structure quite striking, especially in all stems more than a year old, but it is manifested in the whole vegetation of the two kinds of plants, and indicates the division of Phænogamous plants into two great classes, recognizable by every eye; which, in their fully developed forms, may be represented, one by the Oak and the other trees of our climate, the other by the Palm (Fig. 166).

185. The difference between the two, as to the structure of their stems, is briefly and simply this. In the first, the woody system is deposited in *annual concentric layers* between a *central pith* and an *exterior bark*; so that a cross-section presents a series of rings or circles of wood, surrounding each other and a distinct pith, and all surrounded by a separable bark. This is the plan, not only of the Oak, but of all the trees and shrubs of the colder climates. In the second, the woody system is not disposed in layers, but consists of separate bundles or threads of woody fibre, &c., running through the cellular system without apparent order; and presenting on the cross-section a view of the divided ends of these threads in the form of dots, diffused through the whole; but with no distinct pith, and no bark which is at any time readily separable.

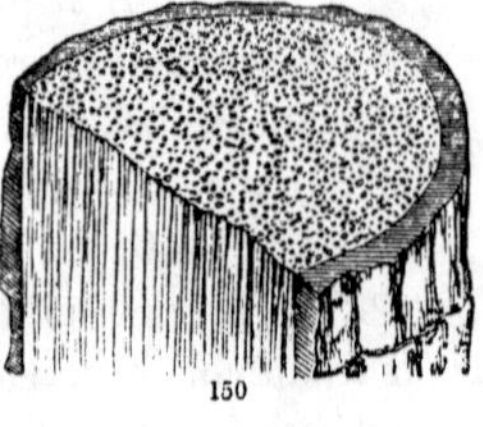

from the wood. The appearance of such a stem, both on the longitudinal and the cross-section, is shown in Fig. 150; it may also be examined in the Cane or Rattan, the Bamboo, and in the annual stalk of Indian Corn or of Asparagus. That of ordinary wood of the first sort is too familiar to need a pictorial illustration.

186. **Exogenous Structure.** The stem, in the first case, increases in diameter by the annual formation of a new layer of wood, which is deposited between the preceding layer and the bark; in other words, the wood increases by annual additions to its outside. Hence, such stems are said to have the EXOGENOUS structure; and the plants whose stems grow in this way are called EXOGENOUS PLANTS, or briefly EXOGENS; that is, as the term literally signifies, *outside-growers.*

187. **Endogenous Structure.** In the second case the new woody matter is intermingled with the old, or deposited towards the centre, which becomes more and more occupied with the woody threads as the stem grows older; and increase in diameter, so far as it depends on the formation of new wood, generally takes place by the gradual distention of the whole, the new wood pushing the old outwards. Accordingly, these stems are said to exhibit the ENDOGENOUS structure or growth; and such plants are called ENDOGENOUS PLANTS, or ENDOGENS; literally, *inside-growers.*

188. The two great classes of Phænogamous plants, indicated by this difference in the stem, are distinguishable even in the embryo state, by differences quite as marked as those which prevail in their whole port and aspect. The embryo of all plants that have endogenous stems bears only a single cotyledon, and therefore sends up but one seed-leaf in germination; hence, Endogens are also called MONOCOTYLEDONOUS PLANTS. The embryo of plants with exogenous stems bears a pair of cotyledons and unfolds a pair of seed-leaves in germination (Fig. 105 – 107); hence, Exogens are likewise called DICOTYLEDONOUS PLANTS.

FIG. 150. Section (longitudinal and transverse) of a Palm-stem.

SECT. V. THE EXOGENOUS OR DICOTYLEDONOUS STEM.

189. SINCE the Exogenous class is by far the larger in every part of the world, and embraces all the trees and shrubs with which we are familiar in the cooler climates, the structure of this kind of stem demands the earlier and more detailed notice. To obtain a true and clear idea of its internal structure, we should commence at its origin and follow the course of development.

190. In the embryo state, or at least at some period antecedent to germination, the rudimentary stem is entirely composed of parenchyma. But as soon as it begins to grow, while the cotyledons only are developing (as in Fig. 106, 107), some of the cells begin to lengthen into tubes, to be marked with transverse bars or spiral lines, and thus give rise to *ducts* or *vessels* (57 – 60); these are grouped as they form into a small and definite number of bundles or threads, say four equidistant ones in the first instance, as in the Sugar Maple: other slender cells of smaller calibre, and destitute of markings, soon appear, surrounding the threads of vessels, and forming the earliest woody tissue. As the rudiments of the next internode and its leaves appear, two or four additional threads of vascular tissue appear in the stem below, in the parenchyma between the earliest ones, and equally surrounded with forming woody tissue. At an early stage, therefore, the developing stem is seen to be traversed by several bundles of woody tissue with some vessels imbedded; and these, as they increase and enlarge, run together so as to make up a woody sheath, or, as seen in the cross-section, a ring, inclosing the central part of the parenchyma within it, and itself inclosed by the external parenchyma. Thus a circle or layer of wood is formed, which is in such a way imbedded in the original homogeneous cellular system as to divide it into two parts; namely, a central portion, which forms the pith, and an exterior zone, which belongs to the bark. The whole is of course invested by the skin or epidermis, which covers the entire surface of the plant. The way in which the layer of wood thus originates is somewhat rudely illustrated by the annexed diagrams (Fig. 151 – 153). The several woody masses, especially in trees and shrubs, are separated from each other by lines or bands of the original cellular tissue, which pass from the pith to the bark, and which necessarily become narrower and more numerous as the woody bundles or wedges increase in size and number. These are the

191. **Medullary Rays**, which form the radiating lines that the cross-section of most exogenous wood so plainly exhibits, especially that of the Oak, Plane, &c. They are the remains of the cellular system of that part of the stem, condensed by the pressure of the woody wedges, or plates, and which serve to keep up the communication between the pith and the bark.

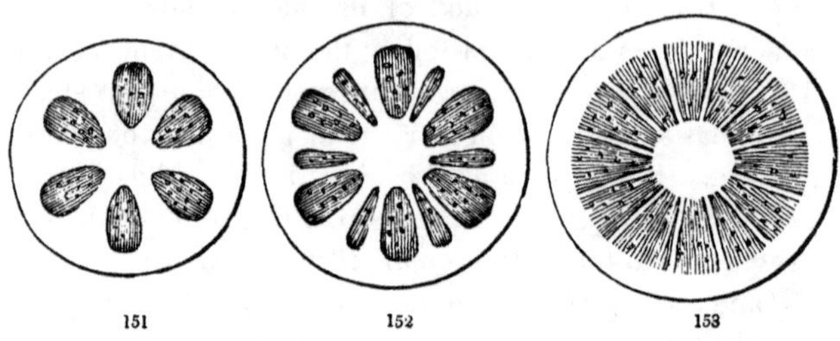

192. **The First Year's Growth** of an exogenous stem accordingly consists of three principal parts; namely, 1st, a central cellular portion, or *Pith;* 2d, a zone of *Wood;* and 3d, an exterior cellular portion, or *Bark.* Fig. 154 represents a section of a woody exogenous stem, a year old, of the natural size. Fig. 155 shows a portion of the same, magnified, so that the different parts may be distinguished, both on the longitudinal and transverse section: and Fig. 156 is a much more magnified view of a slice of the same, reaching from the bark to the pith.

193. **The Pith** (Fig. 155, 156, *a*) consists entirely of soft cellular tissue, or parenchyma (51), which is at first gorged with the nourishing juices of the plant. These are in time exhausted, leaving the older pith dry and light, or mere empty cells, which are of no further use to the plant. Many stems expand so rapidly in diameter during their early growth, that they become hollow, the pith being torn away by the distention, its remains forming a mere lining to the cavity, as in Grasses and other herbs; or else it is separated into horizontal plates, as in the Poke (Phytolacca) and the Walnut. Immediately surrounding the pith, and the very earliest

FIG 151. Plan of a cross-section of a young seedling stem, showing the manner in which the young wood is imbedded in the cellular system.

FIG. 152. The same at a later period, the woody bundles increased so as nearly to fill the circle.

FIG. 153. The same at the close of the season, where the wood has formed a complete circle, separating the pith from the bark, except that they are still connected by narrow portions of the cellular system (the *medullary rays*) which radiate from the pith to the bark.

part of the longitudinal system to appear, is what is called by the superfluous name of

194. **The Medullary Sheath.** This consists merely of the earliest formed vessels, already spoken of (190), and which of course stand in a circle immediately surrounding the pith; but they are seldom, if ever, so numerous as to form a closed layer, or sheath for the pith. More commonly they appear as a few bundles, one at the inner border of each of the larger and earlier woody wedges. They are mostly of the kind named spiral vessels (60), and it is remarkable that this is the only part of an exogenous stem in which spiral vessels ordinarily occur. They may be detected by breaking a woody twig in two, after dividing the bark and most of the wood by a circular incision, and then pulling the ends gently asunder, when their spirally coiled fibres are readily drawn out as gossamer threads. They are shown in place in the vertical section, Fig. 156, *b*.

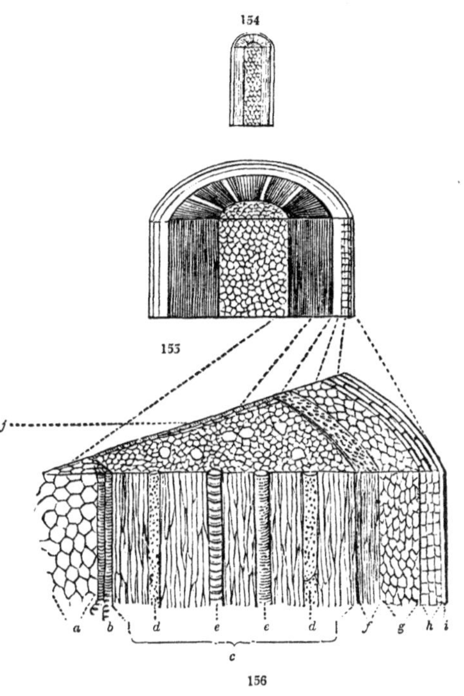

195. **The Wood** (Fig. 156, *c*) consists of proper woody tissue,

FIG. 154. Longitudinal and transverse section of a stem of the Soft Maple (Acer dasycarpum) at the close of the first year's growth; of the natural size.

FIG. 155. Portion of the same, magnified, showing the cellular pith, surrounded by the wood, and that inclosed by the bark.

FIG. 156. More magnified slice of the same, reaching from the bark to the pith: *a*, part of the pith; *b*, vessels of the medullary sheath; *c*, the wood; *dd*, dotted ducts in the wood; *ee*, annular ducts; *f*, the liber, or inner fibrous bark; *g*, the cellular envelope, or green bark; *h*, the corky envelope; *i*, the skin or epidermis; *j*, one of the medullary rays, seen on the transverse section.

among which the vascular is more or less copiously mingled, principally in the form of dotted ducts (*d*), or occasionally some spiral or annular ducts (*e*), &c. The dotted ducts are of so considerable calibre, that they are conspicuous to the naked eye in many ordinary kinds of wood, especially where they are accumulated in the inner portion of each layer, as in the Chestnut and Oak. In the Maple, Plane, &c., they are nearly equably scattered through the annual layer, and are of a size so small that they are not distinguishable by the naked eye.

196. The vertical section in Fig. 156 passes directly through the middle of one of the woody plates that collectively compose the layer; and therefore the medullary rays do not appear. But in the much more magnified Fig. 157, the section is made so as to show the surface of one of these plates, and one of the MEDULLARY RAYS passing horizontally across it, connecting the pith (*p*) with the bark (*b*). These medullary rays form the *silver-grain*, (as it is termed,) which is so conspicuous in the Maple, White Oak, Red Oak, &c., and which gives the glimmering lustre to many kinds of wood when cut in this particular direction. But a section made as a tangent to the circumference, and therefore perpendicular to the medullary rays, brings their ends to view, as in Fig. 158; much

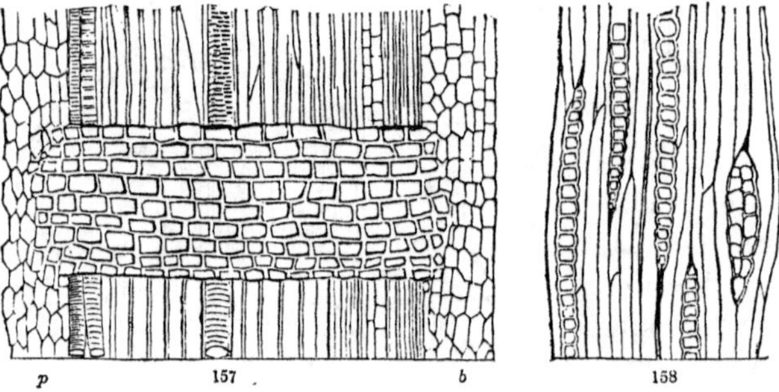

as they appear when seen on the surface of a piece of wood from which the bark is stripped. They are evidently composed of condensed parenchyma merely, and their origin has already been explained (191). They represent the horizontal system of the wood,

FIG. 157. Vertical section through the wood of a branch of the Maple, a year old; so as to show one of the medullary rays, passing transversely from the pith (*p*) to the bark (*b*): magnified. But a section can seldom be made so as to show one unbroken plate stretching across the wood, as in this instance.

FIG. 158. A vertical section across the ends of the medullary rays; magnified.

or the *woof*, into which the vertical woody fibre, &c., or *warp*, is interwoven. The inspection of a piece of oak or maple wood at once shows the pertinency of this illustration.

197. **The Bark**, in a stem of a year old, must next be more attentively considered. At first it consisted of simple cellular tissue, or parenchyma, undistinguishable from that of the pith, except that it assumed a green color when exposed to the light, from the production of *chlorophyll* (87) in its superficial cells. But during the formation of the proper wood, an analogous formation occurs in the bark. The inner portion, next the wood, has woody tissue formed in it, and becomes

198. **The Liber**, or *Fibrous Inner Bark* (Fig. 156, *f*). These fibre-like cells, which give to the inner bark of those plants that largely contain them its principal strength and toughness, are of the kind already described under the name of *bast-cells* or *bast-tissue* (55). They are remarkable for their length, flexibility, and the great thickness of their walls. They are deposited as detached bundles, or in bands separated by extensions of the medullary rays, one accordingly corresponding to each of the woody plates or wedges, or sometimes (as in Negundo, Fig. 159, 160) they are confluent into an unbroken circle round the whole circumference. The liber has received the technical name of ENDOPHLŒUM (literally *inner bark*). The exterior part of the bark, in which no woody tissue occurs, is early distinguishable, in most stems, into two parts, an inner and an outer. The former is

199. **The Cellular Envelope**, or *Green Layer* (Fig. 156, *g*), also called, from its intermediate position, the MESOPHLŒUM. This is composed of loose parenchyma, with thin walls, much like the green pulp of leaves (which last is, indeed, an outlying part of the same system), and containing an equal abundance of chlorophyll. It is the only part of the bark that retains a green color. In woody stems this is covered with

200. **The Corky Envelope**, or EPIPHLŒUM (Fig. 156, *b*), which gives to the twigs of trees and shrubs the hue peculiar to each species, generally some shade of ash-color or brown, or occasionally of much more vivid tints. It is rarely colored green, as in Negundo, where the inner cells contain chlorophyll. It is this tissue, which, taking an unusual development, forms the *cork* of the Cork-Oak, and those corky expansions of the bark which are so conspicuous on the branches of the Sweet Gum (Liquidambar), of

some of our Elms (Ulmus alata and racemosa), &c. It also forms the paper-like, exfoliating layers of Birch-bark. It is composed of laterally flattened parenchymatous cells, much like those of the EPIDERMIS (Fig. 156, *i*), which directly overlies it, and forms the skin or external surface of the stem.

201. To recapitulate the elements which compose the fabric of an exogenous stem of a year old, especially in a woody plant, and at the same time to exhibit them in an accurately drawn, more magnified view, we have, proceeding from the centre towards the circumference, —

I. In the Wood:

1. The *Pith*, belonging to the cellular system (Fig. 159, 160, *p*).

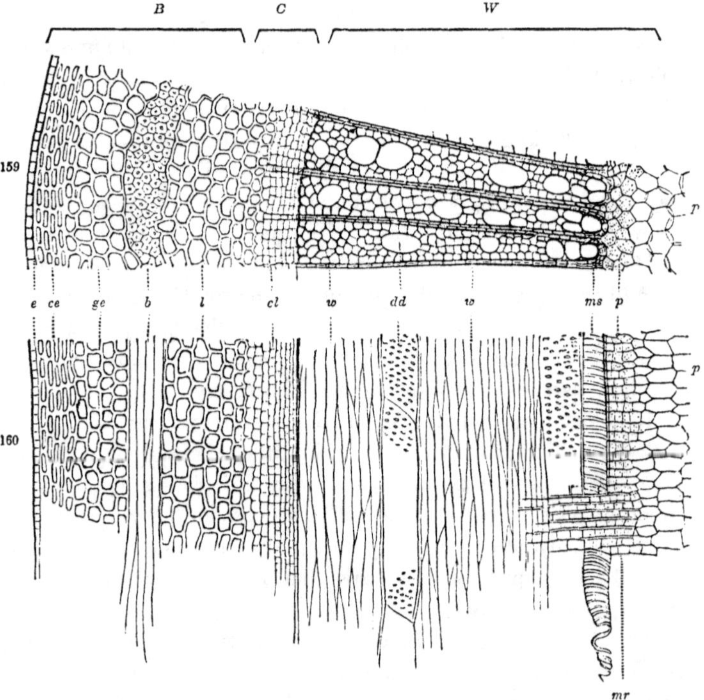

FIG. 159. Portion of a transverse section, and 160, a corresponding vertical section, magnified, reaching from the pith, *p*, to the epidermis, *e*, of a stem of Negundo, a year old: *B*, the bark; *W*, the wood; and *C*, the cambium-layer, as found in February. The references are in the text above; except *mr*, portion of a medullary ray, seen on the vertical section, where it runs into the pith: *dd*, dotted ducts: *cl*, the inner part of the cambium-layer, which begins the new layer of wood. In this tree, we find a thick layer of parenchyma (*l*) inside of the bast-tissue, and therefore belonging to the liber. No bast-tissue is formed in it the second year.

2. The *Medullary Sheath*, *ms*, } which belong to the woody or
3. The *Layer of Wood*, *W*, *w*, } longitudinal system.
4. The *Medullary Rays*, *mr*, a part of the cellular system.

II. In the Bark:

5. The *Liber*, *l*; its bast-tissue, *b*, belongs to the woody system.
6. The *Outer Bark*, belonging wholly to the cellular system, and composed of two parts; namely, 1st, the *Green* or *Cellular Envelope*, *ge*, and 2d, the *Corky Envelope*, *ce*.
7. The *Epidermis*, *e*, or skin, which invests the whole.

202. An herbaceous stem does not essentially differ from a woody one of this age, except that the wood forms a less dense and thinner zone; and the whole perishes, at least down to the ground, at the close of the season. But a shrubby or arborescent stem makes provision for an addition to its fabric the second year, — which may now be considered.

203. **Cambium-layer.** The wedges which constitute the woody layer usually increase in thickness throughout the season, by the continued development of prosenchymatous cells on their outer face, and the medullary rays extend equally by the multiplication of parenchymatous cells: so that there is always a stratum of delicate forming and growing cells interposed between the wood and the bark. This is called the CAMBIUM-LAYER (Fig. 159, 160, *C*). It survives the winter in all exogenous stems capable of more than one year's growth, remaining latent during the suspension of vegetation, and resuming its activity in the spring, to give rise to

204. **The Second Year's Growth in Diameter.** In spring, when vegetation vigorously recommences, and the buds are developing the onward growth of the season, a portion of the sap, charged with mucilage (dextrine, protoplasm, &c.), is at the same time attracted into the cambium-layer, as into every part where growth is going on; and the bark, before adherent, is now readily separable from the wood. To this mucilaginous organizable matter the name of *Cambium* was long ago applied, and hence the forming stratum is termed the cambium-layer; but the latter is only an incipient new woody layer; and the cambium is nothing more than ordinary sap, charged with dissolved assimilated matters, accumulated at the part of the woody stem where further growth alone takes place, and serving as the materials for such growth. It is quite wrong to suppose that there is a real interruption between the wood and the bark at this, or any other period, leaving a space

filled with extravasated sap. A series of delicate slices will at any time show that the bark and the wood are always organically connected, by a very delicate tissue of vitally active, young cells, just in the state in which they multiply by division (26, 32). It is when this process of growth is most rapidly going on, in spring or early summer, and the whole cambium-layer is gorged by the flow of sap, that the bark is so easily *separable ;* but the separation is effected by the rending of a delicate new tissue. The inner portion of this cambium-layer is forming wood ; the outer is forming bark. The cells of the first multiply vertically by division, and then elongate into prosenchyma or woody tissue, a part of them being at the same time commonly transformed into ducts ; thus producing a second layer of wood on the surface of the first, and continuous with the primary layer in the prolongation of the stem and in the branches made the same season. The exterior part of the cambium-layer contributes in much the same way to the thickness of the liber, which therefore grows inversely, or by accessions to its inner face. But the bark exhibits such great diversities in growth and structure, that it cannot well be further considered along with the wood.

205. **Annual Increase of the Wood.** Each successive year a new layer is added to the wood in the same manner ; each layer being, like the first, intersected by the extended medullary rays. A cross-section of such a stem, therefore, exhibits the wood disposed in concentric rings between the bark and the pith ; the oldest lying next the latter, or in the centre, and the youngest occupying the circumference. Each layer being the product of a single year's growth, the age of an exogenous tree may, in general, be correctly ascertained by counting the rings in a cross-section of the trunk. It is obvious, moreover, that the growing parts of an exogenous tree or shrub (and the same applies to the herb) are, — 1. The apex of the stem and branches, by buds, which continue the plant upwards and develope the foliage. 2. The lower extremity of the roots, by which these are advanced from year to year. 3. The cambium-layer, which annually produces a stratum of fresh tissue under the bark, between the buds and the rootlets, over the whole extent of the plant ; its ordinary growth giving rise to new annual layers of wood and inner bark ; while certain cells, taking a special development, form buds and consequently branches in the axils of the leaves, or, adventitiously (152), from other places, or

else, under favoring circumstances, secondary or adventitious roots (130). Lateral buds and roots, although they originate in the cambium-layer, have to grow and break through the bark before they appear externally.*

206. The limits of each year's growth in diameter in exogenous wood are apparent in the cross-section in the form of concentric layers, from two causes, either separate or combined; viz., the greater abundance of ducts in the earlier part of each annual increment, and the smaller size of the woody fibres in the latest growth of the season, which is destitute of ducts, and forms a finer-grained border to the ring. This is well shown in the cross-section of bass-wood, where the ducts compose the greater part of the wood at the inner edge of each layer, and very gradually diminish in number towards the outer edge, which is marked by a thin stratum of minute, laterally flattened wood-cells; — probably a portion of the cambium-layer that took no further growth. This fine exterior border alone marks the layers in white-pine wood, where there are no ducts or other vessels interspersed, and in such wood as that of the Sugar Maple, where the ducts are somewhat equably distributed through the whole breadth of the layer. In oak and chestnut wood, the layers are most strikingly marked, by the accumulation of all the large dotted ducts, here of extreme size and in great abundance, in the inner portion of each layer, where their open mouths on the cross-section are conspicuous to the naked eye, making a strong contrast between the inner porous, and exterior solid part of the successive layers.

207. The annual layers are most distinct in trees of temperate climates like ours, where there is a prolonged period of total repose, from the winter's cold, followed by a vigorous resumption of vegetation in spring. In tropical trees they are rarely so well defined; but even in these climes there is generally a more or less marked annual suspension of vegetation, occurring, however, in the dry and hotter, rather than in the cooler season. There are numerous cases, moreover, in which the wood forms a uniform stratum, whatever be the age of the trunk, as in the arborescent species of Cactus; or where the layers are few and by no means corresponding with the age of the trunk, as in the Cycas.

---

* That peculiar state of the wood of the Sugar Maple, called *Bird's-eye Maple*, is apparently caused by numberless rudimentary adventitious buds, which, failing to grow, have become involved in the woody layers.

208. In many woody climbing or twining stems, such as those of Clematis, Aristolochia Sipho, and Menispermum Canadense, the annual layers are obscurely, if at all, marked, while the medullary rays are unusually broad, and the wood therefore forms a series of separable wedges disposed in a circle around the pith. In the stem of one of our Trumpet-creepers (the Bignonia capreolata) the annual rings, after the first four or five, are interrupted in four places, and here as many broad plates of cellular tissue, belonging properly to the bark, are interposed, passing at right angles to each other from the circumference towards the centre, so that the transverse section of the wood nearly resembles a Maltese cross. But these are all exceptional cases, which scarcely require notice in a general view.

209. The wood of the Pine, Yew, Cypress, and the whole tribe of what are called *Coniferæ*, or cone-bearing trees, is characterized by its uniformity of structure, being formed of a peculiar woody tissue with little or no intermixture of true ducts, and by having the walls of these woody tubes marked with large circular discs, as in Fig. 23 (45, 54).

210. **Sap-wood and Heart-wood.** In the germinating plantlet and in the developing bud, the sap ascends through the whole tissue, of whatever sort; at first through the parenchyma, for there is then no other tissue; and the transmission is continued through it, especially through its central portion, or the pith, in the growing apex of the stem throughout. But in the older parts below, the pith is soon drained of sap by the demand above, and becomes filled with air in its place: thenceforth it bears no part in the plant's nourishment. As soon as wood-cells and ducts are formed, they take an active part in the conveyance of sap; for which their tubular and capillary character is especially adapted. But the ducts in older parts, except when gorged with sap, contain air alone; and the sap now continues to rise only or chiefly through the stem, year after year, to the places where growth is going on, through the proper woody tissue of the wood. In this transmission the new and fresh tissues are the most active. The walls of the cells that compose them soon begin to thicken by internal deposition and by incrustation with mineral matters introduced with the sap (39, 40, 53); and by the formation of new annual layers outside of them, their predecessors are each year removed a step farther from the region of growth; or rather the growing stratum, which connects

the fresh rootlets, that imbibe, with the foliage, that elaborates, the sap, is each year removed farther from them. The latter, therefore, after a few years, cease to convey sap, as they have long before ceased to take part in any vital operations. This older, more solidified, and harder wood, which occupies the centre of the trunk and is the part principally valuable for timber, &c., is called HEART-WOOD, or DURAMEN: while the newer layers of softer, more open and bibulous wood, which is apt to be surcharged with sap, receive the name of SAP-WOOD, or ALBURNUM. The latter name was given by the earlier physiologists in allusion to its white

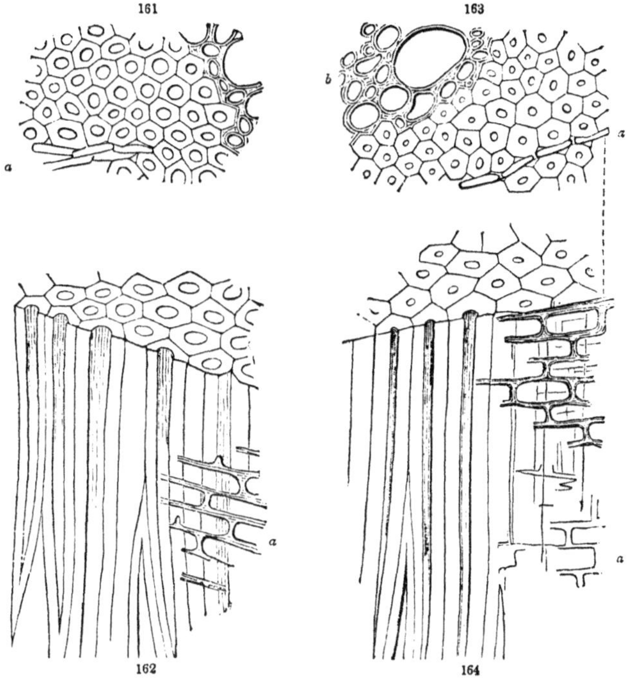

or pale color. In all trees which have the distinction between the sap-wood and heart-wood well marked, the latter acquires a deep-

FIG. 161. Magnified cross-section of a portion of woody tissue of White Oak, a year old. 162. A longitudinal as well as cross section of the same, a little higher magnified. *a, a,* Portions of one of the smaller medullary rays.

FIG. 163. Magnified cross-section of woody tissue from the same stem, taken from a layer of heart-wood, 24 years old: *b,* ducts: *a,* portion of one of the minuter medullary rays. 164. Combined cross and longitudinal section of the same: *a,* tissue of a medullary ray.

er color, and that peculiar to the species, such as the dark brown of the Black Walnut, the blacker color of the Ebony, the purplish-red of Red Cedar, and the bright yellow of the Barberry. These colors are owing to special vegetable products mixed with the incrusting matters ; but sometimes the hue appears to be rather an alteration of the lignine with age. In the Red Cedar, the deep color belongs chiefly to the medullary rays. To show that the older wood-cells are more solidified than the new, the annexed figures are given from corresponding parts of the same trunk of White Oak ; Fig. 161, 162, from sap-wood a year old ; Fig. 163, 164, from a layer of heart-wood twenty-four years old. The walls in both are greatly thickened with lignine ; but in the latter the calibre of a large part of the cells is almost obliterated. In many of the softer woods, there is little solidification in this way, and scarcely any change in color of the heart-wood, except from incipient decay, as in the White Pine, Poplar, Tulip-tree, &c.

211. Each layer of wood, once formed, remains unaltered in dimensions and position, and unchangeable except from internal deposition and from decay. The heart-wood is no longer in any sense a living part of the tree ; it may perish, as it frequently does, without affecting the life of the tree.

212. **The Bark** is much more various in structure and growth than the wood : it is also subject to grave alterations with advancing age, on account of its external position, to distention from the constantly increasing diameter of the stem within, and to abrasion and decay from the influence of the elements without. It is never entire, therefore, on the trunks of large trees ; but the dead exterior parts, no longer distending with the enlarging wood, are gradually fissured and torn, and crack off in layers, or fall away by slow decay. So that the bark of old trunks bears but a small proportion in thickness to the wood, even when it makes an equal annual growth.

213. The three constituent strata (197 – 200), for the most part readily distinguishable in the bark of young shoots, grow independently ; each by the addition of new cells to its inner face, so long as it grows at all. The green layer does not increase at all after the first year or two ; the thickening of the opaque corky layer soon excludes it from the light ; and it gradually perishes, never to be renewed again. The corky layer commonly increases for a few years only, by the formation of new tabular cells : occa-

sionally it takes a remarkable development, the cells swell out into polyhedral shapes, and multiply with unusual rapidity and in great quantities, forming the substance called *Cork*, as in the Cork-Oak. A similar growth occurs on the bark of several species of Elm, of our Liquidambar or Sweet-Gum, &c., producing corky plates on branches. In the Birch, thin annual layers, of very durable nature, are formed for a great number of years: each layer of tabular, firmly coherent cells (Fig. 165, *a*) alternates with a thinner stratum of delicate, somewhat cubical and less compact cells (*b*), which separate into a fine powder when disturbed, and allow the thin, paper-like plates to exfoliate.

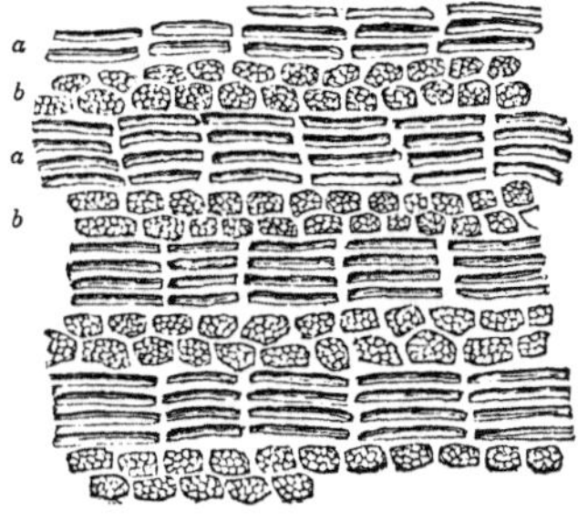

214. The liber, or inner bark (198), continues to grow throughout the life of the tree, by an annual addition from the cambium-layer applied to its inner surface. Sometimes the growth is plainly distinguishable into layers, corresponding with the annual layers of the wood: often, there is scarcely any trace of such layers to be discerned. The liber of the Bass-wood or Linden, and of other trees and shrubs with an evidently fibrous bark, consists of alternate strata of bast-cells * (or of parenchyma abounding with bundles of bast-cells) and of parenchyma alone. In the Sugar Maple, only a small proportion of bast-cells are formed after the first year. In Negundo there is a fine deposit of bast-cells the first year (Fig. 159, well distinguished by their opaline appearance in the magnified cross-section), but they are not again repeated, and the liber ever after consists of parenchyma alone, or with some thin and short prosenchymatous cells intermixed. The brittle liber of the Beech is nearly destitute of bast-cells. So is that

* The name, *liber*, is applied, even by the same author, sometimes to the whole inner bark, of whatever structure, sometimes to its bast-cells alone. It is applied in this work to the inner bark which grows year after year from the cambium-layer, (that is, to all within the green layer,) whether it continues to produce bast-cells or not.

FIG. 165. Transverse section of a minute portion of Birch-bark, the corky layer, highly magnified: *a*, the firm, tabular cells, *b*, the delicate thin-walled cells, in alternate layers.

of the Birch ; but it abounds with clusters of solidified cells, which take their place and exactly imitate ordinary bast-cells on the cross-section (Fig. 18) ; but a longitudinal section exhibits the same appearance, showing that they are globular in shape. In the first year's growth of the stem of Menispermum Canadense, there is a broad arc of bast-cells immediately before each wedge of wood ; in a stem of two or three years this is carried away from the wood by the development of purely cellular bark from the interposed cambium-layer ; it is finally thrown off at the surface, and no more is ever formed. A singular anomaly occurs in a species of Cocculus, where Decaisne has shown that the bast-cells remain connected with the face of the wood, and are covered by its second layer, so as eventually to be found in the interior of the wood. Laticiferous vessels or canals (63) abound in the newer parts of the liber.

215. Sometimes thin plates of delicate cells, like those of cork, are formed in the liber alternately with its proper tissue ; these early give way in the external layers, so that the outer part of the liber, as it grows older, scales off in plates year after year ; as is strikingly the case in the Buttonwood or Plane-tree, in the Shell-bark Hickory, in the Larch, Pine, &c. Even the liber of only one or two years old is thus annually detached in membranous layers or fibrous shreds from the stems of the Currant and Honeysuckle, the Spiræa opulifolia or Nine-Bark, and most strikingly in the Grape-vine. In the latter cases, the green and the corky layers are thrown off the first or second year ; in other cases, they disappear at a later period.

216. Obviously the recent liber and the newer layers of wood, with the interposed cambium-layer, are alone concerned in the life and growth of the tree. The old bark is constantly decaying or falling away from the surface, without any injury to the tree ; while the heart-wood may equally decay within without harm, except by mechanically impairing the strength of the trunk.

217. The crude sap rises to the leaves principally through the newer wood (210). The elaborated sap (79) is returned into the newest bark, thence sent to the cambium-layer, and horizontally diffused through the medullary rays (which may be viewed as inward extensions of the bark) into the sap-wood and all other living parts.

218. The proper juices and peculiar products of plants (80) are accordingly elaborated in the foliage and the bark, especially in the latter. In the bark, therefore, medicinal and other principles

are usually to be sought, rather than in the wood. Nevertheless, as the wood is kept in connection with the bark through the medullary rays, many products which probably originate in the former are found in the wood.

219. Exogenous plants almost always develope axillary buds, and produce branches: hence their stems and branches gradually taper upwards, or are conical.

SECT. VI. THE ENDOGENOUS OR MONOCOTYLEDONOUS STEM.

220. A cursory notice must now be taken of the stem of Endogens (*Inside-growers*), a great class of plants, which, although they have many humble representatives in northern climes, yet only attain their full characteristic development, and display their noble arborescent forms, under a tropical sun. Yet Palms — the type of the class — do extend as far north in this country as the coast of North Carolina (the natural limit of the Palmetto, Fig. 166); while in Europe the Date and the Chamærops have found their way to the warmer parts of the European shore of the Mediterranean. The

166

FIG. 166. The Chamærops Palmetto, in various stages, and the Yucca Draconis.

manner of their growth gives them a striking appearance; their trunks being unbranched cylindrical columns, rising majestically to the height of from thirty to one hundred and fifty feet, and crowned at the summit with an ample plume of peculiar foliage. Their internal structure is equally different from that of ordinary wood.

221. The stem of an Endogen, as already remarked (185), offers no manifest distinction into bark, pith, and wood; and the latter is not composed of concentric rings or layers, nor traversed by medullary rays. But it consists of bundles of woody and vascular tissue, in the form of thick fibres or threads, which are imbedded, with little apparent regularity, in cellular tissue; and the whole is inclosed in an integument which does not strictly resemble the bark of an Exogenous plant; inasmuch as it does not increase by layers, and is never separable from the wood. The fibrous bundles which compose the wood, and which consist of a mass of woody fibres surrounding several vessels, are distributed throughout the cellular system of the stem, most copiously near the circumference, but without being arranged in layers. Each bundle usually contains all the elements of the wood of the exogenous stem, namely, vessels, proper woody tissue, and bast-cells. The bundles may be traced directly from the base of the leaves down through the stem, some of them to the roots in a young plant, while others, curving outwards, lose themselves in the cortical integument, or rind. As the stem increases, new bundles, springing from the bases of more recently developed leaves, are at first directed towards the centre of the stem, along which they descend for some distance, growing more slender in their course, and then, curving outwards, mostly terminate in the rind. It is partly in consequence of the cohesion of these obliquely descending fibres to the false bark, that the latter cannot, as in Exogens, be separated from the wood beneath. The manner in which the woody threads are consequently interwoven is shown in Fig. 167. The Palm-like

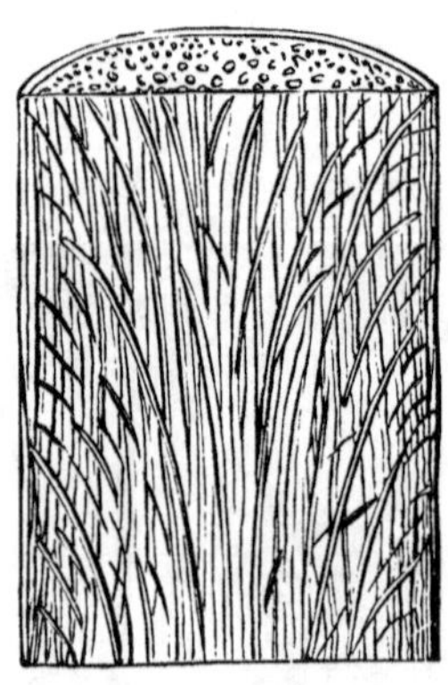

FIG. 167. Vertical and transverse section of a young endogenous stem, to show the curving of the fibres.

Yuccas of the Southern States offer beautiful illustrations of the kind. The appearance on a cross-section of an endogenous stem is shown in Fig. 150. The new woody bundles which are added from year to year, instead of arranging themselves outside the earlier wood and inclosing it, as in Exogens, actually descend more in the centre, and gradually force outward those which were first formed. Such a stem, therefore, instead of having the oldest and hardest wood at the centre and the newest and softest at the circumference, as in ordinary trees, is softest towards the centre and most compact at the circumference. In this way, and by the general growth of the cellular tissue in which the fibro-vascular bundles are imbedded, the stem increases in diameter as long as the rind is capable of distention. In some instances, as in the arborescent Yuccas and the Dracænas or Dragon-trees, the rind remains soft and capable of unlimited expansion; and the woody bundles descend after having reached the circumference, and thus the older stems continue to increase in diameter, much after the manner of an Exogen; but in the Palms, and in most woody Endogens, it soon indurates, and the stem consequently increases no further in diameter. The wood of the lower part of such stem is more compact than the upper, being more filled with woody bundles, the cells of which are lignified by internal deposition; and the rind is harder, from the greater number of ligneous fibres which terminate in it, and from its proper induration. Further increase in diameter being in these cases impossible, and the lower part of the stem becoming at length choked up by the multitude of descending bundles, it appears that the life of such Endogens must be limited.

222. Palms generally grow from the terminal bud alone, and perish if this bud be destroyed; the foliage is also borne in a cluster at the summit of the trunk; which consequently forms a simple cylindrical column. But in some instances two or more buds develope, and the stem branches, as in the Doum-Palm of Upper Egypt, and in the Pandanus, or Screw-Pine (Fig. 117), which belongs to a family closely allied to Palms: in such cases the branches are cylindrical. But when lateral buds are freely developed (as in the Asparagus), or the leaves are scattered along the stem or branches (as in the Bamboo, Maize, &c.), these taper upwards, just as in Exogens.

223. Grasses have endogenous stems, mostly of annual dura-

tion, and which early become hollow in the manner already indicated (193). In several of them, such as the Maize and Sugar-Cane, the stem remains solid; and these furnish good examples of ordinary endogenous structure.

Sect. VII. Of the Theoretical Structure of the Stem, etc.

224. **Origin of the Wood, &c.** We have seen that the plantlet which has as yet developed only one internode and one leaf (188), or one pair of leaves in germination (118), is complete in its parts, being provided with all the organs of vegetation, namely, with root, stem, and foliage. By this time its layer of wood is also manifest (a few vessels being first developed in four or more clusters, around which, principally on the outer side, woody tissue at once begins to appear); and the bark a little later exhibits traces of the elements of its three layers. This nascent wood begins to form early in germination. In a large and highly developed embryo it exists before germination. The conversion of young cells of parenchyma into vessels and wood-cells either commences in the radicle or stem-part and extends upwards into the cotyledons, when the latter are proportionally little developed; or, when they are large in proportion (as in the Almond, Fig. 97), it commences in the cotyledons and grows downwards into the radicle. The wood of the rudimentary stem and that of the leaf or leaves it bears are therefore in connection, are parts of the same system. As the root is produced from the lower end of the radicle (Fig. 107), its forming woody tissue extends downwards into it (the primary vessels, however, commonly developing as ducts instead of spiral vessels), and grow on as that advances by its cellular growth. The leaf or pair of leaves of the second internode by this time begins to appear; in which, or at the base of which, new vascular and woody tissues originate in the same way, extending through the leaf to form its woody system, or framework, making the woody stratum in the second internode of stem as it lengthens, and then contributing to the increase of the wood of the parent internode beneath. This is repeated throughout the whole growth of the season; each internode forming its own woody system, a portion of which appears separately in the leaf, while that in its stem blends with that of the internodes below to form the general zone of wood, in the exogenous stem. It is nearly the same in the en-

dogenous stem, except that the wood forms in separate bundles or threads, and these are commingled through the whole circumference of the young stem, instead of the new wood being constantly applied to the outside of that which was first formed. In the endogenous stem, the individual threads or bundles which form the wood may often be separately traced from the base of the leaf to their termination, at a considerable distance below. In the exogenous stem, their elements are usually confluent and undistinguishable in the common layer.

225. That the wood thus originates in connection with the leaves is shown, — (1.) By tracing the threads of soft woody Endogens, such as Yucca, and some Palms, directly from the base of the leaf into the stem, and thence downward to their termination, towards which they become attenuated, lose their vessels, and are finally reduced to slender shreds of woody tissue. (2.) Because the amount of wood formed in a stem or branch, other things being equal, is in proportion to the number and size of the leaves it bears; its amount in any portion of the branch is in direct proportion to the number of leaves above that portion. Thus, when the leaves are distributed along a branch, it tapers to the summit, as in a common Reed or a stalk of Indian Corn; when they grow in a cluster at the apex, it remains cylindrical, as in a Palm (Fig. 166). Consequently the aggregate diameter of the branches is (*cæteris paribus*) equal to that of the trunk from which they arise; as is beautifully illustrated by the *excurrent* stem of Pines and Firs, (carried directly upwards by the continued growth of the leading shoot, 155,) the diameter of which regularly diminishes as the lateral branches are given off. Consequently the increase of the trunk in diameter directly corresponds with the number and vigor of the branches. The greater the development of vigorous branches on a particular side of a tree, the more wood is formed and the greater the thickness of the annual layers on that side of the trunk. (3.) In a seedling, the wood appears just in proportion as the leaves are developed. (4.) If a young branch be cut off just below a node (141), so as to leave an internode without leaves or bud, no increase in diameter will take place down to the first leaf below. But if a bud be inserted into or ingrafted upon this naked internode, as the bud developes, increase in diameter, with the formation of new wood, recommences.

226. These facts conspire to show, not only the general depend-

ence of the wood on the leaves for its formation, but also that its formation proceeds *from above downwards.** The following are some of the considerations that may be adduced in confirmation of this view: — (1.) When a ligature is closely bound around a growing exogenous stem, the part above the ligature swells; that below does not. Every one may have observed the distortions that twining stems thus accidentally produce upon woody exogenous trunks. On examination, the woody fibres are found to be arrested at the upper margin of the ligature, and thrown into curved and knotted forms; or, where the ligature is spiral, the descending fibres follow the course of the obstruction. (2.) When we girdle the exogenous stem, by removing a ring of bark so as completely to expose the surface of the wood, the part above the ring enlarges in the same manner; that below does not, except by the granulation of cellular tissue, until the incision is healed. (3.) In a graft, the descending wood of the scion may sometimes be traced quite distinct from the stock. (4.) In many cases the fibres of wood are found to curve abruptly round a projection, gradually resuming their perpendicular direction below. Sometimes they take a very sinuous course, when there is no obstruction or evident cause of disturbance; the fibres of adjacent layers even crossing each other at right angles, showing an entire independence of the antecedent layer in their growth. (5.) The wood of the roots is manifestly formed in a descending direction. But it is continuous with that of the stem; and its first layer, the extension of the wood of the radicle into the primary root, agrees in composition with the wood of the succeeding layers in the stem, having no spiral vessels, but only ducts.

227. We have seen (148) that lateral buds develope into branches, just as the original embryo developed into the primary stem. Now the original embryo, or primary bud (144), not only grew upwards to form the stem, but downwards to form the root. Buds grow upwards into branches; have they aught corresponding to the downward growth which in the original stem is represented by the roots? The answer is furnished by those buds which may

---

* There is an article by James Warren, in the first volume of the Memoirs of the American Academy of Arts and Sciences, published in 1785, ingeniously maintaining the downward growth of the wood, apparently from original observations altogether.

be made to grow independently of the parent stem; such, for instance, as the bulblets of the Tiger Lily (Fig. 143), which are merely axillary buds with fleshy scales, and which, when they fall to the ground, or even while yet in their native situation, emit rootlets from their base, whose downward growth is the counterpart of the upward growth of the stem to which the bud gives rise. The same evidence is furnished by those ordinary buds which naturally grow in union with the parent, but which the gardener transfers to the soil in the form of cuttings (which are merely buds with a small piece of the stem), where they throw out roots from the base and grow into independent plants. As the bud, excited by warmth and moisture, developes upwards into a stem, just as it would have done into a branch had it remained in union with the parent, so it strikes root downward from the base of the cutting, and the woody matter of these roots, taken together, may be traced back directly to the bud. The resemblance between the original stem and the branches it bears, therefore, holds good throughout. As the downward growth of the original stem gives rise to roots, so the downward growth of the lateral buds, when they grow in connection with the parent stem, contributes to the wood beneath, and at length to the roots. In layering (167), the gardener well knows that roots strike more readily when an incision is made into the stem where it is covered with the soil. The evident explanation is, that the descending woody growth, arrested by the incision in the cellular callus that forms there, is forced, as it were, to strike at once into the soil, instead of pursuing the longer course through the main trunk to the same ultimate destination. This is the very economy of shrubs and trees which naturally multiply by suckers and stolons; from which the singular Banyan (Fig. 119), that in time spreads into a grove,

"High over-arched, with echoing walks between,"

in no wise differs, except that the roots strike and the whole process goes on high in the open air. In this case, portions of the new wood merely take another and nearer course to the ground in the form of aerial roots, which in time produce additional trunks, instead of continuing their adhesion to the branches, and contributing to the increase in diameter of the main trunk. The additional trunks thus produced, and which eventually, by separation and the decay of the original trunk, may form the stems of independent trees, exactly represent the outer and newer layers of an ordinary tree, the main stem representing the old and decaying centre. Further and very striking illustrations are furnished by those curious stems of Barbacenia, Kingia, and some Lycopodia, in which numerous aerial roots, instead of striking off free from the exterior, descend under the bark or rind, where they are closely pressed together, and form, as it were, coarse threads of wood; but on reaching the ground they assume the appearance and functions of real roots. Every transition is found between this arrangement and that in which they are united and blended with one another in a continuous ligneous tissue.

228. Nevertheless, it is carrying such conclusions much too far to assert, with Thouars and Gaudichaud, that wood is the roots of buds or of leaves, and to insist that each branchlet or branch contributes a distinguishable or definite portion to the trunk below, which is prolonged into a particular root or set of roots. In Palms, indeed, according to the high authority of Martius, there are no other threads of wood in the trunk than those which have proceeded from the bases of the leaves. But in exogenous stems, — of which most is known, — although the principal growth commences and proceeds in the manner above described (224), yet it undoubtedly goes on from year to year by the continual multiplication and growth of cells (32, 203 – 205) over the whole extent of the cambium-layer nearly simultaneously, irrespective, at least in the trunk and roots, of any direct connection with buds or leaves above. The formation of wood is resumed each spring where it was interrupted the previous autumn. This is shown in the case of stumps which have been kept alive for several years, in consequence of the natural ingrafting of some of their roots with the roots of adjacent trees of the same species, and which have continued to form annual layers, although very thin ones, while they survived, notwithstanding they bore no leafy shoots, or scarcely any.* The cambium-

---

* The ascertained fact, that the fibro-vascular tissue of secondary roots originates independently in the parenchyma, adjacent to, but not at first in contact with, the wood of the stem, is decisive against the Thouarsian hypothesis. So also is the case of a tree of Nyssa, found by M. Trécul, in Louisiana, which continued to live after the trunk was extensively stripped of its bark for its whole circumference, not only forming its annual layer (although a very thin one) below the girdled portion, but actually forming new wood in spots on the denuded surface.

layer, however it may have originated in the first instance, blends into a common stratum, which possesses an inherent power of continuing and reproducing itself, while it is nourished by the elaborated sap, which is generally supplied by the foliage above. It is well known that the ascending sap is laterally diffused with great readiness through the whole circumference of the sap-wood; if this be destroyed on one side of the tree, the sap that ascends on the other is equally supplied to all the branches throughout. The branches of each year's growth are, therefore, kept in fresh communication, by means of the newer layers of wood, with the fresh rootlets, which are alone active in absorbing the crude food of the plant from the soil. The fluid they absorb is thus conveyed directly to the branches of the season, which alone develope leaves to digest it. And the food they receive, having been elaborated and converted into organic nourishing matter, is partly expended in the upward growth of new branches, and partly in the downward formation of a new layer of wood, reaching from the highest leaves to the remotest rootlets. These two essential organs, namely, the rootlets which absorb, and the leaves which digest, the plant's nourishment, are, therefore, annually renewed; and, whatever their distance or the age of the tree, are maintained in fresh communication through the new annual layers. As the exogenous tree, therefore, annually renews its buds and leaves, its wood, bark, and roots, — every thing, indeed, that is concerned in its life and growth, — there seems to be no reason, no necessary cause inherent in the tree itself, why it should not live indefinitely. Accordingly, some trees are known to have lived for a thousand years or more; and others are now living which are with high probability thought to be above two thousand years old.* This longevity, however, will not appear surprising when we remember that

229. **The Plant is a Composite Being,** or community, lasting, in the case of a tree especially, through an indefinite and often immense number of generations. These are successively produced, enjoy their term of existence, and perish in their turn. Life passes

---

* The subject of the longevity of trees has been ably discussed by De Candolle, in the *Bibliothèque Universelle* of Geneva, for May, 1831, and in the second volume of his *Physiologie Végétale:* also, more recently, by Professor Alphonse De Candolle. In this country, an article on the subject has appeared in the *North American Review*, for July, 1844.

onward continually from the older to the newer parts, and death follows, with equal step, at a narrow interval; no portion of the tree is now living that was alive a few years ago; the leaves die annually and are cast off, while the internodes or joints of stem that bore them, as to their wood at least, are buried deep in the trunk, under the wood of succeeding generations; converted into heart-wood they are equally lifeless, or perchance decayed, while the bark that belonged to them is in time thrown off from the surface. It is the aggregate, the blended mass alone, that long survives. Plants of single cells are alone perfectly simple, and their existence is extremely short. But the more complex vegetable of a higher grade is not to be compared with the animal of the highest organization, where the offspring always separates from the parent, and the individual is consequently simple and indivisible; while it is truly similar to the branching of arborescent coral, or other compound animals of the lowest grade, where successive generations, though capable of living independently and sometimes separating spontaneously, yet are usually developed in connection, blended in a general body, and nourished more or less in common. Thus the coral structure is built up by the combined labors of a vast number of individuals, — by the successive labors of a great number of generations. The surface or the recent shoots alone are alive; and here life is superficial, all underneath consisting of the dead remains of former generations. The arborescent species are not only lifeless along the central axis, but are dead throughout towards the bottom: as, in a genealogical tree, only the later ramifications are among the living. It is the same with the tree, except that, as the plant imbibes its nourishment principally from the soil through its roots, it makes a downward growth also, and, by constant renewal of fresh tissues (216, 228), maintains the communication between the two growing extremities, the buds and the rootlets. We have seen that branches grow from the parent stem just as this grew from the embryo, only that they are implanted on the main trunk instead of the ground; still they are capable of living as independent individuals, and often do in various ways (as by bulbs, tubers, layers, stolons, offsets, &c.) spontaneously acquire a separate existence. The branches, therefore, or the buds, which are the branches in an earlier stage, are real individuals, which conspire to make up the composite tree. The contrary view would lead to the absurdity of an individual consisting of several genera and species; since the Apple, Pear, Mountain Ash, Quince, Medlar, and Hawthorne may all, by ingrafting, be combined in a single tree. It would also oblige us to consider as a single individual all the plants which have arisen from the mechanical subdivision of an original stem, — for example, perhaps all the Lombardy Poplars in this country, or even a large part of the Potatoes of Europe and America.' While united, however, all the branches are to some extent subordinate to the general whole; so that the term *individual plant* is justly applied to the aggregate stem and branches while they remain united, but no longer.

230. **Phytons.** The analysis of the Phænogamous plant must be carried still further: for a branch, or the simple primary stem itself, is composed of a lineal succession of similar parts, developed one upon the summit of another, each produced by the preceding, and producing that which in turn surmounts it (143); that is, it consists of a series of individual *plantlets* or *plant-elements*, which by their repetition make up the vegetable body. The first of these preëxists in the seed, as the embryo, or initial plantlet (Fig. 105): the downward growth from its lower extremity forms the root (Fig. 107), while from above it gives birth to all the rest, in lineal succession. A name being needful by which to designate this potential plant, the repetition of which makes up the perfect vegetable, that of PHYTON (from the Greek φυτόν, a plant) has been adopted for the purpose.

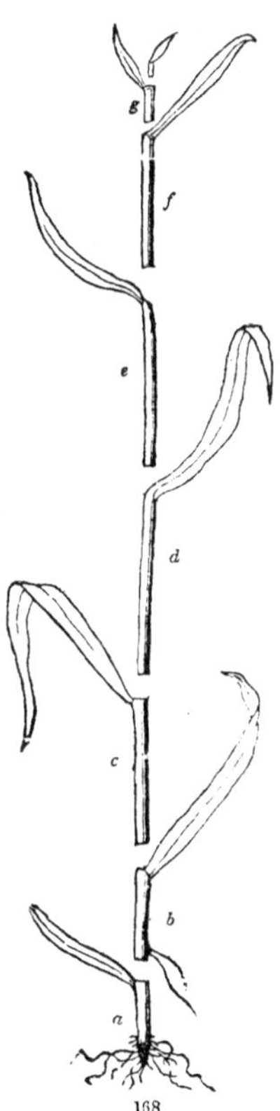

231. The dicotyledonous embryo (Fig. 100, 105) is a double organ, or consists of two simple phytons, with their stem portions united side by side to form the radicle, but each with its own leaf or cotyledon. The monocotyledonous embryo is equivalent to half the dicotyledonous, and

FIG. 168. Diagram to illustrate the development of a Monocotyledonous plant, by superposed phytons; *a*–*g*, the successive phytons, beginning with the first.

therefore exhibits the simplest case. It developes one primary phyton in germination (Fig. 168, *a*), this a second (*b*), this a third (*c*), and so on; each like the preceding, only successively larger and more vigorous as the plant thus multiplies its organs; except that the primary one alone grows downwards into a root in the first instance. But the others mingle their woody tissues with those of the older phytons beneath, and thus draw up their portion of the liquid which the primary root imbibes. They may likewise send forth secondary roots of their own, to establish a direct communication with the soil (as in Fig. 168, *b*). This they commonly do when in contact with the soil (130), and not rarely when raised to some distance above it (131): or they may be made to strike root and live independently, when taken off as cuttings (227). When the dicotyledonous embryo goes on to develope double phytons, like itself, each node bears a pair of leaves (as in Fig. 101 – 104), or, in botanical description, the leaves are said to be *opposite;* as they are in the Maple, in the Mint Family, &c. But quite as frequently the phytons become disjoined or simple after the first or second, each bearing a single leaf only; so that the leaves become *alternate*, just as in those from the monocotyledonous embryo, except that they are there alternate from the very first. This occurs in the Apple, Cherry (Fig. 169), and numberless other instances.

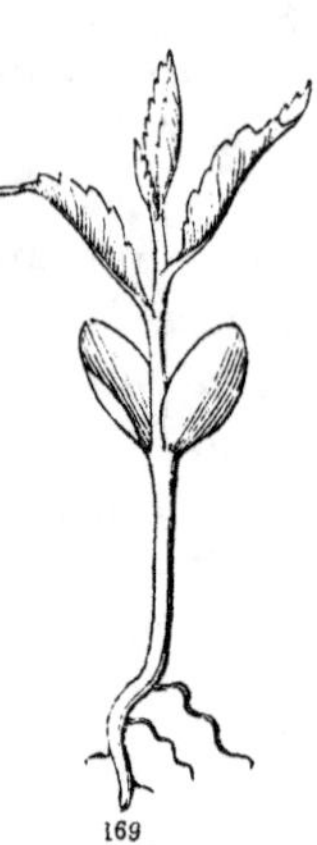

232. The same analysis applies to axillary buds and branches. In most of our trees and shrubs these buds acquire considerable complexity before they begin to unfold (144), and then grow almost simultaneously: but in some of them, as in most annual herbs, the axillary phytons begin to develope one by one.

FIG. 169. Germination of the Cherry; the leaves alternate after the first pair, or cotyledons.

CHAPTER V.

OF THE LEAVES.

Sect. I. Their Arrangement. (Phyllotaxis, etc.)

233. The fundamental organs of the vegetable, namely, the root, stem, and leaves, are so intimately associated and mutually dependent, that the structure and office of no one of them can be separately treated of. The stem, in particular, cannot be understood apart from the leaves. It has accordingly been necessary to anticipate several of the leading points of the present chapter. As to the general office of leaves in the vegetable economy, it has been assumed that the leaf is an apparatus in which, under the agency of sunlight, the sap is digested, and converted into the proper nourishment of the plant (79, 114). As to their situation upon the stem, it has been stated that they invariably arise from the nodes (141), just below the point where buds appear (148). So that wherever a bud or branch is found, a leaf exists, or has existed, either in a perfect or rudimentary state, just beneath it; and buds (and therefore branches), on the other hand, are or may be developed in the axils of all leaves, and do not normally exist in any other situation. And finally, the relation of leaves to the wood and the general structure of the stem has just been noticed (224-231). From its natural connection with that topic, it will be most convenient first to consider their arrangement on the stem. This subject, which has of late been elaborately investigated, has received the name of

234. **Phyllotaxis** (from two Greek words, signifying *leaf-arrangement*). We can here only briefly illustrate the general laws which appear to regulate the arrangement of leaves on the stem, as manifested in the several modes which are of ordinary occurrence.

235. The point of attachment of a leaf (or other organ) with the stem is termed its *insertion*.

236. In botanical descriptions, leaves are said to be *alternate* (149), when there is only one to each node or phyton, as in Fig. 168, in which case the successive leaves are thrown alternately to different sides of the stem: they are said to be *opposite* when each node bears a pair of leaves (149, 231), in which case the two

leaves always diverge from each other as widely as possible, that is, they stand on opposite sides of the stem and point in opposite directions (Fig. 107, 104): or they are *verticillate* or *whorled*, when there are three or more leaves in a circle (*verticil* or *whorl*) upon each node; in which case the several leaves of the circle diverge from each other as much as possible, or are equably distributed around the whole circumference of the axis. The first of the three is the simplest as well as the commonest method, occurring as it does in almost every Monocotyledonous plant (where it is plainly the normal mode, Fig. 168), and in the larger number of Dicotyledonous plants likewise, after the first or second nodes. It should therefore be first examined.

237. **Alternate Leaves.** This general term, which commonly suffices in descriptive botany, obviously comprises a variety of modes. There is, first, the case to which the name is strictly applicable, namely, where the leaves are *alternately disposed on exactly opposite sides of the stem* (as in Fig. 168); the second leaf being thrown to the side farthest from the first, while the third is equally removed from the direction of the second, and is consequently placed directly over the first, the fourth stands over the second, and so on throughout. Such leaves are accordingly *distichous* or *two-ranked*. They form two vertical rows: on one side is the series 1, 3, 5, 7, &c.; on the opposite, the series 2, 4, 6, 8, and so on. This mode occurs in all Grasses, in many other Monocotyledonous plants, and among the Dicotyledonous in the Linden. A second variety of alternate leaves is

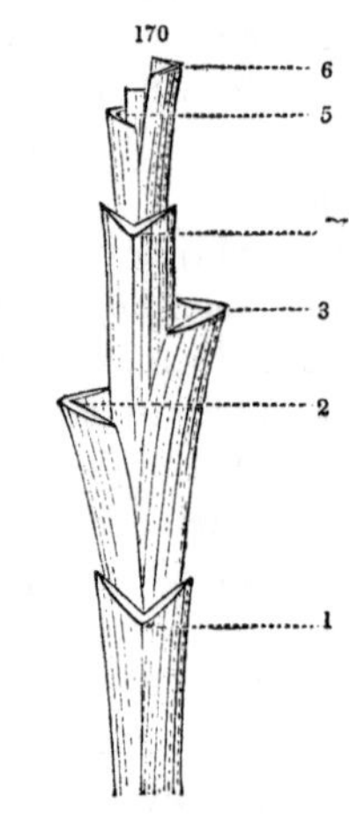

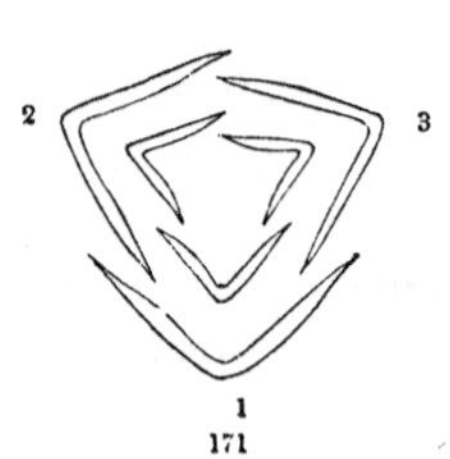

238. The *tristichous* or *three-ranked* arrangement, which is seen in Sedges (Fig. 170) and some other Monocotyledonous plants. Taking any leaf we please to begin with, and numbering

FIG. 170. Piece of a stalk, with the sheathing bases of the leaves, of a Sedge-Grass (Carex Crus-corvi), showing the three-ranked arrangement. 171. Diagram of the cross-section of the same, showing two cycles of leaves.

it 1, we pass round one third of the circumference of the stem as we ascend to leaf No. 2; another third of the circumference brings us to No. 3; another brings us round to a line with No. 1, exactly over which No. 4 is placed. No. 5 is in like manner over No. 2, and so on. They stand, therefore, in three vertical rows, one of which contains the numbers 1, 4, 7, 10; another, 2, 5, 8, 11; the third, 3, 6, 9, 12, and so on. If we draw a line from the insertion of one leaf to that of the next, and so on to the third, fourth, and the rest in succession, it will be perceived that it winds around the stem spirally as it ascends. In the distichous mode (237), the second leaf is separated from the preceding by half the circumference of the stem; and, having completed one turn round the stem, the third begins a second turn. In the tristichous, each leaf is separated from the preceding and succeeding by one third of the circumference, there are three leaves in one turn, or *cycle*, and the fourth commences a second cycle, which goes on in the same way. That is, the *angular divergence*, or size of the arc interposed between the insertion of two successive leaves, in the first is $\frac{1}{2}$, in the second $\frac{1}{3}$, of the circle. These fractions severally represent, not only the angle of divergence, but the whole plan in these two modes; the numerator denoting the number of times the spiral line winds round the stem before it brings a leaf directly over the one it began with; while the denominator expresses the number of leaves that are laid down in this course, or which form each cycle. The two-ranked mode ($\frac{1}{2}$) is evidently the simplest possible case. The three-ranked ($\frac{1}{3}$) is the next, and the one in which the spiral character of the arrangement begins to be evident. It is further illustrated in the next, namely,

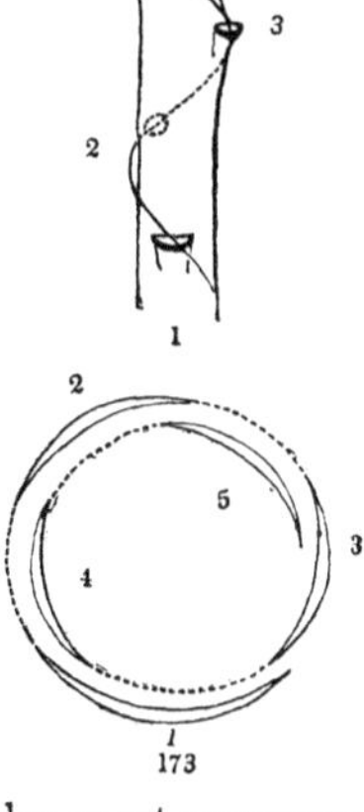

239. The *pentastichous*, *quincuncial*, or *five-ranked* arrangement (Fig. 172). This is much the most common

FIG. 172. Diagram of the five-ranked arrangement of the leaves, as in the Apple-tree; a spiral line is drawn ascending the stem and passing through the successive scars which mark the position of the leaves from 1 to 6. It is made a dotted line where it passes on the opposite side of the stem, and the scars 2 and 5, which come on that side, are made fainter. 173. A plane, horizontal projection of the same; the dotted line passing from the edge of the first leaf to the second, and so on to the fifth leaf, which completes the cycle; as the sixth would come over, or within the first.

case in alternate-leaved Dicotyledonous plants. The Apple, Cherry, and Poplar afford ready examples of it. Here there are five leaves in each cycle, since we must pass on to the sixth before we find one placed vertically over the first. To reach this, the ascending spiral line has made two revolutions round the stem, and on it the five leaves are equably distributed, at intervals of $\frac{2}{5}$ of the circumference. The fraction $\frac{2}{5}$ accordingly expresses the angular divergence of the successive leaves; the numerator indicates the number of turns made in completing the cycle, and the denominator gives the number of leaves in the cycle, or the number of vertical ranks of leaves on such a stem. If we shorten the axis, as it was in the bud, or make a horizontal plan, we have the parts disposed as in the diagram, Fig. 173, the lower leaves being of course the exterior.

240. The eight-ranked arrangement, the next in order, is likewise not uncommon. It is found in the Holly, the Callistemon of our conservatories, the Aconite, the tuft of leaves at the base of the common Plantain, &c. In this case the ninth leaf is placed over the first, the tenth over the second, and so on; and the spiral line makes three turns in laying down the cycle of eight leaves, each separated from the preceding by an arc, or angular divergence of $\frac{3}{8}$ of the circumferance.

241. All these modes, or nearly all of them, were pointed out by Bonnet as long ago as the middle of the last century; but they have recently been extended and generalized, and the mutual relations of the various methods brought to light, by sagacious recent researches, principally those of Schimper and Braun. If we write down in order the series of fractions which represent the simpler forms of phyllotaxis already noticed, as determined by observation, viz. $\frac{1}{2}$, $\frac{1}{3}$, $\frac{2}{5}$, $\frac{3}{8}$, we can hardly fail to perceive the relation that they bear to each other. For the numerator of each is composed of the sum of the numerators of the two preceding fractions, and the denominator of the sum of the two preceding denominators. (Also the numerator of each fraction is the denominator of the next but one preceding.) We may carry out the series by applying this simple law, when we obtain the further terms, $\frac{5}{13}$, $\frac{8}{21}$, $\frac{13}{34}$, $\frac{21}{55}$, &c. Now these numbers are those which are actually verified by observation, and, with some abnormal exceptions, this series comprises all the cases that occur. These higher forms are the most common where the leaves are crowded on the stem, as in the rosettes of the

Houseleek (Fig. 174), and the scales of Pine-cones (for the arrangement extends to all parts that are modifications of leaves), or where they are numerous and small in proportion to the circumference of the stem, as the leaves of Firs, &c. In fact, when the internodes are long and the base of the leaves large in proportion to the size of the stem, it is difficult, and often impossible, to tell whether the 8th, 13th, or 21st leaf stands exactly over the first. When, on the other hand, the internodes are very short, so that the leaves touch one another, or nearly so, we may readily perceive what leaves are superposed; but it is then difficult to follow the succession of the intermediate leaves. When this cannot be directly done, however, the order may be deduced by simple processes.

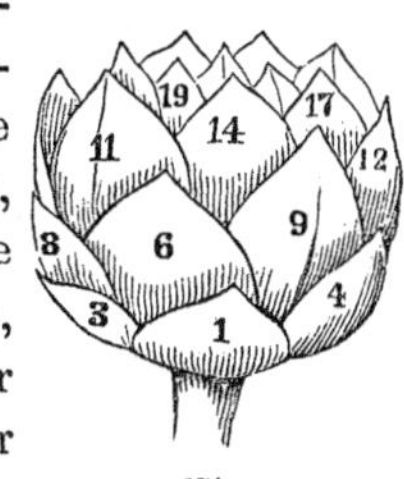

242. Sometimes we can readily count the number of vertical ranks, which gives the denominator of the fraction sought. Thus, if there are eight, we refer the case to the $\frac{3}{8}$ arrangement in the regular series; if there are thirteen, to the $\frac{5}{13}$ arrangement, and so on.

243. Commonly, however, when the leaves are crowded, the vertical ranks are by no means so manifest as two or more orders of oblique series, or *secondary spirals*, which are at once seen to wind round the axis in opposite directions, as in the Houseleek (Fig. 174; where the numbers 1, 6, 11 belong to a spire that winds to the left, 1, 9, 17 to another, which winds to the right, and 3, 6, 9, 12 to still another, that winds in the same direction): they are still more obvious in Pine-cones (Fig. 175, 176). These oblique spiral ranks are a necessary consequence of the regular ascending arrangement of parts with equal intervals over the circumference of the axis: and if the leaves are numbered consecutively, these numbers will necessarily stand in arithmetical progression on the oblique ranks, and have certain obvious relations with the primary spiral which originates them; as will be seen by projecting them on a vertical plane.

244. Take, for example, the quincuncial ($\frac{2}{5}$) arrangement, where, as in the annexed diagram, the ascending spiral, as written on a plane surface, appears in the numbers 1, 2, 3, 4, 5, 6, and

FIG. 174. An offset of the Houseleek, with the rosette of leaves unexpanded, exhibiting the 5-13 arrangement; the fourteenth leaf being directly over the first.

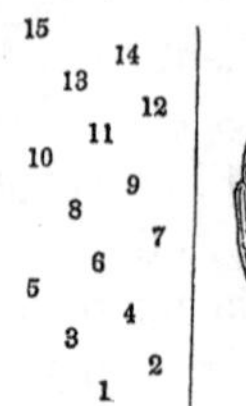

so on: the vertical ranks thus formed, beginning with the lowest (which we place in the middle column that it may correspond with the Larch-cone, Fig. 175, where the lowest scale, 1, is turned directly towards the observer), are necessarily the numbers **1, 6, 11; 4, 9, 14; 2, 7, 12; 5, 10, 15;** and **3, 8, 13.** But two parallel oblique ranks are equally apparent, ascending to the left; viz. **1, 3, 5**, which, if we coil the diagram round a cylinder, will be continued into **7, 9, 11, 13, 15;** and also **2, 4, 6, 8, 10**, which runs into **12, 14**, and so on, if the axis be further prolonged. Here the circumference is occupied by two secondary left-hand series, and we notice that the common difference in the sequence of numbers is two: that is, the number of the parallel secondary spirals is the same as the common difference of the numbers on the leaves that compose them. Again, there are other parallel secondary spiral ranks, three in number, which ascend to the right; viz. **1, 4, 7**, continued into **10, 13; 3, 6, 9, 12**, continued into **15;** and **5, 8, 11, 14**, &c.; where again the common difference, **3**, accords with the number of such ranks. This fixed relation enables us to lay down the proper numbers on the leaves, when too crowded for directly following their succession, and thus to ascertain the order of the primary spiral series by noticing what numbers come to be superposed in the vertical ranks. We take, for example, the very simple cone of the small-fruited American Larch (Fig. 175), which usually completes only two cycles; for we see that the lowest, one intermediate, and the highest scale, on the side towards the observer, stand in a vertical row. Marking this lowest scale 1, and counting the parallel secondary spirals that wind to the left, we find that two occupy the whole circumference. From 1, we number on the scales of that spiral 3, 5, 7, and so on, adding the common difference, **2**, at each step. Again, counting from the base the right-hand secondary spirals, we find three of them, and therefore proceed to number the lowest one by adding this common difference, viz. **1, 4, 7, 10;** then, passing to the next, on which the number **3** has already been fixed, we carry on that sequence, 6, 9, &c.; and on the third, where No. 5 is already fixed, we continue the numbering, 8, 11, &c. This gives us,

FIG. 175. A cone of the small-fruited American Larch (Larix microcarpa), with the scales numbered, exhibiting the five-ranked arrangement, as in the annexed diagram.

in the vertical rank to which No. 1 belongs, the sequence 1, 6, 11, showing that the arrangement is of the quincuncial ($\frac{2}{5}$) order. It is further noticeable, that the smaller number of parallel secondary spirals, 2, agrees with the numerator of the fraction in this the $\frac{2}{5}$ arrangement; and that this number added to that of the parallel secondary spirals which wind in the opposite direction, viz. 3, gives the denominator of the fraction. This holds good throughout; so that we have only to count the number of parallel secondary spirals in the two directions, and assume the smaller number

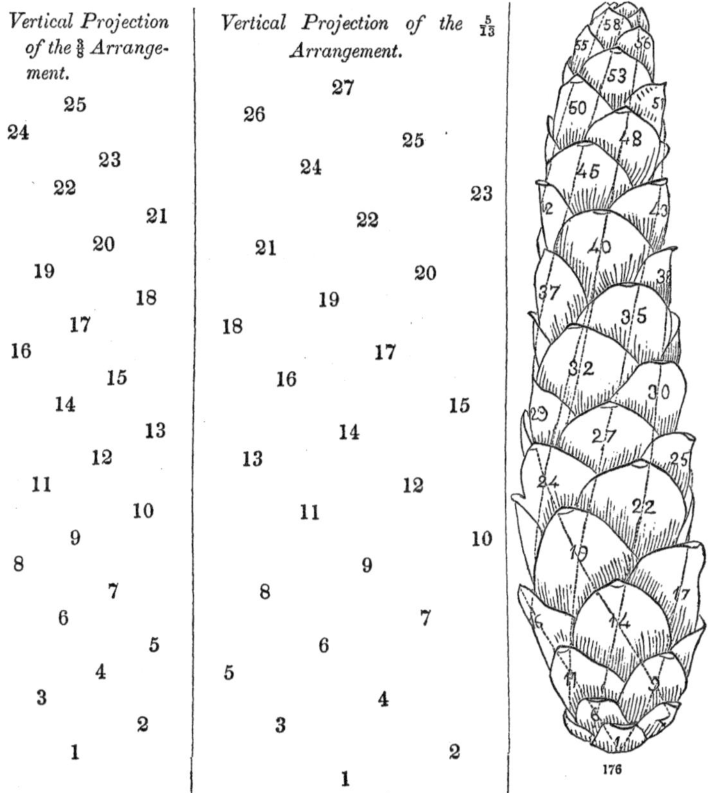

as the numerator, and the sum of this and the larger number as the

FIG. 176. A cone of the White Pine, on which the numbers are laid down, and the leading higher secondary spirals are indicated: those with the common difference 8 are marked by dotted lines ascending to the right; two of the five that wind in the opposite direction are also marked with dotted lines: the set with the common difference 3, in one direction, and that with the common difference 2, in the other, are very manifest on the cone.

denominator, of the fraction which expresses the angular divergence sought. For this we must take, however, the order of secondary spirals nearest the vertical rank in each direction, when there are more than two, as there are in all the succeeding cases.

245. A similar diagram of the $\frac{3}{8}$ arrangement introduces a set of secondary spirals, in addition to the two foregoing, ascending in a nearer approach to a vertical line, and with a higher common difference, viz. 5. There are accordingly five of this sort, viz. those indicated in the diagram by the series **1, 6, 11, 16 ; 4, 9, 14, 19, 24 ; 2, 7, 12, 17, 22 ; 5, 10, 15, 20, 25** ; and **3, 8, 13, 18, 23.** The highest obvious spiral in the opposite direction, viz. that of which the series **1, 4, 7, 10, 13** is a specimen, has the common difference 3, and gives the numerator, and 3 + 5 the denominator, of the fraction $\frac{3}{8}$. The next case, $\frac{5}{13}$, which is exemplified in the rosettes of the Houseleek (Fig. 174) and in the cone of the White Pine (Fig. 176), introduces a fourth set of secondary spirals, eight in number, with the common difference eight, viz. that of which the series **1, 9, 17, 25** is a representative. The set that answers to this in the opposite direction, viz. **1, 6, 11, 16, 21, 26,** with the common difference 5, gives the numerator, and 5 + 8 the denominator, of the fraction $\frac{5}{13}$. We may here compare the diagram with an actual example (Fig. 176) : a part of the numbers are of course out of sight on the other side of the cone. The same laws equally apply to the still higher modes.

246. The order is uniform in the same species, but often various in allied species. Thus, it is only $\frac{2}{5}$ in our common American Larch ; in the European species, $\frac{8}{21}$. The White Pine is $\frac{5}{13}$, as is also the Black Spruce ; but other Pines with thicker cones exhibit in different species the fractions $\frac{8}{21}$, $\frac{13}{34}$, and $\frac{21}{55}$. Sometimes the primitive spiral ascends from left to right, sometimes from right to left. One direction or the other generally prevails in each species, yet both directions are not unfrequently met with even in the same individual plant.

247. But when a branch springs from a stem or parent axis, the spiral is found to be continued directly from the leaves of the stem to those of the branch, so that the leaf from whose axil the branch arises begins the spire of that branch. When the spire of the branch turns in the same direction as that of the parent axis, as it more commonly does, it is said to be *homodromous* (from two Greek words, signifying *like course*) : when it turns in the

opposite direction, it is said to be *heterodromous* (or of *unlike course*).

248. The cases represented by the fractions $\frac{1}{2}$, $\frac{1}{3}$, and $\frac{2}{5}$ are the most stable and certain, as well as the easiest to observe. In the higher forms, the exact order of superposition often becomes uncertain, owing to a slight torsion of the axis, or to the difficulty of observing whether the 9th, 14th, 21st, 35th, or 56th leaf is directly over the first, or a little to the one side or the other of the vertical line. Indeed, if we express the angle of divergence in degrees and minutes, we perceive that the difference is so small a part of the circumference, that a very slight change will substitute one order for another. The divergence in $\frac{5}{13} = 138° 24'$. In all those beyond, it is 137° plus a variable number of minutes, which approaches nearer and nearer to 30′. Hence M. Bravais considers all these as mere alterations of one typical arrangement, namely, with the angle of divergence 137° 30′ 28″, which is irrational to the circumference, that is, not capable of dividing it an exact number of times, and consequently never bringing any leaf precisely in a right line over any preceding leaf, but placing the leaves of what we take for vertical ranks alternately on both sides of this line and very near it, approaching it more and more, without ever exactly reaching it. These forms of arrangement he therefore distinguishes as *curviserial*, because the leaves are thus disposed on an infinite curve, and are never brought into exactly straight ranks. The others are correspondingly termed *rectiserial*, because, as the divergence is an integral part of the circumference, the leaves are necessarily brought into rectilineal ranks for the whole length of the stem. Organic forms and arrangements, it may be observed, always have a degree of plasticity and power of adaptation, even in their numerical relations, which approximate, but are never entirely restricted to mathematical exactness.

249. A different series of spirals sometimes occurs in alternate leaves, viz. $\frac{1}{4}$, $\frac{1}{5}$, $\frac{2}{9}$, $\frac{3}{14}$; and still others have been met with; but these are all rare or exceptional cases, and do not require to be noticed here.

250. **Opposite Leaves** (236). The arrangement of opposite leaves usually follows very simple laws. Almost without exception, the second pair is placed over the intervals of the first, the third over the intervals of the second, and so on. More commonly, as in plants of the Labiate or Mint Family, the successive pairs cross

each other exactly at right angles, so that the third pair stands directly over the first, the fourth over the second, &c., forming four equidistant vertical ranks for the whole length of the stem. In this case the leaves are said to be *decussate*. In other cases, as in the Pink Family, it may often be observed that the successive pairs deviate a little from this line, so that we have to pass several pairs before we find one exactly superposed over the pair we start with. This indicates a spiral arrangement, which falls into some one of the modes already illustrated in alternate leaves, only that here each node bears a pair of leaves.

251. **Verticillate or Whorled Leaves** (236) follow the same modes of arrangement as opposite leaves. Sometimes they *decussate*, or the leaves of one whorl correspond to the intervals of that underneath, making twice as many vertical ranks as there are leaves in the whorl; sometimes they wind spirally, so that each leaf of the whorl belongs to as many parallel spirals, analogous to the secondary spirals in the case of alternate leaves.

252. The opposition or alternation of the leaves is generally constant in the same species, and often through the same family; yet the transition from opposite to alternate leaves upon the same stem is not very rare: it is seen in the common Myrtle, and the Snapdragon. All Exogens, having their cotyledons or embryo leaves opposite, necessarily commence with that mode; many retain it throughout; others change to alternation, either directly in the primordial leaves, or at a later period (231). In Endogens, on the contrary, the first leaves are necessarily alternate (188), and it is very seldom that they afterwards exhibit opposite or whorled leaves.

253. Only one leaf arises from the same organic point. What are called *fascicled* or *tufted* leaves are merely those of an axillary branch, which is so short that the bases of the leaves are in contact. This is plainly seen in the Barberry, where, the primary leaves hardening into a kind of thorn, the bud in its axil developes into a branch, with very slight elongation of the internodes. Of the same nature are the fascicled leaves of the Pine, and, more evidently, of the Larch (Fig. 177), where the whole foliage of such branches

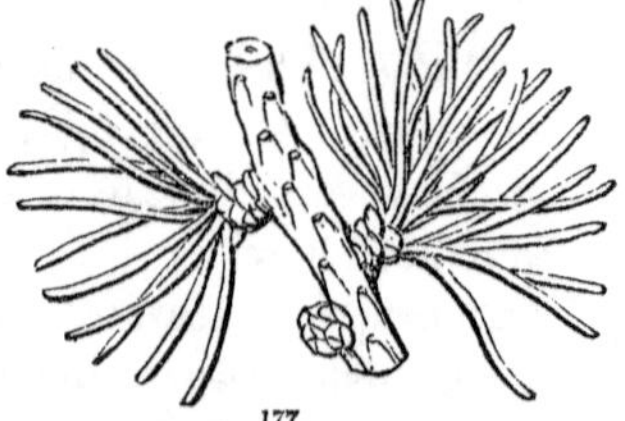

FIG. 177. Clustered or fascicled leaves of the Larch.

is developed without any elongation of the axis. Some of these elongate and grow on through the summer, producing the growth of the season, on which the leaves are distributed so as to show their natural, alternate arrangement.

254. As regards their position on the stem, leaves are said to be *radical*, when they are *inserted* (235) into the stem at or below the surface of the gronnd, so as apparently to grow from the root, as those of the Plantain, Primrose, and of the acaulescent (139) Violets: those that arise along the main stem are termed *cauline;* those of the branches, *rameal;* and those which stand upon or at the base of flower-branches are called *floral;* the latter, however, are generally termed *bracts.*

255. With respect to succession, those leaves which manifestly exist in the embryo are called *seminal;* the first or original pair receiving the name of *Cotyledons* (113), and usually differing widely in appearance from the *ordinary* leaves which succeed them. The earliest ordinary leaves, termed *primordial*, as well as the cotyledons, usually perish soon after others are developed to supply their place.

256. As pertaining to the arrangement of leaves, we should here notice the modes in which they are disposed before expansion in the bud; namely, their

257. **Vernation or Præfoliation.** The latter is the most characteristic name, but the former, given by Linnæus (literally denoting their spring state), is the more ancient and usual. Two things are included under this head: 1st, the mode in which each leaf considered separately is disposed; 2d, the arrangement of the several leaves of the same bud in respect to each other. This last is evidently connected with phyllotaxis, or their position and order of succession on the stem. As to the first, leaves are for the most part either *bent or folded*, or *rolled up* in vernation. Thus, the upper half may be bent on the lower, so that the apex of the leaf is brought down towards the base, as in the Tulip-tree, when the leaves are *inflexed* or *reclinate* in vernation; or the leaf may be folded along its midrib or axis, so that the right half and the left half are applied together, as in the Oak and the Magnolia, when the leaves are *conduplicate;* or each leaf may be folded up a certain number of times like a fan, as in the Maple, Currant, and Vine, when they are said to be *plicate* or *plaited.* The leaf may be rolled either parallel with its axis, or on its axis. In the latter case

it is spirally rolled up from the apex towards the base, like a crosier, or *circinnate*, as in true Ferns (see the young leaves in Fig. 94), and among Phænogamous plants in the Drosera or Sundew. Of the former there are three ways; viz. the whole leaf may be laterally rolled up from one edge into a coil, with the other edge exterior, when the leaves are said to be *convolute*, as in the Apricot and Cherry; or both edges may be equally rolled towards the midrib; either inwards, when they are *involute*, as in the Violet and the Water-Lily; or else outwards, when they are *revolute*, as in the Rosemary and Azalea.

258. Considered relatively to each other, leaves are *valvate* in vernation when corresponding ones touch each other by their edges only, without overlapping: they are *imbricated* when the outer successively overlap the inner, by their edges at least, in which case the order of overlapping exhibits the phyllotaxis, or order of succession and position. In these cases the leaves are plane or convex, at least not much bent or rolled. When leaves with their margins involute are applied together in a circle without overlapping, the vernation is *induplicate*. When in conduplicate leaves the outer successively embrace or sit astride of those next within, the vernation is *equitant*, as the leaves of the Iris at their base: or, when each receives in its fold the half of a corresponding leaf folded in the same manner, the vernation is *half-equitant* or *obvolute*. These terms equally apply to leaves in their full-grown condition, whenever they are then folded or placed so as to overlie or embrace one another. They likewise apply to the parts in the flower-bud, under the name of æstivation or præfloration.

SECT. II. THEIR STRUCTURE AND CONFORMATION.

259. **Anatomy of the Leaf.** The complete leaf consists of the BLADE (*Lamina or Limb*), with its PETIOLE or *Leafstalk*, and at its base a pair of STIPULES. Of these the latter are frequently absent altogether, or else they fall away as the leaf expands: the petiole is very often wanting, when the leaf is *sessile*, or has its blade resting immediately on the stem that bears it. Sometimes, moreover, there is no proper blade or expanded portion, but the whole organ is cylindrical or stalk-like. It is the general characteristic of the leaf, however, that it is an expanded body. Indeed, it may be viewed as a contrivance for increasing the green surface

of a plant, so as to expose to the light and air the greatest practicable amount of parenchyma containing the green matter of vegetation (*chlorophyll*, 87), upon which the light exerts its peculiar action. In a general, mechanical way, it may be said leaves are definite protrusions of the green layer of the bark, expanded horizontally into thin lamina, and stiffened by tough, woody fibres (connected both with the liber, or inner bark, and the wood), which form its framework, *ribs*, or *veins*. Like the stem, therefore, the leaf is made up of two distinct parts, the *cellular* and the *woody*. The cellular portion is the green pulp or parenchyma: the woody, is the skeleton or framework which ramifies among and strengthens the former.

260. The woody or fibrous portion fulfils the same purposes in the leaf as in the stem, not only giving firmness and support to the delicate cellular apparatus, but also serving for the conveyance and distribution of the sap. The subdivision of these *ribs*, or *veins*, of the leaf, as they are not inappropriately called, continues beyond the limits of unassisted vision, until the bundles or threads of woody tissue are reduced to nearly separate fibres, ramified throughout the green pulp, so as to convey to every portion the sap it consumes.

261. The cellular portion, or parenchyma, of the leaf is not a structureless, pulpy mass, such as it appears to the naked eye. The *chlorophyll* (87), to which the green color is entirely owing, and which consists of innumerable rounded globules, is all inclosed in cells of lax parenchyma (51); and these cells are not heaped promiscuously, but exhibit a regular arrangement; upon a plan, too, which varies in different parts of the leaf, according to the different conditions in which it is placed.

262. Leaves are almost always expanded horizontally, so as to present one surface to the ground and the other to the sky; and the parenchyma forms two general strata, one belonging to the upper and the other to the lower side. The microscope displays a manifest difference in the parenchyma of these two strata. That of the upper stratum is composed of one, two, three, or several compact layers of oblong cells, placed endwise, or with their long diameter perpendicular to the surface; while that of the lower is very loosely arranged, leaving numerous vacant spaces between the cells; and when the cells are oblong, their longer diameter is parallel with the epidermis. This is shown in Fig. 7, which represents a magnified section through the thickness (perpendicular

to the surface) of a leaf of the Star-Anise of Florida; where the upper stratum of parenchyma consists of only a single series of perpendicular cells. Also in Fig. 178 (after Brongniart), which represents a similar view of a thin slice of a leaf of the Garden Balsam. Fig. 179 represents a similar section through the thickness of a leaf of the White Lily; where the upper stratum is composed of only one compact layer of vertical cells. The parenchyma is alone represented; the woody portion, or veins, being left out. The structure shows why the upper surface of leaves is of a deeper green than the lower.

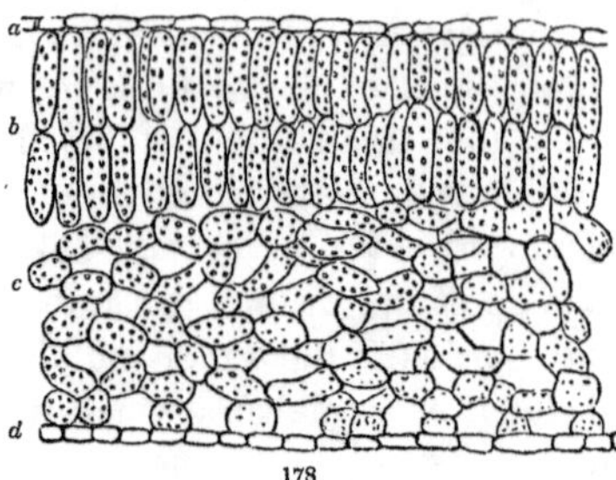

263. The object which this arrangement subserves will appear evident, when we consider that the spaces between the cells, filled with air, communicate freely with each other throughout the leaf, and also with the external air (by means of holes in the epidermis presently to be described); and when we consider the powerful action of the sun to promote evaporation, especially in dry air, and that the thin walls of the cells, like all vegetable membrane, allow of the free escape of the contained moisture by transudation. The compactness of the cells of that stratum which is presented immediately to the sun, and their vertical elongation, so that each shall expose the least possible surface, obviously serve to protect the loose parenchyma beneath from the too powerful action of direct sunshine. This provision is the more complete in the case of plants indigenous to arid regions, where the soil is usually so parched during the dry season, that, for a long period, it affords only the scantiest supply of moisture to the roots. Compare, in this respect, the leaf of the Lily (Fig. 179), where the upper stratum contains but a single layer of barely oblong cells, with that of the Oleander (which is obliged to stand a season of drought), the upper stratum of which consists of two layers of long and narrow vertical cells as closely compacted as possible (Fig. 184). So different is the organization of the two strata, that a leaf soon perishes if reversed so as to expose the lower surface to direct sunshine.

FIG. 178 Magnified section through the thickness of a leaf of the Garden Balsam: *a*, section of the epidermis of the upper surface; *b*, of the upper stratum of parenchyma; *c*, of the lower stratum; *d*, of the epidermis of the lower surface.

264. A further and more effectual provision for restraining the perspiration of leaves within due limits is found in the epidermis, or skin, that invests the leaf, as it does the whole surface of the vegetable, and which is so readily detached from the succulent leaves of such plants as the Stone-crop and the Live-for-ever (Sedum) of the gardens. The EPIDERMIS (69) is composed of small cells belonging to the outermost layer of cellular tissue, with the pretty thick-sided walls very strongly coherent, so as to form a firm membrane. Its cells usually contain no chlorophyll. In ordinary herbs that allow of ready evaporation, this membrane is made up of a single layer of cells; as in the Lily, Fig. **179**, and the Balsam, Fig. 178. It is composed of two layers in cases where one might prove insufficient; and in the Oleander, besides the provision already described, the epidermis consists of three layers of very

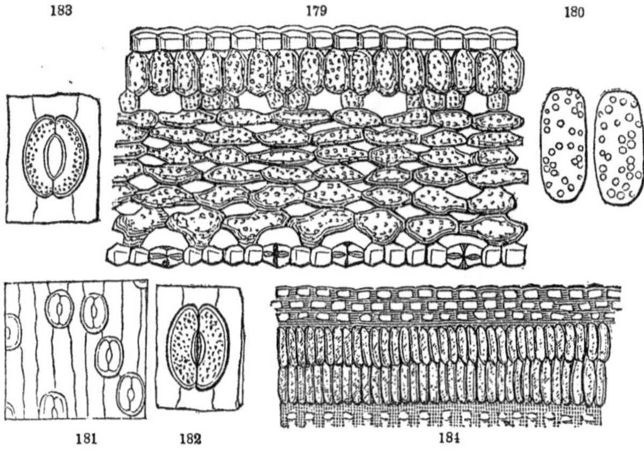

thick-sided cells (Fig. 184). It is generally thick, or hard and impermeable, in the firm leaves of the Pittosporum, Laurustinus, &c., which will thrive, for this very reason, where other plants are liable to perish, in the dry atmosphere of our rooms in winter.

FIG. 179. Magnified section through the thickness of the leaf of the White Lily, showing the parenchyma, and the epidermis of both surfaces; the lower pierced with stomata. (After Brongniart.) 180. Two of the cells of the upper stratum of parenchyma, detached and more magnified, showing the contained grains of chlorophyll.

FIG. 181. Magnified view of the 10,000th part of a square inch of the epidermis of the lower surface of the White Lily, with the stomata, or breathing pores, it bears. These are unusually large in the Lily. One is shown more magnified in Fig. 182: and widely open in Fig. 183.

FIG. 184. Magnified perpendicular section through the thickness of the epidermis and upper stratum of parenchyma in the leaf of the Oleander (after Brongniart); showing the epidermis of three layers of thick-sided cells, and the upper parenchyma of very compact vertical cells.

265. In such firm leaves, especially, the walls of the epidermal cells are soon thickened by secondary deposition (39), especially on the superficial side. This is well seen in the epidermis of the Aloe, and in other fleshy plants, which bear severe drought with impunity: in Fig. 185, it is shown, at *a*, in the rind of a Cactus, where the green layer of the whole stem answers the purpose of the leaves. Sometimes an exterior layer of this superficial deposit in the epidermis, or a secretion from it, may be detached in the form of a continuous, apparently structureless membrane, which Brongniart and succeeding authors have called the CUTICLE. That it may shed water readily, the surface of leaves is commonly protected by a very thin varnish of wax, or else with a *bloom* of the same substance in the form of a whitish powder, which easily rubs off (86), as familiarly seen in a cabbage-leaf.

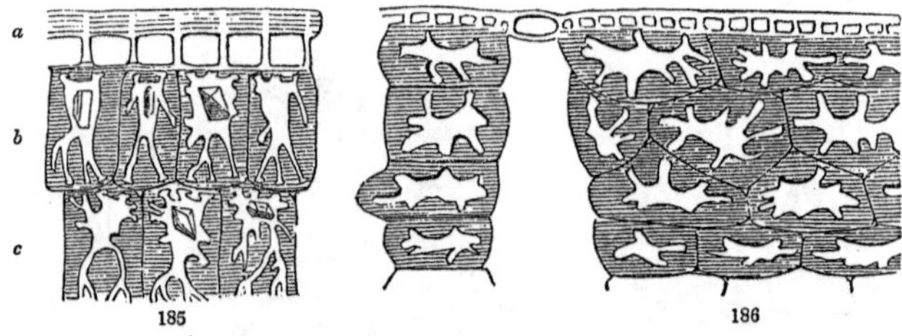

266. A thickening deposit sometimes takes place in the cells of parenchyma immediately underneath the epidermis, especially in the Cactus Family, where the once thin and delicate walls of the cells become excessively and irregularly thickened, so as doubtless greatly to obstruct or arrest all exhalation through the rind. Something like this choking of the cells must commonly occur with age in most leaves, particularly those that live for more than one season.

267. But the multiplication of these safeguards against exhalation might be liable to defeat the very objects for which leaves are principally destined. Evaporation from the parenchyma of the leaves is essential to the plant, as it is the only method by which its excessively dilute food can be concentrated. Some arrange-

FIG. 185. Magnified slice of the epidermis and superficial parenchyma of a Cactus, after Schleiden; exhibiting the epidermis greatly thickened by a stratified deposition in the cells: and the cells of the parenchyma likewise nearly filled with an incrusting deposit. The deposition in such cases is always irregular, leaving canals or passages which nearly connect the adjacent cells. Several of the cells contain crystals (91).

FIG. 186. Similar section from another species of Cactus, passing through one of the stomata, and the deep intercellular space beneath it.

ment is requisite that shall allow of sufficient exhalation from the leaves while the plant is freely supplied with moisture by the roots, but restrain it when the supply is deficient. It is clear that the greatest demand is made upon the leaves at the very period when the supply through the roots is most likely to fail: for the summer's sun, which acts so powerfully on the leaves, at the same time parches the soil upon which the leaves (through the rootlets) depend for the moisture they exhale. So long as their demands are promptly answered, all goes well. The greater the force of the sun's rays, the greater the speed at which the vegetable machinery is driven. But whenever the supply at the root fails, the foliage begins to flag and droop, as is so often seen under a sultry meridian sun; and if the exhaustion proceeds beyond a certain point, the leaves inevitably wither and perish. Some adaptation is therefore needed, analogous to the governor in machinery, or the self-acting valve, which shall regulate the exhalation according to the supply. Such an office is actually fulfilled by

268. **The Stomata,** *Stomates*, or *Breathing-pores* (70). Through the orifices which bear this name, exhalation principally takes place, in all ordinary cases, where the epidermis is thick and firm enough to prevent much escape of moisture by direct transudation. The stomata (Fig. 181–183, 187) are situated so as to open directly into the hollow chambers, or air-cavities, which pervade the parenchyma (Fig. 179, 186), especially the lower stratum; so as to afford free communication between the external air and the whole interior of the leaf. The perforation of the epidermis is between two (or rarely four) small and delicate cells, which, unlike the rest of the epidermis, usually contain some chlorophyll, and in other respects resemble the parenchyma beneath. Their exact mechanism is not very well made out; but it appears that, when moist, these hygrometric cells become turgid, and in elongating diverge or curve outwardly in their middle, where they do not cohere, so as to open a free communication between the outer air and the interior of the leaf. When dry, they incline to

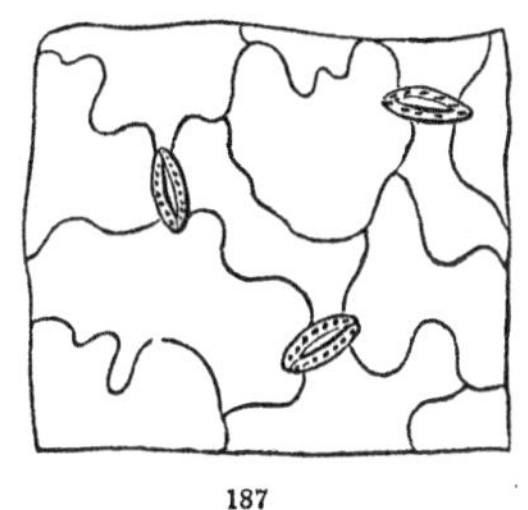

FIG. 187. A highly magnified piece of the epidermis of the Garden Balsam, with three stomata (after Brongniart).

shorten and straighten, so as to bring their sides into contact and close the orifice completely. This structure is sufficiently illustrated in the figures referred to, and especially in those of the Lily, where the stomata are unusually large and easy of examination. The action and use of this mechanism will readily be understood. So long as the leaf is in a moist atmosphere, and is freely supplied with sap by the stem and roots, the cells that guard the orifice are expanded, and the open stomata allow the free escape of moisture by evaporation. But when the supply fails, and the parenchyma begins to be exhausted, the guardian cells, at least equally affected by the dryness, quickly collapse, and by closing these thousands of apertures check the drain the moment it becomes injurious to the plant.

269. As a general rule, the stomata wholly or principally belong to the epidermis of the lower surface of the leaf: the mechanism is too delicate to work well in direct sunshine. The position of the stomata, and the loose texture of the lower parenchyma, require that this surface should be shielded from the sun's too direct and intense action; and show why leaves soon perish when artificially reversed, and prevented from resuming (as otherwise they spontaneously will) their natural position. This general arrangement is variously modified, however, under peculiar circumstances. The stomata are equally distributed on the two sides of those leaves, of whatever sort, which grow in an erect position, or present their edges, instead of their surfaces, to the earth and sky (294), and have the parenchyma of both sides similarly constituted, sustaining consequently the same relations to light. In the Water-Lilies (Nymphæa, Nuphar), and other leaves which float upon the water, the stomata all belong to the upper surface; and all leaves growing under water, where there can be no evaporation, are destitute, not only of stomata, but usually of a distinct epidermis also.

270. The number of the stomata varies in different leaves from 800 to about 170,000 on the square inch of surface. In the Apple, there are said to be about 24,000 to the square inch (which is under the average number, as given in a table of 36 species by Lindley); so that each leaf of that tree would present about 100,000 of these orifices. From their great numbers, they are doubtless fully adequate to the office that is attributed to them, notwithstanding their minute size. Their size varies so greatly in different plants, that no safe inference can be drawn of the comparative amount of

exhalation in different leaves from the mere number of their stomata. When the stomata are not all restricted to the lower surface, still the greater portion usually occupy this position. Thus, the leaf of Arum Dracontium is said to have 8,000 stomata to a square inch of the upper surface, and twice that number in the same space of the lower. The leaf of the Coltsfoot has 12,000 stomata to a square inch of the lower epidermis, and only 1,200 in the upper. That of the White Lily 60,000 to the square inch on the lower surface, and perhaps 3,000 on the upper.

271. At the points on the surface of the developing leaf where stomata are about to be formed, one of the epidermal cells early ceases to enlarge and thicken with the rest, but divides into two (in the manner formerly described, 32), forming the two guardian cells: as they grow, the two constituent portions of their common partition separate, leaving an interspace or orifice between. In some cases, each new cell divides again, when the stomata are formed of four cells in place of two.

272. Succulent or fleshy plants, such as those of the Cactus tribe, Mesembryanthemums, Sedums, Aloes, &c., are remarkable for holding the water they imbibe with great tenacity, rather in consequence of the thickness of the epidermis, or from the deposit which early accumulates in the superficial cells of the parenchyma (265), than from the want of stomata. The latter are usually abundant,* but they seem to remain closed, or to open less than in ordinary plants, except in young and growing parts. Hence the tissue becomes gorged as it were with fluid, which is retained with great tenacity, especially during the hot season. They are evidently constructed for enduring severe droughts; and are accordingly found to inhabit dry and sunburnt places, such as the arid plains of Africa, — the principal home of the Stapelias, Aloes, succulent Euphorbias, &c., — or the hottest and driest parts of our own continent, to which the whole Cactus Family is indigenous. Or, when such plants inhabit the cooler temperate regions, like the Sedums and the common Houseleek, &c., they are commonly found in the most arid situations, on naked rocks, old walls, or sandy plains, exposed to the fiercest rays of the noonday sun, and

* The thickened epidermis of the fleshy leaves of the Sea-Sandwort (Honkenya) is provided with an abundance of large stomata, on the upper as well as the lower face. But this plant, though very fleshy, grows in situations where its roots are always supplied with moisture.

thriving under conditions which would insure the speedy destruction of ordinary plants. The drier the atmosphere, the greater their apparent reluctance to part with the fluid they have accumulated, and upon which they live during the long period when little or no moisture is yielded by the soil or the air. Their structure and economy fully explain their tolerance of the very dry air of our houses in mid-winter, when ordinary thin-leaved plants become unhealthy or perish.

273. Sometimes the leaves of succulent plants merely become obese or misshapen, like those of the Ice-plant and other species of Mesembryanthemum, &c.: sometimes they are reduced to triangular projections or points, or are perfectly confounded with the unusually developed green bark of the stem, which fulfils their office, as in the Stapelia and most Cacti.

274. **The Development of Leaves** proceeds from the apex (which first appears, in the form of a little tumor or papilla) towards the base, which is later eliminated from the axis. The apex is pushed forward by the formation and growth of the parts beneath: after the blade has shaped itself, the rudiment of the petiole, if there is to be any, begins to be visible, and this grows in like manner from the apex downwards, the lower part of it being the last formed: its growth subsequent to its formation is greater in proportion to its original size than that of any other part of the leaf. The sheath at the base (as in most Monocotyledons), or the stipules (304, which principally belong to Dicotyledons), are at first continuous with the blade, or divided from it by a mere constriction: the formation and elongation of the petiole soon separate them. The stipules, remaining next the axis or source of nourishment, undergo a rapid development early in the bud, so that, at a certain stage, they are often larger than the body of the leaf, and they accordingly form in such cases the teguments of the bud. Divided or lobed and compound leaves are simple at the commencement, but the lobes are very early developed; they grow in respect to the axis of the leaf nearly as that grew from the axis of the plant, and in the compound leaf at length isolate themselves, and are often raised on footstalks of their own. Commonly the upper lobes or leaflets are first formed, and then the lower: but in those of the Walnut and Ailanthus, and other large compound leaves, new leaflets continue to be produced from the apex, even after the lowermost are nearly full grown. In the earliest stage leaves con-

sist of parenchyma alone: the fibro-vascular tissue which makes the ribs, veins, or framework appears later. No good researches have yet been made into the mode and order of its production.

275. **The Forms of Leaves** are almost infinitely various. These afford some of the readiest, if not the most certain, marks for characterizing species. Their principal modifications are therefore classified, minutely defined, and embodied in a system of nomenclature which is equally applicable to other parts of the plant, and which as an instrument is indispensable to the systematic botanist. The numerous entirely unconnected technical terms which have gradually accumulated from the infancy of the science, and have multiplied with its increasing wants, are mostly quite arbitrary, or have been suggested by real or fancied resemblances of their shapes to natural or other objects. This arbitrary nomenclature, which formerly severely tasked the memory of the student, was reduced by De Candolle to a clear and consistent system, based upon scientific principles, and of easy application. The fundamental idea of the plan is, that the almost infinite varieties in the form and outline of leaves may be deduced from the different modes and degrees in which the woody skeleton or framework of the leaf is expanded or ramified in the parenchyma. Upon this conception our following sketch is based; in which we endeavor to introduce and define the more important terms of the nomenclature of leaves. It should be kept in mind, however, that this system is partly if not altogether empirical, and is not to be taken as an explanation of the actual formation of the leaf; but rather as an account of the mutual adaptation and corespondence of the outlines and the framework of leaves. For the parenchyma is developed, and the form of the leaf is often fixed, before the framework has an existence. The latter, therefore, cannot have determined the outline or shape of the organ. The distribution of the veins or fibrous framework of the leaf in the blade is termed its

276. **Venation.** The veins are distributed throughout the lamina in two principal modes. Either the vessels of the petiole divide at once, where they enter the blade, into several veins, which run parallel with each other to the apex, connected only by simple transverse veinlets (as in Fig. 201); or the petiole is continued into the blade in the form of one or more principal or coarser veins, which send off branches on both sides, the smaller branchlets uniting with one another (*anastomosing*) and forming a kind

of network; as in Fig. 191, 199. The former are termed *parallel-veined*, or commonly *nerved* leaves; the veins in this case having been called nerves by the older botanists, — a name which it is found convenient to retain, although of course they are in no respect analogous to the nerves of animals. The latter are termed *reticulated* or *netted-veined* leaves.

277. Parallel-veined or nerved leaves are characteristic of Endogenous plants; while reticulated leaves are almost universal in Exogenous plants. We are thus furnished with a very obvious, although by no means absolute, distinction between these two great classes of plants, independently of the structure of their stems (185).

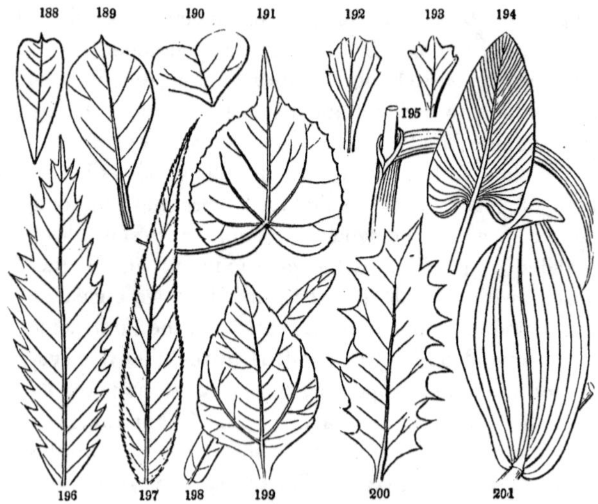

278. In reticulated leaves, the coarse primary veins (one or more in number), which proceed immediately from the apex of the petiole, are called *ribs;* the branches are termed *veins*, and their subordinate ramifications, *veinlets.* Very frequently, a single strong rib (called the *midrib*), forming a continuation of the petiole, runs directly through the middle of the blade to the apex (Fig. 196, 197, &c.), and from it the lateral veins all diverge. Such leaves are termed *feather-veined* or *pinnately veined;* and are subject to various modifications, according to the arrangement of the veins and veinlets; the primary veins sometimes passing straight from

FIG. 188-201. Various forms of simple leaves.

the midrib to the margin, as in the Beech and Chestnut (Fig. 196); while in other cases they are divided into veinlets long before they reach the margin. When the midrib gives off a very strong primary vein or branch on each side above the base, the leaf is said to be *triple-ribbed*, or often *tripli-nerved*, as in the common Sunflower (Fig. 199); if two such ribs proceed from each side of the midrib, it is said to be *quintuple-ribbed*, or *quintupli-nerved*.

279. Not unfrequently the vessels of a reticulated leaf divide at the apex of the petiole into three or more portions or ribs of nearly equal size, which are usually divergent, each giving off veins and veinlets, like the single rib of a feather-veined leaf. Such leaves are termed *radiated-veined*, or *palmately veined;* and, as to the number of the ribs, are called three-ribbed, five-ribbed, seven-ribbed, &c. (Fig. 191, 203, 209). Examples of this form are furnished by the Maple, the Gooseberry, the Mallow Family, &c. Occasionally the ribs of a radiated-veined leaf converge and run to the apex of the blade, as in Rhexia and other plants of the same family, thus resembling a parallel-veined or nerved leaf; from which, however, it is distinguished by the intermediate netted veins. But when the ribs are not very strong, such leaves are frequently said to be nerved, although they branch before reaching the apex.

280. According to the theory of De Candolle (275), the shape which leaves assume may be considered to depend upon the distribution of the veins, and the quantity of parenchyma; the general outline being determined by the division and direction of the veins; and the form of the margin, (whether even and continuous, or interrupted by void spaces or indentations,) by the greater or less abundance of the parenchyma in which the veins are distributed. This view is readily intelligible upon the supposition that a leaf is an expansion of soft parenchyma, in which the firmer veins are variously ramified. Thus, if the principal veins of a feather-veined leaf are not greatly prolonged, and are somewhat equal in length, the blade will have a more or less elongated form. If the veins are very short in proportion to the midrib, and equal in length, the leaf will be *linear* (as in Fig. 198); if longer in proportion, but still equal, the leaf will assume an *oblong* form (Fig. 200), which a slight rounding of the sides converts into an *oval* or *elliptical* outline. If the veins next the base are longest, and especially if they curve forward towards their extremities, the leaf assumes a *lanceolate* (Fig. 197), *ovate* (Fig. 199), or some inter-

mediate form. On the other hand, if the veins are more developed beyond the middle of the blade, the leaf becomes *obovate* (Fig. 189), or *cuneiform* (Fig. 192). In radiated or palmately veined leaves (Fig. 202–204), where the primary ribs are divergent, an orbicular or roundish outline is most common, and indeed is universal when the ribs are of equal strength. Some of the ribs or their ramifications being directed backwards, a recess, or *sinus*, as it is termed, is produced at the base of the leaf, which, taken in connection with the general form, gives rise to such terms as *cordate* or *heart-shaped* (Fig. 191), *reniform* or *kidney-shaped* (Fig. 202), &c., when the posterior portions are rounded; and those of

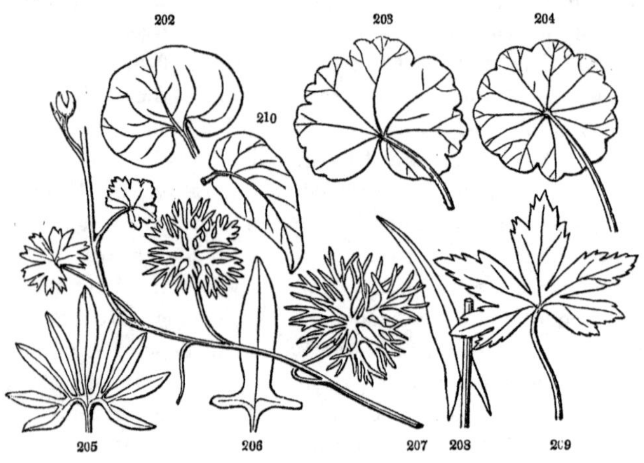

sagittate or *arrow-headed* (Fig. 208), and *hastate* or *halberd-shaped* (Fig. 206), when they are produced into angles or lobes. The margins of the sinus are sometimes brought into contact, when they are frequently united; for whenever soft cellular parts are in close contact at an early period of their development, they are very apt to cohere and grow together. In this case the leaf becomes *peltate* or *shield-shaped* (Fig. 204); the blade being attached to the petiole, not by its apparent base, but by some part of the lower surface. Two or three common species of Hydrocotyle plainly exhibit the transition from common radiated leaves into the peltate form. Thus, the leaf of H. Americana (Fig. 203) is roundish-reniform, with an open sinus at the base; while in H. inter-

FIG. 202–210. Forms of simple, chiefly radiated-veined leaves.

rupta and H. umbellata (Fig. 204), the margins have grown together so as to obliterate the sinus, and an orbicular peltate leaf is produced. In nerved leaves, when the nerves run parallel from the base to the apex, as in Grasses (Fig. 195), the leaf is necessarily linear, or nearly so; but when they are more divergent in the middle, or towards the base, the leaf becomes oblong, oval, or ovate, &c. (Fig. 201). In one class of nerved or parallel-veined leaves, the simple veins or nerves arise from a prolongation of the petiole in the form of a thickened midrib, instead of the base of the blade, constituting the *curvinerved* leaves of De Candolle. This structure is almost universal in the Ginger tribe, the Arrowroot tribe, in the Banana, and other tropical plants; and our common Pontederia, or Pickerel-weed (Fig. 194), affords an illustration of it, in which the nerves are curved backwards at the base, so as to produce a cordate outline.

281. As to the margin and particular outline of leaves, they exhibit every gradation between the case where the blade is *entire*, that is, with the margin perfectly continuous and even (as in Fig. 201), and those where it is cleft or divided into separate portions. The convenient hypothesis of De Candolle connects these forms with the abundance or scantiness of the parenchyma, compared with the divergence and the extent of the ribs or veins; on the supposition that, where the former is insufficient completely to fill up the framework, *lobes*, *incisions*, or *toothings* are necessarily produced, extending from the margin towards the centre. Thus, in the white and the yellow species of Water Ranunculus, there appears to be barely sufficient parenchyma to form a thin covering for each vein and its branches (Fig. 207, the lowest leaf); such leaves are said to be *filiformly dissected*, that is, *cut into threads;* the nomenclature in all these cases being founded on the convenient, but incorrect supposition, that a leaf originally entire is cut into teeth, lobes, divisions, &c. If, while the framework remains the same as in the last instance, the parenchyma be more abundantly developed, as in fact happens in the upper leaves of the same species when they grow out of water, and is shown in the same figure, they are merely *cleft* or *lobed.* If these *lobes* grow together nearly to the extremity of the principal veins, the leaf is only *toothed*, *serrated*, or *crenated;* and if the small remaining notches were filled with parenchyma, the leaf would be *entire.* The study of the development of leaves, however, proves that the

parenchyma grows and shapes the outlines of the organ in its own way, irrespective of the framework, which is, in fact, adapted to the parenchyma rather than the parenchyma to it. The principal terms which designate the mode and degree of division in simple leaves may now be briefly explained, without further reference to this or any other theory.

282. A leaf is said to be *serrate*, when the margin is beset with sharp teeth which point forwards towards the apex (Fig. 196); *dentate*, or *toothed*, when the sharp salient teeth are not directed towards the apex of the leaf (Fig. 200); and *crenate*, when the teeth are rounded (Fig. 203, 204). A slightly waved or sinuous margin is said to be *repand;* a strongly uneven margin, with alternate rounded concavities and convexities, is termed *sinuate* (as in the Oak). When the leaf is irregularly and sharply cut deep into the lamina, it is said to be *incised;* when the portions, or *segments*, are more definite, it is said to be lobed; and the terms *two-lobed*, *three-lobed*, *five-lobed*, &c., express the number of the segments. If the incisions extend about to the middle of the blade, or somewhat deeper, the leaf is said to be *cleft;* and the terms *two-cleft*, *three-cleft*, &c. (or in the Latin form, *bifid*, *trifid*, &c.), designate the number of the segments: or when the latter are numerous or indefinite, the leaf is termed *many-cleft*, or *multifid*. If the segments extend nearly, but not quite, to the base of the blade or the midrib, the leaf is said to be *parted* (Fig. 209): if they reach the midrib or the base, so as to interrupt the parenchyma, the leaf is said to be *divided;* the number of *partitions* or *divisions* being designated, as before, by the terms *two-*, *three-*, *five-parted*, or *two-*, *three-*, *five-divided*, &c.

283. As the mode of division always coincides with the arrangement of the primary veins, the lobes or incisions of feather-veined, are differently arranged from those of radiated or palmately veined leaves: in the latter, the principal incisions are all directed to the base of the leaf; in the former, towards the midrib. These modifications are accurately described by terms indicative of the venation, combined with those that express the degree of division. Thus, a feather-veined (in the Latin form, a *pinnately veined*) leaf is said to be *pinnately cleft* or *pinnatifid*, when the sinuses reach halfway to the midrib; *pinnately parted*, when they extend almost to the midrib; and *pinnately divided*, when they reach the midrib, dividing the parenchyma into separate portions. A few

subordinate modifications are indicated by special terms: thus, a pinnatifid or pinnately parted leaf, with regular, very close and narrow divisions, like the teeth of a comb, is said to be *pectinate;* a feather-veined leaf, more or less pinnatifid, but with the lobes decreasing in size towards the base, is termed *lyrate*, or *lyre-shaped* (Fig. 212); and a lyrate leaf with sharp lobes pointing towards the base, as in the Dandelion (Fig. 213), is called *runcinate.* A palmately veined leaf is in like manner said to be *palmately cleft*, *palmately parted*, *palmately divided*, &c. (Fig. 207, 209), according to the degree of division. The term *palmate* was originally employed to designate a leaf more or less deeply cut into about five spreading lobes, bearing some resemblance to a hand with the fingers spreading; and it is still used to designate a palmately lobed leaf, without reference to the depth of the sinuses. A palmate leaf with the lateral lobes cleft into two or more segments, is said to be *pedate* (Fig. 205), from a fancied resemblance to a bird's foot. By designating the number of the lobes in connection with the terms which indicate their extent and their disposition, botanists are enabled to describe all these modifications with great brevity and precision. Thus, a *palmately five-parted* leaf is one of the radiated-veined kind, which is divided almost to the base into five segments: a *pinnately five-parted* leaf is one of the feather-veined kind cut into five lobes (two on each side, and one terminal), with the sinuses extending almost to the midrib: and the same plan is followed in describing cleft, lobed, or divided leaves.

284. The segments of a lobed or divided leaf may be again divided, lobed, or cleft, upon the same principle as the leaf itself, and the same terms are employed in describing them. Sometimes both the primary, secondary, and even tertiary divisions are defined by a single word or phrase; as *bipinnatifid* (Fig. 214), *tripinnatifid*, *bipinnately parted*, *tripinnately parted*, *twice palmately parted*, &c.

285. Parallel-veined or nerved leaves may be expected to present entire margins, and this in fact almost universally occurs when the nerves are convergent (Fig. 201). Such leaves are often lobed or cleft when the principal nerves diverge greatly, as in the Dragon Arum; but the lobes themselves are entire. So, also, ribbed leaves are mostly entire, when the ribs converge to the apex: and leaves which exhibit a well-marked marginal vein (the

falsely ribbed leaves of Lindley), into which the lateral veinlets are confluent (as in all Myrtaceous plants), are also entire.

286. There are a few terms employed in describing the apex of a leaf, which may be here enumerated. When a leaf terminates in an acute angle, it is said to be *acute* (Fig. 199, 208): when the apex is an obtuse angle, or rounded, it is termed *obtuse* (Fig. 194, 198): an obtuse leaf, with the apex slightly indented or depressed in the middle, is said to be *retuse*, or, if more strongly notched, *emarginate* (Fig. 188): an obovate leaf with a wider and more conspicuous notch at the apex is termed *obcordate* (Fig. 190), being a cordate or heart-shaped leaf inverted. When the apex is, as it were, cut off by a straight transverse line, the leaf is said to be *truncate:* when abruptly terminated by a small projecting point, it is *mucronate* (Fig. 188, 189): and when an acute leaf has a narrowed and prolonged apex, or tapers to a point, it is *acuminate*, or *pointed*, as in Fig. 191.

287. All these terms are equally applicable to expanded surfaces of every kind, such as petals, sepals, &c.: and those terms which are used to describe the modifications of solid bodies, such as stems and stalks, are equally applicable to leaves when they affect similar shapes, as they sometimes do.

288. The whole account, thus far, relates to Simple Leaves, namely, to those which have a blade of one piece, however cleft or lobed, or, if divided, where the separate portions are neither raised on stalklets of their own, nor *articulated* (by a joint) with the main petiole, so that the pieces are at length detached from it. The distinction, however, cannot be very strictly maintained; there are so many transitions between simple and

289. **Compound Leaves** (Fig. 211, 215–221). These have the blade divided into entirely separate pieces; or, rather, they consist of a number of blades, borne on a common petiole, usually supported on stalklets of their own, between which and the main petiole an articulation or joint is formed, more or less distinctly. These separate blades are called Leaflets: they present all the diversities of form, outline, or division, which simple leaves exhibit; and the same terms are employed in characterizing them. Having the same nature and origin as the lobes or segments of simple leaves, they are arranged in the same ways on the common petiole. Compound leaves accordingly occur under two general forms, the *pinnate*, and the *palmate*, otherwise called *digitate.*

The *pinnate* form is produced when a leaf of the pinnately veined sort becomes compound; that is, the leaflets are situated along the sides of the common petiole. There are several modifications of the pinnate leaf. It is *abruptly pinnate*, when the leaflets are even in number, and none is borne on the very apex of the petiole or its branches, as in Cassia; and also in the Vetch tribe, where, however, the apex of the petiole is generally prolonged into a tendril (Fig. 216). It is *impari-pinnate*, or pinnate with an odd leaflet, when the petiole is terminated with a leaflet (Fig. 215, 220). There are some subordinate modifications; such as *lyrately pinnate*, when the blade of a lyrate leaf (Fig. 212) is completely divided, as in Fig. 220; and *interruptedly pinnate*, when some minute leaflets are irregularly intermixed with larger ones, as is also shown to some extent in the figure last cited. The number of leaflets varies from a great number to very few. When reduced to a small number, such a leaf is said to be *pinnately seven-*, *five-*, or *tri-foliolate*, as the case may be. A pinnate leaf of three or five leaflets is often called *ternate* or *quinate;* which terms, however, are equally applied to a palmately compound leaf, and also, and more appropriately, to the case of three or five simple leaves

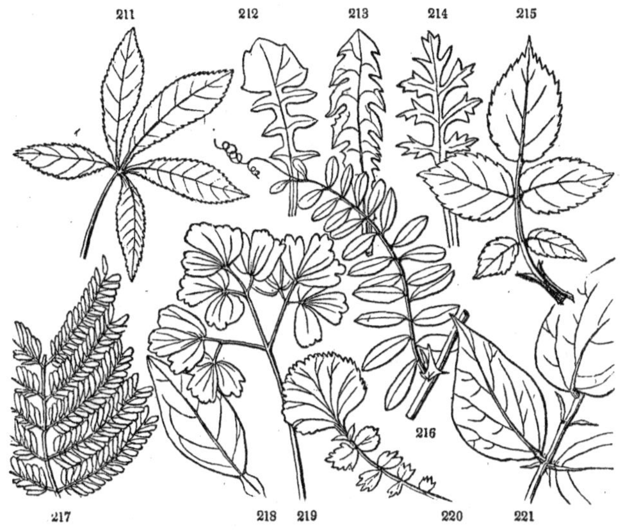

FIG. 211 - 221. Compound and lobed leaves.

growing on the same node. A *pinnately trifoliolate* leaf (Fig. 221) is readily distinguished by having the two lateral leaflets attached to the petiole at some distance below its apex, and by the joint which is observable at some point between their insertion and the lamina of the terminal leaflet. Such a leaf may even be reduced to the paradoxical case of a single leaflet; as in the Orange (Fig. 218), and frequently in one variety of Rhynchosia tomentosa; which is distinguished from a really simple leaf by the joint at the junction of the partial with the general petiole.

290. The *palmate* or *digitate* form is produced when a leaf of the palmately veined sort becomes compound; in which case the leaflets are necessarily all attached to the apex of the common petiole, as in the Horsechestnut and Buckeye (Fig. 211). Such leaves of three, five, or any definite number of leaflets are termed *palmately* (or *digitately*) *trifoliolate*, *five-foliolate*, &c. A leaf of two leaflets, which rarely occurs, is *unijugate* (one-paired) or *binate*. By this nomenclature, the distinction between pinnately and palmately compound leaves is readily kept up.

291. The stalk of a leaflet is called a *partial petiole* (*petiolula*); and the leaflet thus supported is *petiolulate.*

292. The partial petioles may bear a set of leaflets instead of a single one, when the leaf becomes *doubly* or *twice* compound. Thus a pinnate leaf again compounded in the same way becomes *bipinnate*, or if still a third time divided it is *tripinnate*, &c. In these cases the main divisions or branches of the common petiole are called *pinnæ*. So a trifoliolate leaf twice compound becomes *biternate;* or thrice, *triternate*, &c. When the primary division is digitate, the secondary division is often pinnate, thus combining the two modes in the same leaf. A leaf irregularly or indeterminately several times compounded, in whatever mode, is *decompound.*

293. The blade of a leaf is almost always symmetrical, that is, the portions on each side of the midrib or axis are similar; but occasionally one side is more developed than the other, when the leaf is *oblique*, as is strikingly the case in the species of Begonia (Fig. 210) now common in gardens.

294. The blade is also commonly horizontal, presenting one surface to the sky, and the other to the earth; in which case the two surfaces differ in structure (262) as well as in appearance, each being fitted for its peculiar offices: if artificially reversed, they spontaneously resume their natural position, or soon perish if pre-

vented from doing so. But in *erect* and *vertical* leaves, the two surfaces are equally exposed to the light, and are similar in structure and appearance. In such erect leaves as those of Iris, it is what corresponds to the lower surface of ordinary leaves that is presented to the air; for the leaf is folded together lengthwise and consolidated while in the nascent state, so that the true upper surface is concealed in the interior, except near the base, where they override each other in the *equitant* manner (258). True vertical leaves, which present their edges instead of their surfaces to the earth and sky, generally assume this position by a twisting of the base or of the petiole; as is strikingly seen in a large number of New Holland trees of the Myrtle Family, now common in greenhouses.

295. Leaves assume extraordinary appearances when they become *succulent*, as in the different species of Mesembryanthemum (Ice-plant), &c., and no less so when, on the contrary, producing little or no green parenchyma, they become scale-like, as in Beech-drops, Monotropa, and other parasitic plants; where they do not perform the ordinary office of leaves. Not unlike these are the altered or degenerate leaves that form the integuments of scaly buds (146). The primary leaves on every shoot of the Pine are merely thin and dry scales; from the axils of which the ordinary foliage is developed in fascicles of needle-shaped leaves (253).

296. Leaves which grow under water are often nearly or quite destitute of parenchyma; as in Ranunculus Purshii (Fig. 207), and Ranunculus aquatilis, Bidens Beckii, Myriophyllum, &c. A very remarkable instance of the kind occurs in Ouvirandra fenestralis, a South African aquatic plant, with nerved leaves, which exhibit a complete framework or skeleton, while the parenchyma is entirely wanting. In the Barberry some of the summer leaves harden as they grow into compound or branching spines (Fig. 222).

222

297. When the blade of the leaf is wanting, its office is sometimes performed by the petiole, or by the stipules.

FIG. 222. A summer shoot of the Barberry, showing a lower leaf in the normal state; the next partially, those still higher completely, transformed into spines.

298. **The Petiole**, or *Leafstalk*, is usually either round, or half-cylindrical and channelled on the upper side. But in the Aspen, it is strongly flattened at right angles with the blade, so that the slightest breath of air puts the leaves in motion. It is not unfrequently furnished with a leaf-like border, or wing; which, in the Sweet Pea of the gardens, extends downward along the stem, on which the leaves are then said to be *decurrent;* or the stalk or stem thus bordered is said to be *alate* or *winged.* In many Umbelliferous plants, the base of the petiole is dilated into a broad and membranaceous inflated sheath; and in a great number of Endogenous plants, especially in Grasses, the petiole consists of a *sheath*, embracing the stem, which in the true Grasses is furnished at the summit with a membranous appendage, in some sort equivalent to the stipules, called the *ligule* (Fig. 195). In the proper Pea tribe, the apex of the petiole is often changed into a tendril (Fig. 216); and in one plant of that tribe (Lathyrus Aphaca), the whole petiole becomes a tendril, the office of the leaf being fulfilled by a pair of large stipules. Sometimes, as in one section of Astragalus, the petioles harden into spines after the leaflets fall off.

299. The woody and vascular tissue runs lengthwise through the petiole, in the form usually of a definite number of parallel threads, to be ramified in the blade. The ends of these threads are apparent on the base of the leafstalk when it falls off, and on the scar left on the stem, as so many round dots (Fig. 130, 127, *b*), of a uniform number and arrangement in each species. Sometimes they are so close as to be confluent into a continuous line or bundle.

300. **Phyllodia** (Fig. 226, 227). Occasionally the woody system spreads and the whole petiole dilates into a kind of blade, traversed by ribs, mostly of the parallel-veined kind. In these cases the proper blade of the leaf is commonly abortive or disappears; this substitute, called a PHYLLODIUM (meaning a leaf-like body), taking its place. These phyllodia constitute the whole foliage of the numerous Australian Acacias. Here they are at once distinguished from leaves with a true blade by being entire and parallel-veined; while their proper leaves, as the primordial ones uniformly appear in germination, and also later ones in casual instances, are compound and netted-veined. They are also recognized by their uniformly vertical position, presenting their margins instead of their surfaces to the earth and sky; and they sometimes bear a true compound lamina at the apex, as in Fig. 227. These

Acacias, with the Myrtaceous trees that have leaves with a proper

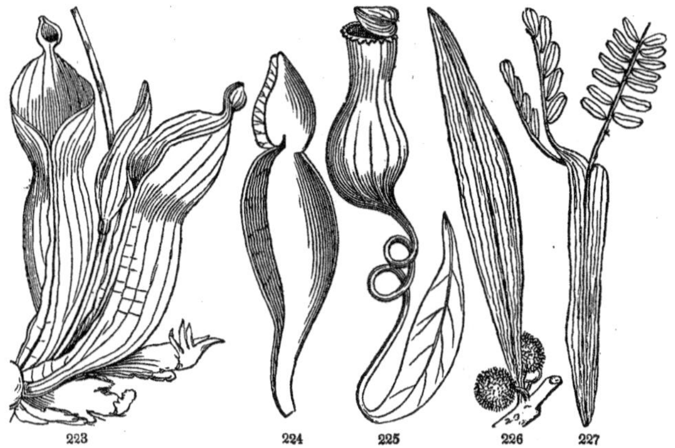

blade which becomes vertical by a twist (294), compose more than half of the forests of New Holland, and give to them a prevailing and very peculiar feature, and an unusual distribution of light and shade; the cause of which was detected by the scrutinizing glance of Robert Brown.

301. In the Dionæa, or Venus's Fly-catcher (Fig. 228), the proper lamina, or blade of the leaf, is the terminal portion, fringed with stiff bristles, which closes suddenly and with considerable force when the upper surface is touched. This is borne on a dilated, foliaceous body, which may be held to represent the petiole; but it is horizontally expanded

FIG. 223. Pitchers of Heliamphora; 224, of Sarracenia purpurea; 225, of Nepenthes. 226. A phyllodium of a New Holland Acacia. 227. The same, bearing a reduced compound blade.

FIG. 228. A plant of Dionæa muscipula, reduced in size.

and netted-veined. Still more singular modifications of the leaf are met with in the form of

302. **Ascidia**, or *Pitchers* (Fig. 223 – 225). These occur in several plants of widely different families. If we conceive the margins of the dilated petiole of Dionæa to curve inwards until they meet, and cohere with each other, there would result a leaf not unlike that of Sarracenia purpurea, the common Pitcher-plant or Sidesaddle Flower of the Northern United States (Fig. 224), in which, accordingly, the tube or pitcher may be considered as the petiole, and the hood at the summit as the lamina. This view is confirmed by a new Pitcher-plant of the same family (Heliamphora, Fig. 223), recently discovered by Mr. Schomburgk in the mountains of British Guiana, and described by Mr. Bentham; in which the margins of the dilated petiole are not always united quite to the summit, and the lamina is represented by a small concave terminal appendage. In the curious Nepenthes (Fig. 225), the petiole is first dilated into a kind of lamina, then contracted into a tendril, and finally dilated into a pitcher, containing fluid secreted by the plant itself; the orifice being accurately closed by a lid, which is from analogy supposed to represent the real blade of the leaf.

303. The cohesion of the edges of a leaf with each other, or with neighboring organs, is by no means infrequent; since all parts or organs of a plant which are contiguous at the time of their development are liable to become ingrafted or to cohere together. This is illustrated by the formation of peltate leaves (Fig. 203, 204), and likewise by what are termed *perfoliate* leaves; whether formed by the union of the bases of a pair of opposite sessile leaves (*connate-perfoliate*), as in Silphium perfoliatum, Triosteum perfoliatum, the upper pairs of the Honeysuckle, &c.; or consisting of a single clasping leaf, the posterior lobes of which encompass the stem and cohere on the opposite side, as is seen in Bupleurum rotundifolium, Uvularia perfoliata, and Baptisia perfoliata (Fig. 229).

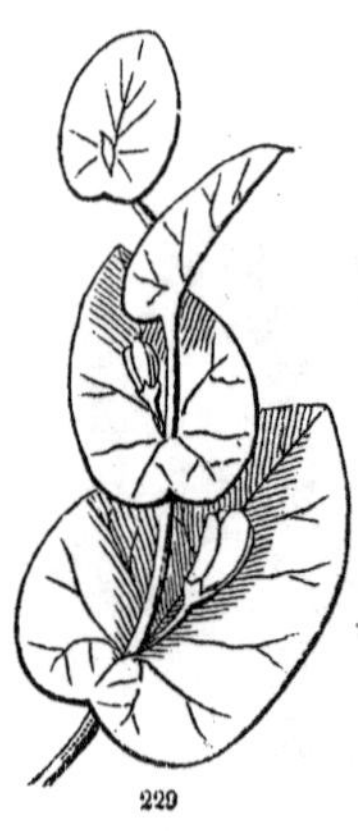

FIG. 229. Perfoliate leaves of Baptisia perfoliata.

304. **Stipules** (259) are lateral appendages of leaves, usually in the form of small foliaceous bodies, situated on each side of the base of the petiole (Fig. 215, &c.). They are not found at all in a great number of plants; but their presence or absence is usually uniform throughout each natural order. They commonly have the texture, color, and venation of leaves, and are subject to similar modifications. Like leaves, they are sometimes membranaceous or scale-like, and sometimes transformed into spines, &c.; and they have also a strong tendency to cohere with each other, or with the base of the petiole. Thus, in the Clover, the Strawberry, and the Rose (Fig. 215), a stipule adheres to each side of the base of the petiole; in the Plane-tree, they are free from the petiole, but cohere by their outer margins, so as to form an apparently single stipule opposite the leaf. In other cases, both margins are united, forming a sheath around the stem, just above the leaf: these are called *intrafoliaceous* stipules, or, when membranaceous, as in Polygonum (see Ord. Polygonaceæ), they have been termed *ochreæ*. When opposite leaves have stipules, which is not very common, they usually occupy the space between the petioles on each side, and are termed *interpetiolar*. The stipules of each leaf (one on each side), being thus placed in contact, frequently unite, so as to form apparently but a single pair of stipules for each pair of leaves; instances of which are very common in the order Rubiaceæ.

305. When leaves are furnished with stipules, they are said to be *stipulate:* when destitute of these appendages, *exstipulate.* They are sometimes present in young leaves only; as in the Beech, the Fig, and the Magnolia (Fig. 130, 131), where they form the covering of the buds, but fall away as these expand.

306. The leaflets of compound leaves are sometimes provided with small stipules (*stipelles*) of their own, as in the Bean (Fig. 221); when they are said to be *stipellate.*

Sect. III. The Death and Fall of the Leaves; Exhalation, etc.

307. While the axis, or portion of each phyton that belongs to the stem, is permanent during the life of the individual plant, the leaf lasts only for a limited period, and is thrown off, or perishes and decays, after having fulfilled its temporary office.

308. **Duration of Leaves.** In view of their duration, leaves are

called *fugacious*, when they fall off soon after their first appearance; *deciduous*, when they last only for a single season; and *peristent*, when they remain through the cold season, or other interval during which vegetation is interrupted, and until after the appearance of new leaves, so that the stem is never leafless; as in *Evergreens*.

309. Leaves last for a single year only in many Evergreens, as well as in deciduous-leaved plants; the old leaves falling soon after those of the ensuing season are expanded, or, if they remain longer, ceasing to bear any active part in the economy of the vegetable, and soon losing their vitality altogether. In Pines and Firs, however, as in many other evergreen trees and shrubs, although there is an annual fall of leaves while the growth of the season is taking place, yet these were the produce of some season earlier than the last; and the branches are continually clothed with the foliage of from two to five, or even eight or ten, successive years. On the other hand, it is seldom that all the leaves of an herb endure through the whole growing season, but the earlier foliage near the base of the stem perishes and falls, while fresh leaves are still appearing at the summit. In our deciduous trees and shrubs, however, the leaves of the season are mostly developed within a short period, and they all perish nearly at the same time. They are not destroyed by frost, as is commonly supposed; for they begin to languish, and often assume their autumnal tints (as happens with the Red Maple especially), or even fall, before the earlier frosts; and when vernal vegetation is destroyed by frost, the leaves blacken and wither, but do not fall off entire, as in autumn. Some leaves are cast off, indeed, while their tissues, at least at the base of the petiole, have by no means lost their vitality. Death is often rather a consequence than the cause of the fall. Others die and decay on the stem without falling, as in Palms and most Endogens; or else the dead leaves may hang on the branches through the winter, as in the Beech and some kinds of Oak, to fall when the new buds expand, the following spring. We must therefore distinguish between the death and the fall of the leaf.

310. **The Fall of the Leaf** is owing to an organic separation, through an *articulation*, or joint, which forms between the base of the petiole and the surface of the stem on which it rests. The formation of the articulation is a vital process, a kind of disintegration of a transverse layer of cells, which cuts off the petiole by a

regular line, in a perfectly uniform manner in each species, leaving a clean scar (Fig. 127, 130) at the insertion. The solution of continuity begins in the epidermis, where a faint line marks the position of the future joint while the leaf is still young and vigorous: later, the line of demarcation becomes well marked, internally as well as externally; the disintegrating process advances from without inwards, until it reaches the woody bundles; and the side next the stem, which is to form the surface of the scar, has a layer of cells condensed into what appears like a prolongation of the epidermis, so that, when the leaf separates, "the tree does not suffer from the effects of an open wound." "The provision for the separation being once complete, it requires little to effect it; a desiccation of one side of the leafstalk, by causing an effort of torsion, will readily break through the small remains of the fibro-vascular bundles; or the increased size of the coming leaf-bud will snap them; or, if these causes are not in operation, a gust of wind, a heavy shower, or even the simple weight of the lamina, will be enough to disrupt the small connections and send the suicidal member to its grave. Such is the history of the fall of the leaf. We have found that it is not an accidental occurrence, arising simply from the vicissitudes of temperature and the like, but a regular and vital process, which commences with the first formation of the organ, and is completed only when that is no longer useful; and we cannot help admiring the wonderful provision that heals the wound even before it is absolutely made, and affords a covering from atmospheric changes before the part can be subjected to them." * Leaves fall by an articulation in most Exogenous plants, where the insertion usually occupies only a moderate part of the circumference of the stem, and especially in those with woody stems which continue to increase in diameter. When they are not cast off in autumn, therefore, the disruption inevitably takes place the next spring, or whenever the circumference further enlarges. But in most Endogenous plants, where the leaves are scarcely, if at all, articulated with the stem, which increases little in diameter subsequently to its early growth, they are not thrown off, but simply wither and decay; their dead bases or petioles being often persistent for a long time.

311. **The Death of the Leaf,** however, in these and other cases, is

---

* Dr. Inman, in Henfrey's *Botanical Gazette*, 1. p. 61.

still to be explained. Why have leaves such a temporary existence? Why in ordinary cases do they last only for a single year, or a single summer? The answer to this question is to be found in the anatomical structure of the leaf, and the nature and amount of the fluid which it receives and exhales. The water continually absorbed by the roots dissolves, as it percolates the soil, a small portion of earthy matter. In limestone districts especially, it takes up a sensible quantity of carbonate and sulphate of lime, and becomes *hard*. It likewise dissolves a smaller proportion of silex, magnesia, potash, &c. A part of this mineral matter is at once deposited in the woody tissue of the stem (210); but a larger portion is carried into the leaves (40, 92), where, as the water is exhaled or distilled perfectly pure, all this earthy substance must be left behind to incrust the delicate cells of the parenchyma, much as the vessels in which water is boiled for culinary purposes are in time incrusted with an earthy deposit. This earthy incrustation, in connection with the deposition of organic solidified matter (39), gradually chokes the tissue of the leaf, obstructs the exhalation, and finally unfits it for the performance of its offices. Hence the fresh leaves most actively fulfil their functions in spring and early summer; but languish towards autumn, and ere long inevitably perish. Hence, although the roots and branches may be permanent, the necessity that the leaves should be annually renewed. But the former are, in fact, annually renewed likewise; and life abandons the annual layers of wood and bark almost as soon as it does the leaves they supply (216, 217, 228), and for similar reasons; although their situation is such that they become part of a permanent structure, and serve to convey the sap even when no longer endowed with vitality.

312. The general correctness of this view may be tested by direct microscopical observation. In Fig. 185, 186, some superficial parenchyma thus obstructed by long use is represented; and similar illustrations may be obtained from ordinary leaves. That this deposit consists in great part of earthy matter is shown by carefully burning away the organic materials of an autumnal leaf over a lamp, and examining the ashes by the microscope; which will be found very perfectly to exhibit the form of the cells. The ashes which remain when a leaf or other vegetable substance is burned in the open air represent the earthy materials which it has accumulated. A vernal leaf leaves only the minutest quantity of ash-

es; an autumnal leaf yields a very large proportion, — from ten to thirty times as much as the wood of the same species; although the leaves contain the deposit of a single season only, while the heart-wood is loaded with the accumulations of successive years.*

313. **Exhalation from the Leaves.** The quantity of water exhaled from the leaves during active vegetation is very great. In one of the well-known experiments of Hales, a Sunflower three and a half feet high, with a surface of 5,616 square inches exposed to the air, was found to perspire at the rate of twenty to thirty ounces avoirdupois every twelve hours, or seventeen times more than a man. A Vine, with twelve square feet of foliage, exhaled at the rate of five or six ounces a day; and a seedling Apple-tree, with eleven square feet of foliage, lost nine ounces a day. The amount varies with the degree of warmth and dryness of the air, and of exposure to light; and is also very different in different species, some exhaling more copiously even than the Sunflower. But when we consider the vast perspiring surface presented by a large tree in full leaf, it is evident that the quantity of watery vapor it exhales must be immense. This exhalation is dependent on the capacity of the air for moisture at the time, and upon the presence of the sun; often it is scarcely perceptible during the night. The Sunflower, in the experiment of Hales, lost only three ounces in a warm, dry night, and underwent no diminution during a dewy night.

314. **Rise of the Sap.** Now this exhalation by the leaves requires a corresponding absorption by the roots. The one is the measure of the other. If the leaves exhale more in a given time than the roots can restore by absorption from the soil, the foliage droops; as we see in a hot and dry summer afternoon, when the drain by exhalation is very great, while a further supply of moisture can hardly be extorted from the parched soil; — as we observe also in a leafy plant newly transplanted, where the injured rootlets are not immediately in a fit condition for absorption. Ordinarily, however, exhalation by the leaves and absorption by the roots are in

* The dried leaves of the Elm contain more than eleven *per cent.* of ashes, while the wood contains less than two *per cent.;* those of the Willow, more than eight *per cent.*, while the wood has only 0.45; those of the Beech, 6.69, the wood only 0.36; those of the (European) Oak, 4.05, the wood only 0.21; those of the Pitch-Pine, 3.15, the wood only 0.25 *per cent.* Hence the decaying foliage in our forests restores to the soil a large proportion of the inorganic matter which the trees from year to year take from it.

direct ratio to each other, and the loss sustained by the leaves is immediately restored (by endosmosis, 37) through the ascent of the sap from the branches, the latter being constantly supplied by the stem; so that, during active vegetation, the sap ascends from the remotest rootlets to the highest leaves, with a rapidity corresponding to the amount of exhalation. The action of the leaves is, therefore, the principal mechanical cause of the ascent of the sap. This is beautifully illustrated when a graft has a different time of leafing from that of the stock upon which it is made to grow, the graft wholly regulating the season or temperature at which the sap is put in motion, and controlling the habits of the original stock. Also by introducing the branches of a tree into a conservatory during winter; when, as their buds expand, the sap in the trunk without is set unseasonably into motion to supply the demand.

315. During the summer's vegetation, while the sap is consumed or exhaled almost as fast as it enters the plant, no considerable accumulation can take place: but in autumn, when the leaves perish, the rootlets, buried in the soil beyond the influence of the cold, which checks all vegetation above ground, continue for a time slowly to absorb the fluid presented to them. Thus the trunks of many trees are at this season gorged with sap, which will flow from incisions made into the wood. This sap undergoes a gradual change during the winter, and deposits its solid matter in the tubes and cells of the wood. The absorption recommences in the spring, before new leaves are expanded to consume the fluid; the soluble matters in the tissue of the stem are redissolved, and the trunk is consequently again gorged with sap, which will flow, or *bleed*, when wounded. But when the leaves resume their functions, or when flowers are developed before the leaves appear, as in many forest-trees, this stock of rich sap is rapidly consumed, and the sap will no longer flow from an incision. It is not, therefore, at the period when the trunk is most gorged with sap, in spring and autumn, but when least so, during summer, that the sap is probably most rapidly ascending.

CHAPTER VI.

OF THE FOOD AND NUTRITION OF PLANTS.

SECT. I. THE GENERAL PHYSIOLOGY OF VEGETATION.

316. THE Organs of Vegetation or Nutrition (those by which plants grow and form their various products) having now been considered, both separately and to some extent in their combined action, we are prepared to take a comprehensive survey of the general phenomena and results of vegetation; to inquire into the elementary composition of plants, the nature of the food by which they are nourished, the sources from which this food is derived, and the transformations it undergoes in their system, chiefly in the leaves. It is in vegetable digestion, or, to use a better term, in *assimilation*, that the essential nature of vegetation is to be sought, since it is in this process alone that mineral, unorganized matter is converted into the tissue of plants and other forms of organized matter (12, 15, 16). From this point of view, therefore, the reciprocal relations and influences of the mineral, vegetable, and animal kingdoms may be most advantageously contemplated, and the office of plants in the general economy of the world best understood. This portion of general physiology is intimately connected with chemistry, and some knowledge of that science is requisite for the due comprehension of the subject, especially in relation to its exceedingly important applications to agriculture and horticulture. We are here restricted to the bare statement of the leading facts which are thought to be established, and the more important deductions which may be drawn from them; omitting, for the most part, to adduce the evidence by which these general propositions are supported.

317. Although the organs of vegetation have been considered anatomically and morphologically, or in view of their structure and development, still the leading points of their physiology, or connected action in the maintenance of the life and growth of the plant, have from time to time been explained or assumed.

318. The functions of nutrition, which, in the higher animals, comprise a variety of distinct processes, are reduced to the greatest

degree of simplicity in vegetables. *Imbibition*, *assimilation*, *growth*, and perhaps *secretion*, apparently include the whole.

319. Plants absorb their food, entirely in a liquid or gaseous form, by imbibition, according to the law of endosmosis (37), through the walls of the cells that form the surface, principally those of the newest roots and their fibrils (120). The fluid absorbed by the roots, mingled in the cells with some previously assimilated matter they contain in solution (27, 79), is diffused by exosmosis and endosmosis from cell to cell, aided by the capillary action of the fibro-vascular tissue of the wood, through the newer parts of which the sap principally rises in stems of some age (210, 217); and is attracted into the leaves (or to other parts of the surface of the plant exposed to the air and light) by the exhalation which takes place from them (314), and the consequent inspissation of the sap. Here, exposed to the light of the sun, the crude sap is assimilated, or converted into organizable matter (79), with the evolution of oxygen gas into the air; and, thus prepared to form vegetable tissue or any organic product, the elaborated fluid is attracted into growing parts by endosmosis, in consequence of its consumption and condensation there, or is diffused through the newer tissues. The fluids are transferred from place to place by permeation and diffusion, according to a simple physical law. There is no movement in plants of the nature of the circulation in animals (37). Even in the so-called vessels of the latex there is merely a mechanical flow from the turgid tubes towards the place where the liquid is escaping when wounded, or from a part placed under increased pressure (63). The only circulation, or directly vital movement of fluid, in vegetable tissue, is that of rotation, or the system of currents in or next the layer of protoplasm in young and active cells (36): this movement is confined to the individual cell, and can have no influence in the transference of the sap from cell to cell. Respiration is likewise a function of animals alone. What is so called in vegetables is connected with assimilation, and is of entirely different physiological significance, as will presently be shown. None of the secretions of plants appear, like many of those of animals, to play any part, at least any essential part, in nutrition. Many, if not all of them, are purely chemical transformations of the general assimilated products of plants,—are *excretions* rather than secretions (80).

320. The appropriation of assimilated matter in vegetable growth,

and the production and multiplication of cells, which make up the fabric of the plant, have already been treated of (25 – 39). We have now only to consider what the food of plants is, whence it is derived, and how it is elaborated.

Sect. II. The Food and the Elementary Composition of Plants.

321. **The Food** and the elementary composition of plants stand in a necessary relation to each other. Since it is not to be supposed that plants possess the power of creating any simple element, whatever they consist of must have been derived from without. Their *composition* indicates their *food*, and *vice versâ*. If we have learned the chemical composition of a vegetable, and also what it gives back to the soil and the air, we know consequently what it must have derived from without, that is, its *food*. Or, if we have ascertained what the plant takes from the soil and air, and what it returns to them, we have learned its chemical composition, namely, the difference between these two. And when we compare the nature and condition of the materials which the plant takes from the soil and the air with what it gives back to them, we may form a correct notion of the influence of vegetation upon the mineral kingdom. By considering the materials of which plants are composed, we may learn what their food must necessarily contain.

322. **The Constituents of Plants** are of two kinds; the *earthy* or *inorganic*, and the *organic*. It has been stated (40, 91) that various earthy matters, dissolved by the water which the roots absorb, are drawn into the plant, and at length deposited in the wood, leaves, &c. These form the ashes which are left on burning a leaf or a piece of wood. Although these mineral matters are often turned to account by the plant, and some of them are necessary in the formation of certain products, (as the silex which gives needful firmness to the stalk of Wheat, and the phosphates which are found in the grain,) yet none of them are essential to simple vegetation, which may, and sometimes does, proceed without them. These materials, the presence of which is in some sort accidental, though in certain cases essential, are distinguished as the *earthy*, or *mineral*, or *inorganic constituents* of plants. This class may be left entirely out of view for the present. But the analysis of any newly formed vegetable tissue, or of any part of the plant,

such as a piece of wood, after the incrusting mineral matter has been chemically removed, invariably yields but three or four elements. These, which are indispensable to vegetation, and make up at least from eighty-eight to ninety-nine *per cent.* of every vegetable substance, are termed the *universal, organic constituents* of plants. They are Carbon, Hydrogen, Oxygen, and Nitrogen (10) The proper vegetable structure, that is, the tissue itself, uniformly consists of only three of these elements, namely, carbon, hydrogen, and oxygen. These are absolutely essential and universal; while the fourth, nitrogen, is an essential constituent of the protoplasm, which plays so important a part in the formation of the cells (27), and of certain vegetable products.

323. **The Organic Constituents.** These four elements must be furnished by the food upon which the vegetable lives; — they must be drawn from the soil and the air; in some cases, doubtless, from the latter source, as in Epiphytes, or Air-plants (132), but generally and principally by absorption through the roots. The plant's nourishment is wholly received either in the gaseous or the liquid form; for the leaves can imbibe air or vapor only (262 – 268), while the tissue of the rootlets is especially adapted to absorb liquids, and is incapable of taking in solid matter, however minutely divided (Fig. 108 – 110).

324. In whatever mode imbibed, evidently the main vehicle of the plant's nourishment is *water*, which as a liquid bathes its roots, and in the state of vapor continually surrounds its leaves. We have seen how copiously water is taken up by the growing plant, and have formed some general idea of its amount by the quantity that is exhaled unconsumed by the leaves (313). But pure water, although indispensable, is insufficient for the nourishment of plants. It consists of oxygen and hydrogen; and therefore may furnish, and doubtless does principally furnish, these two essential elements of the vegetable structure. But it cannot supply what it does not itself contain, namely, the carbon and nitrogen which the plant also requires.

325. Yet the question arises, whether the water which the plant actually imbibes contains in fact a quantity of these remaining elements. Though pure water cannot, may not *rain-water* supply the needful carbon and nitrogen? It is evident that, if the water which in such large quantities rises through the plant and is exhaled from its leaves contain even a very minute quantity of these

ingredients, in such a form that they may be detained when the superfluous water is exhaled, this might furnish the whole organic food of the vegetable; since the plant may condense and accumulate the carbon and nitrogen, just as the extremely minute quantity of earthy matter which the water contains is in time largely accumulated in the leaves and wood.

326. As respects the nitrogen, nearly seventy-nine *per cent.* of the atmosphere consists of this gas in an uncombined or free state, that is, merely mingled with oxygen. And, being soluble to some extent in water, every rain-drop that falls through the air absorbs and brings to the ground a minute quantity of it, which is therefore necessarily introduced into the plant with the water which the roots imbibe. This accounts for the free nitrogen which is always present in plants.

327. The plant also receives, probably, a larger portion of its nitrogen in the form of ammonia (or hartshorn), a compound of hydrogen and nitrogen, which is always produced when any animal and almost any vegetable substance decays, and which, being very volatile, must continually rise into the air from these and other sources. Besides, it appears to be formed in the atmosphere, through electrical action in thunderstorms (in the form of nitrate of ammonia). The extreme solubility of ammonia and all its compounds prevents its accumulation in the atmosphere, from which it is greedily absorbed by aqueous vapor, and brought down to the ground by rain. That the roots actually absorb it may be inferred from the familiar facts, that plants grow most luxuriantly when the soil is supplied with substances which yield much ammonia, such as animal manures; and that ammonia may be detected in the juices of almost all plants. Rain-water, therefore, contains the third element of vegetation, namely, nitrogen, both in a separate form and in that of ammonia.

328. The source of the remaining constituent, carbon, is still to be sought. Of this element plants must require a copious supply, since it forms much the largest portion of their bulk. If the carbon of a leaf or of a piece of wood be obtained separate from the other organic elements, — which may be done by charring, that is, by heating it out of contact with the air, so as to drive off the oxygen, hydrogen, and carbon, — although a small part of the carbon is necessarily lost in the operation, yet what remains perfectly preserves the shape of the original body, even to that of its most

delicate cells and vessels. With the exception of the ashes, this consists of carbon, or charcoal, amounting to from forty to sixty *per cent.*, by weight, of the original material. Carbon is itself a solid, absolutely insoluble in water, and therefore incapable of assumption by the plant. The chief, if not the only fluid compound of carbon which is naturally presented to the plant, is that of carbonic acid gas, which consists of carbon united with oxygen. This gas makes up on the average one two-thousandth of the bulk of the atmosphere; from which it may be directly absorbed by the leaves. But, being freely soluble in water up to a certain point, it must also be carried down by the rain and imbibed by the roots. The carbonic acid of the atmosphere is therefore the great source of carbon for vegetation.

329. It appears, then, that the atmosphere — considering water in the state of vapor to form a component part of it — contains all the essential materials for the growth of vegetables, and in the form best adapted to their use, namely, in the fluid state. It furnishes water, which is not only food itself, inasmuch as it supplies oxygen and hydrogen, but is likewise the vehicle of the others, conveying to the roots what it has gathered from the air, namely, the requisite supply of nitrogen, either separately or in the form of ammonia, and of carbon in the form of carbonic acid.

330. These essential elements, the whole *proper food* of plants, *may be* absorbed by the leaves directly from the air, in the state of gas or vapor. Doubtless most plants actually take in a portion of their food in this way, at least when other supply is arrested. Drooping foliage may be revived by sprinkling with water, or by exposure to a moist atmosphere. A vigorous branch of the common Live-for-ever (Sedum Telephium), or of many such plants, it is well known, will live and grow for a whole season when pinned to a dry, bare wall; and the Epiphytes, or Air-plants (132), as they are aptly called, must derive their whole sustenance immediately from the air; for they have no connection with the ground.

331. But the peculiar office of leaves is something different from that of absorbing nourishment. As a comprehensive statement, leaving extraordinary cases out of view, it may be said that plants, although they derive their food from the air, receive it chiefly through their roots. The *aqueous vapor*, condensed into rain or dew, and bringing with it to the ground a portion of *carbonic acid*, and of *nitrogen* or *ammonia*, &c., supplies the appropriate food of the plant to the rootlets. Imbibed by these, it is conveyed through the stem and into the leaves, where the now superfluous water is restored to the atmosphere by exhalation,* while the residue is converted into the proper nourishment and substance of the vegetable.

332. The atmosphere is therefore the great storehouse from which vegetables derive their nourishment; and it might be clearly shown that all the constituents of plants, excepting the small earthy portion that many can do without, have at some period formed a part of the atmosphere. The vegetable kingdom represents an amount of matter, which the force of organization has withdrawn from the air, and confined for a time to the surface.

333. Does it therefore follow, that the soil merely serves as a foothold to plants, and that all vegetables obtain their whole nourishment directly from the atmosphere? This must have been the case with the first plants that grew, when no vegetable or animal matter existed in the soil; and no less so with the first vegetation that covers small volcanic islands raised in our own times from the sea, or the surface of lava thrown from ordinary volcanoes. No vegetable matter is brought to these perfectly sterile mineral soils, except the minute portion contained in the seeds wafted thither by winds or waves. And yet in time a vast quantity is produced, which is represented not only by the existing vegetation, but by the mould that the decay of previous generations has imparted to the soil. We arrive at the same result by the simple experiment of causing a seed of known weight to germinate on powdered flints, watered by rain-water alone. When the young plant has

* The water exhaled may be again absorbed by the roots, laden with a new supply of the other elements from the air, again exhaled, and so on; as is beautifully illustrated by the cultivation of plants in closed *Ward cases*, where plants are seen to flourish for a long time with a very limited supply of water, every particle of which (except the small portion actually *consumed* by the plants) must pass repeatedly through this circulation. This vegetable microcosm well exhibits the actual relations of water, &c. to vegetation on a large scale in nature; where the water is alternately and repeatedly raised by evaporation and recondensed to such extent that what actually falls in rain is estimated to be reëvaporated and rained down (on an average throughout the world) ten or fifteen times in the course of a year. In this way the atmosphere is repeatedly washed by the rain; and those vapors *washed out* which else by their accumulation would prove injurious to men and animals, and conveyed to the roots of plants, which they are especially adapted to nourish.

attained the fullest development of which it is capable under these circumstances, it will be found to weigh (after due allowance for the silex it may have taken up) perhaps fifty or one hundred times as much as the original seed. There can be no question as to the source of this vegetable matter in all these cases. *The requisite materials exist in the air. Plants possess the peculiar faculty of drawing them from the air. The air must have furnished the whole.* This conclusion is amply confirmed by a great variety of familiar facts; such as the accumulation of vegetable matter in peat-bogs, and of mould in neglected fields, in old forests, and generally wherever vegetation is undisturbed. Since this rich mould, instead of diminishing, regularly increases with the age of the forest and the luxuriance of vegetation, the trees must have drawn from the air, not only the vast amount of carbon, &c. that is stored up in their trunks, but an additional quantity which is imparted to the soil in the annual fall of leaves, &c.

334. Still it by no means follows, that each plant draws all its nourishment directly from the air. This unquestionably happens in some of the special cases just mentioned; with Air-plants, and with those that first vegetate on volcanic earth, bare rocks, naked walls, or pure sand. But it is particularly to be remarked, that only certain tribes of plants will continue to live under such circumstances, and that none of the vegetables most useful as food for man or the higher animals will thus thrive and come to maturity. In nature, the races of plants that will grow at the entire expense of the air, such as Lichens, Mosses, Ferns, and certain succulent tribes of Flowering plants, gradually form a soil of vegetable mould during their life, which they increase in their decay; and the successive generations live more vigorously upon the inheritance, being supported partly upon what they draw from the air, and partly upon the ancestral accumulation of vegetable mould. Thus, each generation may enrich the soil, even of those plants that draw largely upon vegetable matter thus accumulated; for it annually restores a portion by its dead leaves, &c., and when it dies it bequeathes to the soil, not only all that it took from it, but all that it drew from the air. It is in this way that the lower tribes and so-called useless plants create a soil, which will in time support the higher plants of immediate importance to man and the higher animals, but which could never grow and perfect their fruit, if left, like their humble but indispensable predecessors, to derive an unaided subsistence directly from the inorganic world. While it is strictly true, therefore, that all the organic elements have been originally derived from the air, it is not true that what is contained in almost any given plant, or in any one crop, is immediately drawn from this source. A part of it is thus supplied, but in proportions varying greatly in different species and under different circumstances. Undisturbed vegetation consequently tends always to enrich the soil. But in agriculture the crop is ordinarily removed from the land, and with it not only what it has taken from the earth, but also what it has drawn from the air; and the soil is accordingly impoverished. Hence the farmer finds it necessary to follow the example of nature, and to restore to the land, in the form of manure, an amount substantially equivalent to what he takes away.

335. The mode in which vegetable mould is turned to account by growing plants has not yet been sufficiently investigated. According to Liebig, the decaying vegetable matter is not employed until it has been resolved into its original inorganic elements, namely, into water, carbonic acid, ammonia, &c.; which, slowly absorbed by the water that percolates the soil, are imbibed by the roots. Others suppose that a portion of the food which plants derive from decaying vegetable matter may consist of soluble, still organic compounds. The economy of the greenless parasitic plants (135) is adduced in confirmation of this view: but these are nourished by the foster plant just as its own flowers are nourished. Decisive evidence to the point is furnished by Fungi, the greater part of which live upon decaying organic matter, and have not the power of forming organizable products from inorganic materials; and there is reason to think, that at least one Phænogamous plant (our Monotropa, 137, Fig. 812) lives in much the same way.

336. **The Earthy Constituents.** The mineral substances which form the inorganic constituents of plants (322) are furnished by the soil, and are primarily derived from the slow disintegration and decomposition of the rocks and earths that compose it.* These are dissolved, for the most part, in very minute proportions, in the water which percolates the soil (aided, as to the more insoluble earthy salts, by the carbonic acid which this water contains), and with this

* According to Liebig, the quantity of potash contained in a layer of soil formed by the disintegration of 40,000 square feet of the following rocks, &c.,

water are taken up by the roots. However minute their proportion in the water which the roots imbibe, the plant concentrates and accumulates them, as it does its most dilute inorganic food, by the constant exhalation of the water from the leaves, until they amount to an appreciable quantity, often to a pretty large percentage, of the solid matter of the vegetable. As might be expected (**311**), the leaves contain a much larger amount of ashes, or earthy matter, than the wood, but the trunk more than the branches (**210**). Herbaceous plants also accumulate more than trees in proportion to their weight when dry.*

337. The ashes left after combustion are mostly composed of the "alkaline chlorides, with the bases of potash and soda, earthy and metallic phosphates, caustic or carbonate of lime and magnesia, silica, and oxides of iron and of manganese. Several other substances are also met with there, but in quantities so small that they may be neglected." Different species growing in the same soil appear to take in some portion of all such materials that are

---

to the depth of twenty inches, is as follows. This quantity of Felspar (a large component of granite, &c.) contains . . . 1,152,000 lbs.

Clinkstone, ...	from 200,000 to 400,000 "
Basalt, ...	" 47,500 " 75,000 "
Clay-slate, ...	" 100,000 " 200,000 "
Loam, ...	" 87,000 " 300,000 "

The silex yielded to the soil by the gradual decomposition of granite and other rocks is in the form of a silicate of potash or other alkali, which, though insoluble in pure water, is slowly acted upon and dissolved by the united action of water and carbonic acid, or more largely by water impregnated with carbonate of potash, which is abundantly liberated during the natural decomposition of these rocks.

* The subjoined results, selected from Boussingault, exhibit in a tabular form the relative quantities of organic and inorganic constituents in several kinds of herbage, compared, in several cases, with the root or grain. The water was previously driven off by desiccation.

	Leaves of Mangel-Wurzel.	Root of Mangel-Wurzel.	Potato-tops.	Potatoes.	Pea-straw.	Peas.	Clover-hay.	Wheat-straw.	Wheat.
Carbon,	38.10	42.75	44.80	43.72	45.80	46.06	47.53	48.48	46.10
Hydrogen,	5.10	5.77	5.10	6.00	5.00	6.09	4.69	5.41	5.80
Oxygen,	30.80	43.58	30.50	44.88	35.57	40.53	37.96	38.79	43.40
Nitrogen,	4.50	1.66	2.30	1.50	2.31	4.18	2.06	0.35	2.27
Ashes,	21.50	6.24	17.30	3.90	11.32	3.14	7.76	6.97	2.43
	100.00	100.00	100.00	100.00	100.00	100.00	100.00	100.00	100.00

naturally presented to them in solution, but not, however, in the same proportions, nor in any close proportion to the relative solubility of these several substances: while, on the other hand, the same species in different localities, under generally similar circumstances, and also each of its particular parts or organs, contains, or tends to contain, the same mineral constituents in nearly the same proportion. One base, however, is often substituted for another, equivalent for equivalent, as magnesia for lime, soda for potash. The roots, therefore, appear to have a certain power of selection in respect to these mineral materials. Nor is it a valid objection to this view, that they absorb poisons which destroy them. These are either organic products, such as opium; or else are corrosive substances, such as sulphate of copper, which disorganize the rootlets, and are then indiscriminately imbibed by mere capillary attraction. For mutilated roots or stems absorb all dissolved materials of the proper density that are presented to them, not only in much larger quantity (so long as the cut is fresh) than do the uninjured rootlets, but almost indifferently, and in the same proportion that they absorb the water they are dissolved in.

338. In the ashes, only the salts which resist the action of heat, such as the phosphates, sulphates, and hydrochlorates, are in the state in which they existed in the plant itself. A great part of the bases were combined with organic acids, formed in the plant, and most largely with the oxalic (90, 91): these compounds are by incineration, or by subsequent exposure to the air, principally converted into carbonates.

339. It being indispensable that a plant should find in the soil such mineral matters as are necessary to its growth or perfect development, we are enabled to understand why various species will only flourish in particular soils or situations; why plants which take up common salt, &c. are restricted to the sea-shore and to the vicinity of salt-springs; why numerous weeds which grow chiefly around dwellings, and follow the footsteps of man and the domestic animals, flourish only in a soil abounding in nitrates (their ashes containing a notable quantity either of nitrate of potash or of lime); why the Vine requires alkaline manures, to replace the large amount of tartrate of potash which the grapes contain; and why Pines and Firs, the ashes of which contain very little alkali, will thrive in the thinnest and most sterile soil, while the Beech, Maple, Elm, &c., abounding with potash, are only found in strong and fertile land.

340. Where vegetation is undisturbed by man, all these needful earthy materials, which are drawn from the soil during the growth of the herbage or forest, are in time restored to it by its decay, in an equally soluble form, along with organic matter which the vegetation has formed from the air. But in cultivation, the produce is carried away, and with it the materials which have been slowly yielded by the soil. "A medium crop of Wheat takes from one acre of ground about 12 pounds, a crop of Beans about 20 pounds, and a crop of Beets about 11 pounds, of phosphoric acid, besides a very large quantity of potash and soda. It is obvious that such a process tends continually to exhaust arable land of the mineral substances useful to vegetation which it contains, and that a time must come, when, without supplies of such mineral matters, the land would become unproductive from their abstraction. . . . . . In the neighborhood of large and populous towns, for instance, where the interest of the farmer and market-gardener is to send the largest possible quantity of produce to market, consuming the least possible quantity on the spot, the want of saline principles in the soil would very soon be felt, were it not that for every wagon-load of greens and carrots, fruit and potatoes, corn and straw, that finds its way into the city, a wagon-load of dung, containing each and every one of these principles locked up in the several crops, is returned to the land, and proves enough, and often more than enough, to replace all that has been carried away from it." * The loss must either be made up by such equivalent return, or the land must lie fallow from time to time until these soluble substances are restored by further disintegration of the materials of the soil: or meanwhile the more exhausting crops may be alternated with those that take least from the soil and most from the air; or

---

* Boussingault, *Economie Rurale;* from the Engl. Trans., p. 493. Further: — "It may be inferred that, in the most frequent case, namely, that of arable lands not sufficiently rich to do without manure, there can be no continuous [independent] cultivation without annexation of meadow; in other words, one part of the farm must yield crops without consuming manure, so that this may replace the alkaline and earthy salts which are constantly withdrawn by successive harvests from another part. Lands enriched by rivers alone permit of a total and continued export of their produce without exhaustion. Such are the fields fertilized by the inundations of the Nile; and it is difficult to form an idea of the prodigious quantities of phosphoric acid, magnesia, and potash, which, in a succession of ages, have passed out of Egypt with her incessant exports of corn." — p. 503.

with one which, like clover, although it takes up 77 pounds of alkali per acre, may be consumed on the field, so as to restore most of this alkali in the manure for the succeeding crop.

341. It has been asserted that the advantage of preceding a wheat crop by one of Leguminous plants (such as Peas, Clover, Lucerne, &c.), or of roots or tubers, is owing to the fact, that these leave the phosphates, &c. nearly untouched for the wheat which is to follow, and which largely abstracts them. The results of Boussingault's experiments and analyses show that these products are far from having the deficiency of phosphates which was alleged. "For example, beans and haricots take 20 and 13.7 pounds of phosphoric acid from every acre of land; potatoes and beet-root take 11 and 12.8 pounds of that acid, exactly what is found in a crop of wheat. Trefoil is equally rich in phosphates with the sheaves of corn that have gone before it." * His further researches seem to show that these crops exhaust the soil less than the cereal grains, in part at least, on account of the large quantity of organic matter, rich in nitrogen, which they leave to be incorporated with the soil. The theory of rotation in crops, founded by De Candolle on the assumption that excretions from the roots of a plant accumulated in the soil until in time they became injurious to that crop, but furnished appropriate food for a different species, is entirely abandoned as an explanation; and even the fact that

* Boussingault, *l. c.*, p. 497.— Subjoined is a table, from the same work, of *the percentage of Mineral Substances taken up from the soil by various plants grown at Bechelbronn.*

Substances which yielded the Ashes.	Acids.			Chlorine.	Lime.	Magnesia.	Potash.	Soda.	Silica.	Oxide of Iron, Alumina, &c.	Charcoal, Moisture, and Loss.
	Carbonic.	Sulphuric.	Phosphoric.								
Potatoes,	13.4	7.1	11.3	2.7	1.8	5.4	51.5	traces	5.6	0.5	0.7
Mangel-Wurzel,	16.1	1.6	6.1	5.2	7.0	4.4	39.0	6.0	8.0	2.5	4.2
Turnips,	14.0	10.9	6.0	2.9	10.9	4.3	33.7	4.1	6.4	1.2	5.5
Potato-tops,	11.0	2.2	10.8	1.6	2.3	1.8	44.5	traces	13.0	5.2	7.6
Wheat,	0.0	1.0	47.0	traces	2.9	15.9	29.5	traces	1.3	0.0	2.4
Wheat-straw,	0.0	1.0	3.1	0.6	8.5	5.0	9.2	0.3	67.6	1.0	3.7
Oats,	1.7	1.0	14.9	0.5	3.7	7.7	12.9	0.0	53.3	1.3	3.0
Oat-straw,	3.2	4.1	3.0	4.7	8.3	2.8	24.5	4.4	40.0	2.1	2.9
Clover,	25.0	2.5	6.3	2.6	24.6	6.3	26.6	0.5	5.3	0.3	0.0
Peas,	0.5	4.7	30.1	1.1	10.1	11.9	35.3	2.5	1.5	traces	2.3
French beans,	3.3	1.3	26.8	0.1	5.8	11.5	49.1	0.0	1.0	traces	1.1
Horse beans,	1.0	1.6	34.2	0.7	5.1	8.6	45.2	0.0	0.5	traces	3.1

such excretions are formed, at least to any considerable extent, is not made out. That they could accumulate and remain in the soil without undergoing decomposition is apparently impossible.

SECT. III. ASSIMILATION, OR VEGETABLE DIGESTION, AND ITS RESULTS.

342. WE have reached the conclusion, that the universal food of plants is rain-water, which has absorbed some carbonic acid gas and nitrogen (partly in the form of ammonia or of other compounds) from the air, or dissolved them from the remains of former vegetation in the soil, whence it has also taken up a variable (yet more or less essential) quantity of earthy matter.

343. This fluid, imbibed by the roots, and carried upwards through the stem, receives the name of *sap* or *crude sap* (79). During its ascent, its properties are often more or less altered, chiefly by dissolving the soluble organized matter it meets with; thus becoming sweet in the Maple, &c., and acquiring different sensible properties in different species. This dissolved portion is already elaborated food, and may therefore be immediately employed in vegetable growth. But the crude sap itself is merely raw material, unorganized, mineral matter, as yet incapable of forming a part of the living structure. Its conversion into organized matter constitutes the process of

344. **Assimilation** (12, 15), or what, from an analogy with an animal function, is usually termed *Vegetable Digestion*. To undergo this important change, the crude sap is attracted into the leaves, or other green parts of the plant, which constitute the apparatus of vegetable digestion, where it is exposed to the light of the sun, under which influence alone can this change be effected. Under the influence of solar light, the fabric is itself constructed, and the *chlorophyll*, or green matter of plants, upon which, or in connection with which, the light exerts its wonderful action, is first developed. When plants are made to grow in insufficient light, as when potatoes throw out shoots in cellars, this green matter is not formed. When light is withdrawn, it is soon decomposed; as we see when Celery is blanched by heaving the soil around its stems. So, also, the naturally greenless leaves of plants parasitic upon the roots or stems of other species (135) have no direct power of assimilation, but feed upon and grow at the ex-

pense of already assimilated matter. But all green parts of plants, such as the cellular outer bark of most herbs, act upon the sap in the same manner as leaves, even supplying their places in plants which produce few or no leaves, as in the Cactus, &c. Under the influence of light, an essential preliminary step in vegetable digestion is accomplished, namely, the concentration of the crude sap by the evaporation or exhalation of the now superflous water, the mechanism and various consequences of which have already been considered (**267**, **313**).

345. We have only to consider the further agency of light in the process of vegetable digestion itself, namely, its action in the leaf upon the concentrated sap. Here it accomplishes two perfectly unparalleled results, which essentially characterize vegetation, and upon which all organized existence absolutely depends (**1**, **16**). These are, — 1st. *The chemical decomposition of one or more of the substances in the sap which contain oxygen gas, and the liberation of this oxygen at the ordinary temperature of the air.* The chemist can in certain cases liberate oxygen gas from its compounds, but only with the aid of powerful reagents, or of a heat equal to that of red-hot iron. 2d. *The transformation of this mineral food, this inorganic into organic matter, — the organized substance of living plants, and consequently of animals.* These two operations, although separately stated to convey a clearer idea of the results, are in fact but different aspects of one great process. We contemplate the first, when we consider what the plant gives back to the air; — the second, when we inquire what it retains as the materials of its own growth. The concentrated sap is decomposed; the portion which is not required in the growth of the plant is returned to the air; and the remaining elements are at the same time rearranged, so as to form peculiar *organic* products.

346. The principal material given back to the air, in this process, is oxygen gas,* that element of our atmosphere which alone

---

* A small proportion of nitrogen gas is likewise almost constantly exhaled from the leaves; but this appears to come from the nitrogen which the water imbibed by the roots had absorbed from the air (326), and which passes off unaltered from the leaves when this water is evaporated, or from the air which the rootlets directly absorb to some extent. In the course of vegetation, no more nitrogen is given out than what is thus taken in, and probably not so much. So that the exhalation of nitrogen may be left out of the general view of the changes which are brought about in vegetation.

renders it fit for the breathing and life of animals. That the foliage of plants in sunshine is continually yielding oxygen gas to the surrounding air has been familiarly known since the days of Ingenhouss and Priestley, and may at any moment be verified by simple experiment. The readiest way is, to expose a few freshly gathered leaves to the sunshine in a glass vessel filled with water, and to collect the air-bubbles which presently arise while the light falls upon them, but which cease to appear when placed in shadow. This air, when examined, proves to be free oxygen gas. In nature, diffused daylight produces this result; but in our rude experiments, direct sunshine is generally necessary. What is the source of this oxygen gas, which is given up to the air just in proportion to the vigor of assimilation in the leafy plant, or, in other words, to the consumption of crude sap?

347. To take for illustration the case which exhibits the general result (and whether this is actually attained at one operation, or not, does not affect the view), and enables us directly to contrast the *materials* with the principal *product* of vegetation, we will suppose the plant is assimilating its food immediately into *Cellulose*, or the substance of which its tissue consists (27). This matter, when in a pure state, and free from incrusting materials, has a perfectly uniform composition in all plants. It is composed of carbon, hydrogen, and oxygen, of which the latter two exist in the same proportions as in water.* It may therefore be said to consist of carbon and the elements of water. These materials are necessarily furnished by the plant's food. But the universal food of the plant, that which is only and absolutely essential to bare vegetation (324, 329), is carbonic acid and water. If this be decomposed in vegetation, and the carbonic acid give up its oxygen, there remain carbon and water, or rather the elements of water, — the very composition of *cellulose* or vegetable tissue. Doubtless, then, the oxygen which is rendered to the air in vegetation comes from the carbonic acid which, as we have seen (328), the plant took from the air.

348. This view may be confirmed by direct experiment. We have seen that many plants *must*, and all *may*, imbibe the whole or a part of their food directly from the air into their leaves (132,

---

* Cellulose is chemically composed of 12 equivalents of Carbon, 10 of Hydrogen, and 10 of Oxygen, viz. C^{12}, H^{10}, O^{10}.

330). All leafy plants doubtless obtain a part of their carbonic acid in this way. It is accordingly found, that when a current of carbonic acid is made slowly to traverse a glass globe containing a leafy plant exposed to the full sunshine, the carbonic acid disappears, and an equal bulk of oxygen gas supplies its place. Now, since carbonic acid gas contains just its own bulk of oxygen, it is evident that what has thus been decomposed in the leaves has returned all its oxygen to the air. Plants take carbonic acid from the atmosphere, therefore (directly or indirectly); they retain its carbon; they give back its oxygen.*

349. But cellulose, being the final, insoluble product of vegetation appropriated as tissue, cannot itself be formed in the first instance. The materials from which it is deposited, and which we actually find in the elaborated sap, are *Dextrine* or *Vegetable Mucilage* (81, 83), sugar (84), &c. The first of these is probably directly produced in assimilation. Its chemical composition is the same as that of pure cellulose: it consists, not only of the same three elements, but of the same elements in exactly the same proportion. Dextrine, vegetable mucilage, &c., are the primary, as yet unappropriated materials of vegetable tissue, or unsolidified cellulose, and their production from the crude sap is attended with the evolution of the oxygen which was contained in the carbonic acid of the plant's food, as already stated.† Nor is the result in any

* At least, the result is *as if* the oxygen exhaled were all thus detached from the carbon of the carbonic acid. Just this amount is liberated, and the facts obviously point to the carbonic acid as its real source. But on the other hand, it appears unlikely that a substance which holds oxygen with such strong affinity as carbon should yield the whole of it under these circumstances: and water is certainly decomposed, with the evolution of oxygen, in the formation of a class of vegetable products soon to be mentioned; besides, Edwards and Colin have shown that water is directly decomposed during germination. Still, as no one supposes that the residue after the liberation of oxygen *is* carbon and water, but only the three elements in the proportions which would constitute them, it amounts to nearly the same thing whether we say that *the oxygen of the carbonic acid*, or *an amount of oxygen equivalent to that of the carbonic acid, derived partly from it and partly from the water*, is liberated in such cases. That Schleiden should assert that the oxygen liberated comes from the decomposition of water alone, shows gross carelessness, or an ignorance of arithmetic as well as chemistry, which is the less excusable in one who, in a scientific treatise, habitually applies opprobrious epithets to a great part of his fellow-laborers.

† The result is just the same, if, with Henfrey, we suppose that the mat-

respect altered when *Starch* is produced. In that case, the direct product of assimilation in the form of dextrine, instead of being immediately appropriated in growth, is solidified in the starch-grains, and in that compact and temporarily insoluble form accumulated as the ready prepared materials of future growth (81). So, also, when *Inuline* is formed instead of starch, as in the roots of Elecampane (Inula Helenium) and the Dahlia, and the tubers of the Jerusalem Artichoke: here the dextrine is solidified into a substance intermediate in its properties between dextrine and cellulose, which is closely analogous to starch, and subservient to the same purpose. Notwithstanding the difference in their properties and chemical reactions, these various products are strictly *isomeric*, that is, they consist of the same elements, combined in the same proportions; and physiologically they are merely different states of one and the same thing. Dextrine is the most soluble state (dissolving freely in cold water), and that originally formed in assimilation in the foliage: starch and inuline are two temporarily solidified states, and cellulose is the ultimate and usually permanent insoluble condition. Accordingly, whenever the materials of growth are supplied from such accumulations of nourishment, as especially from the seed in germination, from fleshy roots (128), rootstocks (174), tubers (175), &c., the starch or inuline is dissolved in the sap, being spontaneously reconverted into dextrine, &c., and attracted in this liquid state into the growing parts, where it is transformed into cellulose, and becomes a portion of the permanent vegetable fabric.

350. Assimilated matter also occurs in the sap under the still more soluble form of *Sugar* (84). If we suppose this to be a direct product of the assimilation of carbonic acid and water, the amount of oxygen gas exhaled will be just the same as before. For *sugar* has the same elementary composition as dextrine, starch, and cellulose, with the addition of one equivalent of water in the case of cane-sugar, and of three more in that of grape-sugar.* If, as is more probable, sugar is a subsequent transformation of dextrine, then the latter has only to appropriate some water. In the forma-

---

ters acted upon in assimilation are at first as much deoxidized as in chlorophyll, since these general products of vegetation have immediately to absorb oxygen enough to bring them to the form of dextrine, starch, cellulose, &c.

* The formula for cane-sugar is C^{12}, H^{11}, O^{11}; for grape-sugar, C^{12}, H^{14}, O^{14}.

tion of all these products, therefore, the same quantity of carbonic acid is consumed, and all its oxygen restored to the air.* It is more and more evident, therefore, that, by just so much as plants grow, they take carbonic acid from the air, they retain its carbon, and return its oxygen.

351. In the production of that modification of cellúlose called *Lignine* (41), which forms a secondary deposit thickening the walls of the cells, and which abounds in wood, if this be really a simple product, and not a mixture, not only must a larger amount of carbonic acid be decomposed, but a small portion of water also, with the liberation of its oxygen. For the composition attributed to it shows that it contains less oxygen than would suffice to convert its hydrogen into water.† This excess of hydrogen, and the still larger excess of carbon, renders those woods that abound with incrusting deposit, other things being equal, more valuable as fuel than those of which the issue in great part consists of proper cellulose, as has already been stated.

352. The whole class of fatty substances, including the *Oils*, *Wax*, *Chlorophyll* (85 - 87), and most of the products of their alter-

---

* Since all these *neutral ternary substances* are identical, or nearly so, in elementary composition, and since, with the same amount of carbon, derived from the decomposition of carbonic acid, the plant can form them all, notwithstanding the great difference in their external characters, it will no longer appear so surprising that they should be so readily convertible into each other in the living plant, and even in the hands of the chemist. But the chemistry of organic nature exceeds the resources of science, and constantly produces transformations which the chemist in his laboratory is unable to effect. The latter can change starch into dextrine, and dextrine into sugar; but he cannot reverse the process, and convert sugar into dextrine, or dextrine into starch. In the plant, however, all these various transformations are continually taking place. Thus, the starch deposited in the seed of the Sugar-cane, as in all other Grasses, is changed into sugar in germination; and the sugar which fills the tissue of the stem at the time of flowering is rapidly carried into the flowers, where a portion is transformed into starch and again deposited in the newly-formed seeds. And although the chemist is unable to transform starch, sugar, &c. into cellulose, yet he readily effects the opposite change, by reconverting woody fibre, &c. (under the influence of sulphuric acid) into dextrine and sugar. The plant does the same thing in the ripening of fruits, during which a portion of tissue is often transformed into sugar. Starch-grains and cellulose never can be formed artificially, because they are not merely organizable matter, but have an organic structure, or are the result of growth.

† According to Payen, lignine, separated as much as possible from cellulose, consists of Carbon 53.8, Hydrogen 6.0, and Oxygen 40.2 per cent., $= C^{35}, H^{24}, O^{20}$.

ation, contain, a few of them no oxygen at all (such as caoutchouc and some oils), and all of them less oxygen than is requisite to convert their hydrogen into water. In their direct formation, therefore, not only all the oxygen of the carbonic acid has been given out, but also a portion belonging to the water. If formed by a further deoxidation of neutral ternary products, as chlorophyll from starch (87) and wax from sugar (86), the same result is attained as respects the liberation of oxygen gas, but by two or more steps instead of one. The *Resins*, doubtless, are not direct vegetable products, but originate from the alternation and partial oxidation of the essential oils. *Balsams*, which exude from the bark of certain plants, are natural solutions of resins in their essential oils, as rosin, or pine-resin, in the oil of turpentine.

353. An opposite class, the *Vegetable Acids* (90), contain more oxygen than is necessary for the conversion of their hydrogen into water, but less than the amount which exists in carbonic acid and water. Indeed, the most general vegetable acid, the *oxalic* (which is formed artificially by the action of nitric acid on starch), has no hydrogen, except in the atom of water that is connected with it. These acids are sometimes formed in the leaves, as in the Sorrel, the Grape-vine, &c., but usually in the fruit. If produced directly from the sap, as is probably the case in acid leaves, only a part of the oxygen in the carbonic acid which contributes to their formation would be exhaled. But if they are formed from sugar, or any other of the general products of the proper juice, the absorption of a portion of oxygen from the air would be required for the conversion; and this absorption takes place (at least in some cases) when fruits acquire their acidity. Even their formation by the plant, therefore, is attended by the liberation of oxygen gas, though less in quantity than in ordinary vegetation.

354. There is still another class of vegetable products of universal occurrence, and, although comparatively small in quantity, of as high importance as those which constitute the permanent fabric of the plant; namely, the *neutral quaternary organic compounds*, of which nitrogen is a constituent (79). These are mutually convertible bodies, related to each other as dextrine and sugar to starch and cellulose, and playing the same part in the animal economy that the neutral ternary products do in the vegetable. To the basis or type of these azotized products Mulder has given the name of *Proteine* (27): hence they are sometimes collectively

called proteine compounds. In their production from the crude sap, the ammonia, or other azotized matter it contains, plays an essential part; and oxygen gas is restored to the air from the decomposition of all the carbonic acid concerned and a part of the water.*

355. In living cells the proteine exists as azotized mucilage, and forms the *protoplasm* or vitally active lining which may be said to give origin to the vegetable structure, since the cellulose is deposited under its influence to form the permanent walls or fabric of the cells, as has already been explained (27 – 32). When the cells have completed their growth and transformation, the protoplasm abandons them, being constantly attracted onwards into forming and growing parts, where it incites new development. For this azotized matter has the remarkable peculiarity of inducing chemical changes in other organic products, especially the neutral ternary bodies, causing one kind to be transformed into another, or even the decomposition of a part into alcohol, acetic acid, and finally into carbonic acid and water (as in germination, &c.), — itself remaining the while essentially unaltered.

356. The constant attraction of the protoplasm from the completed into the forming parts of the plants explains how it is, that so small a percentage of azotized matter should be capable of playing such an all-important part in the vegetable economy. It does its work with little loss of material, and no portion of it is fixed in the tissues. At least, the little that remains in old parts is capable of being washed out, showing that it forms no integral part of the fabric. It explains why the heart-wood of trees, especially the most solidified kinds, yields barely a trace of nitrogen, while the sap-wood yields an appreciable amount, and the cambium-layer

---

* The chemical changes have been tabulated thus: —

The materials:	C.	H.	N.	O.	From which are formed the product:	C.	H.	N.	O.
74 of Water,		74		74	1 of Proteine,	48	36	6	14
94 of Carbonic acid,	94			188	4 of Cellulose,	48	40		40
2 of Carbonate of ammonia,	2	2	6	4	212 of Oxygen liberated,				212
	96	76	6	266		96	76	6	266

It seems now to be conceded, that proteine, as well as all its transformations or states, contains also, as essential constituents, a minute quantity of sulphur and phosphorus, one or both (10).

and all parts of recent formation, such as the buds, young shoots, and rootlets, always contain several hundredths of it. This gives the reason, also, why sap-wood is so liable to decay (induced by the proteine), and the more likely in proportion to its newness and the quantity of sap it contains, while the perfectly lignified heart-wood is so durable. Following this course, we find that the azotized matter rapidly diminishes in the stem and herbage during flowering, while it accumulates in the forming fruit, and is finally condensed in the seeds (which have a larger percentage than any other organ), ready to subserve the same office in the development of the embryo plant it contains.*

357. When wheat-flour, kneaded into dough, is subjected to the prolonged action of water, the starch is washed away, and a tenacious, elastic residue, the *Gluten* of the flour, which gives it the capability of being raised, contains nearly all the proteine compounds of the seed, mixed with some fatty matters (which may be removed by alcohol and ether) and with a little cellulose. The azotized products constitute from eight to thirty *per cent.* of the weight of wheat-flour; the proportion varying greatly under different circumstances, but always largest when the soil is best supplied with manures that abound in nitrogen. The gluten is not itself a simple quatenary principle; but is a mixture of four nearly isomeric bodies of this sort, distinguished by chemists under the names *Fibrine* (identical in nature with that which forms the muscles of animals), *Albumen* (of the same nature as animal albumen), *Caseine* (identical with the curd of milk), and *Glutine.* In beans and all kinds of pulse, or seeds of Leguminous plants, the large proportion of azotized matter principally occurs in the form of *Legumine*, a form nearly intermediate in character between albumen and caseine.

358. Having now noticed all the principal products of assimilation in plants, at least those concerned in nutrition, as compared with the inorganic materials from which they must needs be formed, we may the more clearly perceive, that the principal result of vegetation, as concerns the atmosphere, from which plants draw their food, consists in the withdrawal of water, of a little am-

* The cotyledons of peas and beans, according to Mr. Rigg, contain from 100 to 140 parts, and the plumule about 200 parts, of nitrogen, to 1,000 parts of carbon.

monia, and of a large proportion of carbonic acid, with the restoration of oxygen. The latter is a constant effect of vegetation and the measure of its amount. As respects the tissue of the plant, which makes up almost the whole bulk of a tree or other vegetable fabric, the sole consequences of its formation upon the air are the withdrawal of a small quantity of water, and of a large amount of carbonic acid gas, and the restoration of the oxygen of the latter. In the formation of the azotized products, a portion of ammonia or of some equivalent compound of nitrogen is also withdrawn. It is true, indeed, that leaves decompose carbonic acid only in daylight; and that they sometimes impart a quantity of carbonic acid to the air in the night, especially when vegetation languishes, or even take from it a little oxygen. But this does not affect the general result, nor require any qualification of the general statement. The work simply ceases when light is withdrawn. The plant is then merely in a passive state. Yet, whenever exhalation from the leaves slowly continues in darkness, the carbonic acid which the water holds necessarily flies off with it, during the interruption to vegetation, into the atmosphere from which the plant took it. So much of the crude sap, or raw material, merely runs to waste. Furthermore, it must be remembered, that the decomposition of carbonic acid in vegetation is in direct opposition to ordinary chemical affinity; or, in other words, that all organized matter is in a state corresponding to that of unstable equilibrium. Consequently, when light is withdrawn, ordinary chemical forces may perhaps to some extent resume their sway, the oxygen of the air combine with some of the newly deposited carbon to reproduce a little carbonic acid, and thus demolish a portion of the rising vegetable structure which the setting sun left, as it were in an unfinished or unstable state. This is what actually takes place in a dead plant at all times, and whenever an herb is kept in prolonged darkness; chemical forces, exerting their power uncontrolled, demolish the whole vegetable fabric, beginning with the chlorophyll (as we observe in blanching Celery), and at length resolve it into the carbonic acid and water from which it was formed. But this must all be placed to the account of *decomposing*, not of *growing* vegetation; and even if it were a universal phenomenon, which is by no means the case,* would not affect the

* In repeating the old experiments upon this subject with due precautions, and with improved means of research, it is found that many ordinary plants,

general statement, that, *by so much as plants grow*, they decompose carbonic acid and give its oxygen to the air; or, in other words, purify the air.

359. Every six pounds of carbon in existing plants has withdrawn twenty-two pounds of carbonic acid gas from the atmosphere, and replaced it with sixteen pounds of oxygen gas, occupying the same bulk. To form some general conception of the extent of the influence of vegetation upon the air we breathe, therefore, we should compute the quantity of carbon, or charcoal, that is contained in the forests and herbage of the world, and add to the estimate all that exists in the soil as vegetable mould, peat, and in other forms; all that is locked up in the vast deposits of coal (the product of the vegetation of bygone ages); and, finally, all that pertains to the whole existent animal kingdom; — and we shall have the aggregate amount of a single, though the largest, element which vegetation has withdrawn from the atmosphere. By

---

when in full health and vigorous vegetation, *impart no carbonic acid to the air during the night.* — See Pepys, in *Philosophical Transactions*, for 1843. — They deteriorate the air only in their decay, and in peculiar processes, distinct from vegetation and directly the reverse of assimilation; as in germination, for instance, where, as will hereafter be explained, the proteine induces the decomposition of a portion of the store of assimilated matter, in order that the rest may be brought into a serviceable condition. For at the beginning, it must be recollected, the plant or the shoot grows, not by assimilation, but by consuming and appropriating a store of nourishment which was assimilated by the parent. The evolution of carbonic acid by plants, therefore, which has so long been taken for granted, and misinterpreted, has no existence as a general phenomenon. And it is by a false analogy that this *loss* which plants sustain in the night has been dignified with the name of *vegetable respiration*, and vegetables said to vitiate the atmosphere, just like animals, by their respiration, while they purify it by their digestion. If, indeed, this were a constant function, in any way contributing to maintain the life and health of the plant, it might be properly enough compared with the respiration of animals, which is itself a decomposing operation. But this is not the case. And herein is a characteristic difference between vegetables and animals: the tissues of the latter continue to live and act through the lifetime of the animal, and therefore require constant interstitial renewal by nutrition, new particles replacing the old, which are removed and restored to the mineral world by *respiration:* while in plants there is no such renewal, but the fabric, once completed, remains unchanged, ceases to be nourished, and consequently soon loses its vitality; while new parts are continually formed farther on to take their places, to be in turn abandoned. Plants, therefore, having no decomposition and recomposition of any completed fabric, cannot have the function of respiration.

multiplying this vast amount of carbon by sixteen, and dividing it by six, we obtain an expression of the number of pounds of oxygen gas that have in this process been supplied to the atmosphere.

360. Rightly to understand the object and consequences of this immense operation, which has been going on ever since vegetation began, it should be noted, that, so far as we know, vegetation is the only operation in nature which gives to the air free oxygen gas, that indispensable requisite to animal life. There is no other provision for maintaining the supply. The prevailing chemical tendencies, on the contrary, take oxygen from the air. Few of the materials of the earth's crust are saturated with it; some of them still absorb a portion from the air in the changes they undergo; and none of them give it back in the free state in which they took it,—in a state to support animal life,—by any known natural process, at least upon any considerable scale. Animals all consume oxygen at every moment of their life, giving to the air carbonic acid in its room; and when dead, their decaying bodies consume a farther portion. Decomposing vegetable matter produces the same result. Its carbon, taking oxygen from the air, is likewise restored in the form of carbonic acid. Combustion, as in burning our fuel, amounts to precisely the same thing; it is merely rapid decay. The carbon which the trees of the forest have been for centuries gathering from the air, their prostrate decaying trunks may almost as slowly restore to the air, in the original form of carbonic acid. But if set on fire, the same result may be accomplished in a day. All these causes conspire to rob the air of its life-sustaining oxygen. The original supply is indeed so vast, that, were there no natural compensation, centuries upon centuries would elapse before the amount of oxygen could be so much reduced, or that of carbonic acid increased, as to affect the existence of the present races of animals. But such a period would eventually arrive, were there no natural provision for the decomposition of the carbonic acid constantly poured into the air from these various sources, and for the restoration of its oxygen. We have seen that vegetation accomplishes this very result. The needful compensation is therefore found in the vegetable kingdom. While animals consume the oxygen of the air, and give back carbonic acid which is injurious to their life, this carbonic acid is the principal element of the food of vegetables, is consumed and decomposed by them, and its oxygen restored for the use of animals.

Hence the perfect adaptation of the two great kingdoms of living beings to each other; — each removing from the atmosphere what would be noxious to the other; — each yielding to the atmosphere what is essential to the continued existence of the other.*

361. The relations of simple vegetation, under this aspect, to the mineral kingdom on the one hand and the animal kingdom on the other, are simply set forth in the first part of the diagram placed at the close of this chapter.

362. But, besides this remotely essential office in purifying the air, the vegetable kingdom renders to the animal another service so immediate, that its failure for a single year would nearly depopulate the earth; namely, in providing the necessary food for the whole animal kingdom. It is under this view that the grand office of vegetation in the general economy of the world is to be contemplated. Plants are the sole producers of nourishment. They alone transform mineral, chiefly atmospheric materials, they condense air, into organized matter. While they thus *produce* upon a vast scale, they *consume* or destroy comparatively little; and this never in proper vegetation, but in some special processes hereafter to be considered (370). Often when they appear to consume their own products, they only transform and transfer them (128, 174), as when the starch of the Potato is converted into new shoots and foliage.

363. Animals *consume* what vegetables produce. They themselves produce nothing directly from the mineral world. The herbivorous animals take from vegetables the organized matter which they have produced; — a part of it they consume, and in respiration restore the materials to the atmosphere from which plants derived them, in the very form in which they were taken, namely, as carbonic acid and water. The portion they accumulate in their tissues constitutes the food of carnivorous animals; who consume and return to the air the greater part during life, and the

* It is plain, however, that, while the animal kingdom is entirely dependent on the vegetable, as no function of animals restores to the atmosphere the oxygen they consume, yet the latter is, in a good degree at least, independent of the former, and might have existed alone. The decaying races of plants, giving back their carbon to the air and to the soil (333) would furnish food for their successors. And since all the carbonic acid which animals render to the air in respiration they have derived from their vegetable food, it would in time have found its way back to the air, for the use of new generations of plants, without the intervention of animals. At most, they merely expedite its return.

remainder in decay after death. The atmosphere, therefore, out of which plants create nourishment, and to which animals as they consume return it, forms the necessary link between the animal and vegetable kingdoms, and completes the great cycle of organic existence. Organized matter passes through various stages in vegetables, through others in the herbivorous animals, and undergoes its final transformations in the carnivorous animals. Portions are consumed at every stage, and restored to the mineral kingdom, to which the whole, having accomplished its revolution, finally returns.

364. Plants not only furnish all the materials of the animal fabric, but furnish each principal constituent ready formed, so that the animal has only to appropriate it. The food of animals is of two kinds; — 1. that which serves to support respiration and maintain the animal heat; 2. that which is capable of forming a portion of the animal fabric, of its flesh and bones. The ternary vegetable products furnish the first, in the form of sugar, vegetable jelly, starch, oil, &c., and even cellulose; substances which, containing no nitrogen, cannot form a part of the animal frame, but, conveyed into the blood, are decomposed in respiration, the carbon and the excess of hydrogen combining with the oxygen of the air, to which they are restored in the form of carbonic acid and water. Any portion not required by the immediate demands of respiration is stored in the tissues in the form of fat, (which the animal may either accumulate directly from the oily and waxy matters in its vegetable food, or produce by an alteration of the starch and sugar,) as a provision for future use; any deficiency subjects the tissues themselves, or the proper supporting food, to immediate decomposition in respiration. The quaternary or azotized products furnish the proper materials of the animal frame, the fibrine, caseine, albumen, &c. being directly appropriated from the vegetable food to the blood, muscles, &c.; while a slight transformation of them gives origin to *gelatine*, of which the sinews, cartilages, and the organic part of the bones, consist. The earthy portion of the bones, the iron in the blood, and all the saline ingredients of the animal body (with the exception of common salt, which is sometimes taken directly from the mineral kingdom), are drawn from the earthy constituents (336) of the plants upon which the animal feeds. The animal merely appropriates and accumulates these already organizable materials, changing them, it may be, little by little, as he destroys them, but rendering them all back (those

of the first class through the lungs, of the second through the kidneys) finally to the earth and air, from which, and in the condition in which, the vegetable took them.

365. The relations of vegetation to the mineral and animal kingdoms, as especially concerns the elaboration of the constituents of the animal body, are shown in the second part of the subjoined diagram.

Diagram illustrating the Mutual Relations of the Three Kingdoms of Nature.

I. *Vegetable Fabric.*

Mineral Kingdom.		Vegetable Kingdom.		Animal Kingdom.		Mineral Kingdom.
WATER	Oxygen, Hydrogen	Oxygen, Hydrogen, Carbon	= Cellulose, Starch, Sugar, &c.	Consumed by animals and in respiration returned to the air, as	Oxygen, Hydrogen	WATER.
CARBONIC ACID	Carbon, Oxygen				Carbon, Oxygen	CARBONIC ACID.

II. *Complete Vegetation.*

Mineral Kingdom.		Vegetable Kingdom.		Animal Kingdom.		Mineral Kingdom.
AMMONIA	Hydrogen, Nitrogen	Hydrogen, Nitrogen, Hydrogen, Oxygen, Carbon	= Proteine and Cellulose	the first forming Albumen, Fibrin (Muscle), Gelatine (Sinews), Caseine (Curd), &c. — returned as Urea, and finally resolved into Carbonate of Ammonia and Water,	Hydrogen, Nitrogen	AMMONIA.
WATER	Hydrogen, Oxygen				Hydrogen, Oxygen	WATER.
CARBONIC ACID	Carbon, Oxygen				Carbon, Oxygen	CARBONIC ACID.

CHAPTER VII.

OF FLOWERING AND ITS CONSEQUENCES.

366. PLANTS have thus far been considered only as respects their *Organs of Vegetation;* — those which essentially constitute the vegetable being, by which it grows, deriving its support from the surrounding air and soil, and converting these inorganic materials into its own organized substance. As every additional supply of nourishment furnishes materials for the development of new branches, roots, and leaves, thus multiplying both those organs which receive food and those which assimilate it, it would seem that, apart from accidents, the increase and extension of plants would be limited only by the failure of an adequate supply of nourishment. After a certain period, however, varying in different species, but nearly constant in each, a change ensues, which controls this otherwise indefinite extent of the branches, and is attended with very important results. A portion of the buds, instead of elongating into branches, are developed in the form of FLOWERS; and the nourishment which would otherwise contribute to the general increase of the plant, is partially or wholly expended in their production, and in the maturation of the *fruit* and *seeds* (110). So far as we know, the sole office of the flower and fruit in the vegetable economy is the production of seed. Hence they are termed ORGANS OF REPRODUCTION (115).

367. **Flowering an Exhaustive Process.** Plants begin to bear flowers at a nearly determinate period for each species; which is dependent partly upon constitutional causes that we are unable to account for, and partly upon the requisite supply of nutritive matter in their system. For, since the flower and fruit draw largely upon the powers and nourishment of the plant, while they yield nothing in return, fructification is an exhaustive process, and a due accumulation of food is requisite to sustain it.* Annuals flower

* When the branch of a fruit-tree, which is sterile or does not perfect its blossoms, is *ringed* or *girdled* (by the removal of a narrow ring of bark), the elaborated juices, being arrested in their downward course, are accumulated in the branch, which is thus enabled to produce fruit abundantly; while the shoots that appear below the ring, being fed chiefly by the crude ascending sap,

in a few weeks or months after they spring from the seed, when they have little nourishment stored up in their tissue; and their lives are destroyed in the process (127): biennials flower after a longer period, rapidly exhausting the nourishment accumulated in the root during the previous season, and then perishing (128); while shrubs and trees do not commence flowering until they are sufficiently established to endure it. The exhaustion consequent upon flowering, however, is often exhibited in fruit-trees, which, after producing an excessive crop (especially of late fruits, such as apples), sometimes fail to bear the succeeding year. When the crop of one year is destroyed, the nourishment which it would have consumed accumulates, and the tree may bear more abundantly the following season, and so on alternately from year to year.

368. The actual consumption of nourishment in flowering may be shown in a variety of ways; as by the rapid disappearance of the farinaceous or saccharine store in the roots of the Carrot, Beet, &c., when they begin to flower, leaving them light, dry, and empty; and from the rapid diminution of the sugar in the stalk of the Sugar-cane (as also in that of Maize) at the same period. The stalks are therefore cut for making sugar just before the flowers expand, as they then contain the greatest amount of saccharine matter.

369. The consequences of this exhaustion upon the duration of plants have already been adverted to. They are further illustrated by the facility with which annuals may be changed into biennials, or their life prolonged indefinitely by preventing their flowering; while they perish whenever they bear flowers and seed, whether during the first or any succeeding year. So, a common annual Larkspur has given rise to a double-flowered variety in the gardens, which bears no seed, and has therefore become a perennial. So, also, cabbage-stumps, which are planted for seed, may be made to bear heads the second year by destroying the flower-shoots as they arise; and the process may be continued from year to year, thus converting a biennial into a kind of perennial plant. The effect of flowering upon the longevity of the individual is strikingly shown

do not bear flowers, but push forth into leafy branches. So the flowers of most trees and shrubs that bear large or fleshy fruit are produced from lateral buds, resting directly upon the wood of the previous year, in which a quantity of nutritive matter is deposited. So, also, a seedling shoot, which would not flower for several years if left to itself, blossoms the next season when inserted as a graft into an older trunk, from whose accumulated stock it draws.

by the Agave, or Century-plant, — so called because it flowers in our conservatories only after the lapse of a hundred, or at least a great number of years; although, in its native sultry clime, it generally flowers when five or six years old. But whenever this occurs, the sweet juice with which it is filled at the time (which by fermentation forms *pulque*, the inebriating drink of the Mexicans) is consumed at a rate answering to the astonishing rapidity with which its huge flower-stalk shoots forth (24), and the whole plant inevitably perishes when the seeds have ripened. So, also, the Corypha, or Talipot-tree, a magnificent Oriental Palm, which lives to a great age and attains an imposing altitude (bearing a crown of leaves, each blade of which is often thirty feet in circumference), flowers only once; but it then bears an enormous number of blossoms, succeeded by a crop of nuts sufficient to supply a large district with seed; while the tree immediately perishes from the exhaustion consequent upon this over-production.

370. Flowering and fruiting, then, draw largely upon the plant's resources, while they give back nothing in return. In these operations, as also in germination, vegetables act as true consumers (like animals, 363), decomposing their own products, and giving back carbonic acid and water to the air, instead of taking these materials from the air. It is in flowering that they actually consume most. In fruiting, although the plant is robbed of a large quantity of nourishment, this is mostly accumulated in the fruit and seed, in a concentrated form, for the future consumption, not of the parent plant, but of the new individual inclosed in the seed. As we may treat of the latter elsewhere, we have here to contemplate only the real and immediate consumption of nourishment by the flower.

371. This is shown by the action of flowers upon the air, so different from that of leaves. While the foliage withdraws carbonic acid from the air, and restores oxygen (346, 358), flowers take a small portion of oxygen from the air, and give back carbonic acid. While leaves, therefore, purify the air we breathe, flowers contaminate it; though, of course, only to a degree which is relatively and absolutely insignificant.

372. **Evolution of Heat.** When carbon is consumed as fuel, and by the oxygen of the air converted into carbonic acid, an amount of heat is evolved, directly proportionate to the quantity of carbon consumed, or of carbonic acid produced. Precisely the same amount is more slowly generated during the gradual decomposition

of the same quantity of vegetable matter by decay, — a heat which is employed by the gardener when he makes hot-beds of tan, decaying leaves, and manure, — or by the breathing of animals, where it maintains their elevated temperature (364). The consumption of a given amount of carbon and hydrogen, under whatever form, and whether slowly or rapidly, generates in all cases the very same amount of heat. Now, since flowers consume carbon and produce carbonic acid, acting in this respect like animals, they ought to evolve heat in proportion to that consumption. This, in fact, they do. The evolution of heat in blossoming was first observed by Lamarck, about seventy years ago, in the European Arum, which, just as the flowers open, "grows hot," as Lamarck stated, "as if it were about to burn." It was afterwards shown by Saussure in a number of flowers, such as those of the Bignonia, Gourd, and Tuberose, and the heat was shown to be in direct proportion to the consumption of the oxygen of the air, or, in other words, of the carbon of the plant. The increase of temperature, in these cases, was measured by common instruments. But now that thermo-electric apparatus affords the means of measuring variations inappreciable by the most delicate thermometer, the heat generated by an ordinary cluster of blossoms may be detected. The phenomenon is most striking in the case of some large tropical Aroideous plants, where an immense number of blossoms are crowded together and muffled by a kind of hood, or spathe (390), which confines and reverberates the heat. In some of these, the temperature rises at times to twenty or even fifty degrees (Fahrenheit) above that of the surrounding air.*

373. The source of the heat in flowering is therefore evident. As to its object, we cannot say whether its production is the immediate end in view, and the plant burns some of its carbon merely

---

* This increase of temperature occurs daily from the time the flowers open until they fade, but is most striking during the shedding of the pollen. At night, the temperature falls nearly to that of the surrounding air; but in the course of the morning the heat comes on, as it were, like a *paroxysm of fever*, attaining the maximum, day after day, very nearly at the same hour of the afternoon, and gradually declining towards evening. In ordinary cases, the heat of flowering is absorbed by the vaporization of the sap and the exhalation of oxygen by the foliage (besides, a large amount is absorbed from the solar radiation and rendered latent in the process of assimilation); so that the actual temperature of a leafy plant in summer is lower than that of the atmosphere.

as fuel, or whether the evolution of heat and the formation of carbonic acid are incidental consequences of certain necessary transformations. We have remarked that the principal *consumption* takes place in the flower; and that a store is laid up in the fruit and seed. But much even of this is consumed, with the evolution of heat, when the seed germinates. By a not very violent metaphor it may be said, therefore, that in the Century-plant (369), which, after living a hundred years, consumes itself in producing and giving life to its offspring, who literally rise from its ashes, we have the realization of the fabled Phœnix!

374. **Plants need a Season of Rest.** There is another condition, which, if not essential to the production of flowers, exerts an important influence. When plants are in continual and luxuriant growth, rapidly pushing forth leafy branches, they are not apt to produce flower-buds. Our fruit-trees, in very moist seasons, or when cultivated in too rich a soil, often grow luxuriantly, but do not flower. The same thing is observed when our Northern fruit-trees are transported into tropical climates. On the other hand, whatever checks this continuous growth, without affecting the health of the individual, causes blossoms to appear earlier and more abundantly than they otherwise would. It is for this reason that transplanted fruit-trees incline to flower the first season after their removal, though they may not blossom again for several years. A season of comparative rest is essential to the transformation by which flowers are formed. It is in autumn, or at least after the vigorous vegetation of the season is over, that our trees and shrubs, and most perennial herbs, produce the flower-buds of the ensuing year.

375. The requisite annual season of repose, which in temperate climes is attained by the lowering of the temperature in autumn and winter, is scarcely less marked in many tropical countries, where winter is unknown. But the result is brought about, in the latter case, not by cold, but by excessive heat and dryness. The Cape of Good Hope, the Canary Islands, and the southern part of California, may be taken as illustrations. In the Canaries, the growing season is from November to March, — the winter of the northern hemisphere, — their winter also, as it is the coolest season, the mean temperature being 66° Fahr. But the rains fall regularly and vegetation is active; while in summer, from April to October, it very seldom rains, and the mean temperature is as high

as 73°. During this dry season, when the scorching sun reduces the soil nearly to the dryness and consistence of brick, ordinary vegetation almost completely disappears; and the Fig-Marigolds, Euphorbias, and other succulent plants, which, fitted to this condition of things, alone remain green, not unaptly represent the Firs and other evergreens of high northern latitudes. The dry heat there brings about the same state of vegetable repose as cold with us. The roots and bulbs then lie dormant beneath the sun-burnt crust, just as they do in our frozen soil. When the rainy season sets in, and the crust is softened by moisture, they are excited into growth under a diminished temperature, just as with us by heat; and the ready-formed flower-buds are suddenly developed, clothing at once the arid waste with a profusion of blossoms. The vegetation of such regions consists mainly of succulents, which are able to live through the drought and exposure; of bulbous plants, which run through their course before the drought becomes severe, then lose their foliage, while the bud remains quiescent, safely protected under ground until the rainy season returns; and of annuals, which make their whole growth in a few weeks, and ripen their seeds, in which state the species securely passes the arid season. A season of interruption to growth, produced either by cold or dryness, occurs, in a more or less marked degree, through every part of the world.

376. These considerations explain the process of *forcing* plants, and other operations of horticulture, by which we are enabled to obtain in winter the flowers and fruits of summer. The gardener accomplishes these results principally by skilful alterations of the natural period of repose. He gives the plant an artificial period of rest by dryness at the season when he cannot command cold, and then, by the influence of heat, light, and moisture, which he can always command, causes it to grow at a season when it would have been quiescent. Thus he retards or advances, at will, the periods of flowering and of rest, or in time completely inverts them.

CHAPTER VIII.

OF THE INFLORESCENCE.

377. **Inflorescence** is the term used to designate the arrangement of flowers upon the stem or branch. The flower, like the branch, is evolved from a bud. Flower-buds and leaf-buds are often so similar in appearance, that it is difficult to distinguish one from the other before their expansion. The most conspicuous parts of the flower are so obviously analogous to the leaves of a branch, that they are called in common language the leaves of the flower. Such a flower as the double Camellia appears as if composed of a rosette of white or colored leaves, resembling, except in their color and great delicacy, the clusters of leaves which crown the offsets of such plants as the Houseleek (Fig. 174), &c. We may therefore naturally consider a flower-bud as analogous to a leaf-bud; and a flower, consequently, as analogous to a short, leafy branch.

378. This analogy is confirmed by the position which flowers occupy. Whatever views may be entertained respecting the nature of flowers, it is certain that they appear at the same situations as ordinary buds, and at no other. They have the same relation to the stem or flower-stalk which bears them, that leaf-buds have to the stem or branch from which they arise; that is, they occupy the extremity of the stem or branch, and the axil of the leaves (144, 148). Consequently, the arrangement of the buds governs the whole arrangement of the blossoms, as well as that of the branches. The flower-stalk is merely the last term of ramification. The almost endless variety of modes in which flowers are clustered upon the stem, many of them exhibiting the most graceful of natural forms, all implicitly follow the general law which has controlled the whole development of the vegetable from the beginning. We have, throughout, merely buds terminating the stem and branches, and buds from the axil of the leaves.

379. The simplest kind of inflorescence is, of course, that of a *solitary* flower,—a single flower-stalk bearing a single flower; as in Fig. 249 and Fig. 229. The naked stalk which supports the flower is termed the PEDUNCLE. If the flower is not raised on a proper stalk, it is said to be *sessile*.

380. In both of the examples just adduced, the flower is *solitary;* but there is a difference in one respect. In Fig. 249, the flower terminates the stem; it stands in the place of a terminal bud. In Fig. 229, it arises from the axil of a leaf, or represents an axillary bud. These two cases, in fact, exhibit the two types (reduced to the greatest simplicity), to the one or the other of which all the forms of inflorescence belong.

381. We may begin with the second of these plans; in which the flowers all spring from axillary buds; while the terminal bud, developing as an ordinary branch, continues the stem or axis indefinitely. For the stem in such case may continue to elongate, and produce a flower in the axil of every leaf, until its powers are exhausted (Fig. 230). This gives rise, therefore, to what is called

382. **Indefinite or Indeterminate Inflorescence.** The primary axis is here never terminated by a flower; but the secondary axes (from axillary buds) are thus terminated. Before we enumerate the various forms of inflorescence of this class, a few terms must be defined which necessarily come into use in distinguishing the parts of a flower-cluster. The *primary axis*, or general stalk which bears the whole cluster of flowers, retains the name of PEDUNCLE (379), while the *secondary axes*, which form the partial flower-stalks and support each a single blossom, now receive the name of PEDICELS. These, being axillary branches, must of course be subtended each by a leaf, or else will show the scar left by its fall. The leaves of an inflorescence, however, are usually reduced in size, or changed in appearance, so as to be quite unlike the ordinary leaves of the plant: they are called sometimes *floral leaves*, or more commonly BRACTS. The bracts are often reduced to a minute size, so as to escape ordinary notice: they very frequently fall off when the flower-bud in their axil expands, or even still earlier; and sometimes, as in the greater part of the Mustard Family, they altogether fail to appear. The portion of the general stalk along which flowers are borne is called the *axis of the inflorescence*, and sometimes, especially when covered with sessile flowers, the RHACHIS (from its resemblance or analogy to the backbone).

383. The various forms of indefinite inflorescence which in descriptive botany are distinguished by special names, as might be expected, run into one another through endless intermediate gradations. In nature they are not so absolutely fixed as in our written definitions; and whether this or that name should be used in a

particular case is often a matter of fancy. The subjoined account of the principal kinds will at the same time bring to view the connection between them.

384. **A Raceme** is formed when the primary axis continues to lengthen, and the flowers, singly produced from the axil of each bract, are supported on pedicels of their own, as in Fig. 230. The flowers and fruit of the Currant, Barberry, and wild Black Cherry (Fig. 236) furnish most familiar examples. The lowest flowers of a raceme, being evidently the oldest, are the first to expand, and the others follow in regular succession, from the base to the summit. Indeed, the lower flowers often produce, or (as in the Snowberry, Symphoricarpus racemosus) even ripen, their fruit, before the summit has ceased to grow and develope new flowers.

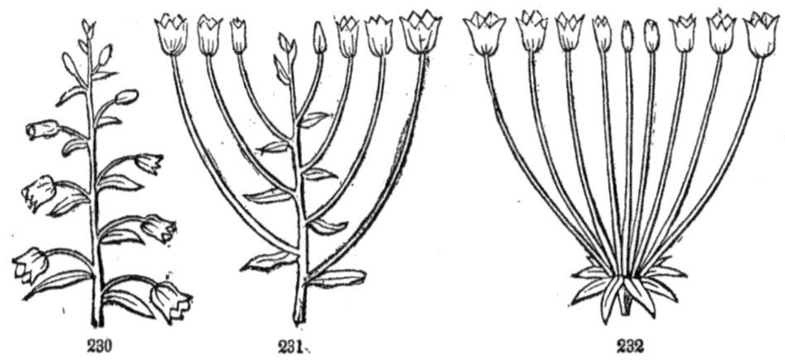

385. **A Corymb** (Fig. 231, 239) is the same as a raceme, except that the lower pedicels are elongated, so as to form a level-topped or slightly convex bunch of flowers; as in the Hawthorn, &c.

386. **An Umbel** (Fig. 232) differs from a corymb only in having all the pedicels arising from the same apparent point; the general peduncle, in this case, bearing several flowers without any perceptible elongation of the internodes of the axis of inflorescence. The Primrose and the Milkweed afford familiar examples of the simple umbel.

387. A corymb being evidently the same as a raceme with a short main axis, and an umbel the same as a corymb with a still shorter axis, it is evident that the outer flowers of an umbel or corymb correspond to the lowermost in the raceme, and that these will first expand, the blossoming proceeding regularly from the base to the apex, or (which is the same thing from the) circumfer-

FIG. 230 - 232. Diagrams of a simple raceme, corymb, and umbel.

ence to the centre. This mode of development uniformly takes place when the flowers arise from axillary buds; on which account the indefinite mode of inflorescence is also called the *centripetal*.

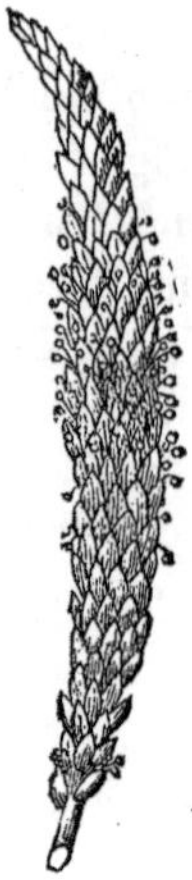

388. In all the foregoing cases, the flowers are raised on stalks, or pedicels. When these are wanting, or very short, the *spike* or the *head* is produced.

389. **A Spike** is the same as the raceme, except that the flowers are sessile, or destitute of any apparent pedicels; as in the Plantain (Fig. 233). It is an indeterminate or centripetal inflorescence, with the primary axis elongated, and the secondary axes not at all elongated, but terminated at their very origin by a flower. Two varieties of the spike have received independent names, viz. the *Spadix* and the *Ament*.

233

390. **A Spadix** is a fleshy spike enveloped by a large bract or modified leaf, called a SPATHE, as in Calla palustris (Fig. 234), the cultivated Calla Æthiopica, Arum triphyllum, or Indian Turnip (Fig. 235), and the Skunk Cabbage (see Araceæ).

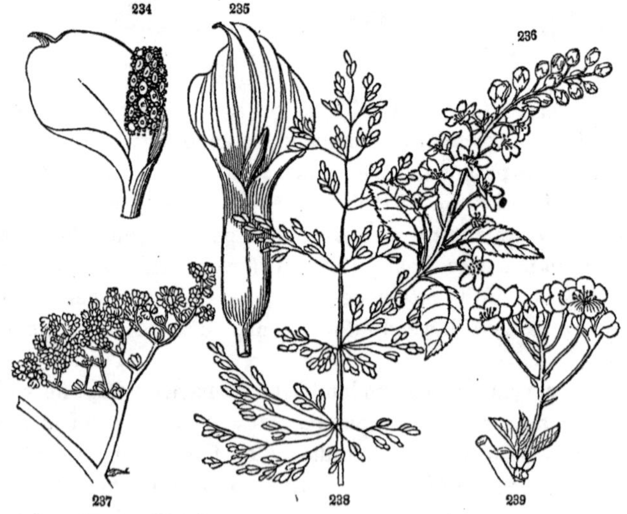

391. **An Ament, or Catkin,** is merely that kind of spike with scaly

FIG. 233. Young spike of Plantago major.

FIG. 234 - 239. Forms of inflorescence. 234, 235. Spadix of Calla and of Arum, with the spathe, 236. A raceme. 237. A cyme. 238. A panicle. 239. A corymb.

bracts borne by the Birch, Poplar, Willow, and, as to one of the two sorts of flowers, by the Oak, Walnut, and Hickory, which are accordingly called *amentaceous trees*. Catkins usually fall off in one piece, after flowering or fruiting, especially the sterile catkins.

392. **The Head, or Capitulum,** is a globular cluster of sessile flowers, like that of the Button Bush, the balls of the Buttonwood or Plane-tree, &c. It is a many-flowered centripetal inflorescence, in which neither the primary axis nor the secondary axes are at all lengthened. We may conceive it to originate, either from the non-development of the pedicels of an umbel (Fig. 232), or the non-elongation of the axis of a spike. In other words, the head differs from a spike only in its shortness. So what is at first a head frequently elongates into a spike as it grows older; as in many species of Clover, &c. In all these forms, the blossoms necessarily expand from the base to the apex, or from the circumference to the centre (387).

393. The base both of the head and the umbel is frequently furnished with a number of imperfect leaves or bracts, crowded together, or forming a whorl (236, Fig. 232), termed an INVOLUCRE. The involucre assumes a great variety of forms; sometimes resembling a calyx; and sometimes (as in Cornus Florida, or the common Dogwood, and C. Canadensis, Fig. 240) becoming petal-like, and much more showy than the blossom itself. It is, however, distinguished from the calyx or corolla by including a number of flowers. Sometimes, however, as in the Mallow Family and Hibiscus, the involucre forms a kind of outer calyx to each flower.

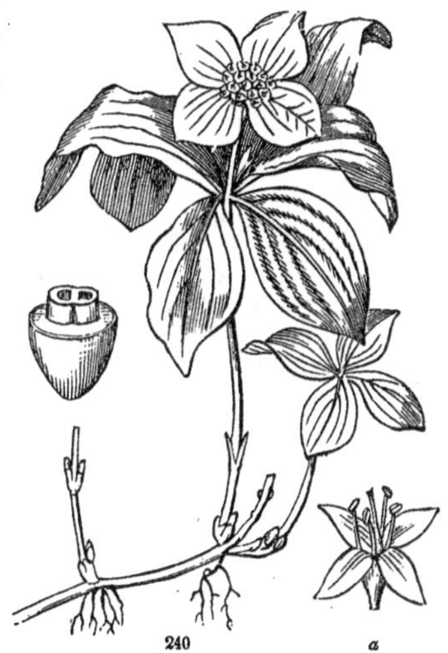

394. The axis, or rhachis (382), of a head is called the RECEPTACLE. Frequently, instead of being globular or somewhat pro-

FIG. 240. Cornus Canadensis; with its petal-like four-leaved involucre surrounding a head of flowers: *a*, a separate flower from the head, enlarged.

longed, it is flat or depressed, and dilated horizontally, so as to allow a large number of flowers to stand on its level or merely convex surface ; as in the Sunflower, and in similar plants. What were called *compound flowers* by the older botanists, such as the Sunflower, Aster, Marigold, &c., are heads of this kind, containing a smaller or larger number of flowers, crowded together on the receptacle (or dilated branch), and surrounded by an involucre. Not unfrequently the separate flowers are also subtended by bracts ; as in the Sunflower, Rudbeckia, Coreopsis, &c., when these receive the name of PALEÆ, or CHAFF. (See Ord. Compositæ.)

395. The FIG presents a case of very singular inflorescence (Fig. 241, 242), where the flowers apparently occupy the inside instead of the outside of the axis, being inclosed within the fleshy receptacle, which is hollow and nearly closed at the top. The magnified slice (Fig. 243) shows that the inner surface is lined, not with mere seeds, as is commonly supposed, but with a multitude of small blossoms. The *fig* is therefore something like a mulberry (Fig. 244), or a pine-apple, turned inside out.

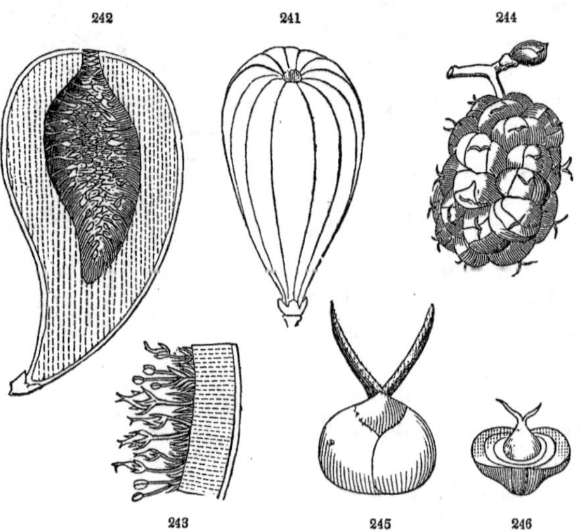

396. In all the cases yet mentioned, the flower-clusters are simple ; the ramification not passing beyond the first step ; the lateral

FIG. 241. A Fig. 242. A vertical section. 243. A thin slice from the same, magnified.

FIG. 244. The Mulberry in fruit. 245. One of the component flowers, magnified. 246. One of the flowers with a section of the juicy floral envelopes.

buds being at once terminated by a single flower. But the lateral flower-stalks may themselves branch, just as ordinary branches give rise to branchlets; when the inflorescence becomes compound. The modifications produced by a second branching of the inflores cence are readily understood. If the branches of a raceme are prolonged, and bear other flowers on pedicels similarly arranged, a *compound raceme* is produced; or if the flowers are sessile, a *compound spike* is formed. A corymb, the branches of which are similarly divided, forms a *compound corymb;* and an umbel, where the branches (often called *rays*) bear smaller umbels at their apex, is termed a *compound umbel;* examples of which occur in almost all the species of the Family Umbelliferæ, which is so named because all its plants bear umbels. For these secondary umbels, a good English name has been employed by Dr. Darlington, that of UMBELLETS. Their involucre, when they have any, is distinguished from that of the principal umbel by the name of INVOLUCEL.

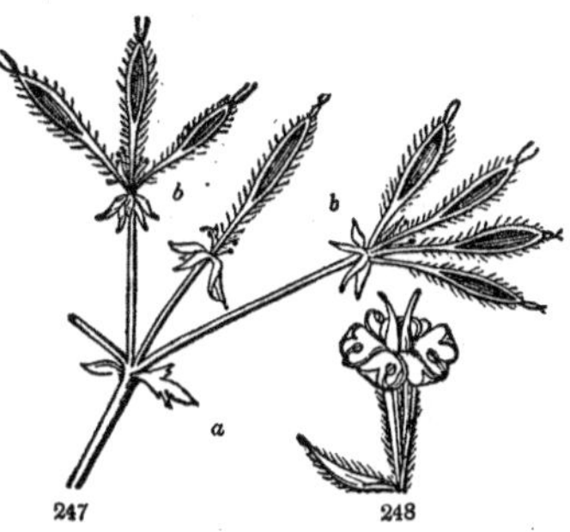

397. It is often necessary to distinguish between the bracts on the branches of the inflorescence, and those at the base of the primary branches; in which case the former are termed BRACTEOLES, or BRACTLETS; but there is no real limit, either between bractlets and true bracts, or between bracts and true leaves.

398. When the inflorescence is compound, it is readily seen that two or more modes of inflorescence may be combined; the first ramification following one plan, and the subdivision another. The combination is usually expressed by a descriptive phrase, as "spikes racemose, or racemed," "heads corymbose," &c. The combination of the raceme and the corymb or the cyme gives rise to a form of inflorescence which has a technical name, viz.: —

399. **The Panicle.** This is formed when the secondary axes of a raceme branch in a corymbose manner, as in numerous Grasses (Fig. 238), or those of a corymb divide in the manner of a raceme. And the name is loosely applied to almost any open and

FIG. 247. Compound umbel (in fruit) of Osmorhiza longistylis: *a*, the involucre; *b*, *b*, involucels. 248. A separate flower enlarged, with its subtending bract of the involucel.

more or less elongated inflorescence which is irregularly branched twice, thrice, or a greater number of times.

400. **A Thyrsus** is merely a compact panicle of a pyramidal, oval, or oblong outline; such as the cluster of flowers of the Lilac and Horsechestnut, a bunch of grapes, &c.

401. **Definite or Determinate Inflorescence.** In this class, the flowers all represent terminal buds (380). The primary axis is directly terminated by a single flower-bud, as in Fig. 249, and its growth is of course arrested, as it is now incapable of any further elongation. In this way we have a solitary terminal flower. Further growth can take place only by the development of secondary axes from axillary buds. These may develope at once as peduncles, or as leafy branches; but they are in either case arrested, after more or less elongation, by a flower-bud, just as the primary axis was (Fig. 250). If further development ensues, it is by the production of branches of the third order, from the axils of leaves or bracts on the branches of the second order (Fig. 251); and so on. Hence this mode of inflorescence is said to be *definite* or *determinate*, in contradistinction to the indeterminate mode, already treated of (382, &c.), where the primary or leading axes elongate indefinitely, or merely cease to grow from the failure of nourishment, or some other extrinsic cause. The most common and most regular cases of determinate inflorescence occur in opposite-leaved plants, for obvious reasons; and such are accordingly chosen for the subjoined illustrations. But the Rose, Potentilla, and Buttercup furnish familiar examples of the kind in alternate-leaved plants.

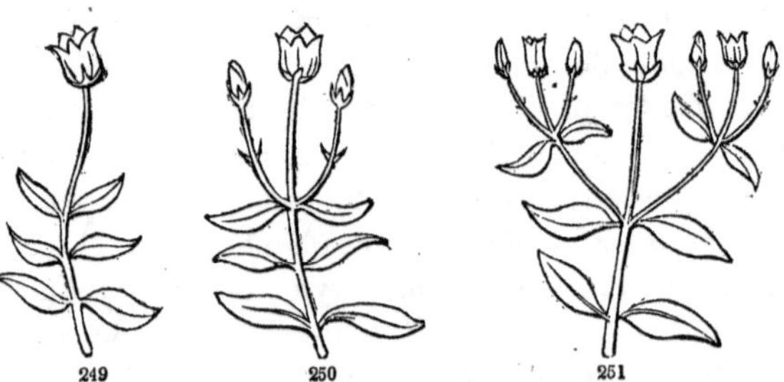

402. The determinate mode of inflorescence assumes forms which closely imitate the various forms of the indeterminate kind, already described, with which they have been confounded, and

FIG. 249-251. Diagrams of regular forms of determinate or centrifugal inflorescence.

on this account have failed to receive distinctive names. When, for example, all the secondary axes connected with the inflorescence are arrested by terminal flowers without any onward growth except what forms their footstalks or pedicels, and these are nearly equal in length, a raceme-like inflorescence is produced, as in Fig. 252. When the flowers are developed in this way, with scarcely any pedicels, the spike is imitated. These are essentially distinguished from the true raceme and spike, however, by the reverse order of development of the blossoms ; the terminal and then the upper ones opening earliest, and the others expanding in succession from above downwards ; while the blossoming of the raceme proceeds from below upwards. Or when, by the elongation of the lower secondary axes, a corymb is imitated, the flowers are found to expand in succession from the centre towards the circumference of the flat-topped cluster, while the contrary occurs in the corymb. That is, while the order in indeterminate inflorescence is centripetal (387), that of the determinate mode is *centrifugal.* When the determinate inflorescence assumes the corymbose form, which it more commonly does, it has a distinctive name, viz. : —

403. **The Cyme.** This is a flat-topped, rounded or expanded inflorescence, whether simple or compound, of the determinate class ; of which those of the Laurustinus, Elder, Dogwood, and Hydrangea are fully developed and characteristic examples. More commonly it is from the upper axils alone that the flower-bearing branches successively proceed, as indicated in Fig. 249 – 251. In more compound and compact cymes (Fig. 237), such as those of the Laurustinus, Dogwood, &c., the leaves or bracts are usually minute, rudimentary, or abortive, and all the numerous flower-buds of the cluster are fully formed before any of them expand ; and the blossoming then runs through the whole cluster in a short time, commencing in the centre of the cyme, and then in the centre of each of its branches, or CYMULES, and thence proceeding centrifugally. But in the Chickweeds (Fig. 253), in Hypericum, and many such like plants, the successive production of the branches and the evolution of the flowers, beginning with that which arrests the growth of the primary axis, go on gradually through the

FIG. 252. Definite inflorescence imitating a raceme.

whole summer, until the powers of the plant are exhausted, or until all the branchlets or peduncles are reduced to single internodes, or pedicels without any leaves, bracts, or bractlets, when no further development can take place. Such cases enable us to study the determinate inflorescence to advantage, and to follow the successive steps of the ramification by direct observation.

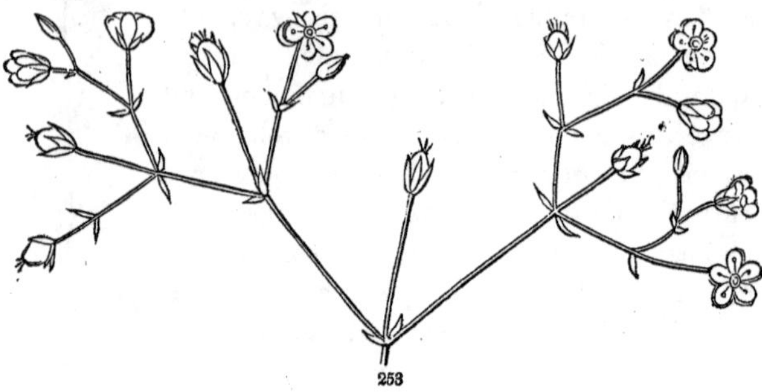

404. **The Fascicle** is a densely crowded cyme, with the flowers almost sessile, or on short peduncles of nearly equal length; as in the Sweet William.

405. **A Glomerule** is a cyme condensed into a head or short spike. It is to the cyme what the capitulum is to the corymb or umbel.

406. There are several abnormal modifications of inflorescence, especially of the determinate or centrifugal kind, arising from irregular development, or the suppression of parts, such as the non-appearance sometimes of the central flower, or often of one of the lateral branches at each division; as in the ultimate ramifications of Fig. 253, where one of the lateral pedicels is wanting. When this deviation is completely carried out, that is, when one of the side branches regularly fails to appear, the cyme is apparently converted into a kind of one-sided raceme, and the flowers seem to expand from below upwards, or centripetally. The diagram, Fig. 254, when compared with Fig. 253, explains this anomaly. The place of the axillary branch which fails to develope at each ramification is indicated by the dotted lines. Cases like this occur in several Hypericums, and in some other opposite-leaved plants. An analogous case occurs in many alternate-leaved plants; where the stem, being terminated by a flower, is continued by a branch from the axil of the uppermost leaf or bract: this, bearing a flow-

FIG. 253. The open, progressively developed cyme of Arenaria stricta.

er, is similarly prolonged by a secondary branch, that by a third, and so on; as is shown in the diagram, Fig. 255. Such forms of inflorescence, which we may observe in Drosera, and in most Sedums and Boraginaceæ, imitate the raceme so nearly, that they have commonly been considered as of that kind. They are distinguishable, however, by the position of the flowers *opposite* the leaf or bract, or at least out of its axil; while in the raceme, and in every modification of centripetal inflorescence, the flowers necessarily spring from the axils of the bracts. But if the bracts disappear, as they commonly do in the Forget-me-not, &c., the true nature of the inflorescence is not readily made out. The undeveloped summit is usually coiled in a spiral or *circinate* (257) manner, gradually unrolling as the flowers grow and expand, and becoming straight in fruit. On account of this coiled arrangement, such cymes or false racemes are said to be *helicoid*, or *scorpioid*.

407. The cyme, raceme, head, &c., as well as the one-flowered peduncle, may be produced, either at the extremity of the stem or leafy branch (*terminal*), or in the axil of the leaves (*axillary*). The case of a peduncle opposite a leaf, as in the Poke (see Ord. Phytolaccaceæ), the Grape-vine, &c., is just that illustrated in Fig. 255, except that in these cases the peduncles bear a cluster of flowers instead of a single one. The tendrils of the vine (Fig. 134) occupy the same position, and are of the same nature, so that they are not incorrectly said to be sterile and modified peduncles. In a growing Grape-vine, it is evident that the uppermost tendril really terminates the stem; and that the latter is continued by the growth of the axillary bud, situated between the petiole and the peduncle; the branch thus formed, assuming the same direction as the main stem, and appearing to be its prolongation, throws the peduncle or tendril to the side opposite the leaf.

408. The *extra-axillary* peduncles of most species of Solanum are to be similarly explained. They are really terminal pedun-

FIG. 254, 255. Plan of two modifications of helicoid cymes or false racemes.

cles, which have become lateral by the evolution of a branch from an axil below, which takes the direction of the main stem, so as to form an apparent continuation of it. This has been explained on the supposition of the cohesion of the base of an axillary peduncle with the stem; which could well apply only to those cases where the peduncle is in the same vertical line as the leaf beneath. Such peduncles may sometimes come from extra-axillary accessory buds, such as those shown in Fig. 133.

409. In the Linden (see Ord. Tiliaceæ) the peduncle appears to spring from the middle of a peculiar foliaceous bract. But this is rather a bractlet, inserted on the middle of the peduncle, and decurrent down to its base, just as many leaves are decurrent on the stem (298) in Thistles, &c.

410. A peduncle which arises from the stem at or beneath the surface of the ground, as in the Primrose, the Daisy, the so-called stemless Violets, &c., is called a *radical peduncle*, or a SCAPE.

411. A combination of the two classes of inflorescence is not unusual, the general axis developing in one way, but the separate clusters of flowers in the other. Thus the heads of all the Compositæ (such as Thistles, Asters, &c.) are *centripetal*, the flowers expanding regularly from the margin or circumference to the centre; while the branches that bear the heads are developed in the *centrifugal* mode, the central heads first coming into flower.

412. This is exactly reversed in all Labiatæ (plants of the Mint tribe); where the stem grows on indefinitely in the centripetal mode, bearing axillary clusters of flowers in the form of a general raceme or spike, which blossoms from below upwards; while the flowers of each cluster form a cyme, and expand in the centrifugal manner. These cymes, or *cymules*, of Labiatæ are usually close and compact, and being situated one in each axil of the opposite leaves, the two together frequently form a cluster which surrounds the stem, like a whorl or verticil (as in the Catnip and Horehound): hence such flowers are often said to be *whorled* or *verticillate*, which is not really the case, as they evidently all spring from the axils of the two leaves. The apparent verticil of this kind is sometimes termed a VERTICILLASTER.

413. True whorled flowers occur only in some plants with whorled leaves, as in Hippuris and the Water Milfoil.

CHAPTER IX.

OF THE FLOWER.

Sect. I. Its Organs, or Component Parts.

414. Having glanced at the circumstances which attend and control the production of flowers, and considered the laws which govern their arrangement, we have next to inquire what the flower is composed of.

415. **The Flower** (110, 111) assumes an endless variety of forms in different species, so that it is very difficult properly to define it. The name was earliest applied, as it is still in popular language generally applied, to the delicate and gayly colored leaves or petals, so different from the sober green of the foliage. But the petals, and all these bright hues, are entirely wanting in many flowers, while ordinary leaves sometimes assume the brilliant coloring of the blossom. The stamens and pistils are the characteristic organs of the flower; but sometimes one or the other of these disappear from a particular flower, and both are absent from full *double* Roses, Camellias, &c., in which we have only a regular rosette of delicate leaves. This, however, is an unnatural state, the consequence of protracted cultivation.

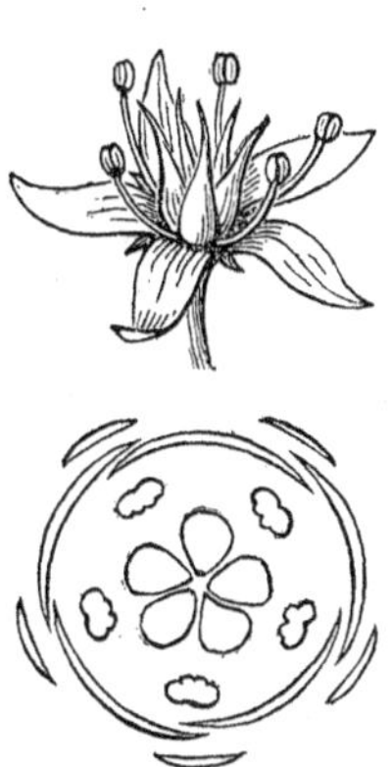

416. *A complete flower* consists of the essential organs of reproduction (viz. stamens and pistils, 110), surrounded by two sets of leaves or envelopes which protect them. (Fig. 256.) The latter are of course *exterior* or *lower* than the former, which in the bud they inclose.

417. **The Floral Envelopes**, then, are of two sorts, and occupy two circles, one above or within the other. Those of the lower circle, the exterior envelope in the flower-bud, form the Calyx: they com-

FIG. 256. The complete flower of a Crassula. 257. Diagram of its cross-section in the bud, showing the relative position of its parts. The five pieces of the exterior circle are sections of the sepals; the next, of the petals; the third, of the stamens through their anthers; the innermost, of the five pistils.

monly exhibit the green color and have much the appearance of ordinary leaves. Those of the inner circle, which are commonly of a more delicate texture and brighter color, and form the most showy part of the blossom, compose the COROLLA. The several parts or leaves of the corolla are called PETALS: and the leaves of the calyx take the corresponding name of SEPALS. One of the five sepals of the flower represented in Fig. 256 is separately shown in Fig. 258; and one of the petals in Fig. 259. The calyx and corolla taken together, or the whole floral envelopes, whatever they may consist of, are sometimes called the PERIANTH (*Perianthium* or *Perigonium*).

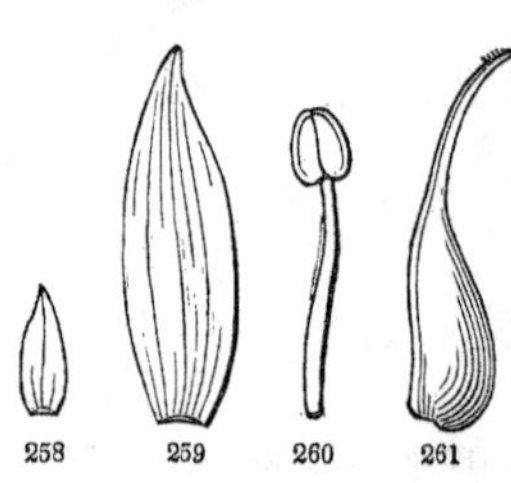

418. **The Essential Organs** of the flower are likewise of two kinds, and occupy two circles or rows, one within the other. The first of these, those next within the petals, are the STAMENS (Fig. 260). A stamen consists of a column or stalk, called the FILAMENT (Fig. 262, *a*), which bears on its summit a rounded body, or case, termed the ANTHER (*b*), filled with a powdery substance called POLLEN, which it discharges through one or more slits or openings. The older botanists had no general term for the stamens taken collectively, analogous to that of corolla for the entire whorl of petals, and of calyx for the whorl of sepals. A name has, however, recently been proposed for the *staminate system* of a flower, which it is occasionally convenient to use; that of ANDRŒCIUM.

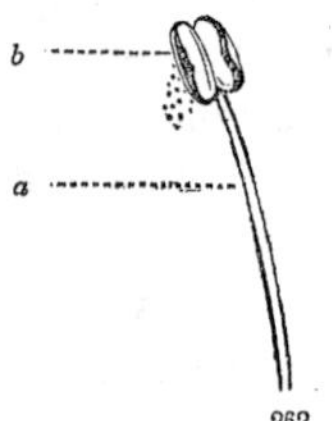

419. The remaining, or seed-bearing organs, which occupy the centre or summit of the flower, to whose protection and perfection all the other parts of the flower are in some way subservient, are termed the PISTILS. To them collectively the name of GYNÆCIUM has been applied. One of them is separately shown in Fig. 261. This is seen more magnified and cut across in Fig. 263; and a different one, longitudinally divided, so as to exhibit the whole length of its cavity, or *cell*, is represented in Fig. 264.

FIG. 258. A separate sepal; 259, a petal; 260, a stamen; and 261, a pistil from the flower of Fig. 256.

FIG. 262. A stamen, with the anther (*b*) discharging its pollen: *a*, the filament.

420. A pistil is distinguished into three parts; namely, the OVARY (Fig. 264, *a*), the hollow portion at the base which contains the OVULES, or bodies destined to become seeds; the STYLE (*b*), or columnar prolongation of the apex of the ovary; and the STIGMA (*c*), a portion of the surface of the style denuded of epidermis; sometimes a mere point or a small knob at the apex of the style, but often forming a single or double line running down a part of its inner face, and assuming a great diversity of appearance in different plants.

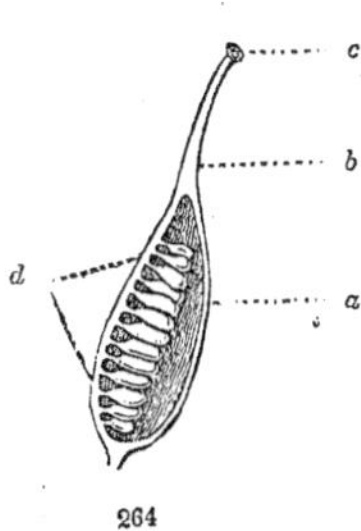

264

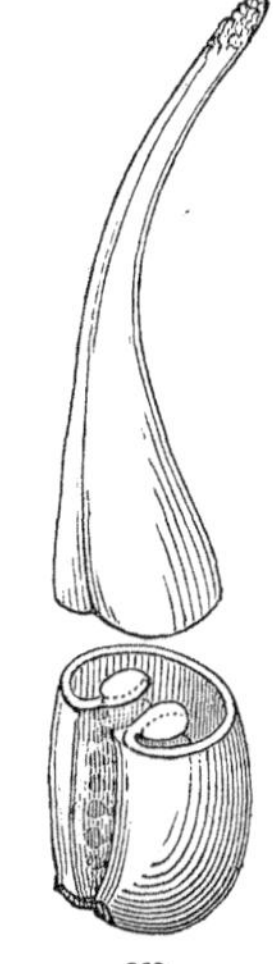
263

421. All the organs of the flower are situated on, or grow out of, the apex of the flower-stalk, into which they are said, in botanical language, to be *inserted*, and which is called the TORUS, or RECEPTACLE. This is the axis of the flower, to which the floral organs are attached (just as leaves are to the stem); the calyx at its very base; the petals just within or above the calyx; the stamens just within the petals; and the pistils within or above the stamens.

422. Such is the structure of a complete and regular flower; which we take as the *type*, or standard of comparison. The calyx and corolla are termed *protecting organs*. In the bud, they envelope the other parts: the calyx sometimes forms a covering even for the fruit; and when it retains its leaf-like texture and color, it assimilates the sap of the plant with the evolution of oxygen gas, in the same manner as do true leaves: the corolla elaborates honey or other secretions, for the nourishment, as is supposed, of the stamens and pistils. Neither the calyx nor corolla is essential to a flower, one or both being not unfrequently wanting. The stamens and pistils are, however, *essential organs*,

FIG. 263. A pistil of Crassula, like that of Fig. 261, but more magnified, and cut across through the ovary, to show its cell, and the ovules it contains. At the summit of the style is seen a somewhat papillose portion, destitute of epidermis, extending a little way down the inner face: this is the stigma.

FIG. 264. Vertical section of a pistil, showing the interior of its ovary, *a*, to one side of which are attached numerous ovules, *d*: above is the style, *b*, tipped by the stigma, *c*.

20

since both are necessary to the production of seed. But even these are not always both present in the very same flower; as will be seen when we come to notice the diverse forms which the blossom assumes, and to compare them with our pattern flower.

Sect. II. The Theoretical Structure or General Morphology of the Flower.

423. To obtain at the outset a correct idea of the flower, it is needful here to consider the relation which its organs sustain to the organs of vegetation. Taking the blossom as a whole, we have recognized, in the chapter on Inflorescence (377), the identity of flower-buds and leaf-buds as to situation, &c. Flowers, consequently, are at least analogous to branches, and the leaves of the flower to ordinary leaves.

424. But the question which now arises is, whether the leaves of the stem and the leaves and the more peculiar organs of the flower are not *homologous* parts, that is, parts of the same fundamental nature, although developed in different shapes that they may subserve different offices in the vegetable economy; — just as the arm of man, the fore-leg of quadrupeds, the wing-like fore-leg of the bat, the true wing of birds, and even the pectoral fin of fishes, all represent one and the same organ, although developed under widely different forms and subservient to more or less different ends. The plant continues for a considerable time to produce buds which develope into branches. At length it produces buds which expand into blossoms. Is there an entirely new system introduced when flowers appear? Are the blossoms formed upon such a different plan, that the general laws of vegetation, which have sufficed for the interpretation of all the phenomena up to the inflorescence, are to afford no further clew? Or, on the contrary, now that peculiar results are to be attained, are the simple and plastic organs of vegetation — the stem and leaves — developed in new and peculiar forms for the accomplishment of these new ends? The latter, doubtless, is the correct view. The plant does not produce essentially new kinds of organs to fulfil the new conditions, but adopts and adapts the old. Notwithstanding these new conditions and the successively increasing difference in appearance, the fundamental laws of vegetation may be traced from the leafy branch into and through the flower.

425. In vegetation no new organs are introduced to fulfil any particular condition, but the common elements, the root, stem, and leaves, are developed in peculiar and fitting forms to subserve each special purpose. Thus, the same organ which constitutes the stem of an herb, or the trunk of a tree, we recognize in the trailing vine, or the twiner, spirally climbing other stems, in the straw of Wheat and other Grasses, in the columnar trunk of the Palm, in the flattened and jointed Opuntia, or Prickly Pear, and in the rounded, lump-like body of the Melon-Cactus. So, also, branches harden into spines in the Thorn, or, by an opposite change, become flexible and attenuated tendrils in the Vine, and runners in the Strawberry; or, when developed under ground, they assume the aspect of creeping roots, and sometimes form thickened rootstalks, as in the Calamus, or tubers, as in the Potato. But the type is readily seen through these disguises. They are all mere modifications of the stem. The leaves, as we have already seen, appear under a still greater variety of forms, some of them as widely different from the common type of foliage as can be imagined; such, for example, as the thickened and obese leaves of the Mesembryanthemums; the intense scarlet or crimson floral leaves of the Euchroma, or Painted Cup, of the Poinsetia of our conservatories, and of several Mexican Sages; the tendrils of the Pea tribe; the pitchers of Sarracenia (Fig. 223), and also those of Nepenthes (Fig. 225), which are leaf, tendril, and pitcher combined. The leaves also appear under very different aspects in the same individual plant, according to the purposes they are intended to subserve. The first pair of leaves, or cotyledons, when gorged with nutritive matter for the supply of the earliest wants of the embryo plant, as in the Bean and Almond (Fig. 97), would seem to be peculiar organs. But when they have discharged this special office in germination, by yielding to the young plant the store of nourishment with which they are laden, they throw off their disguise, and assume, with more or less distinctness, the color and appearance of ordinary foliage; while in other cases, as in the Convolvulus, &c., they are green and foliaceous from the first. As the stem elongates, the successive leaves vary in form or size, according to the varying vigor of vegetation. In our trees, we trace the last leaves of the season into bud-scales; and in the returning spring we may often observe the innermost scales of the expanding leaf-buds to resume, the first perhaps im-

perfectly, but the ensuing ones successfully, the appearance and the ordinary office of leaves (146).

426. The analogies of vegetation would therefore suggest, that, in flowering, the leaves, no longer developing as mere foliage, are now wrought into new forms, to subserve peculiar purposes. In the chapter on Inflorescence, we have already shown that the arrangement and situation of flowers upon the stem conform to this idea. In this respect, flowers are absolutely like branches. The aspect of the floral envelopes favors the same view. We discern the typical element, the leaf, in the calyx; and again, more delicate and refined, in the petals. In numberless instances, we observe a regular transition from ordinary leaves into sepals, and from sepals into petals. And, while the petals are occasionally

green and herbaceous, the undoubted foliage sometimes assumes a

FIG. 265. Open flower, with a flower-bud and leaf of the White Water-Lily (Nymphæa odorata); the inner petals passing into stamens. 266. A flower with all the parts around the pistil cut away except one of the petaloid stamens, one intermediate, and one proper stamen. 267. An inner petal, with the imperfect rudiments of an anther at the tip. 268. Transverse section of an ovary.

delicate texture and the brightest hues (425). The perfect gradation of leaves or bracts into sepals is extremely common. The transition of sepals into petals is exemplified in almost every case where there are more than two rows of floral envelopes; as in the Magnolia, and especially in the White Water-Lily, the Illicium, or Star Anise of the Southern States, and the Calycanthus, or Carolina Allspice, which present several series of floral envelopes, all nearly alike in color, texture, and shape; but how many of the innermost are to be called petals, and how the remainder are to be divided between sepals and bracts, is entirely a matter of arbitrary opinion. In fact, the only real difference between the calyx and corolla is, that the former is the outer, and the latter an inner series of floral envelopes. Sometimes the gradation extends one step farther, and exhibits an evident transition of petals into stamens; showing that these are of the same fundamental nature as the floral envelopes, which are manifestly traceable back to leaves. The White Water-Lily (Fig. 265) exhibits this latter transition, as evidently as that of sepals into petals. Here the petals occupy several whorls, and, while the exterior are nearly undistinguishable from the calyx, the inner are reduced into organs which are neither well-formed petals nor stamens, but intermediate between the two. They are merely petals of a smaller size, with their summits contracted and transformed into imperfect anthers, containing a few grains of pollen: those of the series next within are more reduced in size, and bear perfect anthers at the apex; and a still further reduction of the lower part of the petal completes the transition into stamens of ordinary appearance.

427. Transitions, or intermediate states, between petals and stamens occur in numerous cases. These two are not only adjacent organs, but they appear to have very intimate relations, to which we may allude in another place. But similar transitions between such specialized, and as it were antagonistic, organs as the stamens and the pistils would not be expected normally to occur; nor is there any such regular instance known. Yet they are not unfrequently met with in *monstrous* blossoms, as occasionally in the Oriental Poppy in gardens, in the Houseleek, and in certain Willows.* These are monsters it is true; but the study of monstrosities often throws much light upon the regular structure.

* Thus two Apple-trees in Ashburnham, Massachusetts, annually produce

428. The regular transformation, or *metamorphosis* (if we may use that somewhat ambiguous term), takes an upward course, from leaves into sepals, from sepals into petals, and from the latter into stamens, or even into pistils. We trace the typical leaf forward into the floral envelopes, and thence into the essential organs of the blossom. Now if these organs be, as it were, leaves developed in peculiar states, under the controlling agency of a power which has overborne the ordinary forces of vegetation, they must always have a tendency to develope in their primitive form, when the causes that govern the production of blossoms are interfered with during their formation. They may then reverse the spell, and revert into some organ below them in the series, as from stamens into petals, or pass at once into the state of ordinary leaves. That is, organs which from their position should be stamens or pistils may develope as petals or floral leaves, or in the form of ordinary leaves. Such cases of *retrograde metamorphosis* frequently occur in cultivated flowers, and occasionally in some spontaneous plants.

429. Thus we meet with *the actual reconversion of what should be a pistil into a leaf* very frequently in the double garden Cherry, either completely (Fig. 269), or else incompletely, so that the resulting organ (as in Fig. 270) is something intermediate between the two. *The change of what should be stamens into petals* is of common occurrence in what are called *double* and *semi-double* flowers of the gardens; as in Roses, Camellias, Carnations, &c. When such flowers have many stamens, these disappear as the supernumerary petals increase in number; and the various bodies that may be often observed, intermediate between perfect stamens (if any remain) and the outer row of petals, — from imperfect petals with a small lamina tapering into a slender stalk, to

flowers in which the petals are replaced by five small foliaceous bodies, like the sepals, and in place of stamens there are ten separate and accessory pistils, inserted on the throat of the calyx. For an account of this case, and for a good suite of specimens, I am indebted to Dr. E. Leigh, of Townsend, Mass.

FIG. 269. A small leaf in place of a pistil from the centre of a flower of the double Cherry. 270. An organ intermediate between a leaf and a pistil, from a similar flower.

FIG. 271. Leaflet of a Bryophyllum, developing buds along its margins.

those which bear a small distorted lamina on one side and a half-formed anther on the other, — plainly reveal the nature of the transformation that has taken place. The garden Columbine often affords beautiful illustrations of this kind. Carried a step farther, the pistils likewise disappear, to be replaced by a rosette of petals, as in double Buttercups. This increase in the number of the petals of double flowers is not altogether at the expense of the stamens and pistils. In such cases the petals themselves are prone to *double*, or to multiply in number.

430. In full double Buttercups we may often notice a tendency of the rosette of petals to turn green, or to retrograde still farther into foliaceous organs. And there is a monstrous state of the Strawberry blossom, well known in Europe, in which all the floral organs revert into green sepals, or imperfect leaves. The annexed illustration (Fig. 272) exhibits a similar retrograde metamorphosis in a flower of the White Clover, where the calyx, pistil, &c. are still recognizable, although partially transformed into leaves.

We may observe that the ovary, which has opened down one side, bears on each edge a number of small and imperfect leaves; much as the ordinary leaves, or rather leaflets, of Bryophyllum are apt to develope rudimentary tufts of leaves, or buds, on their margins (Fig. 271), which soon grow into little plantlets. This reversion of a whole blossom into foliaceous parts has been termed *chlorosis*, from the green color thus assumed.

431. Somewhat different is the retrograde metamorphosis which is occasionally seen in *the production of a leafy branch from the centre of a flower, or of one flower out of the centre of another* (as rose-buds out of roses). Here the receptacle, or axis of the

FIG. 272. A flower of the common White Clover reverting to a leafy branch, after Turpin.

flower, resumes the ordinary growth, or vegetation, of the branch. This more commonly takes place after the formation of the floral envelopes and stamens, but before the pistils appear; as in Fig. **273**. The appearance of a leafy branch from the summit of a pear (as in Fig. **274**) is similarly explained. So, likewise, in very wet and warm springs, some of the flower-buds of the Pear and Apple are occasionally forced into active vegetative growth, so as completely to break up the flower, and change it into an ordinary leafy branch.

274

273

432. In such cases the terminal bud goes on to grow, — contrary to the normal condition, in which the flower arrests all further development of the axis that bears it. An analogous monstrosity sometimes occurs, in which axillary buds (**148**) are developed in the flower. Its organs thus exhibit a distinguishing characteristic of leaves, viz. *the production of buds in their axils;* which develope either as branches or as new axes at once terminated by blossoms. Flowers have thus been met with in the axils of the petals, as in Fig. **275**, and sometimes even in those of the stamens or pistils. Monstrosities of this sort are common in the Rose. Of the same kind are most of those cases in which one or more fruits, such as apples or pears, grow out of another fruit. We have met with flowers

275

FIG. 273. Retrograde metamorphosis of a flower of the Fraxinella of the gardens, from Lindley's Theory of Horticulture; an internode elongated just above the stamens, and bearing a whorl of green leaves.

FIG. 274. A monstrous pear, prolonged into a leafy branch, from Bonnet.

FIG. 275. A flower of the False Bittersweet (Celastrus scandens), producing other flowers in the axils of the petals, from Turpin.

of Clarkia elegans which bore an imperfect blossom in the axil of each petal.

433. The irresistible conclusion from all such evidence is, that the flower is one of the forms — the ultimate form — under which branches appear; that the leaves of the stem, the leaves or petals of the flower, and even the stamens and pistils, are all forms of a common type, only differing in their special development. And it may be added, that in an early stage of development they all appear alike. That which, under the ordinary laws of vegetation, would have developed as a leafy branch, does, in a special case and according to some regular law, finally develope as a flower; its several organs appearing under forms, some of them slightly, and others extremely, different in aspect and in office from the foliage. But they all have a common nature and a common origin, or, in other words, are homologous parts (424).

434. Now, as we have no general name to comprehend all those organs which, as leaves, bud-scales, bracts, sepals, petals, stamens, &c., successively spring from the ascending axis or stem, having ascertained their essential identity, we naturally, and indeed necessarily, take some one of them as the *type*, and view the others as modifications or metamorphoses of it. The leaf is the form which earliest appears, and is the most general of all the organs of the vegetable; it is the form which is indispensable to vegetation in its perfect development, in which it plays, as we have seen, the most important part; it is the form into which all the floral organs may sometimes be traced back by numerous gradations, and to which they are liable to revert when flowering is disturbed and the proper vegetative forces again prevail. Hence the leaf may be properly assumed as the type or pattern, to which all the others are to be referred. When, therefore, the floral organs are called *modified* or *metamorphosed leaves* (terms which we have avoided almost entirely, as liable to convey an erroneous impression), it is not to be supposed that a petal has ever actually been a green leaf, and has subsequently assumed a more delicate texture and hue, or that stamens and pistils have previously existed in the state of foliage; but only that what is fundamentally one and the same organ developes, in the progressive evolution of the plant, under each or any of these various forms. When the individual organ has once fairly begun to develope, its destiny is fixed.

435. The theory of vegetable morphology may be expressed in other, and more hypothetical or transcendental forms. We have

preferred to enunciate it in the simplest and most general terms. But, under whatever particular formula expressed, its adoption has not only greatly simplified, but has thrown a flood of light over the whole of Structural Botany, and has consequently placed the whole logic of Systematic Botany upon a new and philosophical basis. Our restricted limits will not allow us to trace its historical development. Suffice it to say, that the idea of the essential identity of the floral organs and the leaves was distinctly propounded by Linnæus,* about the middle of the last century. It was newly taught by Caspar Frederic Wolff, about twenty years later, and again, after the lapse of nearly twenty years more, by the celebrated Goethe, who was entirely ignorant, as apparently were his scientific contemporaries, of what Linnæus and Wolff had written on the subject. Goethe's curious and really scientific treatise was as completely forgotten or overlooked as the significant hints of Linnæus had been. In advance of the science of the day, and more or less encumbered with hypothetical speculations, none of these writings appear to have exerted any influence over the progress of the science, until it had reached a point, early in the present century, when the nearly simultaneous generalizations of several botanists, following different clews, were leading to the same conclusions. Ignorant of the writings of Goethe and Wolff, De Candolle was the first to develope, from an independent and original point of view, the idea of symmetry in the flower; that the plan, or type, of the blossom is regular and symmetrical, but that this symmetry is more or less modified or disguised by secondary influences, giving rise to various deviations, such as those which we are soon to consider. The reason of the prevailing symmetrical arrangement of parts in the blossom has only recently been made apparent, in the investigation of the laws of phyllotaxis (234); from which it appears that the general arrangement of the leaves upon the stem is carried out into the flower.

SECT. III. THE SYMMETRY OF THE FLOWER.

436. **A Symmetrical Flower** is one which has an equal number of parts in each circle or whorl of organs; as, for example, in Fig. 256, where there are five sepals, five petals, five stamens, and five

---

* "Principium *florum* et *foliorum* idem est. Principium gemmarum et foliorum idem est. Gemma constat foliorum rudimentis. Perianthium sit ex connatis foliorum rudimentis," etc. *Philosophia Botanica*, p. 301.

pistils. It is not less symmetrical, although less simple, when there are two or more circles of the same kind of organ; as in Sedum, where there are two sets of stamens, five in each; in the Barberry, where there are two or more sets of sepals, two of petals, and two of stamens, three in each set, &c. *A complete flower* (as already defined, 416) is one that possesses both sorts of floral envelopes, calyx and corolla, and both essential organs, viz. stamens and pistils.

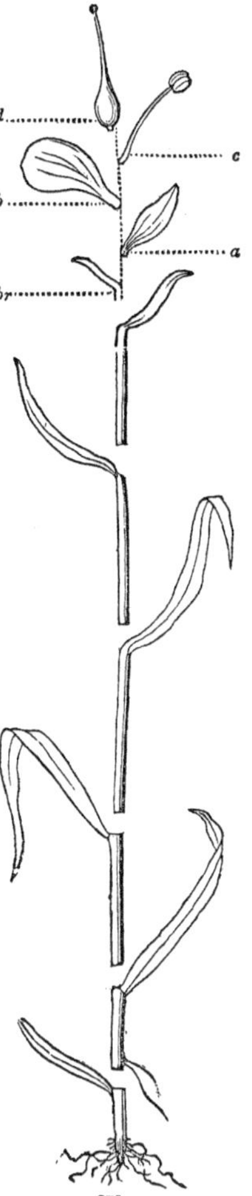

437. The simplest possible complete and symmetrical flower would be one with the calyx of a single sepal, a corolla of a single petal, a single stamen, and a single pistil; such as is represented in the annexed diagram, in connection with the two-ranked arrangement of the leaves (Fig. 276). Each constituent of the blossom represents a phyton, with its stem part reduced to a minimum, and its leaf part developed in a peculiar way, according to the rank it sustains and the office it is to fulfil. That there are short internodes between consecutive organs in the flower is usually apparent on minute inspection of its axis, or receptacle; and some of them are conspicuously prolonged in certain cases. But they are commonly undeveloped, like the axis of a leaf-bud, so that the organs are brought into juxtaposition on a short, mostly conical receptacle, and the higher or later-formed parts are interior or inclosed by the lower.

438. Perhaps the exact case of a flower at once so complete and so simple is not to be met with. For, when the stamens and pistils are thus reduced to the minimum number, the floral envelopes, one or both, commonly disappear, as in the Mare's-tail (Fig. 703). Nor is the production of seed often left to depend upon a single organ;

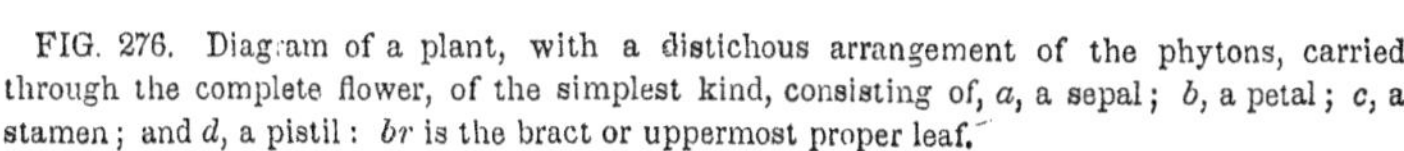
FIG. 276. Diagram of a plant, with a distichous arrangement of the phytons, carried through the complete flower, of the simplest kind, consisting of, *a*, a sepal; *b*, a petal; *c*, a stamen; and *d*, a pistil: *br* is the bract or uppermost proper leaf.

but the essential, and with them the protecting organs, are generally multiplied in each flower, so as greatly to diminish the chances of failure. Thus we find a circle or whorl of each kind of organ, and often two or three circles, or a still larger and apparently indefinite number of parts. In fact, the floral organs usually occur in twos, threes, fours, or fives, and the same number commonly prevails through the several parts of the flower (except when interfered with by some of the disturbing causes hereafter mentioned), which therefore displays a symmetrical arrangement, or a manifest tendency towards it.*

439. Having already noticed the symmetrical arrangement of the foliage (234 – 252), and remarked the transition of ordinary leavesinto those of the blossom (426), we naturally seek to bring the two under the same general laws, and look upon each floral whorl as answering either to a cycle of alternate leaves with their respective internodes undeveloped (237 – 239), or to a pair or verticil of opposite or verticillate leaves (250, 251). Thus, the simplest combination, where the organs are *dimerous*, or in twos, may be compared with the alternate two-ranked arrangement (237), the calyx, the corolla, stamens, &c., each consisting of one cycle of two elements; or else with the case of opposite leaves

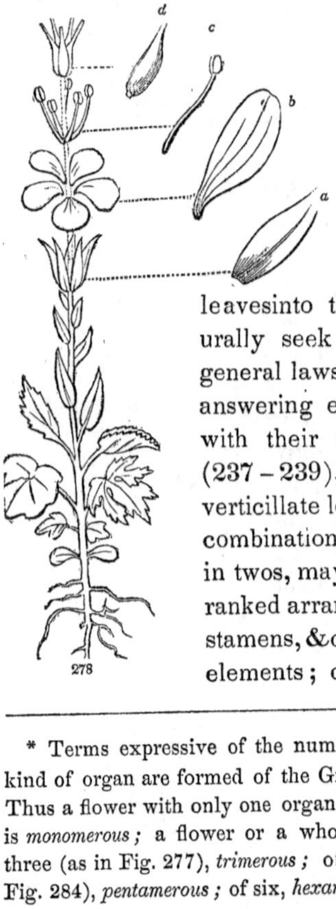

* Terms expressive of the number of parts which compose each whorl or kind of organ are formed of the Greek numerals combined with μέρος, *a part.* Thus a flower with only one organ of each kind, as in the diagram, Fig. 276, is *monomerous;* a flower or a whorl of two organs is *dimerous* (Fig. 298); of three (as in Fig. 277), *trimerous;* of four, *tetramerous* (Fig. 280); of five (as in Fig. 284), *pentamerous;* of six, *hexamerous;* of ten, *decamerous*, &c.

FIG. 277. Parts of a symmetrical trimerous flower (Tillæa muscosa): *a*, calyx; *b*, corolla; *c*, stamens; *d*, pistils.

FIG. 278. Ideal plan of a plant, with the simple stem terminated by a symmetrical pentamerous flower; the different sets of organs separated to some distance from each other, to show the relative situation of the parts; one of each, namely, *a*, a sepal, *b*, a petal, *c*, a stamen, and *d*, a pistil, also shown, enlarged.

(250), when each set would answer to a pair of leaves. So, likewise, the organs of a *trimerous* flower (viz. one with its parts in threes, as in Fig. 277) may be taken, each set as a cycle of alternate leaves of the tristichous mode (171), with the axis depressed, which would throw the parts into successive whorls of threes, or as a proper verticil of three leaves; while those of a *pentamerous* or *quinary* flower (with the parts in fives, as in Fig. 278) would answer to the cycles of the $\frac{2}{5}$ arrangement (239) of alternate leaves, or to proper five-leaved verticils. So the whorls of a tetramerous flower are to be compared with the case of decussating opposite leaves (250) combined two by two, or directly with quaternary verticillate leaves; either of which would give sets of parts in fours.

440. **The Alternation of the Floral Organs.** We learn from observation that the parts of the successive circles of the flower almost universally alternate with each other. The five petals of the flower represented in Fig. 256, for example, are not *opposed* to the five sepals, that is, situated directly above or before them, but *alternate* with, or stand over the intervals between them; the five stamens in like manner alternate with the petals, and the five pistils with the stamens, as is shown in the diagram, Fig. 257. The same is the case in the trimerous flower, Fig. 277; and in fact this is the regular rule, the few exceptions to which have to be separately accounted for.

441. This alternation comports with the more usual phyllotaxis in opposite and verticillate leaves, where the successive pairs decussate, or cross each other at right angles (250), or the leaves of one verticil severally correspond to the intervals of that underneath, making twice as many vertical ranks as there are parts in the whorl (251). The alternation of the floral organs is therefore most readily explained on the assumption that the several circles are true decussating verticils; when it only remains to discover the real connection between the opposite-leaved or verticillate and the spiral phyllotaxis, and to obtain some expression which will harmonize the two modes; both of which are often met with on the same axis. But the inspection of a flower-bud with the parts imbricated in æstivation (492) shows that the several members of the same set do not originate exactly in the same plane. The five petals, for example, in the cross-section of the pentamerous blossom shown in Fig. 257 (and the same arrangement is still more frequently seen in the calyx), are so situated, that two are

exterior in the bud, and therefore inserted lower on the axis than the rest, the third is intermediate, and two others are entirely interior, or inserted higher than the rest. In fact, they exactly correspond with a cycle of the quincuncial, or five-ranked, spiral arrangement, projected on an extremely abbreviated axis, or on a horizontal plane, as is at once seen by comparison with Fig. 172, 173. Also when the parts are in fours, two are almost always exterior in the bud, and two interior. Moreover, whenever the floral envelopes, or the stamens or pistils, are more numerous, so as to occupy several rows, the spiral disposition is the more manifest. It is most natural, accordingly, to assume that the calyx, corolla, stamens, &c. of a pentamerous flower are each a depressed spiral or cycle of the $\frac{2}{5}$ mode of phyllotaxis (239), and those of the trimerous flower are similar spirals of the $\frac{1}{3}$ mode (238). But then the parts of the successive cycles should be superposed, or placed directly before each other on the depressed axis (Fig. 171); whereas, on the contrary, they almost always alternate with each other in the flower, as in the annexed diagram (Fig. 279).

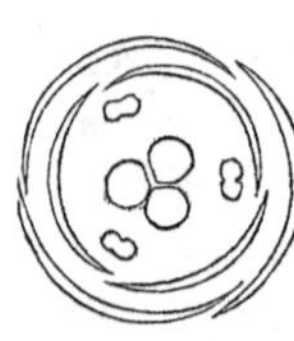
279

442. To reconcile this alternation with the laws of phyllotaxis in alternate leaves, Prof. Adrien de Jussieu has advanced an ingenious hypothesis. He assumes the $\frac{5}{13}$ spiral arrangement (241) as the basis of the floral structure both of the trimerous and pentamerous flower, (at least when the envelopes are imbricated in the bud,) this being the one that brings the successive parts most nearly into alternation, either in threes or in fives; as will readily be observed on inspection of the tabular projection of that mode, given on page 147. The difference between the position of parts in regular alternation, whether in threes or fives, and that assigned by an accurate spiral projection of the $\frac{5}{13}$ mode, is very slight as respects most of the organs, and in none does the deviation exceed one thirteenth of the circumference; — a quantity which becomes nearly insignificant on an axis so small as that of most flowers, especially towards its narrowed apex. Moreover, if the interior organs of a regular and symmetrical flower were thus to originate in the bud nearly in alternation with those that precede them, they would almost necessarily be pushed a little, as they develope, into

FIG. 279. Cross-section of the flower-bud of the trimerous Tillæa, Fig. 277, to show the alternation of parts.

the position of least pressure, and thus fall into these intervals with all the exactness that is actually found in nature. For in these living bodies, endowed as they are with plasticity and a certain power of adaptation to circumstances, the positions assumed are not mathematically accurate ; and the effect of unequal pressure in the bud in throwing the smaller parts more or less out of their normal position may be observed in almost any irregular flower. Moreover, in all the forms of phyllotaxis from $\frac{5}{13}$ onwards, it is doubtful whether what we term vertical ranks are exactly superposed. In tracing them onward to some extent, we perceive indications of a curviserial arrangement, where the superposition is continually approximated, but is never exactly attained.

443. When, therefore, the floral circles consist of parts which are evidently developed in the same horizontal plane (494), they are most simply viewed as decussating verticils, — as formed after the manner of opposite leaves. When they are imbricated in the bud (492), or show in other ways a spiral disposition of parts, we may conceive that the law of alternation is conformed to in the manner which Jussieu has suggested, or in some such way. However explained, we cannot fail to discern an end attained by such arrangement, namely, a disposition of parts which secures the greatest economy of space on an abbreviated axis, and the greatest freedom from mutual pressure.

444. **Position of the Flower as respects the Axis and subtending Bract.** All axillary flowers are situated between a leaf and the stem, or, which is the same thing, between a bract and the axis of inflorescence. These two fixed points enable us to indicate the relative position of the parts of the floral circles with precision. That part of the flower which lies next the leaf or bract from whose axil it arises is said to be *anterior*, or *inferior* (lower) : that which is diametrically opposite or next the axis is *posterior*, or *superior* (upper).* It is important to notice the relative position of parts in this respect. This is shown in a proper diagram by drawing a section of the bract in its true position under the section of the flower-bud, as in Fig. 282 : that of the axis is necessarily diamet-

* As if these were not terms enough, sometimes the organ, or side of the flower, which looks towards the bract, is likewise called *exterior*, and the organ or side next the axis, *interior ;* but these terms should be kept to designate the relative position of the members of the floral circles in æstivation (490).

rically opposite, and its section is sometimes indicated by a dot or small circle. In an axillary tetramerous flower, one of the sepals will be anterior, one posterior, and two lateral, or right and left; as in the annexed diagram of a Cruciferous blossom (Fig. 280);

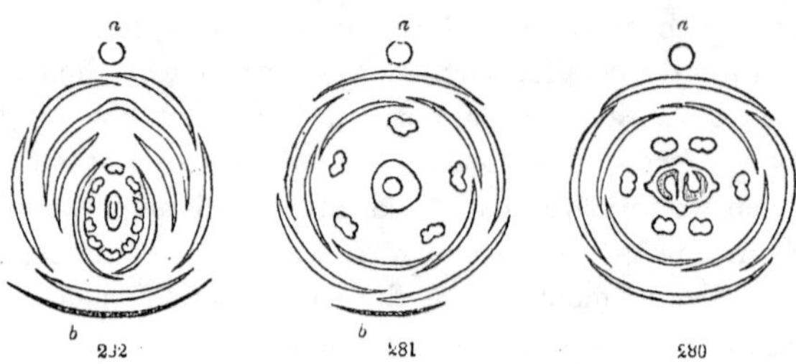

while the petals, alternating with the sepals, occupy intermediate positions, or consist of an anterior and a posterior pair; while the stamens, again, correspond to the sepals in position. A pentamerous axillary flower, having an odd number of parts, will have either one sepal superior or posterior and two inferior or anterior (as in Rhus, Fig. 281), or else, *vice versâ*, with one inferior and two superior, as in Papilionaceous flowers (Fig. 282): in both cases the two remaining sepals are *lateral*. The petals will consequently stand one superior, two inferior, and two lateral, in the last-named case (Fig. 282), and one inferior, two superior, and two lateral in the former (Fig. 281). In terminal flowers (401), the position of parts in respect to the uppermost leaves or bracts should be noted.

SECT. IV. THE VARIOUS MODIFICATIONS OF THE FLOWER.

445. THE complete and symmetrical flowers, with all their organs in the most normal state, that have now been considered, will serve as the type or pattern, with which we may compare the almost numberless variety of forms which blossoms exhibit, and note the character of the differences observed. We proceed upon the supposition, that all flowers are formed upon one comprehensive plan, — a plan essentially consonant with that of the stem or

FIG. 280. Diagram of a Cruciferous flower (Erysimum); *a*, the axis of inflorescence. (The bract is abortive in this, as in most plants of this family.)

FIG. 281. Diagram of a flower of a Rhus, with the axis, *a*, and the bract, *b*, to show the relative position of parts.

FIG. 282. Diagram of a flower of the Pulse tribe, with *a*, the axis, and *b*, the bract.

branch, of which we have shown the flower to be a modified continuation, — so that in the flower we are to expect no organs other than those that, whatever their form and office, answer either to the axis or to the leaves, or, in other words, to phytons (230); so that the differences between one flower and another are to be explained as special deviations from, or circumstantial variations of, one fundamental plan, — variations for the most part similar or analogous to those which are known to occur in the organs of vegetation themselves. Having assumed the type which represents our conception of the most complete, and at the same time the simplest flower, we apply it to all the cases which present themselves; and especially to the elucidation of those blossoms in which the structure and symmetry are masked or obscured; where, like the disenchanting spear of Ithuriel, its application at once reveals the real character of the most disguised and complicated forms of structure.

446. Our pattern flower consists of four circles, one of each kind of floral organ, and of an equal number of parts, successively alternating with one another. It is *complete*, having both calyx and corolla, as well as stamens and pistils (416); *symmetrical*, having an equal number of parts in each whorl (436); *regular*, in having the different members of each circle all alike in size and shape; it has but one circle of the same kind of organs; and moreover, all the parts are distinct or unconnected, so as to exhibit their separate origin from the axis or receptacle of the flower. Our type may be presented under either of the four numerical forms which have been illustrated. That is, its circles may consist of parts in twos (when it is *binary* or *dimerous*), threes (*ternary* or *trimerous*), fours (*quaternary* or *tetramerous*), or fives (*quinary* or *pentamerous*). The first of these is the least common; the trimerous and the pentamerous far the most so. The last is restricted to Dicotyledonous plants, where five is the prevailing number; while the trimerous flower largely prevails in Monocotyledonous plants, although by no means wanting in the Dicotyledonous class, from which Fig. 277 is taken.

447. The principal deviations from the perfectly normal or pattern flower may be classified as follows. They arise, either from, —

1st. The production of one or more additional circles of one or more of the floral organs (*regular multiplication* or *augmentation*):

2d. The production of a pair or a cluster of organs where there should normally be but one, that is, the multiplication of an organ by division (*abnormal multiplication*, also termed *deduplication* or *chorisis*):

3d. The union of the members of the same circle (*coalescence*):

4th. The union of adjacent parts of different circles (*adnation*):

5th. The unequal growth or unequal union of different parts of the same circle (*irregularity*): or,

6th. The non-production or abortion of some parts of a circle, or of one or more complete circles (*suppression* or *abortion*).

7th. To which may be added, the abnormal development of the receptacle or axis of the flower.

448. Some of these deviations obscure the symmetrical structure of the flower; others merely render it irregular, or disguise the real origin of the number of parts. These deviations, moreover, are seldom single; but two, three, or more of them frequently co-exist, so as to realize almost every conceivable variation.

449. Several of these kinds of deviation may often be observed even in the same natural family of plants, where it cannot be doubted that the blossoms are constructed on the same general plan in all the species. Even in the family Crassulaceæ, for example, where the flowers are remarkably symmetrical, and from which our pattern flowers, Fig. 256 and 277, are derived, a considerable number of these diversities are to be met with. In Crassula, we have the completely symmetrical and simple pentamerous flower (Fig. 283, 284), viz. with a calyx of five sepals, a corolla of five petals alternate with the former, an andrœcium (418) of five stamens alternating with the petals, and a gynæcium (419) of five pistils, which are alternate with the stamens; and all the parts are regular and symmetrical, and also distinct and free from each other; except that the sepals are somewhat united at the base, and the petals and stamens slightly connected with the inside of the calyx, instead of manifestly arising from the receptacle or axis, just beneath the pistils. Five is the prevailing or normal number in this family. Nevertheless,

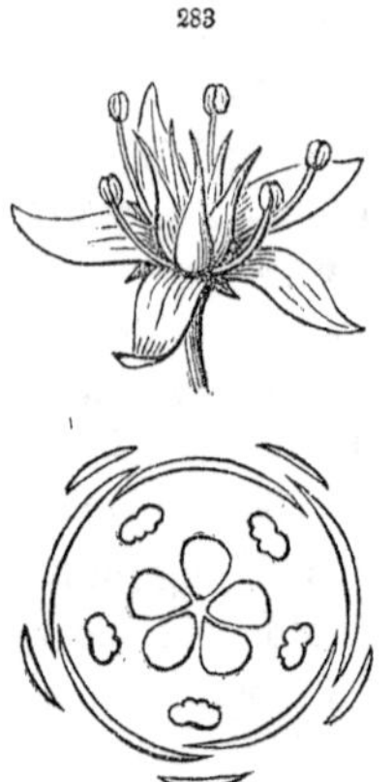

FIG. 283. Flower of a Crassula. 284. Cross-section of the bud.

in the related genus Tillæa, most of the species, like ours of the United States, have their parts in fours, but are otherwise similar, and one common European species has its parts in threes (Fig. 277): that is, one or two members are left out of each circle, which of course does not interfere with the symmetry of the blossom. So in the more conspicuous genus Sedum (the Stonecrop, Live-for-ever, Orpine, &c.) some species are 5-merous, others 4-merous, and several, like our S. ternatum, have the first blossom 5-merous, but all the rest on the same plant 4-merous. But Sedum also illustrates the case of regular augmentation (447, 1st) in its andrœcium, which consists of twice as many stamens as there are members in the other parts; that is, an additional circle of stamens is introduced (Fig. 285, 286), the members of which may be distinguished by being shorter or a little later than those of the primary circle, and also more definitely by their alternation with the primary, which brings them directly opposite the petals. A third genus (Rochea) exhibits the same 5-merous and normal flower as Crassula, except that the contiguous edges of the petals slightly cohere about half their length, although a little force suffices to separate them: in another (Grammanthes, Fig. 287), the petals are firmly united into a tube for more than half their length, and so are the sepals likewise; presenting, therefore, the third of the deviations above enumerated (447). Next, the allied genus Cotyledon (Fig. 288) exhibits in the same flower both this last case of the coalescence of similar parts in its floral envelopes, and an additional circle of stamens, as in Sedum. It likewise presents the next order of deviations, in the adnation of the base of its stamens to the base of the corolla, out of which they apparently arise, as is seen in Fig. 289, where the corolla is laid open and displayed. The pistils, although ordinarily exhibiting a strong tendency to unite, are perfectly distinct in all these cases, and indeed throughout the order, with two exceptions; one of which is seen in Penthorum, where the five ovaries (Fig.

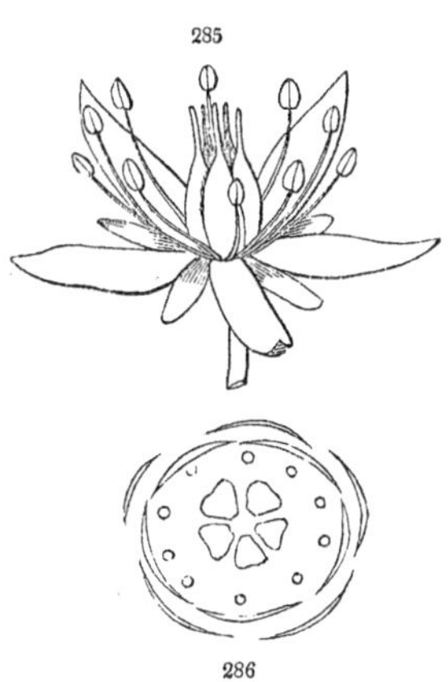

FIG. 285. Flower of a Sedum. 286. Cross-section of the bud.

290) are united below into a solid body, while their summits, as well as the styles, are separate. The same plant also furnishes an example of the non-production (or *suppression*) of one whorl of organs, that of the petals; which, although said to exist in some specimens, are ordinarily wanting altogether. Another instance of increase in the number of parts occurs in the Houseleek (Semper-vivum), in which the sepals, petals, and pistils vary in different species from six to twenty, and the stamens from twelve to forty.

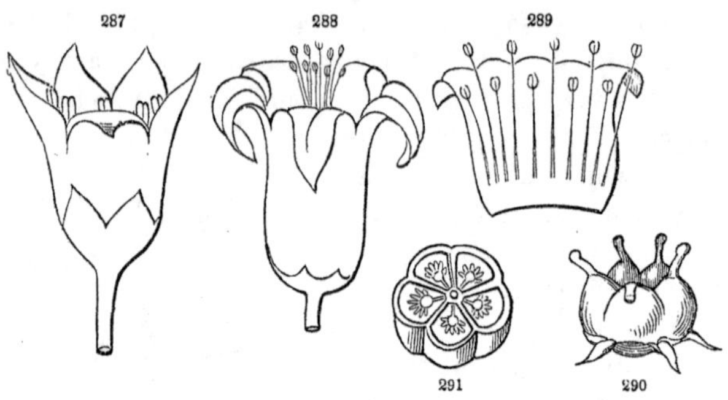

450. Some illustrations of the principal diversities of the flower, as classified above (447), may be drawn at random from different families of plants; and most of the technical terms necessarily employed in describing these modifications may be introduced, and concisely defined, as we proceed. The multiplication of parts is usually in consequence of the

451. **Augmentation of the Floral Circles.** An increased number of circles or parts of all the floral organs occurs in the Magnolia Family; where the floral envelopes occupy three or four rows, of three leaves in each, to be divided between the calyx and corolla, while the stamens and pistils are very numerous, and compactly arranged on the elongated receptacle. The Custard-Apple Family, which is much like the last, has also two circles in the corolla, three petals in each, a great increase in the number of stamens, and, in our Papaw (Fig. 493), sometimes only one circle of pistils, viz. 3, sometimes twice, thrice, or as many as five times that

FIG. 287. Flower of Grammanthes.

FIG. 288. Flower of a Cotyledon. 289. The corolla laid open, showing the two rows of stamens inserted into it.

FIG. 290. The five pistils of Penthorum, united, so as to form a compound ovary. 291. A cross-section of the same.

number. The Water-Lily, likewise, has all its parts increased (Fig. 265), the floral envelopes and the stamens especially occupying a great number of rows ; and the pistils are likewise numerous ; although their number is disguised by a combination, to be hereafter explained. When the sepals, petals, or other parts of the flower are too numerous to be readily counted, or are even more than twelve, especially when the number is inconstant, as it commonly is in such cases, they are said to be *indefinite ;* and a flower with numerous stamens is also termed *polyandrous.*

452. When such multiplication of the floral circles is perfectly regular, the number of the organs so increased is a multiple of that which forms the basis of the flower; but this could scarcely be determined when the numbers are large, as in the stamens of a Buttercup, for example, nor is there much constancy when the whorls of any organ exceed three or four. In such cases, the circles usually appear to run into a continuous spiral, as is plainly seen in the cone of a Magnolia or of a Tulip-tree. The doubling or trebling of any or all the floral circles does not interfere with the symmetry of the flower; but it may obscure it (in the stamens and pistils especially), by the crowding of two or more circles of five members, for example, into what appears like one of ten, or two trimerous circles into what appears like one of six. The latter case occurs in most Endogenous plants.

453. The production of additional floral circles may account for most cases of increase of the normal number of organs, but not for all of them ; unless through the aid of hypotheses that have no intrinsic probability, and are unsupported by any clear analogies drawn from the organs of vegetation, which, it is evident, must give the rule in all questions involving the morphology, or at least the position, of the floral organs. It must, we think, be admitted that certain parts of the blossom are sometimes multiplied by *the production of a pair or a group of organs which occupy the place of one ;* namely, by what has been termed

454. **Chorisis or Deduplication.** The name *dédoublement* of Dunal, which has been translated *deduplication*, literally means *unlining ;* the original hypothesis being, that the organs in question *unline*, or tend to separate into two or more layers, each having the same structure. We may employ the word *deduplication*, in the sense of the doubling or multiplication of the number of parts, without receiving this gratuitous hypothesis as to the nature of the process,

which at best can well apply only to some special cases. The word *chorisis* (χώρισις, the act or state of separation or multiplication), also proposed by Dunal, does not involve any such assumption, and is accordingly to be preferred. By regular multiplication, therefore, we mean the augmentation of the number of organs through the development of additional circles; which does not alter the symmetry of the flower. By chorisis we denote the production of two or more organs in the place of one, through the multiplication of the leaf part of an individual phyton; — a case which may be compared with the multiplication of cells by division (30), and more directly with the division of the blade of a leaf into a number of separate blades or leaflets. Chorisis may take place in two different ways, which are perhaps to be differently explained: in one case, the increased parts stand side by side; in the other, they are situated one before the other. Both cases must evidently disturb or disguise the normal symmetry of the flower.

455. Of the first case, which may be termed *collateral chorisis*, we have a good example in the *tetradynamous* stamens (519) of the whole natural family Cruciferæ. Here, in a flower with symmetrical tetramerous calyx and corolla, we have six stamens; of which the two lateral or shorter ones are alternate with the adjacent petals, as they normally should be, while the four are in two pairs, one pair before each remaining interval of the petals; as is shown in the annexed diagram (Fig. 292). That is, on the anterior and on the posterior side of the flower we have two stamens where there normally should be but a single one, and where, indeed, there is but one in some species of Cruciferæ. Now it occasionally happens that the doubling of this stamen is, as it were, arrested before completion, so that in place of two stamens we see a forked filament bearing a pair of anthers; as is usually the case in several species of Streptanthus (Fig. 293). Here the two stamens which stand in the place of one may be compared with a

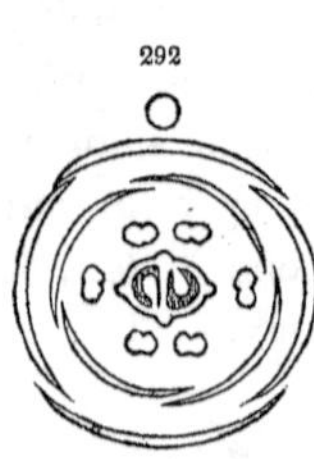

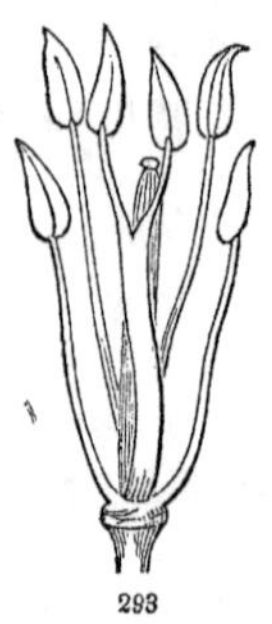

FIG. 292. Diagram of a (tetradynamous) flower of the order Cruciferæ.

FIG. 293. Flower of Streptanthus hyacinthoides, from Texas (the sepals and stamens removed), showing a forked or double stamen in place of the anterior pair.

sessile compound leaf of two leaflets. In the related order Fumariaceæ, each phyton of the andrœcium is *trebled* in the same manner. The circles of the flower in that order are in twos throughout, or dimerous. There is, first, a pair of small scale-like sepals; alternate with these a pair of petals which, in Dicentra, &c. (Fig. 294–296), are saccate or spurred below: alternate and within these there is a second pair of petals (Fig. 297): alternate with these are two clusters of three more or less united stamens, which plainly stand in the place of two single stamens. The arrangement of parts is shown in the annexed diagram (Fig. 298); where the lowest line indicates the subtending bract, and therefore the anterior side of the blossom; the two short lines in the same plane represent the sepals; the two next within, the lateral and exterior petals; those alternate and within these, the inner circle of petals; and alternate with these are the anthers of the two stamen-clusters. The centre is occupied by a section of the pistil, which, as will hereafter be shown, consists of two united. The three stamens are lightly connected in Dicentra (Fig. 296); but in Corydalis and

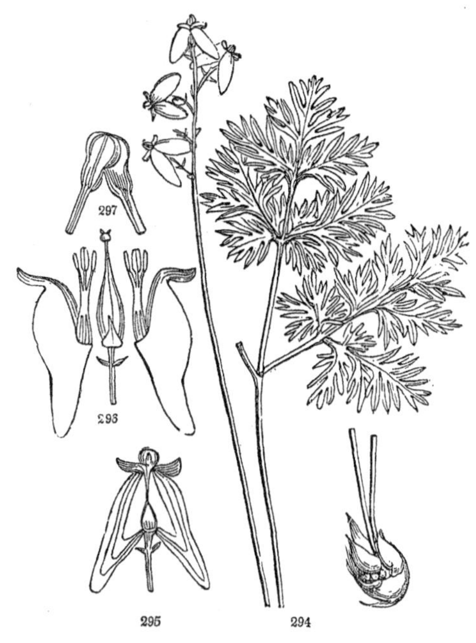

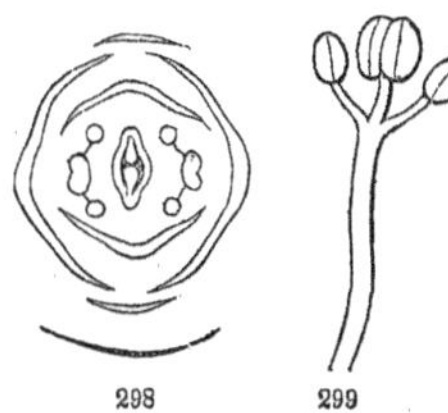

FIG. 294. Dicentra Cucullaria (Dutchman's-breeches), with its kind of bulb, a leaf, and a scape in flower; reduced in size. 295. A flower of the natural size. 296. The same with the parts separated, except the sepals, one of which is seen at the base of the pistil. 297. The inner pair of petals, with their tips coherent.

FIG. 298. Diagram (cross-section) of the similar flower of Adlumia. 299. One of the stamens increased into three by chorisis (the lower part of the common filament is cut away).

Adlumia there is only one strap-shaped filament on each side, which is three-forked at the tip, each fork bearing an anther. One of these trebled stamens is shown in Fig. 299.

456. We have a similar case in some Hypericums and in Elodea (Fig. 300), except that in these, while the floral envelopes are 5-merous, the circles within them are commonly 3-merous. The three members of the andrœcium are normally placed, alternating with the three members of the gynæcium within, and without with three glands, which probably replace an exterior circle of stamens; but each member as it developed has divided above into three stamens (Fig. 301); each anther of which may be viewed as homologous with a leaflet of a trifoliolate leaf (289). In the same way are the false filaments placed between the petals and the real stamens of Parnassia, partly divided into three in our P. Caroliniana (Fig. 305), or into from 9 to 15 shorter glandular lobes in P. palustris. So each cluster of numerous stamens of the polyandrous species of Hypericum (Fig. 553) doubtless arises from the repeated chorisis of a single phyton, and is therefore analogous to a decompound leaf. The actual development of such a cluster from a small protuberance, which in the forming flower-bud stands in the place of a single stamen, and its repeated forkings as it grows, have been traced by Duchatre, particularly in Malvaceous plants.

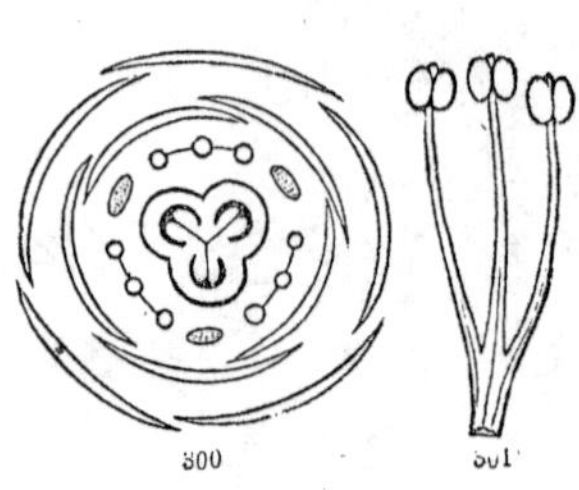

457. Thus far we are sustained by a clear analogy in the organs of vegetation. As the leaf frequently developes in the form of a lobed, divided, or compound leaf, that is, as a cluster of partially or completely distinct organs from a common base, so may the stamen, or even the pistil, become compound as it grows, and give rise to a cluster, instead of completing its growth as a solitary organ: and it appears that the organogeny is strikingly similar in the two cases. Nor is it very unusual for petals to become divided or deeply lobed in the same manner; as, for example, those of Mignonette. In many cases, however, the multiplication takes place in the opposite plane, so that the parts are situated one be-

FIG. 300. Diagram (cross-section) of a flower of Elodea Virginica. 301. One of the three stamen-clusters, consisting of a trebled stamen.

fore the other; — an arrangement which is not known to occur in the leaflets of any compound leaf.

458. Some examples of this *vertical* or *transverse chorisis* may be adduced before we essay to explain them. A common case is that of the *crown*, or small and mostly two-lobed appendage on the inside of the blade of the petals of Silene (Fig. 302) and of many other Caryophyllaceous plants. This is more like a case of real *dédoublement* or *unlining*, a partial separation of an inner lamella from the outer, and perhaps may be so viewed. The stamens sometimes bear a similar and more striking appendage, as in Larrea, for example (Fig. 303), and most other plants of the Guaiacum Family; also in the Dodder (Fig. 930). Let it be noted that in these cases the appendage occupies the inner side of the petal or stamen, and that it is often two-lobed. Again, before each petal of Parnassia (Fig. 305), although slightly if at all united with it, is found a body which in P. palustris is somewhat petal-like, with a considerable number of lobes, and in P. Caroliniana is divided almost to the base into three lobes, which look much like abortive stamens. The true staminneal circle, however, occupies its proper place within these ambiguous bodies, alternate with the petals. We cannot doubt that the former are of the same nature as the scale of the stamens in Larrea, and the crown of the petals of Silene.

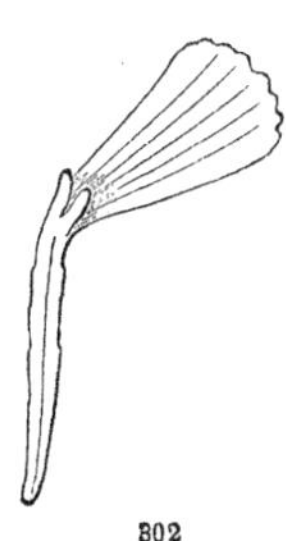

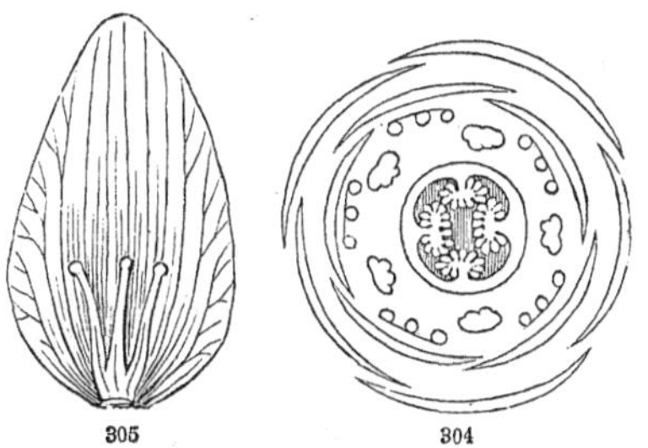

459. It may also be noticed, that, while in collateral chorisis the increased parts are usually all of the same nature, like so many similar leaflets of a compound leaf, in what is called transverse chorisis

FIG. 302. A petal of Silene Pennsylvanica, with its *crown* or appendage.

FIG. 303. A stamen of Larrea Mexicana, with a scale-like appendage cohering with its base on the inner side.

FIG. 304. Diagram (cross-section) of the flower of Parnassia Caroliniana. 305. A petal, with the appendage that stands before it.

22

there is seldom if ever such a division or ramification into homogeneous parts; but the original organ remains, as it were, intact and unmodified, while it bears an appendage of some different appearance or function on its inner face, or at its base on that side. Thus the stamens of Larrea, &c. bear a scale-like appendage; the petals of Sapindus, Cardiospermum, &c., a petaloid scale quite unlike the original petal; the petals of Parnassia, a cluster of bodies resembling sterile filaments united below. In a still greater number of instances, the accession to the petal consists of a real stamen placed before it, and often more or less united with its base, as in the whole Buckthorn Family (Fig. 315), and in the Byttneriaceæ; or of a cluster of stamens, as in the Mallow Family, and indistinctly in most European Lindens; or of such a cluster with a petal-like scale in the midst, as in the American Lindens (Fig. 622, 623, 306). In the first-named cases, the accessory organ developes entire and simple; in the latter, it is multiplied by collateral chorisis.*

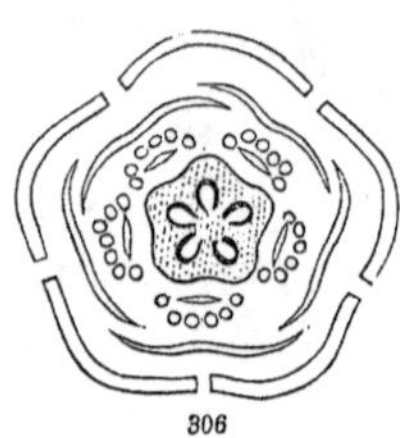

460. A most able writer in a recent number of the Journal of Botany, (with whom we entirely accord as to the nature of collateral chorisis,) "being totally at a loss to find any thing analogous in the ordinary stem-leaves" to this transverse or vertical multiplication of parts, inclines to consider such appendages as those of the petals of Silene, Sapindus, Ranunculus, &c., as deformed glands, and the stamens thus situated, whether singly or in clusters, as developments of new parts in the axil of the petals, &c.† It appears to us, however, that the leaves do furnish the proper analogue of these appendages (especially those of Fig. 302, 303, 305, and the

* For illustrations, and more detailed explanation of these points, the student is referred to the figures and text of *The Genera of the United States Flora Illustrated*, especially to Vol. 2. The opposition of the exterior circle of stamens to the petals in Geranium, &c., we explain in a different way (477).

† Namely, in Hooker's *Journal of Botany and Kew Garden Miscellany*, Dec., 1849, p. 360. — The morphology of true glands is still obscure, notwithstanding the interesting light that is thrown upon them in the article here referred to; and stipules often tend to assume the glandular character, or are reduced to glands, as in Linum.

FIG. 306. Diagram (cross-section) of the unopened flower of the American Linden, to show the scale and the cluster of stamens before each petal.

petaloid scales of Sapindaceæ) in the *ligule* of Grasses (298), and the stipules (304). The former occupies exactly the same position. The latter form an essential part of the leaf (259), and usually develope in a plane parallel with that of the blade, but between it and the axis, particularly when they are of considerable size, and serve as teguments of the bud, as, for example, in Magnolia (Fig. 130) and Liriodendron. The combined intrapetiolar stipules of Melianthus, &c. furnish a case in point, to be compared with the two-lobed internal scale of the stamens in Larrea, the two-cleft adnate appendage of the petals in Caryophylleæ, Sapindus, &c.; and instances of cleft or appendaged stipules may readily be adduced to show that such bodies are as prone to multiplication by division as other foliar parts. The supposition of a true axillary origin of the organs in question, therefore, appears to be gratuitous, and it would certainly introduce needless complexity into the theory of the structure of the flower. Still, as the axillary branch must begin with a single phyton, its development may in the flower be restricted to one phyton (as in the pistillary leaf in the axil of a bract in Coniferæ); thus giving a single axillary organ, which, if it multiply at all as it developes, may do so by collateral chorisis. And, reduced to the simplest case, between the transverse division of a nascent phyton, and the axillar production of a second phyton at an extremely early period in the development of that which subtends it, there is little assignable difference. At present, accordingly, we think that the same generic name may properly enough be employed both for the collateral and the vertical multiplication of organs, where two or more bodies occupy the place of one, carefully distinguishing, however, the two different cases; and also, that a special term is needful for discriminating between such multiplication and that by the regular augmentation of floral organs through the development of additional circles. Nor is a special term the less requisite, at least in systematic botany, because we recognize, in one or both kinds of *chorisis*, processes or modes of division which are common to the floral organs and to the foliage.*

---

* We are aware that Dr. Lindley summarily rejects the whole doctrine of chorisis, or any evolution of two or more bodies in the normal place of one, however explained; and for three reasons, which may be cited from *Introd. to Botany*, 1. p. 333, with a word of comment. "1. There is no instance of

461. **The Coalescence** or union of the parts of the same whorl or set of organs is so frequent, that few cases are to be found in which it does not occur, to a greater or less extent, in some portion of the flower. When the sepals are thus united into a cup or tube, the calyx is said to be *monosepalous*, or, more correctly, *gamosepalous:* when the petals are united, the corolla is said to be *monopetalous*, or *gamopetalous;* the latter being the appropriate term, as it denotes that the petals are combined; but the former is in common use, although strictly incorrect, as it implies that the corolla consists of a single petal. The inappropriate names, in these cases, were given long before the structure was rightly understood. So, also, such a calyx or corolla is said to be *entire*, when the sepals or petals are united to their very summits; as the corolla of Convolvulus (Fig.

---

unlining [read *chorisis*, which Dunal, as quoted by Lindley, proposes to substitute] which may not be as well explained by the theory of alternation." — Not to mention other instances, how is the andrœcium of Fumariaceæ to be explained upon the theory of alternation? If by the hypothesis still reproduced in the *Vegetable Kingdom*, p. 436, we inquire, What analogy warrants the supposition that a stamen or a leaf may split into halves, and the halves unite each with a different filament which has an angular distance of 90 degrees? — "2. It is highly improbable and inconsistent with the simplicity of vegetable structure, that in the same flower the multiplication of organs should arise from two wholly different causes, viz. alternation at one time, and unlining at another. 3. As it is known that in some flowers, where the law of alternation usually obtains, the organs are occasionally placed opposite each other, it is necessary for the supporters of the unlining theory to assume that in such a flower a part of the organs must be alternate and a part unlined, or at one time be all alternate and at another time be all unlined, which is entirely opposed to probability and sound philosophy. See the Camellias figured in the Elements of Botany, p. 76, fig. 156, 157, 158." — In double Camellias the numerous petals of the rosette are in some cases spirally alternate, in others placed opposite each other in five or more ranks. Now, when in the very same species two such different modes of arrangement occur, is it not *a priori* more probable that the two arrangements result from different causes and are governed by essentially different laws? — "4. The examination of the gradual development of flowers, the only irrefragable proof of the real nature of final structure, does not in any degree show that the supposed process of unlining has a real existence." Compare with this the well-stated abstract of Duchatre's memoir on the Morphology and Organogeny of Malvaceæ, which is given in the same work (Vol. 2, p. 70, *et seq.*), and which asserts that the stamens of the Malvaceous flower appear and multiply in a manner wholly conformable to the doctrine of chorisis, as here maintained, and hardly explicable upon any other theory. See, also, several diagrams of the æstivation of flowers of Malpighiaceæ, where the petals extend *within* the outer row of stamens.

921), which thus appears to be one simple organ; or to be *toothed*, *lobed*, *cleft*, or *parted*, according to the degree in which the union is incomplete; this language being employed just as in the case of the division of leaves (281). When the sepals are not united, the calyx is said to be *polysepalous;* and when the petals are distinct, the corolla is said to be *polypetalous;* that is, composed of several petals. Examples of this union of the parts of the same circle have already been shown, as respects the calyx and corolla (Fig. 287), and in the account of what is called the *monopetalous* division of the exogenous natural orders further illustrations are given, exhibiting this union in very different degrees.

462. The union of the stamens occurs in various ways. Sometimes the filaments are combined, while the anthers are distinct. When thus united by their filaments into one set, they are said to be *monadelphous;* as in the Lupine, &c. (Fig. 307). When united by their filaments into two sets, they are *diadelphous* (Fig. 308), as in most plants of the Pea tribe (Leguminosæ), where nine stamens form one set and the tenth is solitary; and in Dicentra (Fig. 296, 299), where the six stamens are equally combined in two sets. When united or arranged in three sets or parcels, they are said to be *triadelphous*, as in the common St. John's-wort (Fig. 553); or if in several, *polyadelphous;* as in other Hypericums, in Tilia, &c. In some of these instances, indeed, the stamens of each group have a common origin, as we suppose (456); still, the same terms are employed in botanical description, under whatever theoretical views. In other cases, the filaments are distinct, or nearly so, and the anthers united into a ring; as in the vast order Compositæ, or class Syngenesia of the Linnæan artificial system; when the stamens are said to be *syngenesious* (Fig. 309, 310). Again, in Lobelia, not only are the anthers syngenesious, but the filaments

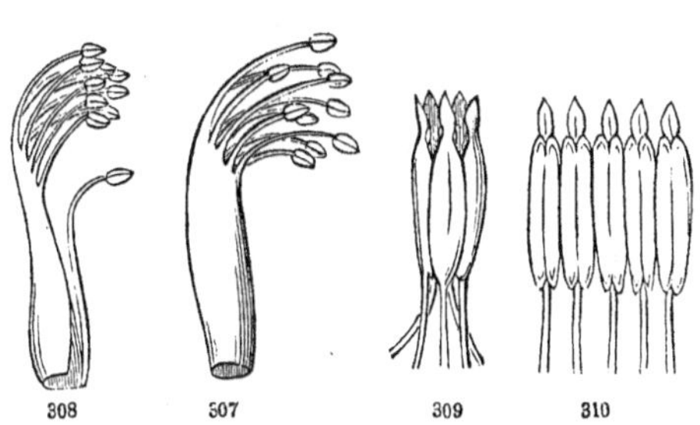

FIG. 307. Monadelphous stamens of a Lupine. 308. Diadelphous stamens (9 and 1) from a papilionaceous flower. Compare with the diagram, Fig. 282.

FIG. 309. Syngenesious stamens of a flower of a Composita. 310. The tube of anthers laid open.

are also combined in a tube for the greater part of their length (Fig. 786). The same thing is seen in the Gourd tribe, where the anthers are sometimes long and sinuous or remarkably contorted, as well as coherent into a mass (Fig. 311–313).

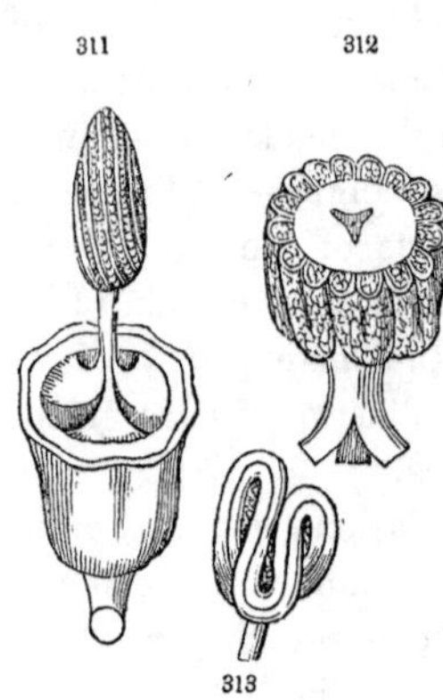

463. The union of the pistils is still more common than that of the stamens. It occurs in every degree, from the partial combination of the ovaries, as in Penthorum (Fig. 290), &c., to their complete union while the styles remain distinct, as in the St. John's-wort (Fig. 554), to the partial union of the latter, as in the Mallow, or to the perfect union of the styles also into a single body, as in Convolvulus (Fig. 922). In some cases, the styles are wholly combined, while the ovaries are only partially so; and in the Milk-weed and Dogbane (Fig. 953), the stigmas are united, while the ovaries are distinct. But the structure of the compound or *syncarpous* pistils will require particular illustration farther on. When there is no such union, but the several organs of the same circle are separate or unconnected, they are said to be *distinct*.

464. The terms *union*, *cohesion*, and the like, must not be understood to imply that the organs in question were first formed as distinct parts, and subsequently cohered. This is seldom the case. The union is congenital; the members of a gamosepalous calyx, a gamopetalous corolla, a monadelphous circle of stamens, or a compound pistil, were connate, and showed their union from the earliest period. The language we use has reference to our idea of these parts, as answering each to a single leaf. We might more correctly say that the several leaves of the same circle have failed to isolate themselves as they grew. The same remark applies to the case of

465. **Adnation**, or the union of different circles of floral organs with one another. This may take place in various degrees. It presents the appearance of one circle or set of parts growing out

FIG. 311. Column of stamens, at once triadelphous and syngenesious, of the Gourd: the floral envelopes cut away. 312. A cross-section of the united anthers, nearly the natural size. 313. A sinuous anther of the Melon.

of another, as the corolla out of the calyx, the stamens out of the corolla, or all of them out of the pistil; and therefore disguises the real origin of the floral organs from the receptacle or axis, in successive series, one within or above the other (42). In the numerous cases where the real origin, or *insertion*, of the floral organs is not obscured by these cohesions, but where they are in appearance as well as in theory inserted on the receptacle, the calyx, corolla, and stamens are said to be *hypogynous*, that is, inserted below the pistils; as in the Buttercup, the Magnolia, in Cruciferous flowers (Fig. 297), &c. The floral organs in such cases are also said to be *free;* which is the term opposed to the adhesion of one organ to another, as that of *distinct* is to the cohesion of the parts of the same whorl or set of organs. Thus, the stamens are said to be *distinct*, when not united with each other, and to be *free*, when they contract no adhesion to the petals, sepals, or pistils; and the same language is equally applied to all the floral organs. The word *connate* (born united) is applied either to the congenital union of homogeneous parts (as when we say that the two leaves of the upper pairs of the Honeysuckle are connate, the sepals or stamens are connate into a tube, or the pistils into a compound pistil), or to the coalescence of heterogeneous parts (as that of the petals with the calyx, or of both with the pistil). But the word *adnate* belongs to the latter case only.

466. When heterogeneous parts are *adnate*, that is, congenitally adherent to each other, some additional technical terms are rendered necessary. Thus two words are used as counterparts of *hypogynous* (under the pistil), and accord with different degrees of adnation, viz. *perigynous* and *epigynous*. The petals and stamens, which almost always accompany each other, are said to be *perigynous* (literally placed around the pistil) when they adhere to the base of the calyx, or in botanical language are inserted on it, either directly, or perhaps more commonly by means of a *disc* or sort of common fleshy base, from the upper surface or edge of which they grow; as in

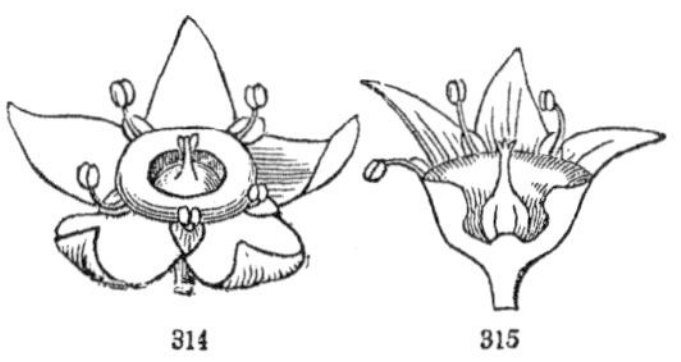

FIG. 314. A flower of Rhamnus alnifolius, showing the perigynous disc, into the margin of which the petals and stamens are inserted. 315. Vertical section through the calyx and the fleshy disc which lines it.

the Cherry, the Buckthorn (Fig. 314, 315), &c. The same term is often applied to the calyx when it is adnate to the base of the ovary, in which case it necessarily carries the petals and stamens with it. Very frequently the calyx invests and coheres with the whole surface of the ovary, so that all the parts of the flower seem to grow out of its summit; as in the Honeysuckle, the Dogwood, (Fig. 240, *a*), the Valerian, &c. The organs which thus apparently arise from the top of the ovary are said to be *epigynous* (literally on the pistil); a case of which is shown in Fig. 316. The earlier botanists called the flower, or calyx, in such cases, *superior*, and the ovary and fruit *inferior*; and when no such combination occurs, the flower, or calyx, &c. was said to be *inferior*, and the ovary *superior*. But these terms are nearly, and should be altogether, superseded by the equivalent and more appropriate expressions of *calyx adherent* in the one case, and *calyx free* in the other; or that of *ovary coherent with the calyx*, and *ovary free from the calyx*, which is the same thing in other words.

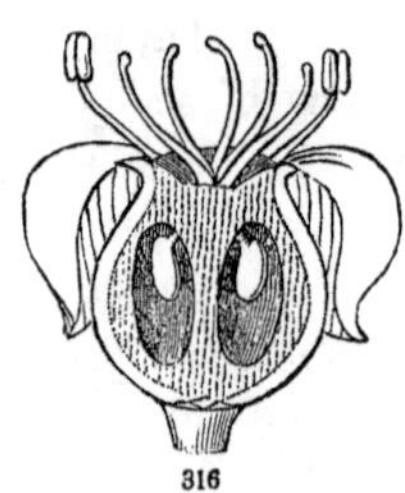

467. The various parts of the flower, thus consolidated, may separate into their integral elements at the point where they become free from the ovary, as in Cornus (Fig. 240); or else remain variously combined; the calyx being frequently prolonged into a tube with which the petals and stamens cohere, as in the Evening Primrose (Ord. Onagraceæ), where the united sepals form a long and slender tube, bearing the petals and stamens on its summit. In most cases, where the corolla is gamopetalous, the stamens continue their adhesion to it; while in the Orchis Family they are free from the corolla, but adherent to the pistil, or *gynandrous*.

468. **Irregularity**, *from unequal development or unequal union.* The Pea tribe affords a familiar illustration of *irregular* flowers arising from the unequal size and dissimilar form of the floral envelopes; especially of the corolla, which, from a fancied resemblance to a butterfly in the flower of the Pea, &c., has been called *papilionaceous*. The petals of such a corolla are distinguished by

FIG. 316. Vertical section through a flower of Aralia nudicaulis, showing the calyx adnate to the whole surface of the compound pistil, on the summit of which the petals and stamens are accordingly inserted.

separate names; the upper one, which is usually most conspicuous, being termed the *vexillum*, *standard*, or *banner* (Fig. 318, *a*); the two lateral (*b*) are called *wings* (*alæ*), and the two lower (*c*), which are usually somewhat united along their anterior edges, and

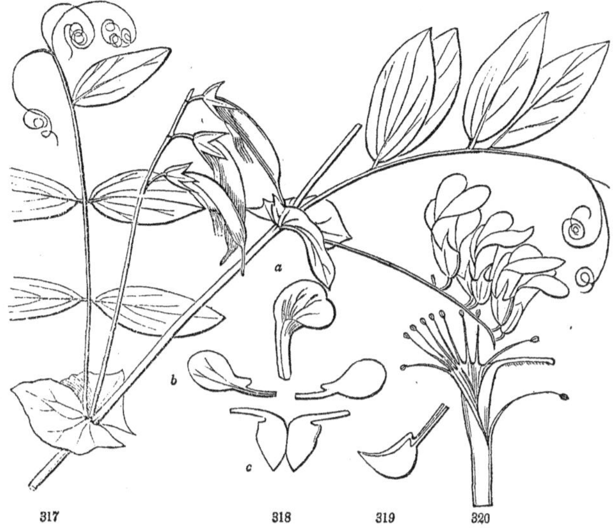

more or less boat-shaped (Fig. 319), together form the *keel* (*carina*). The sepals, which are coalescent below into a cup, are also of unequal size or somewhat unequally united. But here are all the parts of a symmetrical pentamerous calyx and corolla, only they are irregular on account of their unequal size, shape, or union. There is a tendency to become *regular*, however, in some flowers of the same tribe; this is slightly observable in Baptisia (Fig. 321), but is more manifest in Cercis (the Red-bud or Judas-tree), and most of all in Cassia; where the five petals are separate, spreading, and almost alike in size and form. The irregularity of papilionaceous flowers likewise affects the stamens, which, although of symmetrical number, viz. ten, or two circles, are in most cases unequally *diadelphous* (462), nine of them being united by the cohesion of their filaments for the greater part of their length, while the

FIG. 317. A flowering branch of Lathyrus myrtifolius. 318. The corolla displayed: *a*, the vexillum or standard; *b*, the alæ or wings; *c*, the two petals of the carina or keel. 319. The keel-petals in their natural situation. 320. The stamens and pistil, enlarged; the sheath of filaments partly turned back.

tenth (the posterior) stamen is distinct or nearly so (Fig. 320). But in Amorpha (Fig. 323, 324), which belongs to the same family, an approach to regularity is seen in this respect, the ten stamens being united barely at their base; and there is a complete return to regularity in those of Baptisia (Fig. 322), which are perfectly distinct or separate. An example of a different sort of irregular blossom is afforded by the Fumitory Family, the structure of which has already been explained, especially as to the stamens (455, Fig. 296). The floral envelopes of Dicentra are in one view regular, inasmuch as the two members of each circle are alike: but the exterior pair of petals is very unlike the interior pair; and in Corydalis and Fumaria itself one of the exterior petals is unlike the other, rendering the blossom more conspicuously and truly irregular. Here the irregularity is combined with more or less cohesion of the petals; although this union, like that of the two keel-petals of a papilionaceous flower, is not congenital, but occurs subsequently to the development of the organs.

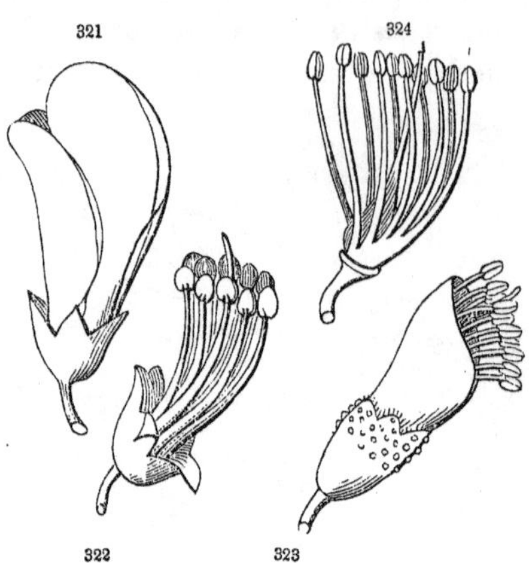

469. There are many other forms of irregular polypetalous blossoms, which we cannot here separately explain, such as that of Polygala, and that of the Larkspur and Monkshood, both of which are farther complicated by the suppression of some organs, as well as by the irregular development of others.

470. Among gamopetalous flowers the most common case of irregularity is that of what are called *bilabiate* (or two-lipped) corollas, which prevail in the Mint Family, and to some extent in several related families. Here the irregularity of form does not arise from the suppression of some of the petals, as might at first

FIG. 321. Papilionaceous flower of Baptisia. 322. The same, with the petals removed, showing the ten distinct stamens.

FIG. 323. Flower of Amorpha. 324. The same, with the solitary petal removed, showing the slightly monadelphous stamens.

sight be supposed, but from their unequal union: the upper lip being formed by the more extensive cohesion of the two upper petals with each other than with the lateral ones; which in like manner unite with the lower petal to form the lower lip (Fig. 367). But, in some such cases, the two upper petals do not cohere with each other as far as they do with the lateral ones, and, being smaller in size, the corolla has the appearance of wanting the upper lip, and shows a deep cleft in its place; as in Teucrium Canadense (Fig. 881). The flowers of Lobelia (Fig. 785) exhibit a striking instance of a similar kind; the two upper petals being united with the lateral (which are still further combined with the lower, to form the lower lip), but wholly unconnected with each other; so that the corolla appears to be split down to the base on the upper side. The ligulate or strap-shaped corollas of Compositæ are evidently formed in the same way, as if by the splitting down of a tubular corolla on one side. In the bilabiate corolla of most Honeysuckles, the upper lip consists of four united petals; the lower of only one (Fig. 743).

471. **Suppression or Abortion.** A complete flower, as already remarked (416), comprises four whorls or sets of organs; namely, calyx, corolla, stamens, and pistils: when any of these are wanting, the flower is said to be *incomplete*. Deviations resulting from the non-production of one or more of the whorls are not uncommon, and may affect any of the floral organs. The calyx, however, is never wanting when the corolla is present, or rather, when the floral envelopes consist of only one whorl of leaves, they are called *calyx*, whatever be their appearance, texture, or color. For since the calyx is frequently delicate and petal-like (in botanical language *petaloid* or *colored*), and the corolla sometimes greenish or leaf-like, the only real difference between the two is, that the calyx represents the outer, and the corolla the inner series; and even this distinction becomes more or less arbitrary when either, or both, of these organs consist of more than one circle. The apparent obliteration of the calyx in some cases is owing to the entire cohesion of the tube with the ovary, and the reduction of the free portion, or limb, to an obscure ring or border, either slightly toothed or entire, as in Aralia (Fig. 316), Fedia (Fig. 764), &c. In Compositæ, the partially obliterated limb of the calyx, when present at all, consists of scales, bristles, or a ring of slender hairs (as in the Thistle), and receives the name of *pappus*.

472. The petals, however, are frequently absent; when the flower is said to be *apetalous*, as in the Anemone (Fig. 325), Clematis, Caltha, &c., in the Crowfoot Family, other genera of which are furnished with both calyx and corolla; as in some species of Buckthorn, while others bear petals; as in our Northern Prickly Ash* (Fig. 641, 642), while the petals are present in the Southern species. They are constantly wanting in a large number of families of Exogenous plants, which on this account form the division *Apetalæ.* When the calyx is present while the corolla is wanting, the flower is said to be *monochlamydeous*, that is with a perianth (417) or floral envelope of only one kind; as in the cases above mentioned. But sometimes both the calyx and the corolla are entirely wanting, as in the Lizard's-tail (Fig. 1021), when the flowers, being destitute of floral envelopes, are termed *achlamydeous.* The essential organs (418) are nevertheless present in these cases, so that the flower is *perfect* (or *bisexual*), although *incomplete.*

473. A still further reduction, however, occurs in many plants; where even these essential organs are not both present in the same flower, but the stamens disappear in some flowers, and the pistils in others. Such flowers are said to be *diclinous*, *unisexual*, or *separated;* that which bears stamens only is termed *sterile* or *staminate*, and that provided with pistils only, *fertile*, or *pistillate.* This separation of the essential organs is very frequently met with where one or both of the floral envelopes are present, as in Menispermum (Fig. 495, 497) and Prickly Ash (Fig. 641, 642); but when these are absent, it presents instances of the greatest possible reduction of which the flower is suscepti-

* In our Northern Zanthoxylum the monochlamydeous perianth which is present may, however, be justly held to be the corolla, and not the calyx, because the five stamens alternate with it, just as they do with the undoubted petals of Z. Carolinianum: in this case, therefore, we may say that the calyx, and not the corolla, is suppressed. See *Gen. Illustr*, 2, p. 148, tab. 156.

FIG. 325. Flower of Anemone Pennsylvanica (apetalous or monochlamydeous).

ble.* An example of the kind is furnished by Ceratiola (Fig. 1036 – 1039), the sterile flowers of which consist merely of a couple of stamens situated in the axil of a bract; and the fertile, of a pistil surrounded by similar bracts. In the Willow (Fig. 326 – 329), which presents a more familiar illustration, the sterile flowers likewise consist of two or three stamens in the axil of bracts, which form a catkin (391); and the fertile, of solitary pistils also subtended by bracts, and disposed likewise in a catkin. That is, the flowers are not only wholly destitute of floral envelopes (unless a little glandular scale on the upper side should be a rudimentary perianth of a single piece), but in one set of blossoms the stamens are also suppressed, and in another, the pistils. The pistillate flowers are reduced to a single pistil. The stamens vary in number in different species, from two to five. If there were only one of the latter, an instance would be afforded of flowers reduced, not merely to one kind of organ, but to a single organ. Now there is one species of Willow, which appears to have a solitary stamen in its staminate flowers,

* Except, perhaps, in what are called *neutral flowers*, such as those which occupy the margin of the cymes of several Viburnums and Hydrangeas, or even the whole cluster in monstrous states, as in the Snowball or Guelder Rose of the gardens (Viburnum Opulus), and the cultivated Hydrangea, which consist of floral envelopes only, with sometimes mere rudiments of stamens or pistils. Of the same kind are the *neutral florets* of Compositæ, such as the marginal flowers, or *rays*, of the Sunflower.

FIG. 326. A catkin of staminate flowers of Salix alba. 327. A single staminate flower detached and enlarged (the bract turned from the eye). 328. A pistillate catkin of the same species. 329. A detached pistillate flower, magnified.

and has therefore been named Salix monandra. But on inspection this seemingly single stamen is found to consist of two united quite to the top (Fig. 330). Here, as in many other cases, the normal condition of the flower is not only much altered by the suppression of some organs, but disguised by the coalescence of those that remain. The blossoms of the Birch are very similar, except that three pistils, the sole representatives of as many flowers, are found under each bract of the fertile catkin.

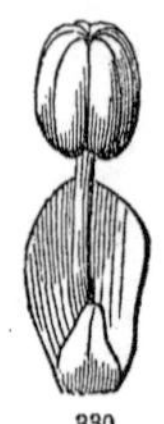

474. When the stamens and pistils are thus separated, the two kinds of blossoms may be borne, either upon different points or branches of the same individual, or upon entirely different individual plants. The flowers are said to be *monœcious* when both kinds are produced by the same individual plants; as in Indian Corn, the Birch, the Oak, Beech, Hazel, Hickory, &c.: and they are called *diœcious* when borne by different individuals; as in the Willow and Poplar, the Sassafras, the Prickly Ash, the Hemp, Hop, &c. In many cases, while some of the flowers are staminate only, and others pistillate only, a portion are perfect, the different kinds occurring either on the same or different individuals; as in most Palms, in many species of Maple, &c.: plants with such flowers are said to be *polygamous*.

475. The term *suppression* in all such cases merely denotes that the parts in question are wholly left out. It is the non-production of some organ or set of organs which forms a component part of our pattern plan of the flower, and which is realized in the complete flower. The term *abortion*, which is often used with exactly the same meaning, is more properly applied to those cases where the organ is deformed or imperfect (where a sterile filament, for example, occupies the position of a stamen), or where a mere rudiment marks the place of a non-developed organ.

476. The suppression or abortion of a whole circle of organs in a symmetrical flower does not destroy its symmetry, if we count the absent members. Thus a monochlamydeous flower, with a single full circle of stamens, usually has the latter placed opposite the leaves of the perianth, that is, of the calyx, the corolla or intervening circle, with the members of which it normally alternates, having failed to appear; as in Comandra (Fig. 1004), Chenopo-

FIG. 330. A staminate flower of Salix purpurea (or monandra), with the stamens coalescent (monadelphous and syngenesious), so as to appear like a single one.

dium, and the Elm (whenever its blossoms have only one set of stamens, Fig. 338).

477. But when, with the abortion of the primary circle, say of the stamens, we have an augmentation of one or more additional circles of the same kind of organ, the law of alternation appears to be violated; the stamens that are present, or the outer circle of them, standing opposite the petals, instead of alternate with them. It is customary to assume this explanation for all cases of the opposition of the stamens to the petals, whether in the Primrose Family, in Claytonia, in the Vine (Fig. 334) and Buckthorn (Fig. 314), or in Byttneriaceæ, &c.: but considerations which have already been adduced indicate a different explanation for many of them (459). It can no longer be deemed sufficient to assume the obliteration of a normal floral circle, and the production of another one, when no traces of the former are to be detected and no clear analogy shown with some strictly parallel instance. But we may confidently apply this view when we find traces of the obliterated or abortive organs, as in the Geranium Family, for example. The pentamerous flower of Geranium (Fig. 633) exhibits ten stamens in two rows, distinguished by their different length, the five of the exterior circle being shorter than the others. One set of these stamens alternates with the petals, the other is opposed to them; which would appear to conform to the law of alternation. But, on closer examination, we see that it is the *inner* circle of stamens that alternates with the petals; those of the *outer* circle stand directly before them. This is a not uncommon case in *diplostemonous* flowers (viz. in those which have twice as many stamens as there are petals or sepals). In this instance the explanation of the anomaly is furnished by the five little bodies, called by the vague and convenient name of *glands*, which stand on the receptacle between the petals and the stamens, and regularly alternate with the former. They accordingly occupy the exact position of the original stamineal circle: wherefore, as situation is the safest guide in determining the nature of organs, we may regard them as the abortive

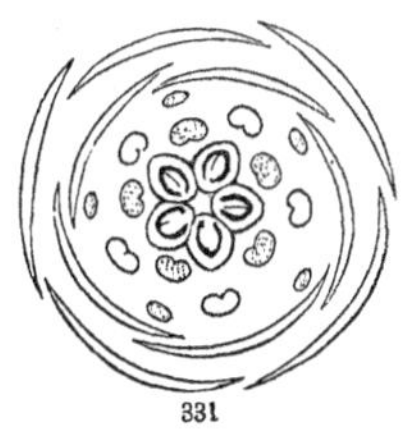

FIG. 331. Diagram (cross-section) of the flower of Geranium maculatum, exhibiting the relative position of parts, especially the glands alternate with the petals, and the two rows of stamens within them.

rudiments of the five proper stamens, which here remain undeveloped. In the annexed diagram (Fig. 331) these are accordingly laid down in the third circle, as five small oval spots, slightly shaded. The actual stamens consequently belong to two augmented circles, those of the exterior and shortor set of which (represented by the larger, unshaded figures), normally alternating with the glands, are of course opposed to the petals, and those of the inner and larger set, normally alternating with the preceding, necessarily alternate with the petals. This view is further elucidated by the closely allied genus Erodium, where all the parts are just the same, except that the five exterior actual stamens are shorter still, and are destitute of anthers; that is, the disposition to suppression which has caused the obliteration of the primary circle of stamens, and somewhat reduced the second in Geranium, has in Erodium rendered the latter abortive also, leaving those of the third row alone to fulfil their proper office. It is just the same in the Flax Family, except that the glands which answer to the primary suppressed stamens are still less conspicuous, and those of the next circle are reduced to very small abortive filaments, or to minute teeth in the ring formed by the union of all the filaments into a cup at the base, leaving five perfect stamens, which, though they alternate with the petals indeed, belong to a third circle (Fig. 332, 333). In a few species of Flax, the second circle of stamens is perfectly obliterated, so that no vestige is to be seen.

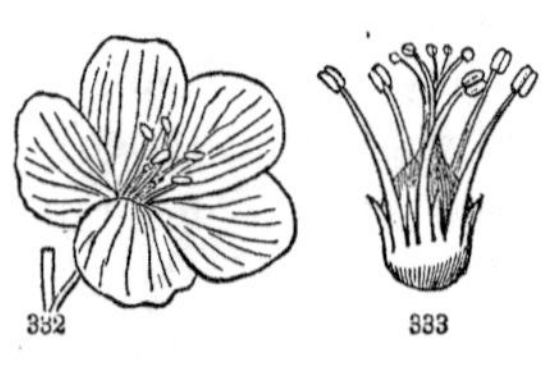

478. The case is different in the Buckthorn Family and in Byttneriaceæ, where we cannot but consider the stamens which alone appear, and stand singly before the petals (with which they are frequently connected at the base), as belonging to the coralline circle (459). Here the symmetrical alternation is interfered with first by chorisis, and then, that process having given an abnormal set of stamens, by the total suppression of the real stamineal circle, as in the Buckthorn Family, &c., or their abortion, and reduction to sterile rudiments, as in many Byttneriaceæ; while in others the genuine circle of stamens appears as an inner

FIG. 332. Flower of Linum perenne. 333. Its stamens and pistils separated: the glands are not represented: the next circle is reduced to minute sterile filaments alternating with the actual stamens.

series. In the same way we incline to explain the opposition of the stamens to the petals in the Grape-vine also (Fig. 334 – 336); inasmuch as the five glands (represented by the small shaded figures in the diagram, Fig. 336) which alternate with the petals clearly belong to a circle within the actual stamens, while there are no vestiges outside of them. The glands, therefore, would seem to represent the proper stamineal circle, in an undeveloped state, reduced to these rudiments or to a lobed disc.

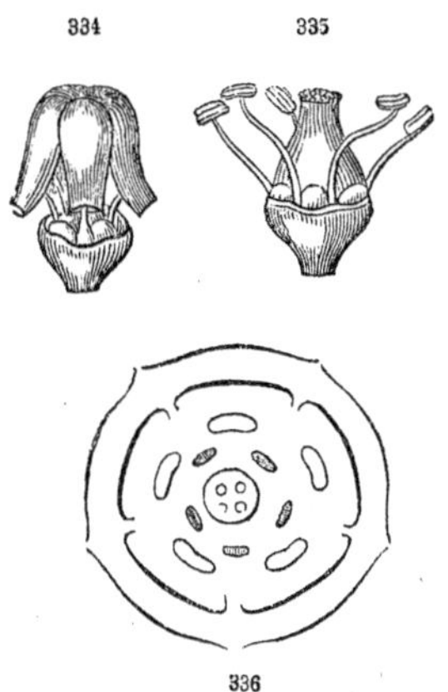

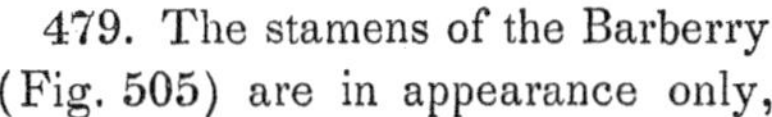

479. The stamens of the Barberry (Fig. 505) are in appearance only, but not really, opposed to the petals, and the petals to the sepals. Here the appearance is caused, not by the suppression, but by the symmetrical augmentation of the floral envelopes and of the stamens. The calyx consists of two alternating circles of sepals, three in each; the corolla of two circles of petals, three in each; the three exterior petals alternating as they should with the inner circle of sepals, and the three interior ones alternating with these. But when the flower opens, the six petals, spreading apparently as one whorl, are necessarily opposed to the six sepals; and the six stamens in two circles, which are still more confluent into one whorl, are equally opposed to these, taken six and six; but they really alternate in circles of threes. In other words, decussating verticils of threes necessarily form six vertical ranks (251, 441). It is just the same in the Lily, Crocus, and most Monocotyledonous plants; where the perianth is composed of six leaves in two circles, and the andrœcium of six stamens in two circles, giving a regular alternation in threes; although, taken as two 6-merous circles, we have a stamen before each leaf of the perianth.

480. The symmetry of the flower is more frequently and seriously obscured by the suppression of a part of the members of the

FIG. 334. Flower of the Grape, casting its petals without expanding them. 335. The same, without the petals: both show the glands distinctly, within the stamens. 336. Diagram of the flower.

same circle, than from any other kind of deviation. The tendency to such obliteration increases as we advance towards the centre of the blossom, owing, doubtless, to the greater pressure exerted on the central parts of the bud, and the progressively diminished space the organs have to occupy on the conical receptacle. So, while the corolla, when present at all, almost always consists of as many leaves as the calyx, the members of the stamineal circle or circles are frequently fewer in number (although from their form they occupy much less room than the petals), and the pistils are still more commonly fewer, excepting where the axis is prolonged for the reception of numerous spiral cycles. Thus, the pistils, which present their typical number in Sedum, and all Crassulaceous plants (Fig. 256, 277, 283 - 290), are reduced to two, or rarely three, in the allied Saxifragaceous Family, while the other floral circles are in fives. So, in Aralia (the Wild Sarsaparilla and Spikenard), the flowers are pentamerous throughout, although the ovaries of the five pistils are united into one (Fig. 316) ; but in Panax, our other genus of the same family, they are reduced to three in the Ground-nut, and to two in the Ginseng, as also in all Umbelliferous plants. Although the pistils are indefinitely augmented in the Rose, Strawberry, and the greater part of Rosaceous plants, or of the normal number five in Spiræa, yet there are only two in Agrimonia, one or rarely two in Sanguisorba, and uniformly one in the Plum and Cherry, although the flowers of the whole order are formed on the pentamerous or sometimes the tetramerous plan, with a strong tendency to augmentation of all the organs. And the Pulse Family has, almost without exception, five members in its floral envelopes, and ten, or two circles, in its stamens, but only a single pistil (Fig. 282). A flower, it may here be added, is *isomerous* (that is, of equal members) when it presents the same number in all its floral circles, — a term therefore equivalent with symmetrical, — and *anisomerous* when the number of parts is different in some of the circles.

481. As to the stamens, it may be remarked that they are usually symmetrical and regular when the floral envelopes are regular (although the common Chickweed and the Maple are exceptions to this rule) ; while they strongly tend to become unsymmetrical by abortion or *irregular* (that is, of unequal size or shape) when the calyx and corolla are irregular, or the whole is oblique in the bud ; the different stamens at the time of their development

being therefore placed in unlike conditions in such cases, so as to favor the growth of some of them, and to arrest or restrain others, either by pressure or by the abstraction of nourishment. Compare in this respect the more or less irregular corolla of Scrophulariaceous plants (Fig. 854–861) with their stamens. The Mullein (Verbascum) is one of the few genera of that family which has as many stamens as there are petals in the composition of its corolla, and sepals in its calyx: but even here they are unequal, and the posterior ones usually bear imperfect or deformed anthers. In other instances, where the five stamens are all present, indeed, the posterior one is either changed into a bearded sterile filament, as in Pentstemon and Chelone, or reduced to a mere rudiment, as in some Snapdragons; or to a deformed filament adherent to the corolla, and bearing a scale-like body in place of the anther, as in Scrophularia. The four remaining perfect stamens, in these cases, and nearly throughout the order, are unequally developed; two of them being longer than the remaining pair; as in Chelone, above cited, in Gerardia, &c.: the same thing is observed in most plants of the related orders Acanthaceæ, Bignoniaceæ, Orobanchaceæ (Fig. 850), Verbenaceæ (Fig. 863–865), and Labiatæ (Fig. 872–884). In such cases, viz. where of four two are long and two are shorter, the stamens are said to be *didynamous*. Not unfrequently, a further suppression takes place, and the two shorter of these stamens either entirely disappear; as in the Sage, Monarda, Lycopus Virginicus, &c. among Labiatæ, and Gratiola Virginica, &c. among the Scrophulariaceæ; or else are reduced to mere sterile filaments, such as those which may commonly be observed in Gratiola aurea, in the Wild Pennyroyal (Hedeoma), and in many other Labiate plants.

482. The obliteration of one or more members of the corolla follows the same laws. The loss of a petal from the circle is a case of irregularity from unequal growth carried to the greatest possible extent, or an arrest of the development of an organ from an early period, and we may sometimes trace the gradation in related plants from the diminution or dwarfing of certain organs to their total suppression. Thus, the papilionaceous corolla (468) of Erythrina herbacea has its five petals, but four of them (all except the posterior or *vexillum*) are small and inconspicuous: in Amorpha (Fig. 323), these same four disappear altogether, and the papilionaceous corolla is reduced to its vexillum alone. In some

cases, the obliteration or diminution may be attributed to local pressure or obstruction of the light, acting uniformly in all instances, from some constant cause. Thus the marginal or ray flowers of the dense head in Compositæ (as in the Aster, Sunflower, Centaurea, &c.) are not only much larger than those of the central or disc flowers, which are much pressed together, but their principal development is external. It is the same in the similar head of the Scabious; where the marginal corollas are not only the larger, but their exterior lobes or petals are much larger than the inner, which are dwarfed, as it were, by the pressure on that side. In other cases, however, we cannot give any such mechanical explanation. In our Buckeyes, for example, the whole five petals are occasionally present, as they are uniformly in the Horsechestnut (another species of the same genus): but more commonly a vacant space marks the place from which the anterior petal has disappeared (Fig. 658). There is also a suppression of two or three stamens out of the two circles of those organs.

483. A few diagrams will exhibit some of the stages of suppression, from the complete and symmetrical to the most reduced condition of the flower. The diagram, Fig. **337**, well enough exhibits the ground-plan of a 5-merous complete flower, symmetrical in all its parts, except that the pistils are reduced from five to two; as in Sullivantia (Fig. 722). Fig. **338** is a diagram of a similar flower, except that the petals are absent (the place they should occupy is denoted by the five dotted lines): this corresponds with the Elm (when pentandrous), and to Chrysosplenium, which is of the same family as Sullivantia, only that there the sepals and stamens are in fours, — one being left out, perhaps we may say, from each circle. Fig. **339** is a ground-plan of the flower of the common Claytonia, or Spring Beauty (Ord. Portulacaceæ), the

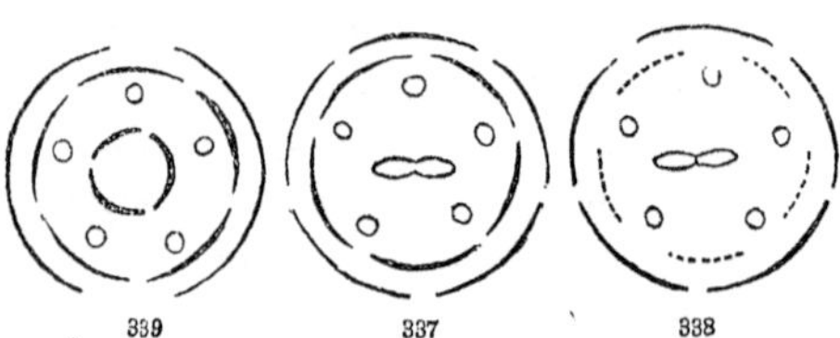

FIG. 337. Ground-plan of the flower of Sullivantia, the united pistils reduced to two.

FIG. 338. Ground-plan of a similar flower when apetalous; the five dotted lines indicating the proper position of the suppressed petals.

FIG. 339. Ground-plan of the flower of Claytonia; the outer lines representing the calyx of two sepals; the next set the corolla of five petals; next are the five stamens *before the petals;* and next the ovary, composed of three parts.

ornament of our vernal woods ; — a complete and regular, but remarkably unsymmetrical blossom, only two of the four circles having the same number of members, and one of those (the stamens) being abnormal in position. There are only two sepals: within these are five petals: within and *opposite* these are five stamens; so that the primary stamineal circle is suppressed, and those present belong to a second circle; or, which is more likely, as they cohere at the base with the claws of the petals, they may arise from a chorisis of the petals themselves: and in the centre there are three pistils with their ovaries combined into one. Further examples will illustrate those graver suppressions which render the flower incomplete, and finally reduce it to a minimum. In the Elm (Fig. 1012), the petals entirely disappear, and the pistils are reduced to two, both of which are abortive in a part of the flowers, and one always disappears in the fertile flowers during the formation of the fruit. The occurrence of numerous cases where parts that actually exist in the pistil at the time of flowering are obliterated in the fruit, justifies the use of the term *suppression* in the case of parts which, though requisite in the ideal plan, are left out in the execution. Our Prickly Ash, as already stated (472), not only wants one circle of floral envelopes altogether (which, however, appears in the species of the Southern States), but, being diœcious (474), the stamens also disappear in all the flowers of one tree, while the pistils are all abortive in those of another individual. In the Blite (Fig. 973, 974), where the plan is trimerous, the petals and two of the stamens are entirely

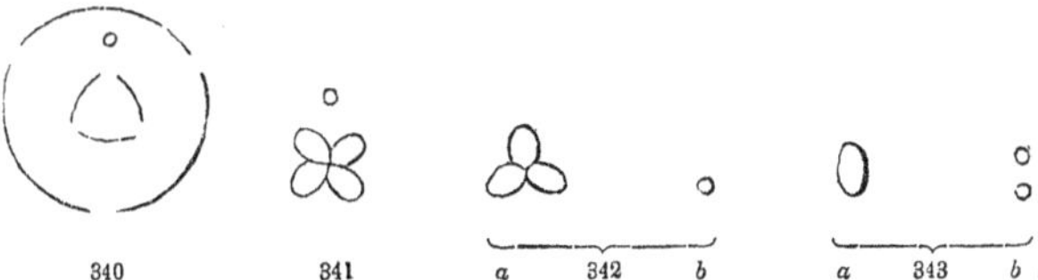

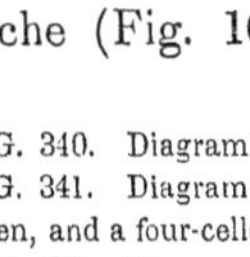

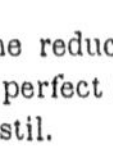

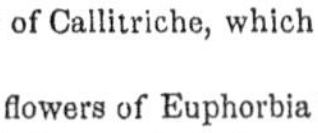

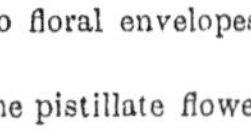

wanting; as the annexed diagram (Fig. 340) shows. In the Callitriche (Fig. 1029 – 1033), where the plan is tetramerous, the

FIG. 340. Diagram of the reduced flower of Blitum.

FIG. 341. Diagram of a perfect flower of Callitriche, which has no floral envelopes, a single stamen, and a four-celled pistil.

FIG. 342. Diagram of the monœcious flowers of Euphorbia: *a*, the pistillate flower reduced to a mere three-celled pistil; and *b*, one of the staminate flowers reduced to a single stamen.

FIG. 343. Diagram of the diœcious flowers of the Willow: *a*, one of the pistillate flowers reduced to a solitary pistil; *b*, a staminate flower reduced to a pair of stamens.

calyx and the corolla wholly disapear, as well as all the stamens but one (Fig. 341); and even this stamen is wanting in some of the flowers on the same stem, while other flowers consist of a single stamen only. This brings us to a case like that of Euphorbia (Fig. 344–348, illustrated by the diagram, Fig. 342), the greatly disguised structure of which would be certainly misapprehended, without special study. Nearly the furthest possible reduction, perhaps, is seen in the Willow (Fig. 326–329), where the staminate and pistillate flowers are distributed to different individual trees, the first reduced usually to a pair of stamens, and the second to a single pistil. The plan is represented in the diagram, Fig. 343.

484. A full illustrative series of almost all the kinds of deviation

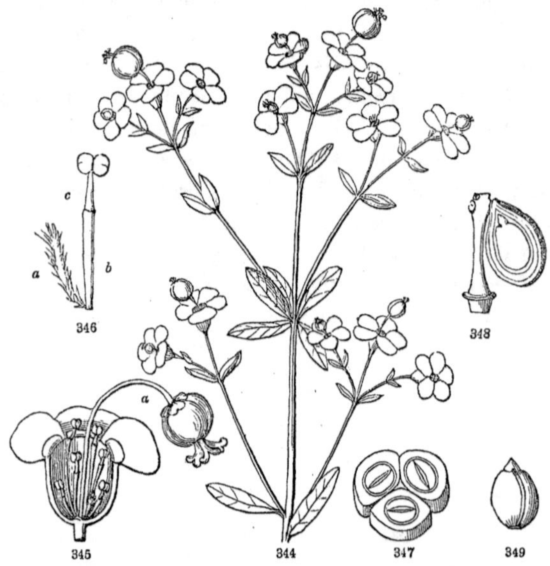

we have mentioned, but especially of simplification through suc-

FIG. 344. Flowering branch of Euphorbia corollata; the lobes of the involucre resembling a corolla. 345. Vertical section of an involucre (somewhat enlarged), showing a portion of the staminate flowers surrounding the pistillate flower (*a*), which in fruit is raised on a slender pedicel. 346. One of the staminate flowers enlarged, with its bract, *a*: *b*, the pedicel, to which the single stamen, *c*, is attached by a joint; there being no trace of floral envelopes. 347. Cross-section of the 3-pistillate fruit. 348. Vertical section of one of the pistils in fruit (the two others having fallen away from the axis), and of the contained seed; showing the embryo lengthwise. 349. A seed.

cessive suppressions, might be drawn from plants of the Euphorbiaceous Family. Among them are complete and perfect flowers, incomplete and perfect flowers, and achlamydeous and separated flowers, both monœcious and diœcious. Of these, the staminate flowers in some species are reduced to a single stamen, either sessile or on a pedicel, in the axil of a bract; and the pistillate, either to one simple pistil, or to a compound pistil formed of two or three simple ones combined. A cluster of such axillary achlamydeous flowers, each of a single stamen, collected at the base of the pedicel of a terminal achlamydeous pistillate flower of three coalescent pistils, and surrounded by an involucre, — the several leaves of which are coalescent below into a kind of cup, — forms the *inflorescence* of Euphorbia, which, until explained by Mr. Brown, was mistaken for a single anomalous blossom (Fig. 344 – 349).

485. Abortive or unusually shaped petals were called NECTARIES by the earlier botanists, whether they secreted honey or had a glandular apparatus, or not. This name was applied to the five spur-shaped petals of the Columbine (Fig. 480, 481), where the floral envelopes are symmetrical and regular, all the petals being alike, although of an extraordinary form; and also to the four reduced and deformed petals of the unsymmetrical and irregular flower of the Larkspur, where two of the petals are spur-shaped and received into the conspicuous spur of the calyx, while the other pair are of a different and more normal form. In the nearly related Aconite, where three of the five petals are obliterated, the two that remain (the *nectaries*, as they have been called) have assumed a shape so remarkable (Fig. 350), that their real nature could only be recognized by the position they occupy. Their appearance is rather that of a deformed stamen. A sterile or deformed stamen, destitute of an anther, or a body that occupies the normal place of a stamen, or is intermediate in appearance and situation between a petal and a stamen, is sometimes called a STAMINODIUM (literally a stamen-like body). Staminodia occur naturally and uniformly in many plants. In cultivated *semi-double* flowers, such transition states are extremely common, as in the Larkspurs, Columbines, &c. of the gardens.

FIG. 350. One of the two deformed, stamen-shaped petals of Aconitum uncinatum.

486. **Abnormal States of the Receptacle** of the flower remain to be mentioned, as obscuring more or less the normal condition, or as giving a singular appearance to the blossom. One of the most remarkable cases of the enlargement of the receptacle is that of the Nelumbium, where it is dilated into a large top-shaped body, nearly inclosing the pistils in separate cavities (Fig. 351). Sometimes it is hollowed out above, as well as dilated, as in the Rose, where the whole receptacle expands into an urn-shaped disc, invested by the adnate tube of the calyx, and bearing the petals and stamens on its border and the numerous pistils on the concave surface (Fig. 684). It is much the same in Calycanthus (Fig. 690 – 695). In Geranium, and many allied plants, the receptacle, which elevates the ovaries more or less, is prolonged between them, and coheres with their styles (Fig. 635). There is nearly a similar prolongation in Euphorbia (Fig. 348). Here there is some development of the axis beyond the proper insertion of the floral organs. Usually the floral internodes remain undeveloped or extremely short, like those of scaly leaf-buds (Fig. 127). But now and then some of them are elongated; as in the Pink and Silene, where the internode between the calyx and the corolla forms a conspicuous stalk, elevating the other parts of the flower in the tube of the calyx; while in many Gentians (Fig. 947) the internode above the circle of stamens is developed, raising the pod on a stalk of its own. This is a common case in the Caper Family; in which the genus Gynandropsis (Fig. 352) exhibits a remarkable development of the whole receptacle. It is enlarged into a flattened disc where it bears the petals, and is then prolonged into a conspicuous stalk which bears the stamens (or rather, perhaps,

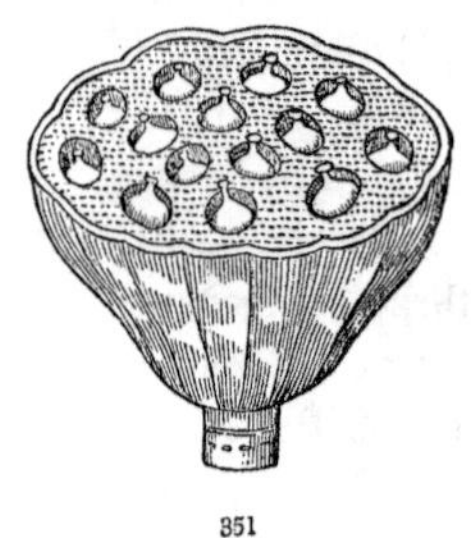

FIG. 351. The enlarged receptacle of Nelumbium.

FIG. 352. Flower of Gynandropsis, showing an elongated receptacle, which separates the different sets of organs.

to which the bases of the stamens are adnate), and then into a shorter and more slender stalk for the pistil; thus separating the four circles or sets of organs, like so many whorls of verticillate leaves.

487. The common name for this kind of stalk, as contradistinguished from the pedicel or stalk of the flower, is the STIPE; and whatever organ or set of organs is thus elevated is said to be *stipitate.* To particularize the portion of the receptacle which is thus developed, the stipe is termed the *Anthophore* when it appears just above the calyx, and elevates the petals, stamens, and pistils; the *Gonophore*, when it supports both the stamens and pistils; and the *Gynophore*, *Gynobase*, or *Carpophore*, when it bears the gynæcium alone. The stalk which sometimes raises each simple pistil of the gynæcium (as in Coptis or the Goldthread) is called a *Thecaphore.* This, however, does not belong to the receptacle at all, but is homologous with the leaf-stalk.*

SECT. V. THE FLORAL ENVELOPES IN PARTICULAR.

488. ALTHOUGH the various organs of the flower have already been connectedly considered under most of their relations, there yet remain some particular points in respect to each of them which require to be separately noticed. It will still be most convenient to treat of the calyx and corolla together, on account of their general accordance in most respects.

489. **Their Development, or Organogeny,** first requires a brief notice. The flower-bud is formed in the same way as the leaf-bud; and what has been stated as to the formation of the leaves of the branch (274) equally applies to the leaves, or envelopes, of the flower. The sepals are necessarily the earliest to appear, which they do in the form of so many cellular tumors or nipples, at first distinct, inasmuch as then their tips only are eliminated from the axis. Each one may complete its development separately, in the

* A few terms which relate to the combination of different kinds of flowers in the same inflorescence, or their corresponding separation, may here be defined. Thus, a head or spike of flowers is said to be *homogamous* when all its blossoms are alike, as in Eupatorium; or *heterogamous* when it includes two or more kinds, as in the Sunflower and Aster. It is *androgynous* when it consists of both staminate and pistillate flowers, as the spikes of many Sedges. When the two kinds of flowers occupy different heads, whether on the same or two different individuals, they are *heterocephalous.*

same manner as an ordinary leaf, (only no petiole is interposed between the blade and the axis,*) when the sepals remain *distinct* (463) or unconnected. Otherwise, the lower and later-eliminated portions of the nascent organs of the circle coalesce as they grow into a ring, which, further developed in union, forms the *cup* or *tube* of the *gamophyllous calyx:* or, in some cases, it would appear that the sepals may at first grow separately, and afterwards, though only at a very early period, coalesce by the cohesion of their contiguous parts. The several parts of an irregular calyx are at first equal and similar; the irregularity is established in their subsequent unequal growth. The petals or parts of the corolla originate in the same way, a little later than the sepals. Their coalescence in the gamopetalous corolla, as far as known, is strictly congenital; the ring which forms its tube appearing nearly as early as the slight projections which become its lobes and answer to the summits of the component petals. The rudiments of the petals are visible earlier than those of the stamens: † but their growth is at first retarded, so that the stamens are earlier completed, and their anthers surpass them, or often finish their growth, while the petals are still minute scales: at length they make a rapid growth, and inclose the organs that belong above or within them. Unlike the sepals in this respect, the base of the petal is frequently narrowed into a portion which corresponds, more or less evidently, to the petiole (the *claw*), which, like the petiole, does not appear until some time after the blade or expanded part; the summit being always the earliest and the base the latest portion formed. As the envelopes of the flower grow and expand, those of each circle adapt themselves to each other in various ways, and acquire the relative positions which they occupy in the flower-bud. Their arrangement in this state is termed

490. Their Æstivation or Præfloration. The latter would be the preferable term; but the former is in common use; the word *Æstivation* (literally the summer state) having been formed for the

---

* At least the case of a petiolate sepal is very rare. The sepals are rather to be compared to bracts, which are mostly sessile, than to ordinary leaves.

† When the stamens, or an exterior set of them, originate by chorisis or deduplication of the petals (459), it appears from the observations of Duchatre that the five protuberances which represent the petals at their first appearance divide transversely, or grow double, the inner half developing into a stamen or a cluster of stamens, the outer into the petal itself.

purpose by Linnæus; — for no obvious reason except that he had already applied the name of *Vernation* (the spring state) to express the analogous manner in which leaves are disposed in the leaf-bud. The same terms are employed, and in nearly the same way, in the two cases, but with some peculiarities. As to the disposition of each leaf taken by itself, the corresponding terms of vernation (257) wholly apply to æstivation; and there are no forms of any consequence to be added, perhaps, except the *corrugate* or *crumpled*, where each leaf is irregularly crumpled or wrinkled, longitudinally or transversely, one or both, as happens in the petals of the Poppy and the Helianthemum, — a case that is not met with in the foliage; the *induplicate*, where the edges are folded inwards, as those of the sepals of Clematis (Fig. 357), — but this, as compared with vernation, is only a modification of the involute; and the *reduplicate*, where the margins are bent outwards instead of inwards, as in the corolla of the Potato, — which is a mere modification of the revolute in vernation.

491. The arrangement in the bud of the several members of the same floral circle in respect to each other is of much importance in systematic botany, on account of the nearly constant characters that it furnishes, and still more in structural botany, from the aid it often affords in determining the true relative superposition or succession of parts on the axis of the flower, by observing the order in which they overlie or envelope each other; for every enveloping part is almost necessarily external to, or of lower insertion than, the part enveloped. The various forms of æstivation that have been distinguished by botanists may be reduced to three essential kinds, namely, the *imbricative*, the *contorted* or *convolutive*, and the *valvular*.*

492. *Imbricative* æstivation, in a general sense, comprises all the modes of disposition in which some members of a floral circle are exterior to the others, and therefore overlie or inclose them in the bud. This must almost necessarily occur wherever the parts are inserted at distinguishably different heights, and is the natural result of a spiral arrangement. The name is most significant

* We should properly say of the *æstivation* that it is *imbricative*, *convolutive* *valvular*, &c., and of the calyx and corolla, or of the sepals, &c., that they are *imbricate* or *imbricated*, *convolute*, *valvate*, &c. in æstivation; but such precision of language is seldom attended to.

when successive leaves are only partially covered by the preceding, as in Fig. 174 – 176; here they manifestly *break joints*, or are disposed like tiles or shingles on a roof, as the term *imbricated* denotes. It is therefore equivalent to the *spiral* arrangement, which word is sometimes substituted for it in æstivation: and, on the other hand, we properly apply the term *imbricated* to any continuous succession of such partly overlying members, as when we say of oppressed and crowded leaves that they are imbricated on the stem, or thus express the whole arrangement of the scales of a bud (Fig. 127), or a bulb (Fig. 141), or of a catkin or cone (Fig. 175). The alternation of the petals with the sepals, &c. necessarily makes the floral envelopes likewise imbricated in the bud, taken as a whole. But in proper æstivation, what we have to designate is the arrangement of the parts of the same floral circle, say the five sepals or the five petals, in respect to each other.

493. Now where the calyx or the corolla exhibits the character of a complete cycle (439) or of a part of a cycle (442) of leaves with the internodes undeveloped, that is, where we may perceive on close inspection that the several members are inserted on the receptacle at unequal heights, this will be manifested in the bud by the relative position of these members: the lower or outer must overlie or inclose the upper or inner. This is just the case in regular imbricative æstivation; where, of five sepals, for example (as in the diagrams, Fig. 300, 281), two will be wholly exterior in the bud, two wholly interior, and one intermediate, namely, covered at one edge by one of the exterior, while its other edge overlies that of one of the inner sepals; — which, on comparison with Fig. 172, 173, will be found to correspond exactly with the $\frac{2}{5}$ or quincuncial arrangement of leaves as presented on a similar ground-plan. Leaves No. 1 and No. 2 are external; No. 3 is internal in respect to these, but external in respect to No. 4, which is two fifths of the circumference distant, and more manifestly to No. 5, which, being separated by an interval of two fifths from the preceding, completes the cycle, and is overlapped by No. 3. In this, the normal and the most common arrangement in the 5-merous flower, the parts are said to be *spirally*, or (with more definiteness as to the numerical kind of spire) *quincuncially imbricated* in æstivation.

494. We have here the advantage of being able to number the successive sepals, or petals, since the third leaf is not only recognizable by its intermediate position, but also indicates the direction in which the spiral turns, as is shown in Fig. 173.

495. The same regularly imbricated arrangement in trimerous flowers gives one exterior, one half interior and half exterior, and one interior member in æstivation, after the order of $\frac{1}{3}$ cycles, as is shown in the diagram, Fig. 353, both for the calyx and corolla; — which compare with Fig. 171, recollecting that the successive cycles are superposed in the foliage, while the floral circles alternate. Regular imbrication in the 4-merous flower gives two outer and two inner members in æstivation (as in the calyx of Cruciferous blossoms, Fig. 280), on the principle of two decussating pairs of leaves (439); or it may sometimes be referable to a modification of some alternate spiral arrangement.

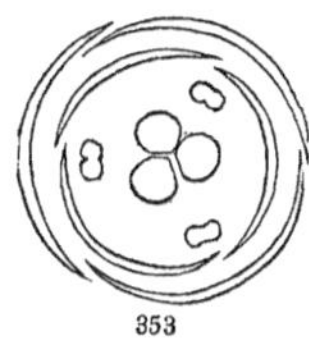

496. The degree of overlapping depends upon the breadth of the parts and the state of the bud; it naturally grows less and less as the bud expands and is ready to open. It is from the full-grown flower-bud, just before *anthesis* (or the opening of the blossom), that our diagrams are usually taken; in which the parts are represented as moderately or slightly overlapping. The same overlapping carried to a greater extent will cause the outer leaf to envelope all the rest, and each succeeding one to envelope those within; as shown in Fig. 354 from one circle of petals of a Magnolia taken in an early state of the bud. Here the mode is just the same as that of Fig. 353. To this, however, has not improperly been applied the name of *convolute*, from its similarity to the convolute vernation of the leaves of the branch (257), similarly rolled up one within the other. But it is practically inconvenient, and wrong in principle, to designate different degrees of the very same mode by distinct names; furthermore, the next general kind of æstivation, when carried to a high degree of overlapping, produces a somewhat similar result; and moreover, it is to this second mode, whatever be its degree, that the name of *convolute* is more commonly applied, in recent systematic botanical writings.

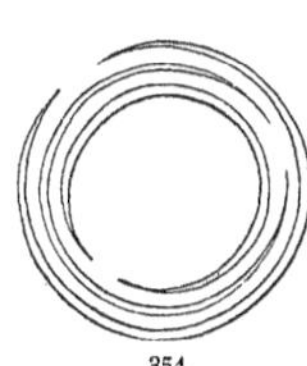

497. There are numerous cases of imbricative æstivation, espe-

FIG. 353. Imbricated æstivation of the calyx and corolla, in a trimerous flower.

FIG. 354. The strongly enveloping imbricative æstivation of the three exterior petals of Magnolia grandiflora.

cially in irregular flowers, where the overlapping of parts does not altogether accord with what must needs be their order of succession on the axis. In the 5-merous calyx and corolla of all truly papilionaceous flowers, for example, one edge of the sepal or the petal No. 2 is placed under, instead of over, the adjacent edge of No. 4, in consequence of which three, instead of only one, of the leaves have one edge covered and the other external; as is shown in Fig. 282. Since, in the corolla of this kind of blossom, the exterior petal, here the vexillum (468), is the larger, and at first embraces all the rest (as is seen in the separate diagram of the corolla, Fig. 359), this modification of imbricative æstivation has received the name of *vexillary*. As nearly the same thing occurs in the Violet, it is probably caused by some slight dislocation that takes place during the early growth of organs in the irregular blossom, which the study of their development should explain. It is not restricted to irregular flowers, however, but occurs as a casual variation, or perhaps more frequently than the quincuncial, in the regular corolla of the Linden (as is shown in Fig. 306). A slight obliquity in the position of the petal No. 2, assumed at an early period, would account for the whole anomaly. That this suggests the true explanation is almost demonstrated by the varying æstivation of the corolla of the Linden; in which the same bunch of blossoms often furnishes instances of regular quincuncial imbrication, of the modification here referred to, and of the similar disposition of the fifth petal, throwing one of its edges outwards also. If the first petal were also to partake of this slight obliquity, the imbricative would be completely converted into what is variously named

498. The *contorted*, *twisted*, or *convolutive* æstivation (Fig. 360, the corolla, and 361). In this mode, the leaves of the circle are all, at least apparently, inserted at the same height, and all occupy the same relative position: one edge of each, being directed obliquely inwards, is covered with the adjacent leaf on that side, while the other covers the corresponding margin of the contiguous leaf on the other side. This is owing to a more or less evident torsion or twisting of each member on its axis early in its development; so that the leaves of the floral verticil, instead of forming arcs of a circle, or sides of a polygon having for its centre that of the blossom, severally assume an oblique direction, by which one edge is carried partly inward and the other outward. This con-

torted æstivation is scarcely ever met with in the calyx, but is very common in the coralla. When this obliquity of position is strong, the petals themselves are usually oblique, or unequal-sided, from the lesser growth of the overlapped side, which is by no means so favorably situated in this respect as is the free external portion, — a case of partial obliteration or dwarfing from pressure. This is well seen in the petals of most Malvaceous plants, to some extent in those of Geranium, Flax, and Wood-Sorrel, and strikingly in those of the St. John's-wort, and in the lobes of the corolla of the Periwinkle (Vinca) and of most other Apocynaceous plants. In the Pink, however, and in many other instances, the petals are symmetrical, although strongly convolute in æstivation. When the petals are broad, this arrangement is frequently conspicuous in the fully expanded flower, as well as in the bud (as in Fig. 365). The convolution in the bud is often so great, that the petals appear as if strongly twisted or rolled up together, each being almost completely overlapped by the preceding, so that they become *convolute* nearly in the sense in which the term is used in vernation; as in the Wallflower (Fig. 360, 361). Although there is some diversity of usage, the terms *convolute* and *contorted* in æstivation are

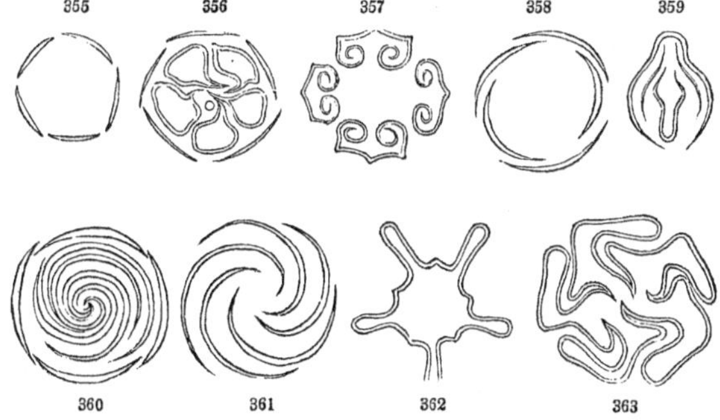

now for the most part employed interchangeably, or nearly so. In

FIG. 355-363. Diagrams of æstivation. (When there are two circles, the outer represents the calyx and the inner the corolla.) 355. Valvate. 356. Valvate calyx; the corolla induplicate or nearly conduplicate. 357. Involute, rather than induplicate, sepals of Clematis. 358. Quincuncially imbricated; the first leaf on the upper side. 359. Vexillary imbricated papilionaceous corolla. 360. Imbricated calyx of Wallflower (two outer and two inner sepals), and within the strongly contorted or convolute corolla. 361. Contorted or convolute corolla, with the petals more expanded. 362. Plaited tube of the corolla of Campanula. 363. Plaited and supervolute corolla of Convolvulus.

Geranium, and in many other cases (as in Fig. 280), we find the prevailingly contorted or convolute æstivation affecting casual transitions to the imbricative mode, corresponding to those already mentioned in the foregoing paragraph.

499. The *valvular* or *valvate* æstivation is that in which the parts of a floral verticil are placed in contact, edge to edge, throughout their whole length, without any overlapping (as in Fig. 355, and the calyx in Fig. 356). Here the members of the circle are strictly verticillate, and stand in an exact circle, no one being in the least degree lower or exterior. The edges of the sepals or petals in this case are generally abrupt, or as thick as the rest of the organ, as is shown in the calyx of the Linden (Fig. 306); by which mark the valvate æstivation may commonly be recognized in the expanded flower. The several parts being all developed under precisely similar conditions in this and the foregoing modes of æstivation, these are naturally and almost without exception restricted to regular flowers alone.

500. By the inflexion of the edges, the strictly valvate æstivation passes by insensible gradations into the *induplicate* (490), as in the calyx of some species of Clematis; a mode which is carried to a maximum in some species of Lysimachia (Fig. 356), where the two edges of the same petal are brought into contact, so as to be *conduplicate*. When the induplicate margins are inrolled, they become *involute* (Fig. 357) in æstivation. On the contrary, the valvate calyx of many Malvaceous plants and the corolla of the Potato blossom have the margins projecting outwards into salient ridges, or are *reduplicate*, in æstivation.

501. The tube of a gamopetalous corolla occasionally exhibits similar ridges or folds, whether salient (as in the bud of some Campanulas, Fig. 362), or reëntering (as in Stramonium): this gives rise to the *plicative*, *plicate*, or *plaited* modification of æstivation. Where the plaits are folded round each other, in a convolutive manner, the æstivation is sometimes termed *supervolutive*, or *supervolute*, as in the Morning-Glory (Fig. 363).

502. The spire in imbricative æstivation, and the order of overlapping in the contorted mode, may turn either from left to right, or from right to left; and the direction is often uniform through the same genus or family, but sometimes diverse in different blossoms on the same plant. In fixing the direction, we suppose the observer to stand before the flower-bud. De Candolle,

indeed, supposes the observer to occupy the centre of the flower, which would reverse the direction; but the former view is generally adopted. The direction is frequently reversed in passing from the calyx to the corolla, — sometimes with remarkable uniformity; while again the two occur almost indifferently in many cases.

503. The *kind* of æstivation, although often the same both in the calyx and corolla, as in Parnassia (Fig. 304) and Elodea (Fig. 300), where both are quincuncially imbricated, is as frequently different; and the difference is often characteristic of families or genera. Thus, the calyx is valvate and the corolla convolute in all Malvaceæ; the calyx imbricated and the corolla convolute in Hypericum, in the proper Pink tribe, &c. Solitary exceptions now and then occur in a family. Thus, the corolla in Rosaceæ is imbricated, so far as known, except in Gillenia, where it is convolute. In general it may be said, that the æstivation of the corolla is more disposed to vary than that of the calyx.

504. **The Calyx.** In treating of the general structure and diversities of the flower, we have already noticed the principal modifications of the calyx and corolla, as well as the terms employed to designate them; which need not be here repeated.

505. The number of sepals that enter into the composition of a calyx is indicated by adjectives formed from the corresponding Greek numerals prefixed to the name; as *disepalous* for a calyx of two sepals; *trisepalous*, of three sepals; *tetrasepalous*, of four; *pentasepalous*, of five; *hexasepalous*, of six sepals; and so on. Very commonly, however, the Greek word for leaves, *phylla*, is used in such composition; and the calyx is said to be *diphyllous*, *triphyllous*, *tetraphyllous*, *pentaphyllous*, *hexaphyllous*, &c., according as it is composed of 2, 3, 4, 5, or 6 leaves or sepals respectively. These terms imply that the leaves of the calyx are distinct, or nearly so. When they are united into a cup or tube, the calyx was by the earlier botanists incorrectly said to be *monophyllous* (literally one-leaved); — a term which we continue to use, guarding, however, against the erroneous idea which its etymology involves, and bearing in mind that the older technical language in botany expresses external appearance, rather than the real structure, as we now understand it. The correct term, *calyx gamophyllous*, is now coming into general use; this literally expresses the true state of the case, and is equivalent to the phrase *sepals united:* the degree of coalescence being indicated by adding "at the base," "to the middle," or "to the summit," as the case may be.

506. Still, in botanical descriptions, it is ordinarily more convenient and usual to regard the calyx as a whole, and to express the degree of union or separation by the same terms as those which designate the degree of division of the blade of a leaf (281–283): as, for example, Calyx *five-toothed*, when the sepals of a pentaphyllous calyx are united almost to the top; *five-cleft*, when united to about the middle; *five-parted*, when they are separate almost to the base; and *five-lobed*, for any degree of division less than five-parted, without reference to its particular extent. The united portion of a gamophyllous calyx is called its *tube;* the distinct portions of the sepals are termed the *teeth*, *segments*, or *lobes*, according to their length as compared with the tube; and the orifice or summit of the tube is named the *throat.* The calyx is said to be *entire* (281), when the leaves of the calyx are so completely confluent that the margin is continuous and even. The terms *regular* and *irregular* (446, 468) are applied to the calyx or corolla separately, as well as to the whole flower. The counterpart to calyx monophyllous or monosepalous in the current glossology is *polyphyllous* or *polysepalous* (viz. of many leaves or sepals). This is equivalent to the phrase *sepals distinct;* and does not mean that they are unusually numerous, or of more than one circle.

507. **The Corolla** has corresponding terms applied to its modifications. When its petals are distinct or unconnected, it is said to be *polypetalous;* when united, at least at the base, *monopetalous*, or more properly *gamopetalous*, as already explained (461). The united portions in the latter case form the *tube* of the corolla, and the distinct parts, the *lobes*, *segments*, &c.; and the orifice is called the *throat*, just as in the calyx. The number of parts that compose the corolla is designated in the manner already mentioned for the calyx; — viz., a corolla of two petals is *dipetalous;* of three, *tripetalous;* of four, *tetrapetalous;* of five, *pentapetalous;* of six, *hexapetalous;* of seven, *heptapetalous;* of eight, *octopetalous*; of nine, *enneapetalous;* of ten, *decapetalous.*

508. Frequently the petals, and rarely the sepals, taper into a stalk or narrow base, analogous to the petiole of a leaf, which is called the *claw* (*unguis*); and hence the petal is said to be *unguiculate* (as in Cruciferous flowers, the Pink, Fig. 302, and Gynandropsis, Fig. 352, &c.); the expanded portion, like that of the leaf, being distinguished by the name of the *lamina*, *limb*, or *blade.*

509. Some kinds of polypetalous flowers receive particular names, from the form or arrangement of their floral envelopes, especially of the corolla. Among the *regular* forms (295) we may mention the *rosaceous* flower, like that of the Rose, Apple, &c., where the spreading petals have no claws, or very short ones; the *liliaceous*, of which the Lily is the type, where the claws or base of the petals or sepals are erect, and gradually spread towards their summits; the *caryophyllaceous*, as in the Pink and Silene, where the five petals have long and narrow claws, which are inclosed in the tube of the calyx; and the *cruciate*, or *cruciform*, which gives name to the Mustard Family (Fig. 525), where the four unguiculate petals, diverging equally from one another, are necessarily disposed in the form of a cross, as in the Mustard, &c. Among the *irregular* polypetalous flowers, which are greatly varied in different families, the *papilionaceous* or *butterfly-shaped* corolla of the Pea tribe has already been described (468).

510. Several forms of the gamopetalous corolla, or gamophyllous calyx, have been distinguished by particular names. These are likewise divided into the *regular*, where their parts are equal in size, or equally united; and the *irregular*, where their size or degree of union is unequal (468). Among the former are the *campanulate* or *bell-shaped*, as the corolla of the Harebell (Fig. 364), which enlarges gradually and regularly from the base to the

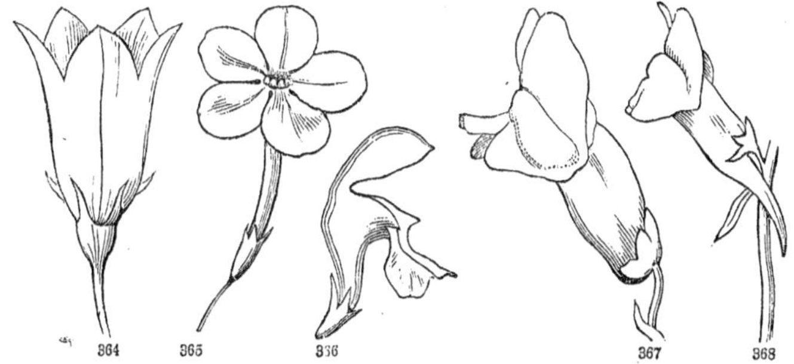

summit; the *infundibuliform*, or *funnel-shaped*, where the tube enlarges very gradually below, but expands widely at the summit,

FIG. 364. Campanulate corolla of Campanula rotundifolia. 365. Salver-shaped corolla of Phlox. 366. Labiate (ringent) corolla of Lamium; a side view. 367. Personate corolla of Antirrhinum. 368. Personate corolla of Linaria, spurred at the base.

as in the corolla of Morning-Glory (Fig. 921) and the Tobacco (Fig. 935); *tubular*, where the form is cylindrical throughout: *hypocrateriform*, or *salver-shaped*, where the limb spreads at right angles with the summit of the more or less elongated tube, as in the corolla of Primula and of Phlox (Fig. 365); and *rotate*, or *wheel-shaped*, when a hypocrateriform corolla has a very short tube, as in the Forget-me-not (Fig. 887) and Bittersweet (Fig. 939).

511. The principal irregular gamopetalous or gamophyllous form that has received a separate appellation is the *labiate* or *bi-labiate*, which is produced by the unequal union of the sepals or petals (470), so as to form an upper and a lower part, or two lips, as they are called, from an obvious resemblance to the open mouth of an animal (Fig. 366). This variety is almost universally exhibited by the corolla of Labiatæ, and very frequently by the calyx also, as in the Sage (Ord. Labiatæ): it likewise occurs in the corolla of most Honeysuckles (Fig. 742, 743), and in the calyx of many papilionaceous flowers. When the upper lip is arched, as in the corolla of Lamium (Fig. 366), it is sometimes called the *galea*, or *helmet*. When the two lips are thus gaping and the throat open, the corolla is said to be *ringent*. But when the mouth is closed by the approximation of the two lips, and especially by an elevated portion or protuberance of the lower, called the *palate*, as in the Snapdragon (Fig. 367) and Toadflax (Fig. 368), the corolla is said to be *personate*, or *masked*.

512. In the Snapdragon, the base of the corolla is somewhat protuberant, or *saccate*, on the anterior side (Fig. 367): in the Toadflax (Fig. 368) the protuberance is extended into a hollow *spur*. A projection of this kind is not uncommon, in various families of plants. One petal of the Violet is thus *spurred* or *calcarate;* so is one of the outer petals in the Fumitory, and each of them in Dicentra (Fig. 295). So, also, one of the sepals is spurred or strongly sac-shaped in the Jewel-weed (Impatiens), the Nasturtium, and the Larkspur; and all five petals take this shape in the Columbine. A monster of the Toadflax is occasionally found, in which the four remaining petals, of the five which enter into its composition, affect the same irregularity, and so bring back the flower to a singular abnormal state of *regularity*. This was called by Linnæus *Peloria;* a name which is now used to designate the same sort of monstrosity in different flowers.

513. The petals are sometimes furnished with appendages on

their inner surface, such as the *crown* at the summit of the claw in Silene (Fig. 302), and the scales similarly situated on the gamopetalous corolla of Myosotis and Symphytum (Fig. 888, 893). The nature of this crown has already been explained (458). Such appendages are sometimes thought to represent an adherent row of abortive stamens or petals.

514. The bodies termed *nectaries* (485) by the old botanists are either petals of unusual form, such as the spurs of the Columbine; or petals passing into stamens, such as the fringe of the Passion-flower; or a deduplication of the petal, as in Parnassia (Fig. 305); or else abortive and transformed stamens, as in Canna. The so-called nectary of Orchidaceous plants is merely a petal, which, being of a different shape from the others, is termed the *labellum*.

515. The duration of the floral envelopes varies greatly in different plants. Sometimes they fall off as the flower opens, or even before expansion, as the calyx of the Poppy and the corolla of the Grape-vine (Fig. 334); when they are said to be *caducous*. More commonly they are *deciduous*, or fall after anthesis but before the fruit forms. When they remain until the fruit is formed or matured, they are *persistent*, which is often the case with the calyx, especially when it has a green color and foliaceous texture. It is occasionally *accrescent*, or takes a further growth during fructification, as in Physalis. When the envelopes persist in a dry or withering state, as the corolla of Heaths, of Campanula, &c., they are said to be *marcescent*.

516. Besides serving as organs of protection, the sepals, when green, assimilate sap, and act upon the air like ordinary foliage (344, 346). The petals, like other uncolored (that is greenless) parts, do not evolve oxygen, but abstract it from the air, and give off carbonic acid; in other words, they decompose assimilated matter, — a process which appears to be needful in flowering, and to subserve some important end at the time (367 – 373). The tissue of a petal is much the same as that of a leaf, except that it is much more delicate, and the fibro-vascular system is reduced to slender bundles of a few spiral vessels, &c., which form its veins.

SECT. VI. THE STAMENS.

517. **The Stamens**, collectively forming the ANDRŒCIUM (418), have been already considered in respect to their component parts,

their nature and symmetry, and their principal modifications as to relative number and disposition. Their absolute number in the flower, it may be remarked, is designated by Greek numerals prefixed to the word used for stamens, as employed by Linnæus in the names of his artificial classes. Thus, a flower with one stamen is said to be *monandrous ;* with two, *diandrous ;* with three, *triandrous ;* with four, *tetrandrous ;* with five, *pentandrous ;* with six, *hexandrous ;* with seven, *heptandrous ;* with eight, *octandrous ;* with nine, *enneandrous ;* with ten, *decandrous ;* with twelve, *dodecandrous ;* and with a greater or indefinite number, *polyandrous.* (See the account of the classes of the Linnæan Artificial System, Part II. Chap. II.)

518. The terms employed to designate their various modifications, most of which have already been incidentally noticed, are likewise derived from the names of Linnæan artificial classes, with the exception of those which relate to their insertion ; such as *hypogynous*, when inserted on the receptacle (466), or, in other words, free from all adhesion to neighboring organs ; *perigynous*, when adherent to the tube of the calyx (as in Fig. 315) ; and *epigynous*, when adherent also to the ovary, and, as it were, raised to its summit (as in Fig. 316). To these may be added the Linnæan term *gynandrous*, expressive of their further cohesion with the style, as in the Orchis Family.

519. As to mutual cohesion, they are *monadelphous* when united by their filaments into one body (as in Fig. 307) ; *diadelphous*, when thus combined in two sets (as in Fig. 308) ; *triadelphous*, when in three sets, as in Hypericum and Elodea (Fig. 300, 301) ; *pentadelphous*, when in five sets, as in our Linden ; and *polyadelphous*, when in several sets, irrespective of the particular number. They are *syngenesious*, when united by their anthers (Fig. 309, 310). As respects inequality of size, they are *didynamous*, when four stamens constitute two pairs of unequal length (481) ; and *tetradynamous*, when six stamens only are present, two of which are shorter than the others, as in Cruciferous flowers (455) ; a case which is sometimes, but less distinctly, seen in the allied Caper Family (Fig. 352). Their complete suppression in some flowers gives rise to such terms as *monœcious*, *diœcious*, and *polygamous*, which have already been defined (473).

520. The proportion of the stamens to the corolla or other floral envelopes is sometimes to be noticed. When they are longer and

protruding, they are said to be *exserted;* when shorter or concealed within, they are *included;* — terms which apply to other organs as well. So of terms which indicate their direction; as *declined*, when curved towards one side of the blossom, as in the Horsechestnut.

521. The stamens are mostly too narrow to furnish any characters of æstivation, except as to the manner in which each one is separately disposed. In this respect they exhibit several varieties, to which the same terms are applied as to the vernation of individual leaves (257).

522. When the stamen is destitute of the filament, or stalk (Fig. 369, *a*), the anther (*b*) is said to be *sessile:* the filament being no more essential to the stamen than the claw is to the petal, or the petiole to the leaf. When the anther is imperfect, abortive, or wanting, the stamen is said to be *sterile*, *abortive*, or *rudimentary;* its real nature being known by its situation.

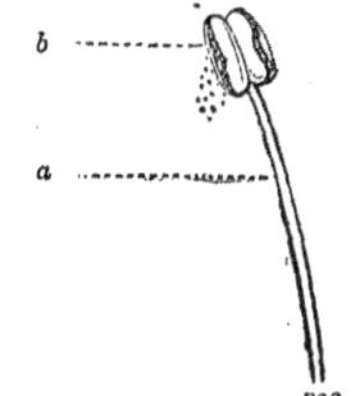

369

523. **The Filament,** although usually slender and cylindrical, or slightly flattened, assumes a great variety of forms: it is sometimes dilated so as to be undistinguishable from the petals, except by its bearing an anther; as in the transition states between the true petals and stamens of Nymphæa (White Water-Lily, Fig. 266, 267). The filament is anatomically composed of a central bundle of spiral vessels or ducts, which represent the fibro-vascular system of the leaf, in the same state as in the petiole, enveloped by parenchyma; the outer stratum of which forms a delicate epidermis.

524. **The Anther** (Fig. 369, *b*), which is the essential part of the stamen, is usually borne on the apex of the filament; and commonly consists of two *lobes*, or *cells* (*thecæ*), placed side by side, and connected by a prolongation of the filament, called the *connectivum*, or *connective.* As the filament answers to the petiole, so the connectivum answers to the midrib of the leaf, and the lobes, or cells, to the blade of the leaf; the portion each side of the midrib forming an anther-lobe. The *pollen*, or powdery substance contained in the anther, originates from a peculiar transformation of the cellular tissue, or parenchyma of the leaf.

525. The attachment of the anther to the filament presents three principal modes. 1st. When the base of the connective exactly corresponds with the apex of the filament and with the axis of the

anther, the latter is termed *innate*, and rests firmly upon the summit of the filament, as in Fig. 370. 2d. When the lobes of the anther adhere for their whole length to a prolongation of the filament, or to a broad connective (whichever it be called), so as to appear lateral, it is said to be *adnate;* as in the Magnolia (Fig. 488). Here the anther must be either *extrorse* or *introrse*. It is *introrse*, or *turned inwards*, when it occupies the inner side of the connective, and faces the pistils, as in Magnolia and the Water-Lily (Fig. 266); but when the anther looks away from the pistils and towards the petals or sepals, it is said to be *extrorse*, or *turned outwards*, as in the Iris, in Liriodendron (Fig. 371), and in Asarum (Fig. 373). 3d. When the anther is fixed by a point to the apex of the filament, on which it lightly swings, it is said to be *versatile;* as in all Grasses, in the Lily, and in the Evening Primrose (Fig. 372), &c. In this case, as in the preceding, the anther is said to be *introrse*, or *incumbent*, when it is turned towards the pistil, which is the most common form; and *extrorse*, when it faces outwards.

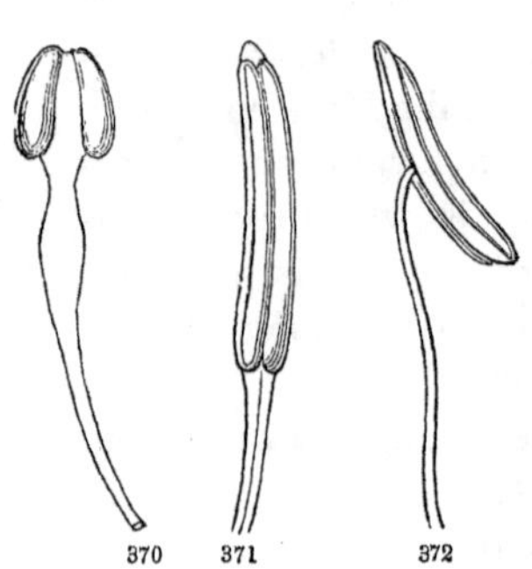

526. The *connective* is frequently inconspicuous or almost wanting, so that the lobes of the anther are directly in contact on the apex of the filament; as in Euphorbia (Fig. 346). It is often produced beyond them into an appendage, as in the Magnolia and Liriodendron (Fig. 371), the Papaw (Fig. 492, where it forms a rounded top), and Asarum (Fig. 373). Appendages or processes from the back of the connective are seen in the stamens of the Violet, and of many Ericaceous plants (Fig. 802 – 804).

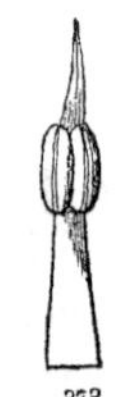

527. Each of the two cells or lobes of the anther is marked with a lateral line or furrow, running from top to bottom; this is the *suture* or *line of dehiscence*, by which the anther opens at maturity, and allows the pollen to fall out (Fig. 369). This line,

FIG. 370. Stamen of Isopyrum biternatum, with an innate anther. 371. Stamen of Liriodendron, or Tulip-tree, with an adnate extrorse anther. 372. Stamen of Œnothera glauca, with the anther fixed by its middle and versatile.

FIG. 373. Stamen of Asarum Canadense, with an adnate anther.

which answers to the margin of the leaf, is exactly lateral in innate anthers, as in Fig. 370; but it looks more or less evidently, and often directly, inward in introrse, and outward in extrorse anthers (Fig. 371, 373).

528. Various deviations from this normal structure of the anther frequently occur; some of which may be cursorily noticed. The opening of the anther, sometimes called its *dehiscence*, does not always take place by a longitudinal fissure for the whole length of the cell. Occasionally the suture opens only at the top, in the form of a chink or pore; as in Pyrola (Fig. 807), Rhododendron, and other Ericaceous plants, and in the Potato, &c. Sometimes the summit of the lobes is prolonged into a tube, which opens by a pore or chink at the apex; as in the Heath and Huckleberry (Fig. 802–804). In the Barberry and other plants of the family (Fig. 507), the Benzoin, &c., nearly the whole face of each anther-cell separates by a continuous line, forming a kind of door, which is attached at the top, and turns back, as if on a hinge: in this case the anthers are said to open by valves. In the Sassafras (Fig. 999), and many other plants of the Laurel Family, each lobe of the anther opens by two such valves, like trap-doors.

529. Sometimes the anthers are one-celled by the suppression of one lobe, being *dimidiate*, or reduced as it were to half-stamens, as in Gomphrena, and some other Amaranthaceous plants; but they more frequently become one-celled by the confluence of the two lobes, and the disappearance of the partition between them. The kidney-shaped one-celled anthers of the Mallow Family may be conceived to arise from the divergence of the base of the two lobes, and their perfect confluence at the apex; and the opening consequently takes place by a continuous sutural line passing round the margin (Fig. 618). A somewhat similar case occurs in Monarda and some other plants of the Mint Family, where only one of the two lobes remains parallel with the filament or connective; while the other, describing a semicircle, is brought into the same vertical line, where it stands bottom upwards; and the two, cohering by their contiguous extremities, become confluent into a single cell, which opens by a continuous straight line from one end to the other. The anther of Teucrium differs from the last chiefly in the enlarged connective, on which the divaricate lobes rest; and the cells, at first distinct, are confluent into one after the anther opens. In the Thyme, the anther-lobes are also

greatly divergent, but are separated by the thickened connective, which in this family is often larger than the cells. In the Sage, the singular elongated connective sits astride the apex of the filament, and bears an anther-cell at each extremity; one of which is perfect and contains pollen, while the other is imperfect or abortive. Illustrations of these diversities will be found under the Ord. Labiatæ. We have no room to pass in review even the more common of the almost endless variations which the anther exhibits.

530. As to its structure, each lobe of the full-grown anther consists of an epidermal membrane, lined with a delicate fibrous tissue, and surrounding a cavity filled with pollen. This fibrous lining, a part of which is shown in Fig. 32, from the anther of Cobæa, is composed of simple or branching attenuated threads or bands, which formed the thickening deposit on the walls of large parenchymatous cells; all the membrane between the bands becoming obliterated as the anther approaches maturity, the latter alone remain, as a set of delicate fibres. This fibrous layer gradually diminishes in thickness as it approaches the line of dehiscence of the cell, and there it is completely interrupted. These very elastic and hygrometric threads lengthen or contract in different ways, according as the anther is dry or moist; which movements, after the pollen has appropriated all the juices of the tissue, aid in the disruption of the anther along the suture, and then favor the egress of the pollen. The walls of many anthers are curved outwards, or completely turned inside out, as in Grasses, by the unlike hygrometric state of the external and the internal layers.

531. Of all the floral organs, the anther shows least likeness to a leaf. Nevertheless, the early development is nearly the same. Like the leaf, the apex is earliest formed, appearing first as a solid protuberance, and the anther is completed before the filament, which answers to the leaf-stalk, makes its appearance. At first, the anther is of a greenish hue, although at maturity the cells assume a different color, more commonly yellow. A transverse section of the forming anther shows four places in which the transformation of the parenchyma into pollen commences, which answer to the centre of the four divisions of the parenchyma of a leaf, viz. the two sides of the blade, each distinguished into its upper and its lower stratum. So that the anther is primarily and typically four-celled; each lobe being divided by a portion of untransformed tissue stretching from the connective to the opposite

side, which corresponds to the margin of the leaf and the line of dehiscence. This appearance is presented by a large number of full-grown anthers: but the partition usually disappears before the anther opens, when each lobe becomes single-celled. The normal anther is consequently considered as two-celled. In Menispermum and Cocculus (Fig. 496), however, the anther is strongly four-lobed externally, and each lobe forms a distinct cell, at maturity. Although the stamens originate a little later than the petals, when these are present, yet they outgrow them at first, and their formation is earlier completed (489).

532. **The Pollen**, contained in the anther, which appears to the naked eye like a mere powder, consists of grains of definite size and shape, which are uniform in the same plant, but often very different in different species or natural families. Although commonly spherical or oval, they are cylindrical in the Spiderwort (Tradescantia), nearly square in Colutea, many-sided in the Teasel, and triangular, with the angles dilated and rounded, in the Evening Primrose (Fig. 419). The most remarkable shape is that of Zostera (a marine aquatic plant), in which the grains consist of long and slender threads, which, as they lie side by side in the anther, resemble a skein of silk. Their surface, although more frequently smooth and even, is banded or crested in many cases; it is reticulated in the Passion-flower, and studded with strong points in Convolvulus purpureus (Fig. 417), or short bristles in the Mallow Family and the Gourd. The color is usually yellow.

533. The grains of pollen are single cells, formed usually in fours, by the division of the living contents of mother cells first into two, and these again into two parts, which, acquiring a layer of cellulose, become four specialized cells, nearly in the manner already described (31, 95). As the pollen completes its growth, the walls of the mother cells are usually absorbed or obliterated, when the grains lie loose in the cell. But sometimes the inclosing cells persist, and collect the pollen-grains into coherent masses of various consistence, as in the Milkweed Family (Fig. 422) and in the Orchis Family (Fig. 1098, 1101). Such pollen-masses are sometimes called *pollinia*. The threads, like cobweb, that are loosely mixed with the pollen of the Evening Primrose (Fig. 700), are vestiges of obliterated mother cells.

534. Not unfrequently the four grains developed in the same cell cohere, more or less firmly, as in most Ericaceous plants; or

grow as one compound grain, without undergoing complete division. The grains of the pollen of the Evening Primrose Family (Fig. 419) thus consist of the rudiments of four, which remain in strict combination; one of them enlarging to form the main body of the grain, while the three others appear as bosses on its angles. Rarely the four cohering grains are placed in the same plane. They usually stand in the same relation to each other as the four angles of a cube. In the Mimosa Family, the division goes farther, and gives rise to eight or sixteen lightly coherent grains in each mass.

535. The pollen-grains have two coats; the exterior of which, called the *extine*, is quite firm and often wax-like, granular, or fleshy; to it the bands, points, or other markings belong. It is a secretion from the inner layer, which is the proper membrane of the cell. This inner coat, named the *intine*, is a thin, uniform, transparent and colorless, highly extensible membrane. It absorbs water rapidly, and when exposed to its action the grain swells and soon bursts, discharging its contents. These contents are a viscid fluid, rich in protoplasm, which often appears slightly turbid under the higher powers of ordinary microscopes, and, when submitted to a magnifying power of three hundred diameters, is found to contain a multitude of minute particles (*fovillæ*) of spherical or oblong form, the larger of which are from the four-thousandth to the five-thousandth of an inch in length, and the smaller only one fourth or one sixth of this size. The smaller exhibit the constant molecular motion of all such minute particles when suspended in a liquid and viewed under a sufficient magnifying power. The larger are some of them of the nature of starch-grains; some consist of oily matter, &c. The pollen of certain aquatic plants — that of Zostera very distinctly — has only a single (the internal or proper) membrane.

536. When wetted, the grains of pollen promptly absorb water by endosmosis (37), and are distended, changing their shape somewhat, and obliterating the longitudinal folds, one or more in number, which many grains exhibit in the dry state. Soon the more extensible and elastic inner coat inclines to force its way through the weaker parts of the exterior, especially at one or more thin points or pores; sometimes projecting so as to form a tube of considerable length, when the absorption is slow and the exterior coating tough. The absorption continuing, the distention soon over-

comes the resistance of the inner coat, which bursts, with the eruption of the contents in a jet. When fresh, living pollen falls upon the stigma, however, which is barely moist, but not wet, it does not burst, but the inner membrane is slowly projected, often through particular points, clefts, or valvular openings of the outer coat, in the form of an attenuated transparent tube (Fig. 416–418), filled with its fluid contents, which penetrates the naked and loose cellular tissue of the stigma, and buries itself in the style (Fig. 419). This, however, is not a mechanical protrusion, but a true growth, depending on nutrition imbibed from the stigma and style. Its further course and the office it subserves will be considered after the structure of the pistil is made known.

Sect. VII. The Pistils.

537. **The Pistils** (419) occupy the centre of the flower, and terminate the axis of growth. Their number is designated by Greek numerals prefixed to the name applied to the pistil from the same language. Thus, a flower with a single pistil is said to be *monogynous ;* with two, *digynous ;* with three, *trigynous ;* with four, *tetragynous ;* with five, *pentagynous ;* with six, *hexagynous ;* with seven, *heptagynous;* with eight, *octogynous;* with nine, *enneagynous ;* with ten, *decagynous ;* and so on: and when more numerous or indefinite, they are termed *polygynous.* (See the Linnæan Orders, 684.)

538. It is comparatively seldom that the pistils are actually equal to the petals or sepals (480) in number; they are sometimes more numerous, and arranged in several rows upon the enlarged or prolonged receptacle, as in the Magnolia, the Strawberry, &c., and perhaps more frequently they are reduced to less than the typical number, or to a single one. Yet often what appears to be a single pistil is not so in reality, but a compound organ, formed by the union of two, three, or a greater number of simple pistils ; as is shown in Fig. 381 – 390.

539. A pistil, as already described (420), is composed of three parts ; the Ovary, or seed-bearing portion ; the Style, or tapering portion, into which the apex of the ovary is prolonged ; and the Stigma, usually situated at the summit of the style, consisting of a part, or sometimes a mere point, of the latter, divested of epidermis, with its moist cellular tissue exposed to the air. The

ovary, which contains the young seeds, or ovules, is of course a necessary part of the pistil: the stigma, which receives from the anthers the pollen (536) by which the ovules are fertilized, is no less necessary: but the intervening style is no more essential to the pistil than the filament is to the stamen, and is therefore not uncommonly wanting. In the latter case, the stigma is *sessile* upon the apex of the ovary. In Tasmannia it actually occupies the side of the ovary for nearly its whole length, and is separated from the line to which the ovules are attached only by the thickness of the walls; and it is nearly the same in our Schizandra (Fig. 375), another plant of the Magnolia Family. The style sometimes proceeds from the side, or even from the apparent base of the ovary; as in the Strawberry.

540. When the pistil is reduced to a single one, or when several coalesce into one, it will necessarily terminate the axis, and appear to be a direct continuation of it. When there are two pistils in the flower, they always stand opposite each other (so that if they coalesce it is by their inner faces); and are either *lateral* as respects the flower, that is, one on the right side and the other on the left, in a plane at right angles to the bract and axis (444), as in the Mustard Family, the Gentian Family, and a few others; or, more commonly, *anterior* and *posterior*, one before the axis and the other before the bract of the axillary flower. When they accord in number with the sepals or petals, they are either opposed to or alternate with them; and the two positions in this respect are sometimes found in nearly related genera, so as to baffle our attempts at explaining the cause of the difference. In Pavonia, for example, the five pistils are opposite the petals; in Malvaviscus and Hibiscus, alternate with them. In Sida, when five, they stand opposite the petals; in Abutilon, opposite the sepals.

541. To attain a correct morphological view of the simple pistil, we must contemplate it as resulting from the transformation of a leaf which is folded inwards, and the margins united; in a manner that will be perfectly evident on comparing Fig. 263 with Fig. 270. The line formed by the union of the margins of the leaf is called the Inner or Ventral Suture, and always looks towards the axis of the flower. This is a true suture, or *seam*, as the word denotes. The opposite line, which answers to the midrib, is sometimes apparent as a thickened line, and is termed the Outer or Dorsal Suture. The surface of the pistil necessarily corresponds

to the lower, and its lining to the upper, surface of a leaf. The stalk of the pistil (487), when it is present, represents the petiole; and a prolongation of the apex of the specialized leaf forms the style. The stigma occupies some portion of what in the style answers to the confluent margins of the transformed leaf (and certainly is not a portion of the midrib, as has been thought); this is evident in Tasmannia, above mentioned, where these margins are actually *stigmatic* for almost their whole length, and in Schizandra, where the stigmatic surface (known by its papillose cells or other surface exposed directly to the air, without any epidermis) begins externally on the ventral edge of the pistil, just above the point where the ovules are attached within (Fig. 375). In the Pæony, in Isopyrum (Fig. 374), and a great number of instances, the stigma consists of two crested ridges or parallel lines running down the inner face of the style; and in a still larger number of cases (as in nearly all Caryophyllaceæ and a part of Malvaceæ), a continuous stigmatic surface extends down this face of the style (Fig. 384). Such unilateral stigmas we accordingly take to be the normal form; and say that, while the united margins of the typical leaf composing the ventral suture are *turned inwards into the cell of the ovary to bear the ovules, in the simple style they are exposed externally to form the stigma.* Where the stigma is terminal, or occupies only the apex of the style, we suppose that these margins are infolded in the style also, and form in its interior the loose conducting tissue through which a communication is established between the terminal stigma and the interior of the ovary. The double nature of the stigma (one lamella of which corresponds to each margin of a leaf) is still evident in the two lobes which the terminal stigma exhibits in many simple pistils, as in Hydrastis (Fig. 376), and Actæa (Fig. 377).

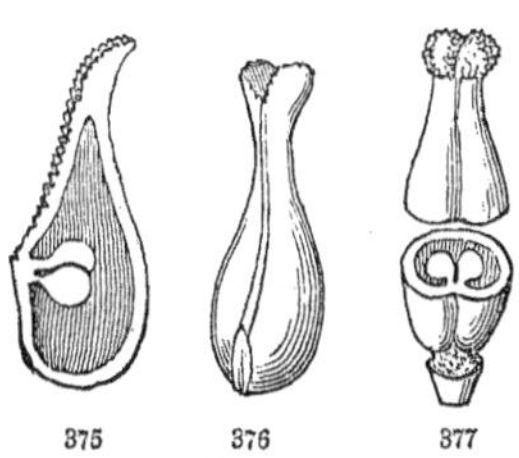

542. The ovary contains only OVULES, or bodies destined to be-

FIG. 374. A ventral view of a pistil of Isopyrum biternatum, showing the double stigma; the ovary cut across, showing the two rows of ovules.

FIG. 375. Vertical section of a pistil of Schizandra coccinea; a side view. 376. Pistil of Hydrastis. 377. Pistil of Actæa rubra, cut across, so as to show the interior of the ovary (the vntral suture turned towards the observer).

come seeds after fertilization (420). These, in all ordinary cases, are borne on the part which represents the margins of the transformed leaf. They are in some sort analogous to buds, which are occasionally developed on the margins of leaves (as in the well-known case of Bryophyllum, Fig. 271). Since both margins of the infolded leaf may bear ovules, the latter are normally arranged in two rows (one for each margin) on the inner or ventral suture; as is seen in Fig. 263, 374, 377. The ovule-bearing portion of the ventral suture, which often forms a ridge or crest projecting more or less into the cavity of the ovary, is named

543. **The Placenta.** As it corresponds with the ventral suture, and is in fact a part of it, or a cellular growth from it, it is always placed next the axis of the flower; as is evidently the case when two, three, or more pistils are present (Fig. 379–383). Each placenta necessarily consists of two parts, one belonging to each of the confluent margins of the transformed leaf. It therefore is frequently two-lobed, or of two diverging lamellæ (Fig. 263). The ovules vary greatly in number; being sometimes very numerous and in several rows on a broad placenta, as in the May-Apple (Podophyllum); sometimes in two normal rows occupying the whole length of the ventral suture, as in the Larkspur, Columbine, Actæa (Fig. 377), &c.; sometimes reduced to one row in appearance, as in the Pea, where on inspection they will be found, however, to be alternately attached to each lamella of the placenta, that is, to each margin of the leaf: again, they occupy only its middle, base, or summit, where they are often reduced to a definite number, to a single pair (Fig. 375), or to a single one (Fig. 316).

544. When the pistils are distinct or uncombined, they are said to be *apocarpous;* when they are united, and form a compound pistil, they are *syncarpous*. We have carefully to distinguish between the *simple pistil*, which represents a single member of the gynæcium (419), and the *compound pistil*, which answers to the whole circle coalescent into one body. To subserve this purpose, botanists have coined the name of

545. **The Carpel or Carpidium.** This name designates an individual member of the gynæcial circle, whether it occur as a separate or simple pistil, or as one of the elements of a compound pistil. It is in the latter case that the name is principally needful. All degrees of union of the carpels may be observed, from the mere cohesion of their contiguous inner angles, to the perfect consolidation of the

ovaries while the styles remain distinct, as in Spergularia (Fig. 387), or of the latter also. Rarely the stigmas or styles are united while the ovaries remain distinct, as in Asclepias and Apocynum (Fig. 953). Numerous illustrations of all the varied forms are given in the systematic part of this volume. The annexed diagrams represent, Fig. 378, 379, three distinct but approximated pistils; Fig. 380, 381, three similar pistils with only their ovaries coalescent; and Fig. 382, 383, three pistils with their styles as well as their ovaries united into one.

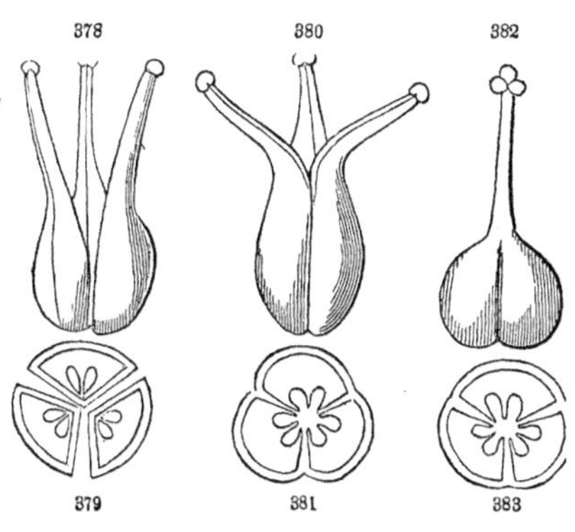

546. **The Compound Pistil.** From these illustrations the regular structure of the compound pistil is readily seen, at least as to the more common and normal case, namely, where the cross-section displays two or more *cells*, or separate cavities. For it is evident that, if the contiguous parts of a whorl of three or more carpels cohere, the resulting compound ovary will have as many cavities, or cells, as there are carpels in its composition, and the placentæ will all be brought together in the axis; as is shown in Fig. 381, 383, in Fig. 291, and in the gynæcium of Fig. 306, as compared with Fig. 284, &c.

547. The partitions, or DISSEPIMENTS, which divide the compound ovary into cells, are evidently composed of the united contiguous portions of the walls of the carpels. These necessarily consist of two layers, one belonging to each carpel; they are always vertical, and are equal in number to the carpels of which the compound pistil is constructed.

548. A single carpel, therefore, has no proper dissepiment. It is, however, sometimes divided by spurious partitions, separating the cavity into separate cells or joints, placed one above another,

FIG. 378. A whorl of three pistils, the line which passes down the inner side representing the ventral suture. 379. A cross-section of their ovaries, showing the two rows of ovules, occupying the inner angle, or ventral suture. 380. A whorl of three pistils, their ovaries united. 381. A cross-section of the same. 382. Three pistils, with their styles also united quite to the summit. 383. A cross-section of the united ovaries.

as in some species of Cassia, in Desmodium, &c. (Fig. 440, 441); or even by a vertical false dissepiment produced by the introflexion of the inner or placental suture, as is partially the case in some species of Phaca and Oxytropis (Fig. 445); or by a projection from the dorsal suture, as in the Flax (Fig. 630, 631), the Service-Berry, and many species of Vaccinium; or by its introflexion, as in Astragalus (Fig. 444).

549. A compound ovary of two cells, or *loculi*, is *bilocular;* of three, *trilocular;* of four, *quadrilocular;* of five, *quinquelocular;* and so on. If of several without reference to the number, it is said to be *plurilocular*, or *multilocular;* the former name being used when the cells are comparatively few, the latter when more numerous. We may, however, have a

550. **Unilocular Compound Pistil,** where the ovary, although composed of two or more carpels, is yet *one-celled*, that is, has a single cavity. The cases of the sort are of two principal kinds, namely, first,

551. **With a free Placenta in the Axis,** as in the Primrose Family (Fig. 825), and in a large part of the Chickweed and Pink Family, as shown in Fig. 384. This is usually explained on the supposition that the dissepiments are obliterated or torn away by the expansion during the growth of the ovary, these alone being wanting to complete the structure of the normal compound ovary already described, as will be seen by comparing the diagram, Fig. 387, with Fig. 383. This is demonstrably the true explanation in the Chickweed and Pink Family; for the dissepiments, or vestiges of them, may be detected at an early stage, and sometimes at the base of the full-grown ovary; while certain plants of the same family, of otherwise identical structure, retain the partitions even in the ripe pod. Other cases, however, especially where there are a few ovules, or even a single one, as in Thrift (Fig. 840), arising from the base of the cell, are more properly referred to the other kind of unilocular compound pistil, namely, that

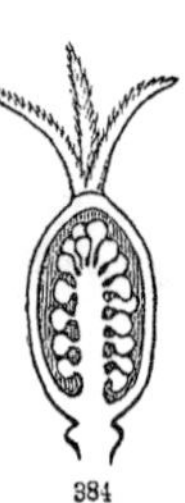

552. **With Parietal Placentation.** If we suppose a circle of three carpellary leaves, with their margins turned inwards, yet not so as to reach the axis, to cohere merely by their contiguous inflexed

FIG. 384. Vertical section through the compound tricarpellary ovary of a plant of the Chickweed Family (Spergularia rubra), showing the free central placenta.

portions, a one-celled tricarpellary ovary would result, with three imperfect dissepiments projecting into the cavity, but not dividing it into distinct cells (as in the diagram, Fig. 385). The placentæ are here borne upon the extremity of the imperfect dissepiments, which, if somewhat prolonged, would meet and unite in the centre, so as to present the regular three-celled structure (as in Fig. 383). This will be evident on comparing the pod of the Common St. John's-wort (Fig. 555), which is completely three-celled with the placentæ united in the axis, with the ovary of another species (Fig. 388), where the three placentæ touch in the centre without cohering, and with the full-grown pod of the last (Fig. 389), where they are drawn asunder by the expansion of the growing pod, and remain attached only to its walls, borne on three slight introflexions, which stand in the place of dissepiments. Parnassia affords a similar instance, only there are usually four such placentæ instead of three (Fig. 304, the centre of which represents a cross-section of the 4-carpellary ovary). These instances bring us to the frequent case in which we may say that the leaves of the gynæcial verticil, placed merely in apposition, as in valvate æstivation (499), directly cohere into one circle by their respective contiguous margins; which, being barely induplicate, form placentæ which are borne directly on the walls. This is shown in the diagram, Fig. 386, representing a cross-section of three carpels thus combined into a compound one-celled ovary, without any appearance of dissepiments. Thus borne upon the walls, instead of in the axis of the compound ovary, the placentæ are said to be *parietal*. Examples of the kind with a tricarpellary ovary are furnished by many

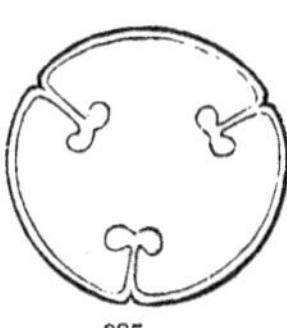

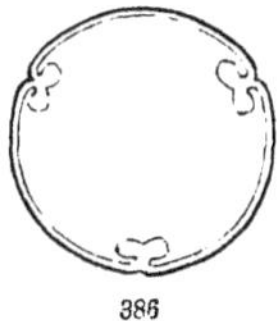

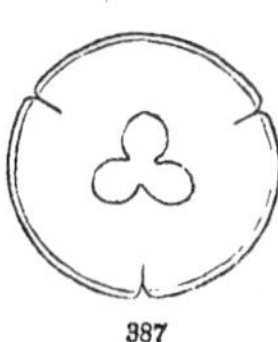

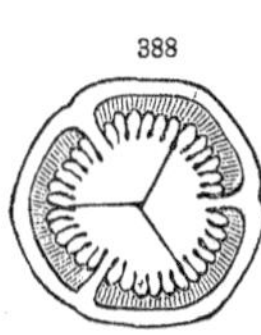

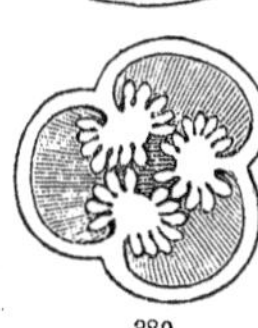

FIG. 385 - 387. Diagrams illustrating parietal and free central placentation. 385. Cross-section of an ovary composed of three united carpels, where the introflexed portions do not reach the centre. 386. Section of a similar ovary, except that the placental margins unite without any introflexion (placentæ strictly parietal). 387. Section of a tricarpellary ovary, with a free central placenta, produced by the obliteration of the dissepiments.

FIG. 388. Magnified cross-section of the ovary of Hypericum graveolens. 389. Enlarged cross-section of the mature pod of the same, where the placentæ become strictly parietal.

Hypericums, by the Violet Family, the Cistus Family (Fig. 548), Drosera (Fig. 390), &c. Also, in an ovary of two carpels, by the Caper Family (Fig. 537), the Fumitory Family (Fig. 298), the Gooseberry (Fig. 711), &c.

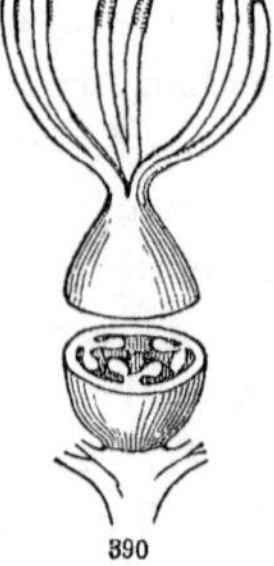

553. An ovary with parietal placentæ is necessarily one-celled; except it be divided by an anomalous partition, such as that of Cruciferous plants, &c.

554. A compound pistil of this kind may have the sutures ovuliferous, or develope placentæ, only at some particular part, as at the summit or the base of the cell; and there few or only solitary ovules may be developed, as in the Thrift (Fig. 840), in Compositæ, &c., which reduces the case to the greatest simplicity. The confluence of two or more basilar parietal placentæ will account for the free central placentation in cases where no dissepiments are discernible at an early period, as in the Primrose Family.

555. It will be seen that parietal placentæ are necessarily double, like the placenta of a simple ovary, or of each carpel of a compound plurilocular ovary; but with this difference, that in these cases the two portions belong to the two margins of the same carpel; while in parietal placentæ they are formed from the coalescent margins of two adjacent carpels. This will readily appear on comparing the diagrams, Fig. 379, 381, with Fig. 385, 386.

556. The number of carpels of which a compound ovary consists is indicated by the number of true dissepiments when these exist (547); or by the number of placentæ, when these are parietal (552); or by the number of styles or stigmas, when these are not wholly united into one body. Thus a simple pistil has a single cell, a single placenta, and a single style. A pistil of two carpels may be two-celled, with two placentæ, two styles, &c.

557. There are, however, some exceptions which qualify these statements: — 1. Each placenta being a double organ (555), it occasionally happens that the two portions are separated more or less, as in Orobanchaceous plants, where a dicarpellary ovary appears on this account to have four parietal placentæ; either approximate in pairs (as in our Cancer-root, Conopholis), or equidistant (as in Aphyllon). 2. Analogous to this is the case where

FIG. 390. Pistil of Drosera filiformibus, with three 2-parted styles; the ovary cut across, showing three parietal placentæ.

the two constituent elements of the stigma (the only essential part of the style) separate into two half-stigmas; a tendency to which is seen in Fig. 376, 377, and which is carried out in most species of Drosera (Fig. 390). The stigma, no less than the placenta, belongs to the margins of the infolded leaf (541), these margins being *ovuliferous* in the ovary and *stigmatiferous* in the style; as Mr. Brown, the most profound botanist of this or any age, has clearly shown. These two constituent portions of the style or stigma are usually combined; but are not unfrequently separate, either entirely or in part, as in Euphorbiaceous plants, in Grasses, and especially in Drosera, where there are consequently twice as many nearly distinct styles as there are parietal placentæ in the compound ovary. If the two component parts of the style of each carpel were reunited into one, in the usual manner, their number would equal the placentæ, and their position would be alternate with the latter. But since, in parietal placentation, each *half-placenta* is confluent, not with its fellow of the same carpel, but with the contiguous *half-placenta* of the adjacent carpel (555), it were surely no greater anomaly for the elements of such *half-stigmas* as those of Drosera (Fig. 390) to follow the same course. This is precisely what takes place in Parnassia, and in other cases where the stigmas are opposite the parietal placentæ; — cases which were thought to be very anomalous, merely on account of the adoption of a false principle (that of the necessary alternation of the stigmas and placentæ), but which are really no more so than the parietal placentation itself. The division of the style in such cases furnishes further examples of collateral chorisis. Sometimes the simple style is repeatedly forked in this way, or cut into a fringe at the summit, as in Turnera, and the short lobes of the compound style in Dionæa. 3. Furthermore, the production of ovules is not always restricted to what answers to the margins of the carpellary leaves. In the Poppy, the whole surface of the long, imperfect partitions is covered with ovules; in Butomus, they are borne over the whole internal face of each carpel, and in the Water-Lilies over the whole surface (Fig. 268), except the inner angle of each cell, where alone they normally belong. Reduced to two in the allied Water-Shield (Brasenia, Fig. 515), the ovules grow from the *dorsal suture*, or the midrib of the carpellary leaf alone! And in Cabomba itself we usually find its three ovules, one in the dorsal and one on the ventral suture, and the third on some variable part of the face of the cell in the vicinity of either suture. In Obolaria, a compound unilocular ovary is ovuliferous over the whole wall of the cell.*

558. When the styles are separate towards the summit, but united below, they are usually described as a single organ; which is said to be *parted*, *cleft*, *lobed*, &c., according to the extent of cohesion. This language was adopted, as in the case of leaves (281) and floral envelopes (461), long before the real structure was understood: but, as it involves an erroneous idea, the expressions, *Styles distinct; united at the base; united to the middle*, or *summit*, &c., as the case may be, should be employed in preference.

559. A few casual exceptions occur to the general rule that ovules and seeds are both produced and matured within an ovary, namely, in a closed carpellary leaf or set of combined carpellary leaves. In the Blue Cohosh, Leontice (Caulophyllum) thalictroides, the ovules rupture the ovary soon after flowering, and the seeds become naked; and in the Mignonette they are imperfectly protected, the ovary being open at the summit from an early period. In all such cases, however, the pistil is formed and the ovules are fertilized in the ordinary way.

560. **Gynæcium of Gymnospermous Plants.** A far more important and remarkable exception is presented by two natural families, the Coniferæ (Pines, Firs, &c., Fig. 391–402), and the Cycadaceæ (Cycas, Zamia, Fig. 403). Here the pistil, as likewise the whole flower, is reduced to the last degree of simplicity; each fertile flower consisting merely of an open carpellary leaf, in place of a pistil, in the form

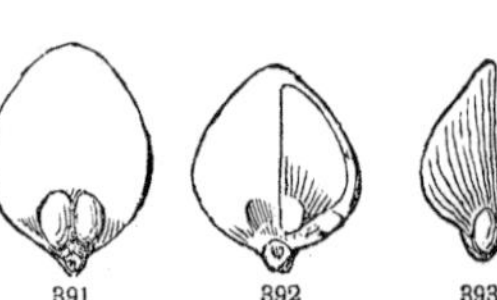

* These various points are elucidated by Mr. Brown, in *Plantæ Javanicæ Rariores*, pp. 107–112, in two notes which apparently are not sufficiently studied by many English botanists. — All placentation is very differently explained by those who adopt the hypothesis of Schleiden and others. According to this new view, as buds regularly arise from the axils of leaves and from the extremity of the stem or axis, and only in some exceptional and abnormal cases from the margins or surface of leaves, so ovules are considered to arise from the axils of the flower, like terminal buds, or from the axils of the carpellary leaves, like axillary buds. Thus, placentæ are supposed to belong to

FIG. 391. A carpellary scale from the ament of a Larch, the upper side turned to the eye, showing the pair of ovules at its base 392. The same in fruit, reduced in size; one of the winged seeds still attached; the other, 393, separated.

of a scale, as in Fig. 391, or sometimes of a different shape (Fig.

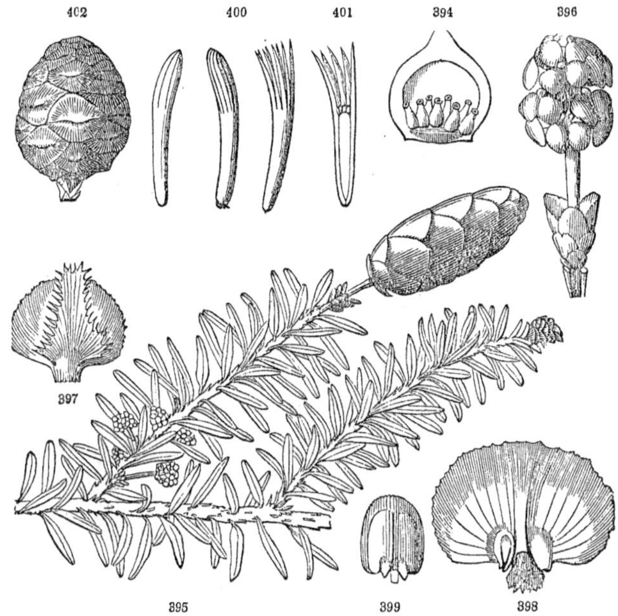

407), which bears two or more ovules upon some part of its mar-

---

the axis, and not to the carpellary leaves; and a one-celled ovary, with one or more ovules arising from the base of the cell, would nearly represent the typical state of the gynæcium. This theory, which the intelligent student may easily apply in detail, offers the readiest explanation of free central placentation, especially in such cases as Primula, &c., where not the slightest trace of dissepiments is ever discoverable. It must be admitted that the monstrosities which occur in Primula, and some other plants with free central placentation, favor this new view. It is also perfectly applicable to ordinary central placentation; where we have only to suppose the cohesion of the inflexed margins of the carpellary leaves with a central prolongation of the axis or receptacle which bears the placentæ. But in case of parietal placentation,

FIG. 394. Carpellary scale of Cupressus sempervirens (the true Cypress), seen from within, and showing the numerous orthotropous ovules that stand on its base. 395. Branch of Abies Canadensis (Hemlock Spruce), with lateral staminate flowers, and a fertile strobile. 396. Staminate ament, magnified. 397. Carpellary scale of a fertile ament, with its bract. 398. Similar fertile scale, more magnified and seen from within; showing the two ovules adherent to its base: one of them (the left) laid open. 399. The scale in front, nearly of the natural size, its inner surface occupied by the two seeds. 400. Polycotyledonous embryos of Abies and Cypress. 401. Vertical section of one. 402. Strobile of Taxodium distichum (Suborder Cupressineæ).

gin or upper surface. The ovules, therefore, instead of being in-

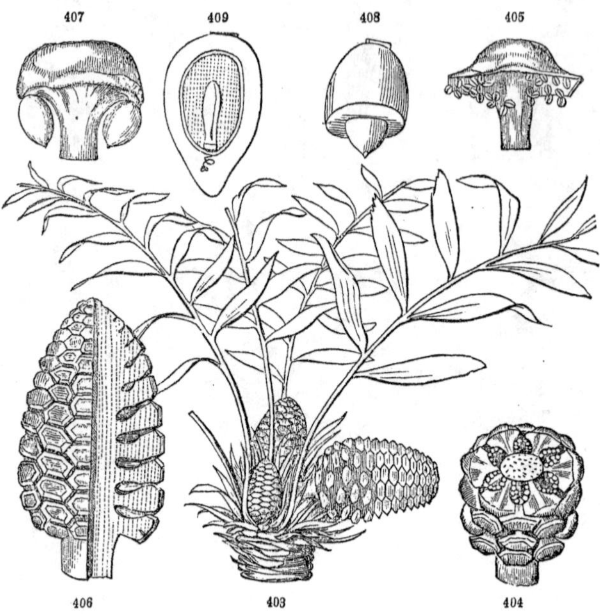

closed in an ovary, and acted upon by the pollen through the in-

---

the advocates of this theory are obliged to suppose that the axis divides within the compound ovary into twice as many branches as there are carpels in its composition, and that these branches regularly adhere, in pairs, one to each margin of all the carpellary leaves. Its application is attended with still greater difficulties in the case of simple and uncombined pistils, where the ovules occupy the whole inner suture, which are doubtless justly assumed as the regular and typical state of the gynæcium; but to which the new hypothesis can be adapted only by supposing that an ovuliferous branch of the axis enters each carpel, and separates into two parts, one cohering with each margin of the metamorphosed leaf. This view, however, not only appears very improbable, but may be disproved by direct observation, as it has been most completely by those monstrosities in which an anther is changed into a pistil, or even one part of the anther is thus transformed and bears ovules, while the other, as well as the filament, remains unchanged; — a case where the forma-

FIG. 403. Zamia integrifolia (the *Coontie* of Florida). 404. Section of the sterile ament. 405. One of its scales detached, bearing scattered anthers. 406. Fertile ament, from which a quarter-section is removed. 407. A pistillate flower, consisting of two ovules pendent from the thickened summit of the carpellary scale. 408. A drupaceous seed, from which a part of the pulpy outer portion is removed. 409. Vertical section through the seed (of the natural size), showing the pulpy outer coat, the hard inner integument, the albumen, and the embryo.

tervention of a stigma, are naked and exposed, — except as they are more or less covered in Pines, Firs, &c., by the imbrication of the carpellary scales into a sort of ament or cone (as in Fig. 176, &c.), — and are fertilized by the direct application of the pollen. Their seeds, accordingly, are destitute of a pod, or any similar inclosure. On this account they have received the name of GYMNOSPERMOUS PLANTS (111); literally, plants with naked seeds.

SECT. VIII. THE OVULE.

561. **Ovules**, the rudiments of future seeds (420), at first appear like minute pulpy excrescences of the placenta; but long before the flower expands they have acquired a regular, and generally round or oval form. They are attached to the placenta by one extremity, either directly, or by a short stalk called the *Funiculus*, or *Podosperm* (Fig. 413, 414). As to number, they vary from one in each ovary, or in each cell of the compound ovary, to several or many upon each placenta. In the former case, they are said to be *solitary;* in the latter, they are *definite* when their number is uniform and not remarkably great, and *indefinite*, when they are too numerous to be readily counted.

562. As to situation and direction with respect to the cavity that contains them, ovules are said to be *erect* when they arise from the very bottom of the ovary; *ascending*, when fixed to the placenta above the base and directed obliquely upwards; *horizontal*, when they project from the side of the cell, without turning either upwards or downwards (Fig. 263); *pendulous*, when their direction is downwards; and *suspended*, when they arise from the sum-

---

tion of the placenta from a process of the axis is out of the question. This hypothesis is, therefore, entirely untenable as a general theory; and whether it affords a correct explanation of any form of central or basilar placentation must be left for further observation to determine. We will only remark, that even the appearance of a placenta or ovuliferous body in the apparent axil of a carpellary leaf no more proves that the body in question belongs to the axis, than that the appendage before the petals of Parnassia and the American Linden, or the stamen of a Rhamnus or Vitis, represents the axis of a branch instead of a leaf. As to the terminal naked ovule of the Yew, where the structure, on any view, is reduced to the greatest possible simplicity, it is surely as probable that it answers to the earliest formed, or *foliar*, portion of the last phyton, here alone developed, as to the *cauline* part, which so seldom appears in the flower.

mit of the ovary and hang perpendicularly in the cavity (Fig. 316). In the Thrift (Fig. 840), and in the Sumach, the ovule is singularly pendent from an ascending funiculus. These terms are applicable to the seed as well as to the ovule.

563. As to its structure and formation, the ovule appears as a mere excrescence, or papilla, of soft and homogeneous parenchyma, which soon acquires a definite form. This NUCLEUS, as it is called, is the essential part of the organ; in the Mistletoe it actually constitutes the whole, its ovule having no integuments of its own. A hollow place is formed in its interior about the time of flowering, in which the embryo at length appears. Most ovules, however, in the course of their growth acquire an envelope, or more commonly two envelopes. Only one envelope is seen in the ovule of the Walnut, where, after the nucleus is formed and has assumed its ovate shape, a circular ring appears around its base, which gradually enlarges into a sheath, but at length covers it like a sac, which, however, remains open at the apex. This orifice, which leads to the nucleus, and through which, indeed, the nucleus often protrudes, is called the FORAMEN or the MICROPYLE. In far the greater number of cases, a second envelope is formed outside of the first, beginning in the same way, though always later than the inner one, which, however, it eventually overtakes and incloses. The outer envelope, when both are present, becomes the exterior integument or *testa* of the seed; and the inner, its tegmen or inner coat. Mirbel named the exterior coat of the ovule the PRIMINE, and the interior the SECUNDINE, names which are attended with the objection that the secundine or second coat is actually older than the primine or first coat in the order of position. Both sacs are open at the apex, and the summit of the nucleus points

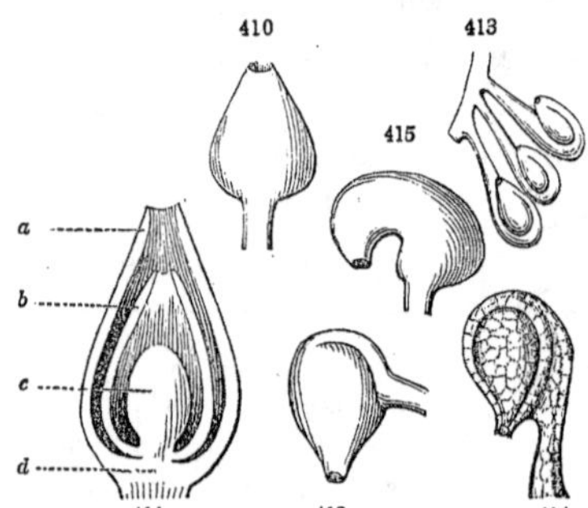

FIG. 410. An orthotropous ovule. 411. Longitudinal section of the same, more magnified: *a*, the primine; *b*, the secundine; *c*, the nucleus; *d*, the chalaza. 412. An amphitropous ovule. 413. Three anatropous ovules, with their funiculi, attached to a portion of the placenta. 414 One of the same, more highly magnified, exhibiting its cellular structure. 415. A campylotroprous ovule.

directly towards the apertures. The orifice or foramen of the exterior integument is called the Exostome (or outer mouth), that of the interior, the Endostome (or inner mouth). The coats of the ovule and the nucleus are distinct and unconnected, except at the base, or point of attachment to the funiculus, where they are all perfectly confluent: this point of union receives the name of the Chalaza (Fig. 411, *d*).

564. Through the funiculus and chalaza the ovule derives its nourishment from the placenta; through the opening at the summit, the nucleus receives the influence of the pollen, which results in the production of the embryo.

565. Our description applies to the complete ovule in its simplest form, where no change in the position of parts takes place during its growth, the chalaza remaining next the placenta, with which the funiculus directly connects it, while the apex, represented by the foramen, or orifice of the coats, is at the opposite extremity (as in Fig. 410). Such an ovule, not being curved or turned from its normal direction, is called *atropous* (literally, not turned), or usually *orthotropous* (straight). This simple orthotropous form occurs in the Cistus Family (Fig. 550), and the Polygonum Family (Fig. 986), and in many others.

566. In the greater number of cases, however, a change of relative position takes place during the development of the ovule; consisting either in its complete inversion upon the funiculus that bears it, so that the orifice or apex is brought down by the side of the stalk and points towards the placenta, while the chalaza looks in the opposite direction (as in Fig. 413, 414, and also in Fig. 263, where such ovules are seen in their natural position in the ovary); or else the ovule curves upon itself, and thus brings down the apex near the funiculus (as in Fig. 416). In the former case, the ovule is *anatropous*, or inverted; in the latter, it is *campylotropous*, or curved. Campylotropous ovules are found in the Mignonette, in all Cruciferous and Caryophyllaceous plants, and in many others; but the anatropous form is by far the most common of all.

567. In *anatropous* ovules, the funiculus coheres firmly with that part of the surface which is applied to it; and in the ripe seed breaks away at the point where it is free from the integument, to which the adherent portion remains attached. The latter receives the name of Rhaphe; and appears in the form of a ridge, cord, or line, passing from the Hilum (as the scar left by the breaking

away of the funiculus from the seed is termed) to the chalaza, maintaining the communication between the interior of the ovule or seed and the placenta. The rhaphe is only found in the anatropous ovule, and serves to distinguish it; since in all others the hilum or scar exactly corresponds to the chalaza, while in this the two occupy opposite extremities of the seed; the chalaza, which is the real base, being by this inversion situated at the apparent apex, while the micropyle, or organic apex, is found next the hilum, or the apparent base. This is perfectly simple on the supposition that an anatropous ovule is produced by the mere adhesion of the funiculus to the whole length of one side of what would otherwise be an orthotropous ovule.*

568. What are called *amphitropous* or *heterotropous* ovules, which are straight, with the chalaza at one end, the micropyle or apex at the other, and the hilum half way between the two (as in Fig. 412), arise from the adhesion of the funiculus for a short distance only, forming a rhaphe of only half the length of the ovule. As the free funiculus in such cases generally diverges at right angles from the axis of the ovule, so that its proper base and apex become lateral, these ovules or seeds are sometimes termed *peltate*, or *transverse*.

569. *Campylotropous* ovules (Fig. 415) differ from the orthotropous in being curved during their development, so that the orifice or apex is brought into juxtaposition with the base; which in this case is both hilum and chalaza.

570. It is important to notice the situation of the orifice, or foramen, of the ovule, as it indicates the future position of the radicle of the embryo (631), which is invariably directed towards the foramen. Its situation with respect to the hilum varies in the different kinds of seeds: in those which arise from orthotropous ovules, it points in the direction exactly opposite the hilum (Fig. 453); in the anatropous form, it is brought close to the hilum, so that it is ordinarily said to point to it (Fig. 454 – 456); in campylotropous seeds, it is also brought round to the hilum; while in the amphitropous, it points in a direction nearly at a right angle with the hilum.

* Thus, in most Cistaceæ, the ovules are orthotropous, but in one small genus (Fumana) the funiculus usually adheres to the side of the ovule, and renders it anatropous. On the contrary, sometimes anatropous ovules become orthotropous in the seed, by the separation of the rhaphe from its face.

SECT. IX. FERTILIZATION AND FORMATION OF THE EMBRYO.

571. THE action of the pollen, by means of which the ovule is fertilized, is now satisfactorily known. The points still controverted mostly relate to the first step in the formation of the embryo.

572. The arrangement and adjustment of parts, mechanical and otherwise, which secure the application of the pollen to the stigma, are so extremely diversified in different plants, that we cannot undertake to give even a general account of them here. The adaptation is sometimes in the relative length of the floral organs in connection with the position of the flower, whether erect, inclined, or nodding; sometimes juxtaposition is effected through transient and often sudden movements, whether mechanical (by elasticity) or spontaneous, which will be mentioned in another place. Frequently the anthers open and the pollen is applied to the stigma while the parts are still approximated in the bud. In monœcious plants the staminate blossoms are commonly situated adjacent to the pistillate, or else raised above them, as in Indian Corn. In diœcious plants, as indeed in a vast number of others, much is left to the action of the winds, or of insects, which convey the pollen from one blossom to another; and the immense abundance of pollen, especially in monœcious and diœcious plants, greatly diminishes the chance of failure. The loose papillæ, or short projecting hairs of the stigma, and especially the viscous fluid which at this time always moistens its surface, serve to retain the grains of pollen on the stigma when they have once reached it. The following brief statement comprises the essential substance of what is known respecting the immediate

573. **Action of the Pollen.** The grain of pollen becomes turgid as it absorbs by endosmosis (37) the viscous moisture of the stigma: its inner membrane consequently extends, breaks through the scarcely extensible outer coat at some one point (or occasionally at two or three points, Fig. 419), and lengthens into a delicate tube, filled with the liquid and molecular matter (fovillæ, 535) that the grain contains. This tube (Fig. 416–419), remaining closed at the extremity, penetrates the loose tissue of the stigma, and is prolonged downwards into the style, gliding along the interspaces between the very loosely disposed cells of the now moist conducting tissue (541), which extends from the stigma to the

cavity of the ovary, and at length reaching the placenta or some other part of the interior of the ovary. This prolongation into a tube, often many hundred times the diameter of the pollen-grain, is a true growth, after the manner of elongating cells (35, 97), nourished by the organizable moisture of the style which it imbibes in its course. Now the orifice of the ovules, or a projection of the nucleus beyond the orifice, is at this time brought into contact with, or proximity to, that portion of the walls of the ovary from which the pollen-tubes emerge; and a pollen-tube thus reaches the nucleus, in which the nascent embryo subsequently appears. In the Gymnospermous plants (Coniferæ and Cycadaceæ, 560), the pollen-tubes grow on and immediately penetrate the nucleus of the ovule, just as they do the stigma in ordinary plants.

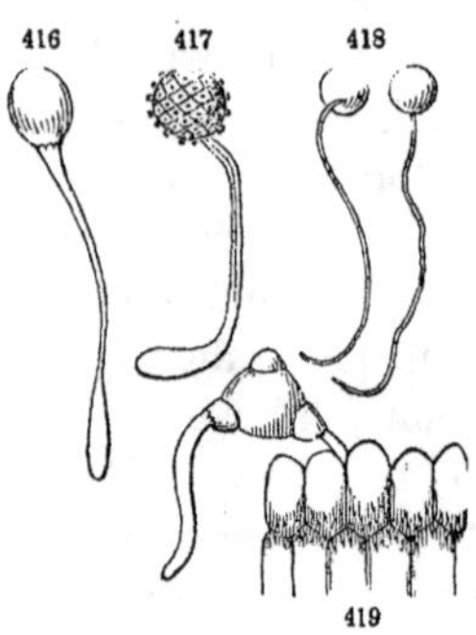

574. The pollen-tubes may be readily inspected under the microscope in many plants; in none more readily than in the Asclepias, or Milkweed, one of the plants in which this subject was so admirably investigated by Mr. Brown. In that family, the pollen-grains of each cell of the anther (Fig. 420) cohere in a mass; and these pollen-masses, dislodged from their cells (Fig. 421, 422), usually by the agency of insects, and brought into proximity with the base of the stigma, protrude their tubes in great abundance, and of a size which renders them visible with a very moderate magnifying power. They may readily be seen to penetrate the base of the stigma, as in Fig. 423, and separate grains with their tubes may be detached from the mass (Fig. 425, 426); but to trace their course down the style (as in Fig. 424), and to their final destination, requires much tact in manipulation and the best means of research. The formation of the pollen-tube commences in some cases almost immediately upon the application of the pollen to the stigma; in many plants it is not perceptible until after the lapse of ten to twenty, or even thirty-six hours. The rate of the growth

FIG. 416. A pollen-grain of Datura Stramonium, emitting its tube. 417. Pollen-grain of a Convolvulus, with its tube. 418. Other pollen-grains, with their tubes, less strongly magnified. 419. A pollen-grain of the Evening Primrose, resting on a portion of the stigma, into which the tube emitted from one of the angles penetrates; the opposite angle also emitting a pollen-tube.

of the pollen-tube down the style is also very various in different plants. In some species, a week or more elapses before they have passed through a style even of a few lines in length. In others, a few hours suffice for their passage through even the longest styles, such as those of Colchicum and Cactus grandiflorus. After the pollen-tubes have penetrated the stigma, the latter dries up, and its tissue begins to wither or die away, as likewise does the body of the pollen-grain, its contents being transferred to the pollen-tube, the lower part of which is still in a growing condition.

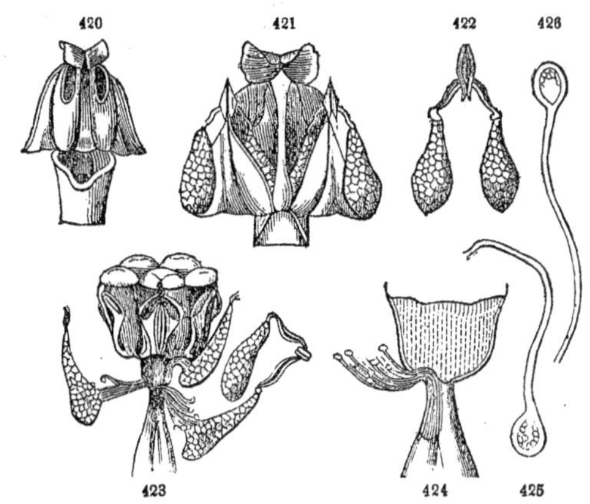

575. **Formation of the Embryo.** Before the pollen-tube reaches the ovule, the nucleus of the latter exhibits a cavity in its interior, towards the apex. In the Mistletoe, this cavity is said to be a mere hollowing out, produced by absorption, and having no evident lining membrane. Usually, however, it results from the special growth of a particular cell, which expands into a bladder or closed sac of considerable size lining the cavity, sometimes remaining inclosed in the tissue of the nucleus towards its summit or orifice, sometimes displacing the upper part of the nucleus entirely, or even projecting through the micropyle. This is the *sac of the amnios* of Mr. Brown, the *embryo-sac* (*sac embryonaire*) of the French botanists.* In this sac the embryo is formed.

* "The ovule is produced by the development of one cell of the placenta,

FIG. 420. A back view of a stamen of the common Milkweed (Asclepias), the appendage cut away. 421. A stamen more magnified, with the two pollen-masses cohering by their *caudicles*, each to a gland from the summit of the stigmatic body, to which a pollen-mass from an adjacent anther is already adherent. 422. A pair of detached pollen-masses (each from a different anther) suspended by their caudicles from the gland. 423. Some of the pollen-masses, with their tubes penetrating the stigma (after Brown). 424. A section through the large stigmatic body and a part of the summit of one of the styles, showing the course of the pollen-tubes. 425, 426. Pollen-grains with their tubes, highly magnified. (The structure of these singular flowers will be more fully explained under the order *Asclepiadaceæ*.)

576. From Linnæus downwards, until recently, it was universally supposed that the embryo originated in the ovule, which was in some way or other fertilized by the pollen. Since the discovery of the pollen-tube in 1824 by Amici, and its actual penetration to the nucleus of the ovule by Mr. Brown, however, the late Professor Horkel and his nephew, Schleiden, — who traced it quite to the embryo-sac, — have propounded a very different view. Schleiden and his followers strongly maintain, as the result of direct observation, that the apex of the pollen-tube itself becomes the embryo; that on reaching the embryo-sac it indents the latter, pushing it forwards so as to reverse a portion on itself, in which cavity the apex of the pollen-tube swells into an oval or globular form, and its contents are transformed into new cells, which, as they grow and multiply, shape themselves into the embryo. Or, according to other observations, it is maintained that the apex of the pollen-tube pierces the embryo-sac and developes into the embryo in its interior, in the manner last stated. It is now unnecessary to adduce the details of the researches, or the theoretical considerations, by which this hypothesis was supported. For, besides the researches of Mirbel, in 1839, the investigations made, between the year 1846 and the present time, by Amici, Mohl, K. Müller, Unger (who had maintained the hypothesis in question), Hoffmeister, Henfrey, and Tulasne, have completely overthrown the foundations on which it rested; by proving, — 1st. That the *embryonal vesicle*, from which the embryo is developed, exists in the embryo-sac, in some cases at least, before the pollen-tube has reached the ovule; so that it cannot owe its origin to the pollen-tube, directly or indirectly, and still less can it be a prolongation of it. 2d. That the end of the pollen-tube is, in many cases, applied to the exterior of the embryo-sac at a point distinguishably, and often considerably, distant from that where the embryo is developed within.*

---

into a cellular body, which essentially consists of a central row of cells, inclosed by a variable number of layers of cells. One of the cells of the central row enlarges and displaces a varying quantity of the rest of the tissue of the ovule. This is the *embryo-sac*." Hoffmeister, as rendered by Henfrey, *Bot. Gazette*, 1, p. 127.

* The latest memoir on this subject, that of Tulasne (in *Ann. Sci. Nat.* for July and August, 1849), is remarkable not only for its thoroughness and its admirable illustrations, but because the author here points out and corrects the error into which he had formerly fallen, which led him to conclude that

577. The general results which all these recent investigations conspire to establish are these: — The pollen-tube, entering the orifice of the ovule, comes directly in contact with the apex of the embryo-sac, penetrating the layer of cells, if there be any, which covers it. Sometimes its extremity slightly indents it; often it glides downwards along the surface of the sac for a little distance; in either case it barely adheres to the membrane, makes no further growth, and after a time begins to wither away. Just before the pollen-tube reaches the embryo-sac, a portion of the protoplasm which the latter contains is attracted into its upper end, next the micropyle. In this protoplasm nuclei appear, usually to the number of three, but sometimes only one, and develope into as many cells. These are the *germinal vesicles*, or *embryonal vesicles*, one of which gives rise to the embryo. The further development of the germinal vesicle begins shortly after the meeting of the pollen-tube with the embryo-sac. The material effect of the former upon the germinal vesicle is supposed to take place by the transudation or endosmotic transference of part of its fluid contents through the membranes of each, and of the embryo-sac between them, into the germinal vesicle, which is thus fertilized; the vitally active contents of two cells of different origin being thus commingled, as in the simpler process of conjugation in the lower Cryptogamous plants (102). Thus endued with new force, the embryonal vesicle, which now adheres to the apex of the embryo-sac, commences an active development; it elongates downwards, or from its free extremity; minute granular matter appears in the interior, which was before perfectly clear and transparent; soon a few transverse partitions are seen, and it is thus converted into a chain of cells, each of which contains a distinct nucleus. This body, which may attain considerable elongation, by the con-

---

the end of the pollen-tube actually penetrates the embryo-sac, and gives rise to the embryonal vesicle. — Hoffmeister asserts (as rendered by Henfrey), that although the pollen-tube generally rests upon the outside of the embryo-sac, yet in a very few isolated cases it perforates it; but "even when the pollen-tube thus penetrates into the interior of the embryo-sac, its end remains perfectly closed, and the membrane of the germinal vesicle quite uninjured: in no case can a direct passage of the contents of one into the other take place. The impregnation is the result solely of an endosmotic exchange of the fluid contents." Henfrey, *Bot. Gazette*, *l. c.*

tinued elongating growth and division of the terminal cell (32 – 34), becomes the *Suspensor*. The lowest of its cells enlarges, and, through cell-formation by division, is converted into a cellular body: this is the nascent *Embryo* (Fig. 430). As it grows it soon

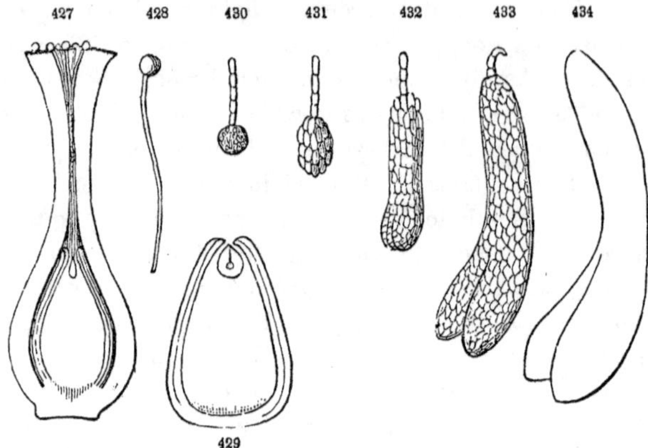

begins to assume its proper form. In a Dicotyledonous plant, as is rudely shown in the accompanying figures, the end farthest from the suspensor begins to be two-lobed (Fig. 432); the lobes increase by ordinary cellular growth, and form the *Cotyledons* (Fig. 433, 434); the opposite extremity is of course the *Radicle*. The suspensor usually disappears before the embryo has attained its full development. A monocotyledonous embryo has this end undivided. In the polycotyledonous embryo of Pines, &c., the budding apex divides successively as it grows into four, six, or more lobes, each of which becomes a cotyledon.*

* The principal points of discordance in the later investigations are connected with the embryonal vesicle. According to Mohl, Henfrey, &c., in Orchis Morio as many as three germinal vesicles exist within the apex of the embryo-sac, anterior to fertilization, as minute free cells, formed from so many nuclei; one (or sometimes more than one) of them enlarges soon after the pol-

FIG. 427. Plan of a vertical section of the pistil of a Polygonum, and of the erect orthotropous ovule it contains, at the period of fertilization: the grains of pollen resting on the stigma have sent their tubes down the style to the mouth of the ovule: and the nascent embryo-sac is seen at the apex of the nucleus. 428. A pollen-grain detached, with its tube. 429. Plan of the vertical section of the ovule more magnified, and at a later period: the nascent embryo with its suspensor seen in the embryo-sac. 430. The nascent embryo with its suspensor, more magnified. 431 – 433. Views of the successive development of the embryo. 434. The embryo as it exists in the seed.

578. Through the fertilization of as many germinal vesicles, two or more embryos are frequently found in the same seed, in the Orange, the Onion, and many other cases. There are generally two embryos in the seed of the Mistletoe; and there is constantly a plurality of embryos in Pines and other Gymnospermous plants (560), though all but one are more commonly abortive or rudimentary.*

579. Contemporaneous with the production of the embryo, a cell-formation takes place in the protoplasm contained in the embryo-sac, soon filling the space with an exceedingly soft and delicate parenchyma, proceeding from the wall of the sac inwards. Sometimes the enlarging embryo, as it grows, obliterates this delicate, half-fluid tissue, is nourished by its contents, and at maturity

---

len-tube has reached the embryo-sac, and developes in the manner above described. According to Hoffmeister, also, in Œnothera two or three germinal vesicles appear a long time before fertilization, from free cell-nuclei, forming so many delicate, free cells, one of which being fertilized developes into the embryo in the manner already described, while the others perish. The results of the more recent researches of Tulasne (upon the embryogeny of Scrophulariaceæ, Campanulaceæ, and Cruciferæ) principally differ in this; — that he was unable to detect any embryonal vesicle before the pollen-tube had penetrated to the embryo-sac; and afterwards he finds only one, mostly of an elongated form, and always from the first attached by one end to the inside of the wall of the embryo-sac, at a point near that to which the pollen-tube is applied externally. He is led to conclude that the embryonal vesicle originates at this point, either from a sort of "dédoublement" of the membrane of the embryo-sac, or from a nucleus adherent there; and he inclines to think that its formation does not precede the conjunction of the pollen-tube with the embryo-sac, but that it is the first visible result of this union. And, calling to mind that Unger met with free cells in the unimpregnated embryo-sac of Hippuris vulgaris, formed from free and floating nuclei, but which were always resolved before the appearance of the real embryonal vesicle, he suggests that the free cells seen by Hoffmeister may be of the same kind. M. Tulasne plausibly considers that the embryo-sac is the cell which receives the fluid of the pollen, and that in its cavity, therefore, the contents of two cells are commingled; the result of which union gives rise to the embryonal vesicle, or potential embryo, endowed from the first with the new specific force which it manifests in its ulterior development. We can only refer the inquirer to this original memoir; an abstract can hardly be made intelligible to the uninstructed reader, without the plates.

* In Coniferæ (at least in the Pines) the embryo is not developed from the embryonal vesicle until long after the cavity of the embryo-sac is filled with the cellular tissue that forms the albumen of the seed; and its formation is in other respects peculiar.

fills the integuments of the seed completely. In other cases, the growth of the embryo in the seed is arrested before it fills the embryo-sac: then this new tissue that surrounds it, solidified by internal deposition, or with its cells filled with starch, &c., becomes permanent, and forms the *albumen* of the seed (627); or sometimes this cellular growth and deposit of nutritive matter take place in the persistent body of the nucleus of the ovule, external to the embryo-sac, as in Nymphæa.

580. With the development of the embryo, the ovule becomes the seed. Its further history should follow that of the fruit.

CHAPTER X.

OF THE FRUIT.

SECT. I. ITS STRUCTURE, TRANSFORMATIONS, AND DEHISCENCE.

581. THE fertilized ovary soon begins to increase in size, and commonly to undergo some change in texture; either becoming dry and membranaceous, crustaceous, or even woody, or else by an opposite change becoming fleshy, pulpy, or juicy: it is now called

582. **The Pericarp**, or *Seed-vessel*. The pericarp and the seeds it incloses together constitute the FRUIT; a term which has a more extensive signification in botanical than in ordinary language; being applied to all mature pistils, of whatever form, size, or texture. The fruit likewise comprises whatever organs may be adnate to the pistils (465). Such incorporated parts, like the fleshy calyx of the Apple and Quince (Fig. 685, 688), sometimes make up the principal bulk of the fruit.

583. It may be remarked that a similar accumulation of fleshy or pulpy matter may take place in adjacent organs wholly unconnected with the pistil; as in the free calyx of the Strawberry Blite (Fig. 993, 995), which becomes greatly thickened, red, and juicy; and in the Wintergreen (Fig. 795–797), where the calyx, at first small and membranaceous, and entirely free from the ovary, gradually enlarges after flowering, and is transformed into a red, pulpy

berry, surrounding the true fruit, which is a small and dry pod. The pulp of the strawberry, moreover, is no part of the proper fruit; but consists of the enlarged and juicy receptacle, or apex of the flower-stalk, bearing the numerous small and dry grains, or true fruits, upon its surface. The bread-fruit and the pine-apple are still more complex, being composed of a whole head or spike of flowers, with their bracts and common receptacle all consolidated into a single fleshy mass. The mulberry is a multiple fruit of the same kind (Fig. 244), in which the component parts may readily be identified. The structure of the fig, which may be likened to a mulberry or a bread-fruit turned inside out, has already been explained (395, Fig. 241 – 243).

584. Under the general name of fruit, therefore, even as the word is used by the botanists, things of very different structure or of different degrees of complexity are confounded. These need to be properly distinguished. For the present, we will consider the fruit in the stricter sense, as consisting of the matured pistil alone, whether simple or compound, either free or in combination with any floral organs, such especially as the tube of the calyx, which, being adnate to the ovary in the flower, is necessarily incorporated with the pericarp in fructification.

585. The pericarp, being merely the matured pistil, should accord in structure with the latter, and contain no organs or parts that do not exist in the fertilized ovary. Some alterations, however, often take place during the growth of the fruit, in consequence of the abortion or obliteration of parts. Thus, the ovary of the Oak (Fig. 1044) consists of three cells, with a pair of ovules in each; but the acorn, or ripened fruit, presents a single cell, filled with a solitary seed. In this case, only one ovule is matured, and two cells and five ovules are suppressed. The ovary of the Horse-chestnut and Buckeye is similar in structure (Fig. 659 – 661), and seldom ripens more than one or two seeds: but the abortive seeds and cells may be detected in the ripe fruit. The ovary of the Birch (Fig. 1053) is two-celled, with a single ovule in each cell: the fruit is one-celled, with a solitary seed; one of the ovules or young seeds being uniformly abortive, while the other in enlarging pushes the dissepiment to one side, so as gradually to close the empty cell (as in Fig. 1056). The Elm presents a similar case (Fig. 1013, 1014); and such instances of suppression in the fruit of parts actually extant in the ovary are not uncommon.

586. On the other hand, the fruit sometimes exhibits more cells than the pistil; as in the two-celled ovary of Datura Stramonium, which soon becomes spuriously four-celled by the projection of the placentæ on one side, so as to reach and cohere with a projection of the dorsal suture on each side. So, also, many legumes are divided transversely into several cells, although the ovary was one-celled with a continuous cavity in the flower.

587. **Ripening.** The growing fruit attracts its food from surrounding parts in the same manner as leaves. When the pericarp preserves its green color and leaf-like texture (as in the Pea, &c.), it is furnished with stomates, and acts upon the air like ordinary leaves. Those which become fleshy or juicy acquire that condition by the accumulation of elaborated sap in their tissue; where it undergoes various transformations, analogous to those which take place in other parts of the plant.

588. Most pulpy fruits are tasteless or slightly bitter during their early growth; at which period their structure and chemical composition are similar to that of leaves, consisting of cellular with some woody tissue; and their action upon the atmosphere is likewise the same (346). In their second stage, they become sour, from the production of acids (353); such as tartaric acid in the grape; the citric in the lemon, orange, and the cranberry; the malic in the apple, gooseberry, &c. At this period they exhale very little oxygen, or even absorb that substance from the surrounding air. The acid increases until the fruit begins to ripen, when it gradually diminishes, and sugar is formed. In the third stage, or that of ripening, the acids, as well as the fibrous and cellular tissues, gradually diminish as the quantity of sugar increases; the latter being produced partly at the expense of the former, by transformations which are very intelligible to the chemist, and which he can partially imitate. A chemical change, similar to that of ripening, takes place when the green fruits are cooked; the acid and the mucilaginous or other products, by the aid of heat reacting upon each other, are both converted into sugar. Mingled with the saccharine matter, a large quantity of *vegetable jelly* (83) is also produced in most acidulated pulpy fruits, existing in the form of *pectine* and *pectic acid.* These arise from the reaction of the vegetable acids during ripening upon the dextrine and other assimilated neutral products accumulated in the fruit.

589. Frequently different parts of the thickness of the pericarp

undergo dissimilar changes during fructification and ripening; the inner portion hardening while the exterior becomes fleshy, or *vice versâ.* When the walls of a pericarp are thus distinguished into two separable portions, the exterior receives the name of EPICARP, or EXOCARP, and the interior that of ENDOCARP. When the exterior part is fleshy or pulpy, as in the peach (Fig. 447) and plum, it is termed the SARCOCARP; and the hard shell or *endocarp* which contains the seed is called the PUTAMEN.

590. Often the walls of the pericarp preserve a nearly uniform texture throughout, becoming either entirely membranaceous, as in many capsules or pods; or fleshy, as in the berry; or indurated throughout, as in the acorn.

591. A part, and in membranaceous or other dry fruits the whole, of the nutritive matter collected in the pericarp is absorbed by the placenta (543) and conveyed to the seed; where the portion which is not consumed in its growth is stored up, either in the embryo or around it, as a provision for its future development in germination.

592. Certain fruits remain closed and entire at maturity, as the acorn, apple, grape, &c.; when they are said to be *indehiscent.* Others separate (wholly or partially) into several pieces, and discharge the seeds; sometimes bursting irregularly, but commonly opening in a uniform and regular manner for each species; these are said to be *dehiscent.*

593. **Dehiscence**, when regular and normal, takes place in a vertical direction, by the opening of one or both sutures (541), or by the disjunction of confluent parts (546). The pieces into which a dehiscent pericarp separates are called its *valves.*

594. A simple carpel dehisces either by the opening of the ventral suture, as in the Columbine, the Peony, &c.; or by the dorsal suture also, as in the Pea and Bean.

595. The dehiscence of a pod which results from the union of two or more carpels may take place by the separation of the constituent carpels from each other, and by the opening of the ventral sutures, as in the Colchicum (Fig. 1115), Rhododendron (Fig. 793), and in the diagram (Fig. 435). In this case, the pericarp splits through the dissepiments; whence the dehiscence is said to be *septicidal.* Sometimes the carpels, although separating from each other in this manner, remain closed or indehiscent, as in the Madder (Fig. 478), the Vervain (Fig. 869), &c.: the separable car-

pels are often termed *cocci;* and the fruit is said to be *dicoccous*, *tricoccous*, &c., according to their number.

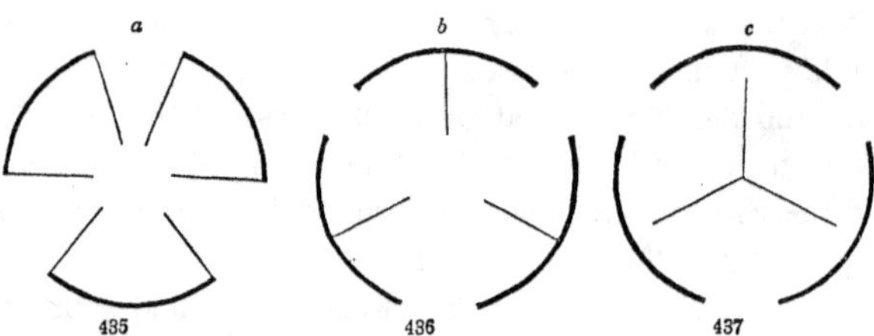

596. Otherwise, the dehiscence may take place by the dorsal suture of each component carpel opening directly into the back of the cells, when the pericarp is more than one-celled; whence this dehiscence is said to be *loculicidal* (as in Fig. 621, 908, 919, and the diagram, Fig. 436). In such cases the dissepiments remain attached to the middle of each valve. In the Helianthemum (Fig. 549), and many other plants, we have an example of loculicidal dehiscence in a one-celled pericarp with parietal placentæ; which in this case are borne directly on the middle of each valve. On the other hand, septicidal dehiscence in a similar pericarp is at once recognizable by the placentæ occupying the margins of the valves.

597. Sometimes the placentæ, being firmly coherent with each other, break away from the dissepiments and remain united in the axis, forming a column, or *columella*, as in Rhododendron (Fig. 793), Polemonium, and Collomia (Fig. 908), &c.

598. Occasionally the dissepiments remain coherent with the axis, while the valves separate from them, as in the Morning-Glory (Fig. 924), and in the diagram, Fig. 437. This modification is termed *septifragal* dehiscence. In like manner, parietal placentæ occasionally separate from the valves, forming what has been termed a *replum;* as in Cruciferous plants, and in the Poppy Family. The same name is applied to the persistent border of the simple pod of Mimosa (Fig. 441).

599. Instead of splitting into separate pieces, the sutures of the pericarp sometimes open for a short distance at their apex only, as in some Chickweeds, and in Tobacco (Fig. 936), and the Primrose (Fig. 826); or by mere points or pores, as in the Poppy.

FIG. 435-437. Diagrams of the dehiscence of capsules (horizontal sections): 435, the septicidal; 436, the loculicidal; 437, the septifragal.

600. In a few cases the opening takes place by a transverse line passing round the pericarp across the sutures, so that the upper part falls off like a lid; as in Anagallis (Fig. 830), the Plantain (Fig. 833), the Henbane (Fig. 941), and the Purslane (Fig. 568). In Jeffersonia, the opening extends only half round the pericarp, and the lid remains attached by the other side, as by a hinge. This anomalous dehiscence is termed *circumcissile* or *transverse.*

Sect. II. The Kinds of Fruit.

601. The various kinds of fruits have been minutely classified and named; but the terms in ordinary use are not very numerous. A rigorously exact and particular classification, discriminating between the fruits derived from simple and from compound pistils, or between those with and without an adnate calyx, becomes too recondite and technical for ordinary use in descriptive botany. Taking first the SIMPLE FRUITS, namely, those that result from single and separate flowers, the principal sorts may be briefly indicated as follows.

602. **A Follicle** is a fruit formed of a single carpel, dehiscing by the ventral suture (541); as in the Larkspur and Columbine (Fig. 483), and the Milkweed (Fig. 460).

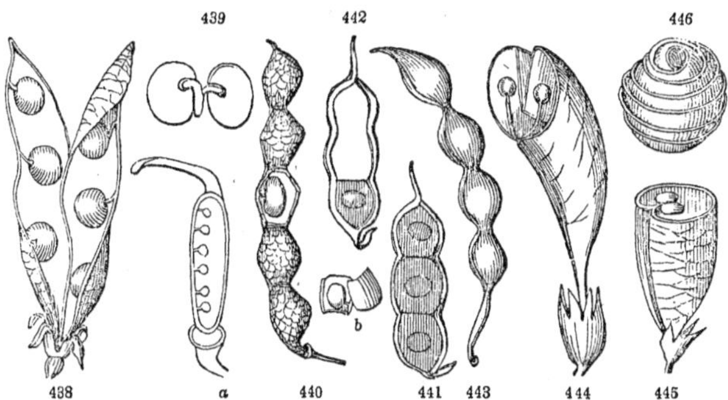

603. **A Legume, or Simple Pod,** is a fruit formed of a single car-

FIG. 438. Open legume of the Pea: *a*, section of the ovary. 439. Embryo, with cotyledons laid open. 440. Loment of Desmodium. 441. Loment of Mimosa: *b*, one of its dehiscent joints which has fallen away from the persisting border or frame (replum), seen in 442. 443. The jointed indehiscent legume of Sophora. 444. A legume of Astragalus, cut across near the summit, to show how it becomes partly or entirely two-celled by the introflexion of the dorsal suture. 445. Similar view of a legume of Phaca, where the ventral suture is somewhat introflexed. 446. A legume of Medicago lupulina, spirally coiled into a globular figure.

pel, and dehiscent by both the ventral and dorsal sutures, so as to separate into two valves; as in the Bean and Pea. The name is extended to the fruit of all Leguminous plants (768), whatever be their form, and whether dehiscent or not. A legume, divided into two or more one-seeded joints, and falling to pieces at maturity, is called a LOMENT, or *lomentaceous* legume. Some of the various kinds of legume are shown in the foregoing figures.

604. **A Drupe, or Stone-Fruit,** is a one-celled, one or two seeded simple fruit which is not dehiscent, with the inner part of the pericarp (*endocarp*, or stone) hard or bony, while the outer (*exocarp*, or *sarcocarp*) is fleshy or pulpy. It is the latter which in our fruits so readily takes an increased development in cultivation. The name is strictly applicable only to fruits of this kind produced by the ripening of a single carpel; as the plum, apricot, peach (Fig. 447), &c.; but is extended in a general way to all one-celled and one or two seeded fruits of similar texture resulting from a compound ovary, and even to those of several bony cells inclosed in pulp, as in the Dogwood (Fig. 240, *b*). The latter, however, are more strictly said to be *drupaceous*, or *drupe-like* fruits.

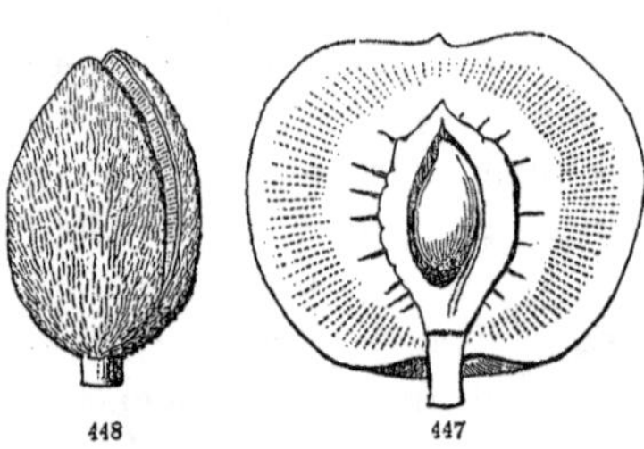

605. **An Achenium** is a small and dry indehiscent one-seeded pericarp, formed of a single carpel; as in the Buttercup, and the allied genera Anemone and Clematis, where they are often terminated by the persistent and often plumose style, in the form of a long tail. In the Rose (Fig. 684), the achenia are borne on the hollow expansion of the receptacle which lines the fleshy tube of the calyx: in Calycanthus the achenia (Fig. 693) are similarly inclosed in a sort of false pod (Fig. 691, 695) of the same nature as the rose-hip, while in the Strawberry (Fig. 678, 679) they are scattered on the surface of the enlarged and pulpy receptacle; where, as in many other cases, they are commonly mistaken for seeds. But they are all furnished with styles, which show their nature; and on cutting them across, we observe the real seed loose

FIG. 447. Vertical section of a peach. 448. An almond; where the exocarp, the portion of the pericarp that represents the pulp of the peach, remains thin and juiceless, and at length separates by dehiscence from the endocarp, or shell.

in the cell. These seed-like fruits were incorrectly called naked seeds by the earlier botanists. The strawberry, raspberry, &c., therefore, taken as a whole, are not simple, but *aggregate fruits.* In the Raspberry and Blackberry (Fig. 680), the achenia are changed into little drupes (604). The name of achenia is also applied to similar one-seeded fruits resulting from a one-celled ovary, even when formed of more than one carpel, and invested by the calyx-tube; as that of the Sunflower and all Composite or Syngenesious plants, where the limb of the calyx, assuming a variety of unusual forms, is termed the *Pappus* (Fig. 776).

606. **A Cremocarp** consists of a pair of achenia placed face to face, and invested by the calyx-tube; which, when ripe, separate from each other, or from a slender central axis, called the *Carpophore;* as in all Umbelliferous plants (Fig. 735 – 737), to which, indeed, the name is restricted. Each separate carpel, or half-fruit, is termed a Hemicarp, or Mericarp, and its inner face the *Commissure.*

607. **A Caryopsis** is a thin and membranaceous pericarp, like an achenium, but adherent to the surface of the seed, so as to be inseparable from its proper covering. The grains of Wheat, Maize, and most Grasses, are examples (Fig. 463 – 465).

608. **A Utricle** is a caryopsis which does not adhere to the seed; or it is an achenium or other one-celled and one-seeded fruit, with a thin and membranous loose pericarp, as in Chenopodium and Amarantus.

609. **A Nut** is a hard one-celled and one-seeded indehiscent fruit, like an achenium, but usually produced from an ovary of two or more cells with one or more ovules in each, all but a single ovule and cell having disappeared during its growth (585); as in the Hazel, Beech, Oak (Fig. 1044), Chestnut, Cocoa-nut, &c. The nut is often inclosed or surrounded by a kind of involucre (393), termed a *Cupule;* as the cup at the base of the acorn, or the burr of the chestnut.

610. **A Samara** is a name applied to a nut, or achenium, having a winged apex or margin; as in the Birch and Elm (Fig. 1014). The fruit of the Maple consists of two united samaræ (Fig. 653).

611. **A Berry** is an indehiscent fruit which is fleshy or pulpy throughout; as the grape, gooseberry (Fig. 707), and persimmon (Fig. 818). The orange, sometimes termed a Hesperidium, is merely a berry with a leathery rind.

612. **A Pome**, such as the apple, pear, and quince (Fig. 685–688), is a fruit composed of two or more papery, cartilaginous, or bony carpels, usually more or less involved in a pulpy expansion of the receptacle or disc, and the whole invested by the thickened and succulent tube of the calyx. It may be readily understood by comparing a rose-hip with a haw, a quince, or an apple.

613. **A Pepo** is an indehiscent fleshy or internally pulpy fruit, composed usually of three carpels, invested by the calyx, and with a firm rind; as the cucumber, melon, and gourd. Its proper structure, which has been variously misconceived, may readily be gathered from a cross-section of a very young melon or gourd (Fig. 449). The three large placentæ project from the axis to the parietes of the cell, where their two constituent parts, more or less separated and recurved, bear the ovules. As the ovary enlarges, the ends of the placentæ usually cohere with the contiguous walls, and the thin dissepiments are at the same time obliterated; so that the fruit presents the deceptive appearance of a three-celled (or, by obliteration of the axis, a one-celled) pericarp, with abnormal parietal placentæ. Sometimes the placentæ are *parietal;* in that case they are revolute without meeting or cohering in the axis.

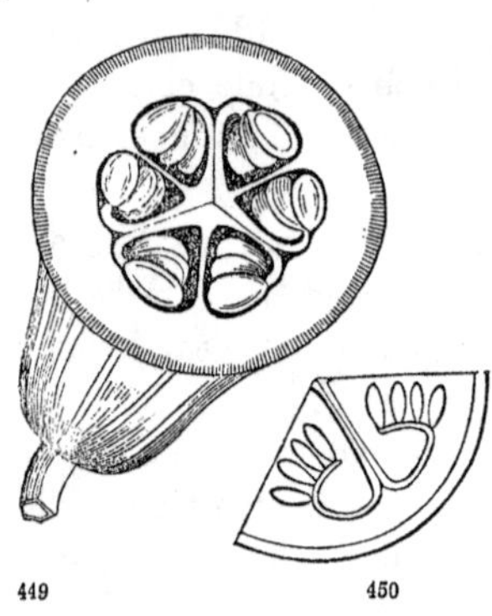

614. **A Capsule** is a general term for all dry and dehiscent pods resulting from a compound ovary, whether opening by valves (593, Fig. 621, &c.), or bursting irregularly, as in Lobelia, or shedding the seeds through chinks or pores, as in the Poppy.

615. **A Silique** is a two-valved capsule, rendered two-celled by a false partition stretched between the parietal placentæ (552), from which the valves separate; as in all Cruciferous plants (Fig. 527), to which family it is confined. A short and broad silique is called a Silicle; as in the Shepherd's Purse or Capsella (Fig. 532).

616. **A Pyxidium, or Pyxis**, is a capsule that opens transversely by a lid or cover, as already explained (600).

617. **Anthocarpous Fruits** are those which, in addition to the peri-

FIG. 449. Section of the ovary of the Gourd; and 450, a diagram of one of its constituent carpels.

carp, have an accessory covering derived from some exterior organ, which, however, does not cohere with the ovary in the fruit; as the nut-like fruit of Mirabilis, the hard outer envelope of which is the indurated and persistent base of the tube of the calyx, which was perfectly free in the blossom. And the berry-like fruit of Shepherdia consists of a fleshy calyx-tube, inclosing a free nut-like pericarp. Instances of this kind are common among what are called

618. **Multiple or Collective Fruits**; or those which result from the aggregation of several flowers into one body or mass. They are, in fact, dense forms of inflorescence, with the fruits or floral envelopes matted together or coherent with each other; as in the pineapple, the mulberry (Fig. 244), &c. The grains of the latter are not the ovaries of a single flower, like those of the blackberry (Fig. 680), but belong to as many separate flowers; and the pulp of these belongs to the floral envelopes instead of the pericarp (583). The fig results from a multitude of flowers concealed in a hollow flower-stalk, if it may be so called, which becomes pulpy and edible (Fig. 241 – 243). Thus the *fruit* seems to grow directly from the branch without being preceded by a flower. In the Partridge-berry (Mitchella repens), and in several species of Lonicera (Fig. 741), the ovaries of two flowers are uniformly united, so as to form a double berry; just as twin apples or cherries are sometimes accidentally produced.

619. **A Cone, or Strobile**, is a collective fruit of the Pine and Cycas Families (Fig. 395, 403); each scale representing an open carpel (375), bearing one or more naked seeds.

620. The *cone* of a Magnolia (Fig. 489) is, however, entirely different, consisting of the numerous aggregated carpels of a single flower, crowded and persistent on an elongated receptacle.

CHAPTER XI.

OF THE SEED.

Sect. I. Its Structure and Parts.

621. **The Seed**, like the ovule (561), of which it is the fertilized and matured state, consists of a Nucleus, usually inclosed within two Integuments.

622. **Its Integuments.** The outer, or proper seed-coat, corresponding to the exterior coat (563) of the ovule, is variously termed the Episperm, Spermoderm, or more commonly the Testa (Fig. 451, *b*). It varies greatly in texture, from membranaceous or papery to crustaceous or bony (as in the Papaw, Nutmeg, &c.), and also in form; being sometimes closely applied (conformed) to the nucleus, and in other cases loose and cellular (as in Pyrola, Fig. 810, and Sullivantia, Fig. 725), or expanded into wings (as in the Catalpa and Bignonia), which render the seeds buoyant, and facilitate their dispersion by the wind; whence winged seeds are only met with in dehiscent fruits. For the same purpose, the testa is sometimes provided with a tuft of hairs at one end, termed a *Coma;* as in Epilobium, Asclepias, or Milkweed (Fig. 963), and Apocynum (Fig. 954). In the Cotton-plant, the whole testa is covered with long wool. It should likewise be noticed, that the integument of numerous small seeds (and also seed-like achenia) is furnished with a coating of small hairs containing spiral threads (one form of which is represented in Fig. 31), and usually appressed and confined to the surface by a film of mucilage. When the seed is moistened, the mucilage softens, and these hairs shoot forth in every direction. They are often ruptured, and the extremely attenuated elastic threads they contain uncoil, and are protruded in the greatest abundance to a very considerable length. This minute mechanism subserves an obvious purpose in fixing these small seeds to the moist soil upon which they lodge, when dispersed by the wind. Under the microscope, these threads may be observed on the seeds of most Polemoniaceous plants, and the achenia of Labiate and Composite plants, as, for example, in many species of Senecio, or Groundsel.

623. The inner integument of the seed, called the Tegmen or

ENDOPLEURA, although frequently very obvious (as in Fig. 451), is often indistinguishable from its being coherent with the testa, or else altogether wanting. Nor when present does it always originate from the secundine or inner coat of the ovule (563). In the Hypericum Family (Fig. 454), in the Pea Family, and probably in a great many other cases, especially where it is tumid or fleshy, or where it adheres firmly to the albumen, it doubtless consists of the remains of the nucleus of the ovule, or of the embryo-sac.

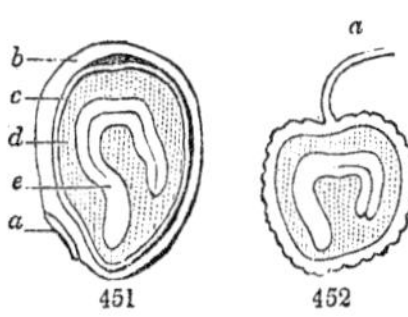

624. The *stalk* of the seed, as in the ovule from which it originated, is called the FUNICULUS (Fig. 452). The scar left on the face of the seed by its separation from the funiculus at maturity is termed the HILUM. The relation of the hilum to the *chalaza*, *micropyle* (563), and other parts of the seed, has been sufficiently indicated when considering the structure of the ovule. The chalaza and rhaphe (567), when present, are commonly obvious in the mature seed, as well as in the ovule (Fig. 455, *b*). The terms *orthotropous*, *anatropous*, *campylotropous*, &c., originally applied to the ovules, are extended to the seeds which result from them; so that we may say, Seeds anatropous, as well as Ovules anatropous, &c.

625. **Aril** (*Arillus*). Some seeds are furnished with a covering, usually incomplete and of a fleshy texture, wholly exterior to their proper integuments, arising from an expansion of the apex of the seed-stalk, or funiculus, or of the placenta itself when there is no manifest seed-stalk. This is called the ARIL. It forms the pulpy envelope of the seed of Podophyllum, Euonymus, and Celastrus, or a mere lateral scale in Turnera, or a tough, lacerated body, known by the name of *mace*, in the Nutmeg. In the White Water-Lily it is a thin, cellular bag, open at the end (Fig. 453). It does not appear in the ovule, but is developed subsequent to fertilization, during the growth of the seed. Of the same nature is the CARUNCLE which grows from the hilum in Polygala, forming a loose lateral append-

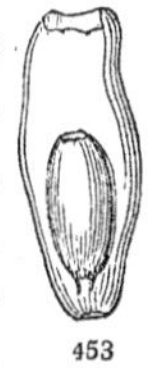

453

FIG. 451. Vertical magnified section of the (anatropous) seed of the American Linden: *a*, the hilum; *b*, the testa; *c*, the tegmen; *d*, the albumen; *e*, the embryo. 452. Vertical section of the (orthotropous) seed of Helianthemum Canadense: *a*, the funiculus.

FIG. 453. Seed of Nymphæa (White Water-Lily), in its membranaceous sac-like aril.

age. Strictly speaking, it is to be distinguished from the STROPHIOLE, the latter being a cellular growth from the micropyle; but the two are not well discriminated. A similar cellular growth takes place on the rhaphe of the Bloodroot, of the Prickly Poppy, and of Dicentra, forming a conspicuous crest on the whole side of the seed.

626. **The Nucleus**, or kernel of the seed, consists of the ALBUMEN, when this substance is present, and the EMBRYO.

627. **The Albumen** (Fig. 451, *d*, 456, *f*) — also variously named the PERISPERM or the ENDOSPERM — which forms the floury part of the seed in our various kinds of grain, consists of whatever portion of the tissue of the ovule persists, and becomes loaded with nutritive matter accumulated in its cells, — sometimes in the form of starch-grains principally, as in wheat and the other cereal grains, sometimes as a continuous, often dense, incrusting deposit, as in the cocoa-nut, the date, the coffee-grain, &c. When it consists chiefly of starch-grains, and may readily be broken down into a powder, it is said to be *farinaceous*, or *mealy*, as in the cereal grains generally, in buckwheat, &c. When a fixed oil is largely mixed with this, it becomes oily, as in the seed of the Poppy, &c.; when more compact, but still capable of being readily cut with a knife, it is fleshy, as in the Barberry, &c.; when it chiefly consists of mucilage or vegetable jelly, as in the Morning-Glory and the Mallow, it is said to be *mucilaginous;* when dense and tough, so as to offer considerable resistance to the knife, as in the Coffee, the Blue Cohosh (Leontice), &c., it is *corneous*, that is, of the texture of horn. Between these all gradations occur. Commonly the albumen is a uniform deposit. But in the nutmeg, as also in the seeds of the Papaw (Fig. 494) and of all plants of the Custard Apple Family, it presents a wrinkled or variegated appearance, owing to numerous transverse divisions, probably caused by inflections of the innermost integument of the seed: in these cases the albumen is said to be *ruminated.*

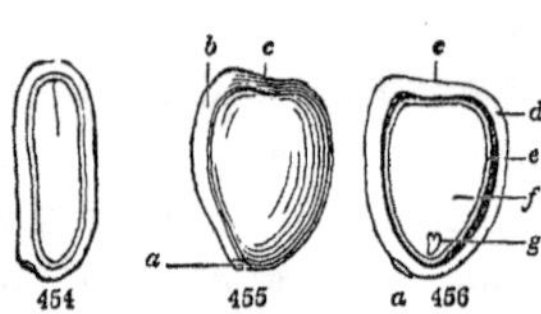

FIG. 454. Vertical section of a seed of Elodea Virginica, showing the two integuments of the seed, and the embryo.

FIG. 455. Seed of Delphinium tricorne (anatropous), enlarged: *a*, the hilum; *b*, the rhaphe; *c*, the chalaza. 456. Vertical section of the same: *c*, the chalaza; *d*, the testa; *e*, the tegmen; *f*, the albumen; *g*, the minute embryo near the hilum, *a*.

628. As already intimated, the albumen may originate from new tissue formed either within the embryo-sac (579), which is probably the more common case; or in the nucleus of the ovule exterior to the embryo-sac, which is certainly the case in the Water-Lily and its allies (the Water-shield, &c., Fig. 518), and in Saururus, for here the thickened embryo-sac persists within or at one extremity of the copious albumen; or both kinds may co-exist. In the first-named case, if any of the proper tissue of the nucleus remains, it is condensed and forms the inner integument of the seed, or becomes confluent with it (623).

629. The office to which the albumen in subservient is the nourishment of the embryo when it begins to develop into a plant. It is a store of nutritive matter, in a very compact or condensed form, accumulated around or next the embryo, which feeds upon it in germination, until it is so far developed that it can obtain and assimilate food for itself (118). The name, therefore, which was applied to it by Gærtner, from its analogy to the albumen or white of the egg of birds, is not inappropriate, although the comparison will not bear to be carried out in detail. As would be expected from its functions, the albumen is the more copious in the seed in proportion as the embryo is smaller and feebler, or less developed. (Fig. 456, compared with Fig. 461, &c.)

630. When the embryo, instead of being arrested in its growth in the seed while yet minute and rudimentary, developes so far as to exhibit its component organs, and form its cotyledons into evident, but usually more or less thickened leaves (as in the Almond, Fig. 457, 458, the Bean, the Maple, Fig. 105, &c.), it absorbs the nutritive matter of the nucleus immediately in the course of its growth; either completely, as in the examples just adduced, or partially, so as to leave a thin albumen (as in Polygala, the Bladder-nut, &c.). In such *exalbuminous* seeds (viz. those entirely destitute of albumen), the requisite store of nourishment, whether of farinaceous, mucilaginous, or oily matter, or frequently of all these kinds combined (as in flax-seed, the walnut, the almond, &c.), is lodged in the embryo, chiefly in the cotyledons, instead of being accumulated around it. Here the embryo occupies the whole cavity, or forms the whole kernel of the seed, and is directly invested by the integuments (Fig. 454, 1047); while in *albuminous* seeds the albumen is interposed between them, at least on one side (Fig. 463, 559), and more commonly on all sides (Fig. 451, 452).

631. **The Embryo,** being an initial plantlet or new individual, is of course the most important part of the seed; and to its production, protection, and support all the other parts of the fruit and flower are subservient. It becomes a plant by the mere development of its parts: it therefore possesses, in a rudimentary or undeveloped state, all the essential organs of vegetation, namely, a root, stem, and leaves, as has already been explained (113, 118, Fig. 105–107). In numerous cases, as in the Maple, the Linden (Fig. 626), and the Convolvulus (Fig. 927), &c., these several parts are perfectly distinguishable in the seed; and the seed-leaves are already foliaceous: sometimes they are large, but thickened by the nourishing matter they contain, as in the Almond (Fig. 457), and the Oak (Fig. 1047). Frequently, however, we only observe an oblong body, cleft or barely two-lobed at one end, as in Fig. 454; but in germination the undivided extremity elongates into a root, the two lobes at the opposite end disclose their real nature by expanding into leaves, and the stem rises between them.

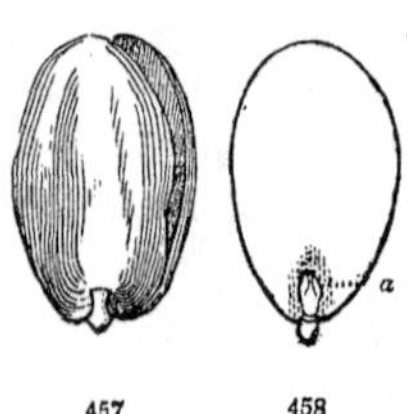

632. The two lobes, or rudiments of the first pair of leaves, are termed Cotyledons; the bud, which, if not actually visible in the seed, as in the Almond (Fig. 458, *a*), appears between them when germination commences, is called the Plumule; and the portion below, which gives rise to the root, is named the Radicle.

633. In these illustrations, we have assumed the embryo with a pair of cotyledons to be the typical, as it is the most common form, occurring as it does in all the families of Exogenous plants (186). Hence the latter are also called Dicotyledonous Plants (188).

634. But in all Endogenous plants only one cotyledon appears, or at least only one on the primary node; if two rudimentary leaves are present, one of them is alternate with the other, and belongs to a second node. Hence Endogens are also termed Monocotyledonous Plants. The monocotyledonous embryo does not usually present the same manifest distinction into radicle, cotyledons, and plumule, as the dicotyledonous; but often appears like a homogeneous and undivided cylindrical or club-shaped body, as in Triglochin (Fig. 460). In this, as in many other

FIG. 457. Embryo (the whole kernel) of the Almond. 458. The same, with one of the cotyledons removed, showing the plumule, *a*.

monocotyledonous embryos, however, a vertical slit, or chink, is observed near the radicular extremity, through which the plumule is protruded in germination. If the embryo be divided parallel with this slit, the plumule is brought into view; as in Fig. 461. If a horizontal section be made at this point (as in Fig. 462), the cotyledon is found to be wrapped around the inclosed plumule, sheathing it, much as the bud and the younger parts of the stem are sheathed by the bases of the leaves in most monocotyledonous plants. The plumule is more manifest in Grasses, especially in the cereal grains, and more complex, exhibiting the rudiments of several concentric leaves, or of a strong bud, previous to germination (Fig. 463 – 465). In many cases, however, no distinction of parts is apparent until germination commences; as in the Onion, the Lily, &c.

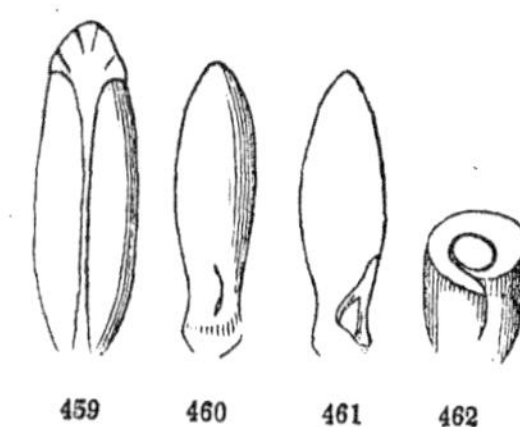

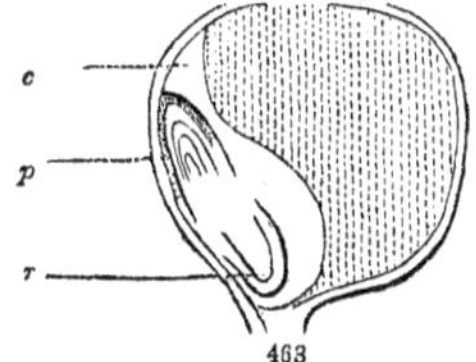

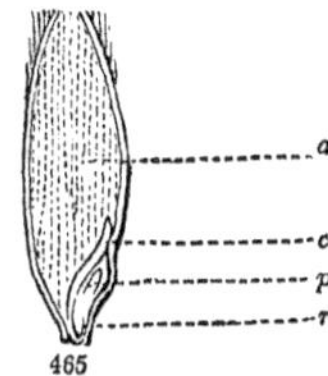

635. The more common of the extremely varied forms under which the embryo occurs may readily be gathered from the numerous illustrations scattered through this volume; which need not be specially enumerated. Its position as respects the albumen, when that is present, is also various. Although more commonly in the axis, it is often *excentric*, or even external to the albumen, as in all Grasses (Fig. 463 – 465), in Polygonum (Fig. 787), &c. When external or nearly so, and curved circularly around the

FIG. 459. Seed of Triglochin palustre; the rhaphe, leading to the strong chalaza at the summit, turned towards the eye. 460. The embryo detached from the seed-coats, showing the longitudinal chink at the base of the cotyledon; the short part below is the radicle. 461. Same, with the chink turned laterally, and half the cotyledon cut away, bringing to view the plumule concealed within. 462. A cross-section through the plumule, more magnified.

FIG. 463. Vertical section of a grain of Indian Corn, passing through the embryo: *c*, the cotyledon; *p*, the plumule; *r*, the radicle. (A highly magnified portion of the albumen, which makes up the principal bulk of the grain, is shown in Fig. 52, p. 57.) 464. Similar section of a grain of Rice. 465. Vertical section of an Oat-grain: *a*, the albumen; *c*, the cotyledon; *p*, the plumule; and *r*, the radicle of the embryo.

albumen, as in Fig. 559, 565, 995, and generally in the families from which these illustrations are taken, it is called *peripheric.* When the embryo is bent so that the radicle is placed against the edges of the cotyledons, the latter are said to be *accumbent* (Fig. 529); or when the radicle rests against the back of one of them (Fig. 538), they are called *incumbent.*

636. The situation of the embryo with respect to the base and apex of the seed is so far uniform, that the radicle always points to the micropyle, as already mentioned. As the nature of the seed may usually, after some practice, be readily determined by external inspection, so the situation of the embryo within, consequently, may often be inferred without actual dissection.

637. The direction of the embryo with respect to the pericarp is also particularly noticed by systematic writers; who employ the terms *ascending*, or *radicle superior*, when the latter points to the apex of the fruit; *descending*, or *radicle inferior*, when it points to its base; *centripetal*, when the radicle is turned towards the axis of the fruit; *centrifugal*, when turned towards the sides; and *vague*, when it bears no evident or uniform relation of the kind to the pericarp.

638. Sometimes the two cotyledons of a dicotyledonous embryo are consolidated or more or less coherent by their contiguous faces into one mass, when they are said to be *conferruminate*, as in the Horsechestnut (Fig. 661).

639. In the Cuscuta, or Dodder, which never produces foliage, the embryo also is entirely destitute of cotyledons (Fig. 122–124). Here these organs are suppressed in an embryo of considerable size; but in most such parasites, the embryo is very minute, as well as reduced to the greatest degree of simplicity, and seems to remain until germination in a very rudimentary state.

640. On the other hand, the embryo assumes the highest complexity in Pines and many other Coniferous plants (400); where the cotyledons as they form are increased in number, from two to four, six, or even fifteen, by collateral chorisis (455); here the embryo is *polycotyledonous.*

Sect. II. Germination.

641. Our narrow limits prevent us from illustrating the various arrangements for the natural dissemination of seeds, which would

form the subject of an interesting chapter; and from considering the circumstances under which the embryo retains its vitality, in many species ordinarily for a few months only, in some perhaps for many centuries.* We must very briefly notice the conditions under which this latent vitality is called into activity, and the embryo is developed into a plant.

642. The conditions requisite to germination are exposure to moisture and to a certain amount of heat, varying from 50° to 80° (Fahrenheit) for the plants of temperate climates, to which must be added a free communication with the air. Direct light, so essential to subsequent vegetation, is unnecessary, if not unfavorable to germination. The degree of heat required to excite the latent vitality of the embryo is nearly uniform in the same species, but widely different in different plants; since the common Chickweed will germinate at a temperature not far above the freezing-point of water, while the seeds of many tropical plants require a heat of 90° to 110° (Fahrenheit) to call them into action, and are often exposed to a considerably higher temperature. Seeds are in the most favorable condition for germination in spring or summer, when loosely covered with soil, which excludes the light while it freely admits the air, moistened by showers, and warmed by the rays of the sun. The water which is slowly absorbed softens all the parts of the seed; the embryo swells, and bursts its envelopes; the radicle is protruded, and, taking a downward direction, fixes itself in the soil; while the other extremity elongates in the opposite direction, bringing the cotyledons (except when these remain under ground, as in the Pea, the Horsechestnut, Wheat, &c.) and

* It is well known that seeds which have been kept for sixty years have germinated; and it seems that grains of wheat, taken from ancient mummies under circumstances which leave little doubt of their high antiquity, have been made to germinate; but in these cases there are several sources of possible deception. Dr. Lindley records the remarkable case of some Raspberries, "raised in the garden of the Horticultural Society from seeds taken from the stomach of a man, whose skeleton was found thirty feet below the surface of the earth, at the bottom of a barrow which was opened near Dorchester. He had been buried with some coins of the Emperor Hadrian; *and it is therefore probable that the seeds were sixteen or seventeen hundred years old.*" Most seeds, when buried deep in the soil, where they are subject to a uniform and moderate temperature, and removed from the influence of the air and light, are in a favorable state for the preservation of vitality, and will germinate when brought to the surface after a long interval.

the plumule, or growing apex of the young stem, to the surface, when the primordial leaves expand in the air. As soon as the root and leaves are developed, each in their appropriate medium, the process of germination is finished; and the plant, deriving through them its nourishment, continues to grow in the manner already described (113).

643. The nourishment which the embryo requires during germination is furnished by the starch, &c. of the albumen (627), when this substance is present in the seed; or by starchy or other matter accumulated in its own tissue (630). But as starch is insoluble in cold water, certain chemical changes are necessary to bring it into a fluid state, so that it may nourish the embryo. These changes are incited by the proteine compounds, or neutral azotized products (354), which are largely accumulated in the seed, whether in the albumen or in the embryo itself (356), and which here, as elsewhere, take the initiative in all the transformations of vegetable matter (27). Here, just as in growth from a bulb or tuber, the changes essentially consist in the transformation of the starch, first into dextrine, or gum, and thence into sugar (350), a part of which is destroyed by resolution, first into acetic acid, and finally into carbonic acid and water, with the abstraction of oxygen from the air, and the evolution of heat (372), while the remainder is rendered directly subservient to the growth of the plantlet. The reason why light, so essential to subsequent growth, impedes or prevents incipient germination, becomes evident when we remember that it incites the decomposition of carbonic acid, and the fixation of carbon by the plant (344 – 350); while germination is necessarily attended by an opposite transformation, namely, the destruction of a portion of organized matter, with the evolution of carbonic acid.

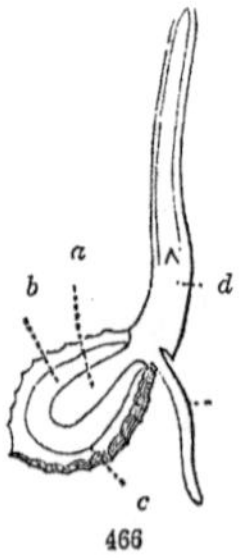

644. In most Dicotyledonous plants, the cotyledons rise out of the ground, and perform more or less perfectly the office of leaves, until those of the plumule expand (Fig. 100 – 107): but when the cotyledons are very thick and fleshy, as in the Horsechestnut, the Pea, the Oak, &c., they serve merely as reservoirs of nourishment, and remain under ground, that is, are *hypogæous* in germination, the first leaves

466

FIG. 466. The germinating seed of Scirpus, a Monocotyledonous plant: *a*, the cotyledon, remaining within the albumen, *b*, inclosed in the pericarp, *c*; from which the plumule (*d*) elongates.

which appear being those of the plumule. This is also the case in all Monocotyledonous plants; in which the cotyledon remains within the integuments of the seed, while the radicle and plumule together pass out at or near the micropyle, as shown in the germinating seed of Scirpus (Fig. 466).

645. Seeds may casually germinate while attached to the parent plant, especially such as are surrounded with pulp, like those of the Cucumber and Melon. The process is liable to commence in wheat and other grain, when protracted warm and rainy weather occurs at the period of ripening; and the albumen becomes glutinous and sweet, from the partial transformation of the starch into gum and sugar. In the Mangrove, which forms dense thickets along tropical coasts, germination commonly takes place in the pericarp while the fruit remains on the tree; and the radicle, piercing the integuments which inclose it, elongates in the air; such a plant being, as it were, *viviparous*. This very naturally takes place, also, in the seeds of *hypogæous* fruits, namely, when the fruit is produced on radical branches, beneath the surface of the soil, as in the Peanut, in Amphicarpæa, Polygala polygama, and many other plants.

CHAPTER XII.

OF REPRODUCTION IN CRYPTOGAMOUS OR FLOWERLESS PLANTS.

646. The general morphology of these simpler forms of vegetation has been very briefly adverted to (Chapter II.) in sketching the progressive development of plants, from those of a single cell or a simple congeries of cells up to those which exhibit the completed type of vegetation. Taken collectively, we distinguish this lower series of the vegetable kingdom by negative characters only; saying that they do not bear true *flowers* (consisting essentially of stamens and pistils), and accordingly do not produce *seeds*, or bodies consisting of a distinguishable embryo plantlet, developed in an ovule, through fertilization by means of pollen. Their *spores* (101), or the bodies produced in their fructification by which they are propagated, and which therefore answer to seeds, are single cells,

in most cases. These, as they germinate in the soil, or whatever medium they grow in, undergo a development at the time of their germination which has been compared with that of the embryonal vesicle (577) during its development into the embryo in the ovule of a Phænogamous plant. But the organs of fructification, and the modes in which the spores are produced, are so exceedingly diverse in the different families of Cryptogamous plants, that botanists are as yet unable to reduce them to a common formula or type, as they have done in Phænogamous vegetation. Each great family of the Cryptogamia seems to be formed on a plan peculiar to itself; each presents a special morphology, and has to be independently treated, — with considerable fulness too, and much particularity of illustration, if the subject is to be made intelligible to the unpractised student. Moreover, the functions of the different organs are in some cases as unsettled as their morphology. The leading characters of the several orders of Cryptogamic plants, and the principal terms applied to their different organs, will be succinctly illustrated in the systematic part of this work (927). Here we have only to notice, very briefly, what is known in respect to their fertilization; concerning which some interesting discoveries have recently been made.

646[1]. The truth of the statement that Cryptogamous plants are *flowerless* depends upon what a flower is defined to be. For it is now demonstrated that at least all the higher orders of Cryptogamous plants are provided with two kinds of reproductive organs, which are so far analogous to stamens and pistils, respectively, that the fertilization of one by the other is essential to reproduction. The structure of these organs, however, the process of fertilization, and the resulting product, are widely different from those of Phænogamous plants. In the latter, one of the grains of pollen, which are the essential part of the stamen, by a peculiar growth (573) comes into proximity with and fertilizes an embryonal vesicle (577) contained in an ovule, which is the essential part of a pistil; and the result is the formation of an *embryo*, or plantlet ready formed in the seed. In Cryptogamous plants the fertilizing cells, or seminal filaments, do not resemble pollen in structure and appearance: the large cells that contain them, being analogous to anthers in function merely, are called ANTHERIDIA, or *anther-like* bodies: the cells or organs upon which they act are analogous in function only either to pistils or ovules, and are therefore named

PISTILLIDIA, or *pistil-like* bodies: and the resulting SPORES (109) are formed, usually four together, free in a mother cell in precisely the same manner that pollen-grains are formed in the anther. The spore, moreover, does not grow at once into a plant like the parent, nor is one part of it predetermined to form one organ of the plant and one another, as in the simplest embryo; but in germination it developes, in all the higher Cryptogamous plants, by a somewhat indefinite multiplication and extension of cells, into a cellular structure, or thallus, called by Hofmeister the PRO-EMBRYO; which is of various form in different orders of plants, but always unlike the parent, and from certain cells of which buds or growing points originate and grow into adult plants. The whole process of fertilization and development presents remarkable differences in different orders of the Cryptogamia.

646[2]. **Reproduction in Mosses.** It is in Mosses that the antheridia and pistillidia were first recognized, and are the readiest observed. The antheridia occur either in the axils of the leaves, or collected into a head at the summit of the stem. They are found either in the same heads as the pistillidia, or in distinct heads on the same individuals (monœcious), or on separate individuals (diœcious). The antheridium (Fig. 1162), is merely a cylindrical or club-shaped sac, composed of a single layer of cells, united to form a delicate membrane; within which are developed vast numbers of minute, very delicate cells, completely filling the sac. The sac bursting at its apex when mature, the delicate vesicles are discharged. These at their first formation contain only an amorphous substance, which turns yellow on the application of iodine: but when mature, a slender filament, thickened at one end and tapering off to a fine point at the other, may be seen through the transparent walls, spirally coiled up in the interior of each vesicle. When these vesicles are extruded in water under the microscope, the contained filaments may be seen to execute lively movements, wheeling round and round in the vesicle, or, when disengaged from the latter, and assuming a corkscrew form, at the same time advancing forward, the thin end of the filament almost always preceding. Minute observation, which is very difficult, both from the rapidity of the motion (which, however, is arrested by poisons), and from the great delicacy of the whole structure, shows that the movements arise from two long and extremely delicate cilia, attached to the tapering end of the filament. The filament itself exhibits no independent motion. The resemblance of these moving filaments to the so-called spermatozoa in animals is manifest. The pistillidia of Mosses (Fig. 1161), which appear at the same time as the antheridia, and often mixed with them, are flask-shaped bodies (like ovaries in shape), with long necks (resembling a style), composed of a single cellular membrane. The neck is perforated by an open canal, leading to the enlarged cavity below, at the base of which a single cell, projecting free into the open space, is the germ of the future capsule or *sporangium* (105), in which a great number of spores are formed. The antheridia are supposed to fertilize the pistillidia by means of the spiral filaments, which are assumed to penetrate the canal of the neck of the pistillidium, and to reach the cell which is afterwards developed into the sporangium or fruit. No such process of fertilization has actually been observed in Mosses: but it is well known that no fruit is produced by plants that bear antheridia alone, and none by the plants that bear only pistillidia unless those with antheridia occur in the vicinity. The spores of Mosses are single cells, with a double coat, like a pollen-grain. In germination, the inner or proper membrane of the spore swells and protrudes, from any part of its surface favorably situated, a tubular process, which forms partitions as it elongates and branches, giving rise to a *pro-embryo* or rudimentary plantlet, which resembles a branched Conferva. Certain cells of its various branches, taking a special development, produce buds, which are soon covered with a tuft of rudimentary leaves, and grow up into the leafy stems of the perfected plant. Here a single spore, or rather the pro-embryo developed from it, gives rise at once to a number of individuals.

646[3]. **Reproduction in Hepaticæ** appears to be affected in a manner physiologically similar to that of the Mosses, especially in those which resemble Mosses in their vegetation. Their antheridia are filled with vesicles containing the same active spiral filaments. In some of the frondose kinds the pistillidia are more like those of the Ferns; but they do not exhibit the remarkable peculiarity of the latter family, which may now be described.

646[4]. **Reproduction in Ferns.** In Mosses, as in all Phænogamous plants, the organs of fructification occur as the last stage of the vegetable development, the perfecting of the seed even involving the death of the individual in numerous cases: and the fertilizing and the fertilized organs are produced at the same time, and the action of one upon the other is immediately succeeded by the full development of the fruit, with its seeds, or spores, as the case may be. But the Ferns, according to recent and most unexpected discoveries, present a very different state of things. Their sporangia (Fig. 1149, 1153), which are not essentially unlike those of Mosses, are, as in the latter, produced and matured on the full-grown plant (usually on the leaves, however, instead of in their axils). But all search for antheridia, whether accompanying the sporangia or upon any other part of the Fern, has been in vain; and consequently the doctrine of sexuality of Cryptogamia, so well established in respect to Mosses, entirely failed in one of the highest Cryptogamic orders. The germination of the spores of Ferns had long since been observed. The process begins in the same manner as in Mosses: but the extremity of the tubular prolongation of the spore, converted by partitions into a row of cells, is developed into an expanded, leaf-like body (the *pro-embryo*, as it is now called), which on a small scale resembles a frondose Liverwort. Upon this body, Nägeli, in 1844, met with moving spiral filaments, like those of the antheridia of Chara, &c. "The announcement of this discovery seemed," as Henfrey remarks, "to destroy all grounds for the assumption of distinct sexes, not only in the Ferns, but in the other Cryptogamia; since it was argued that the existence of these cellular organs producing moving spiral filaments, the so-called spermatozoa, upon the germinating fronds, proved that they were not to be regarded as in any way connected with the reproductive processes. But an essay published by the Count Suminski in 1848 totally changed the face of the question." On the under side of the delicate, Marchantia-like, germinating frond, Suminski found a number of cellular organs of two distinct kinds, namely, antheridia and what he calls "ovules." The former, which are the more numerous, are pedicellated cells on the surface of the germinating frond, in the cavity of which is formed a second cell, filled with minute vesicles containing each a spiral filament coiled up in its interior. The organ bursts at its summit, and discharges the vesicles in a mucilaginous mass; the spiral filaments moving within the vesicles at length make their way out of them and swim about in the water. They resemble those of Mosses, but are flat and ribbon-like, as in Chara, and possess according to Suminski about six, according to Thuret numerous cilia, by whose vibrations they are moved. The so-called "*ovules*" are round cavities in the cellular tissue of the same body, opening on the under side, in the bottom of which is a single globular cell called by Suminski the embryo-sac. Count Suminski asserts that he has even witnessed the process of fertilization in a single instance, observing the entrance of one of the moving filaments into the "ovule." Adopting the Schleidenian doctrine as to the origin of the embryo, he maintains that the vesicular head of a spiral filament penetrates the cell which represents the embryo-sac, and becomes the embryo, producing the first frond and the terminal bud from which the stem of the Fern is developed. These observations and statements have been criticized and extended by Thuret (as to the antheridia), Hofmeister, Schacht, Mettenius, and Merklin (who also testifies to the entrance of spiral filaments into the "ovule"), and in the essential particulars confirmed, except as to the mode of formation of the so-called "embryo" or bud. In Ferns, therefore, it is not the sporangium that is fertilized, nor the spores, but the germinating plantlet formed by the growth of a spore. Without this fertilization, it appears that the pro-embryo of a Fern is incapable of producing leaf-buds and perfecting the development of the plant, as does the Moss: with it, not only buds and the whole vegetation of the individual plant are formed and perfected, but even the fruit and the spores are matured, without further impregnation. While thus in the Ferns the spore forms only the pro-embryo without impregnation, in Mosses it goes on to form the whole leafy stem without impregnation; this operation then taking place, at the same period as in Phænogamous plants, causes the development of the spore-producing part of the plant only. These facts, which may be expressed in various theoretical forms, open some interesting questions and speculations in general physiology.*

646[5]. **Reproduction in Equisetaceæ** is physiologically the same as in Ferns; the antheridia, first detected by Thuret, and so-called "ovules," being produced on an irregular, many-lobed pro-embryo, which results from the germination of the spore.

---

* The English reader is referred to Henfrey's Translation of Mohl's *Anatomy and Physiology of the Vegetable Cell;* and Henfrey's *Report on the Reproduction and supposed Existence of Sexual Organs in the higher Cryptogamous Plants*, in the Report of the British Association for the Advancement of Science, for 1851, reprinted in Silliman's Journal, Vol. 14 and 15; from which the above account has been condensed.

646[6]. **Reproduction in Hydropterides and Lycopodiaceæ** presents still other modifications, not readily explained without many details, and as yet incompletely investigated. Most of these plants produce two kinds of reproductive bodies ; namely, rather large spores, mostly definite in number ; and minute, pollen-like grains, in great abundance. The latter have also been taken for real spores : but it appears that in germination they produce minute vesicles or antheridial cells containing spiral filaments. These doubtless fertilize the minute and transitory pro-embryo formed by the germination of the larger or true spores, on which one or more of the so-called "ovules," and later, "embryos" or growing points, have been detected ; the latter giving immediate origin to the leafy plant. The great difficulty which remains is, that in true Lycopodium only these smaller, pollen-like bodies are produced.

646[7]. **Reproduction of Characeæ.** The two kinds of reproductive organs in Chara have long been recognized, and their relative functions suspected ; the red or orange-colored globule, situated at the base of the conspicuous *nucule* or *sporocarp* (Fig. 1186), having by the earlier botanists been taken for an anther. This is composed of eight shield-shaped valves, containing the coloring matter, and surrounding a cavity into which projects a flask-shaped cell ; and to the apex of this are attached a mass of fine confervoid filaments, divided into a close row of cells, in each of which a spiral filament is developed. These filaments move by means of two very long cilia, attached near one extremity : they escape from the cells after the valves of the globule open, execute very lively movements, and doubtless fertilize the spore-bearing organ, but in what particular manner is not yet well made out. The Charæ are so simple in their organs of vegetation that they have been ranked with the Algæ, and even referred to one of the lowest tribes of that family. But their organs of reproduction ally them rather with the higher Cryptogamia. The remaining, lower forms of Cryptogamous plants have generally been supposed to be strictly asexual, even by those who have maintained the sexuality of Mosses, &c. But very recent researches have now rendered it much more probable that

646[8]. **Reproduction in Thallophytes** generally, both in Lichenes and in the higher grades, at least, of Algæ and Fungi, is effected through the agency of fertilizing cells, or corpuscles of some form, upon the cells in which the spores are produced. The corpuscles of the antheridia do not, indeed, occur in the form of spiral filaments, but are oblong or globular: they are, moreover, motionless, as far as has yet been ascertained, except in the Fucaceæ or olive-colored series of Algæ; in which they execute free and lively movements; and even the cilia by which the motion is effected have been detected by Thuret. As similar, although motionless, corpuscles have been discovered in the Florideæ or Rose-red Algæ, in many Fungi, and in almost all the genera of Lichenes, occurring as a regular part of the structure, and at a certain epoch, there can be little doubt that they subserve similar functions in all these cases; and it is in the highest degree probable that these functions are analogous to those of the spiral filaments of Chara, the Mosses, and the other Cryptogamia of the higher grades. We can here barely refer to those recent memoirs in which these important discoveries are recorded or illustrated, viz.:—Nägeli in *Bot. Zeitung*, 1849. Itsigsohn in *Bot. Zeitung*, 1850. Thuret in *Ann. Sci. Nat.*, 3d ser. 14 & 16, 1850–1. Harvey, *Nereis Bor.-Amer.* in *Smithsonian Contributions*, 1852. Berkeley and Broome in *Report of British Association* for 1851. Tulasne in *Comptes Rendus*, 1851, and *Ann. Sci. Nat.* 17, 1852.

CHAPTER XIII.

OF THE SPONTANEOUS MOVEMENTS WHICH PLANTS EXHIBIT.

647. The facts brought to view in the preceding chapter respecting the production by Cryptogamous plants of minute bodies temporarily endowed with the power of locomotion, leads to the consideration of the various movements which are executed by the ordinary organs of vegetables. Plants, like other living beings, execute certain movements, or changes in the position of their parts, through some inherent powers, which, though far less striking and less varied than in animals, and of a nature different from muscular motion, must not be overlooked.

648. **The Special Directions** which the organs of the plant assume belong to this class of manifestations, although the movements are mostly much too slow to be directly observed. Among these the

most universal are the invariable descent of the root in germination, the ascent of the stem into the light and air, and the turning of branches and the upper surface of leaves towards the light (113, 139, 294). Although these movements are incited by common physical agents, and cannot be the result of any thing like volition, yet all of them are inexplicable upon mechanical principles. Some of them, at least, are spontaneous motions of the plant or organ itself, due to an inherent power, which is merely put in action by light, attraction, or other external influences.

649. The external agencies concerned in the descent of the root and the rise of the stem seem chiefly to be, — 1st, the attraction of the earth acting upon the root; and 2d, the influence of light upon the stem. The influence of gravitation, or of a similar force, was proved by the celebrated experiment of Mr. Knight; who caused the seeds of the Bean to germinate in a quantity of Moss fastened to the circumference of a wheel, which was made to revolve vertically at a rapid rate; where the effect of gravity was replaced by that of centrifugal force. On examination, after some days, the young root and stem were found to have taken the direction of the axis of rotation; the former being turned towards the circumference, and the latter towards the centre of the wheel. The same result took place when the wheel was made to revolve horizontally with considerable rapidity; but when the velocity was moderate, the roots were directed obliquely downwards and outwards, and the stems obliquely upwards and inwards, in obedience both to the centrifugal force and the power of gravitation, acting at right angles to each other. The different behavior of the root and stem is here supposed to depend upon their different mode of growth. The former growing at its extremity only, the soft substance of the growing point was supposed to obey the attraction of gravitation, and curve downwards; while the latter growing by the elongation of a series of internodes already formed, the solid tissues would be unaffected by gravity, which could affect only its nutritive juices, causing their accumulation on the lower side of a stem out of the perpendicular line, which side, thus more actively nourished, would grow more vigorously than the upper, and so cause the stem to turn upwards. There are several objections to this explanation; among them is the fact that the root is capable of penetrating a fluid of greater density than its own substance, such as mercury. That light is the chief cause of the upward direction

of the stem, while it is avoided by the roots, appears from experiments by Schultz and Mohl; who reversed the natural condition, by causing seeds to germinate in Moss, so arranged that the only light they could receive was reflected from a mirror, which threw the solar rays upon them directly from below; in which case it was found that their roots were sent upward into the Moss, contrary to the ordinary direction, and their stems downward towards the light.

650. The Mistletoe obeys the attraction of the trunk or branch upon which it is parasitic (134), just as ordinary plants obey the attraction of the earth; its roots penetrating towards the centre, while the stems grow perpendicular to the surface of the branch, and are therefore placed in various positions as respects the earth. When the germinating seeds of the Mistletoe were fixed on the surface of a cannon-ball, all the radicles were found to be directed towards its centre. A well-devised experiment made by Dutrochet goes to show, that the pointing of the radicle to the adjacent body (and consequently of the germinating root generally towards the earth's centre) is not the result of the immediate attraction of the adjacent body, or of the earth, but is a spontaneous movement due to some internal, vital cause, put in action by the exterior influence. He mounted the seed of a Mistletoe upon one extremity of a very delicately balanced needle, which would turn with the slightest force, and placed it at the distance of half a line from the surface of a large cannon-ball. In germination the radicle directed its point to the ball, and soon came into contact with the surface; but the end of the needle had not moved in the slightest degree towards the ball, as it would have done from a mere exterior attraction. By such experiments Dutrochet has proved that the curvature of the root to grow downward or from the light, and of the stem in the opposite direction, is independent of their growth; and that the cause must be looked for in the cells of the organs themselves. He has attempted the explanation by the aid of endosmosis under different conditions of light, &c.; but without full success, except in showing that curvature is not produced by a contraction of the side which becomes concave, but by the enlargement of that which becomes convex.

651. When the stem has emerged from the earth, it tends to expose itself as much as possible to the light, the growing parts always turning towards the side most strongly illuminated; as is observed when a plant is placed in an apartment lighted from a

single aperture. This is mechanically accounted for by De Candolle, on the supposition, that, as the side upon which the light strikes will fix most carbon by the decomposition of carbonic acid, so its tissue will become more solid than the shady side, and therefore elongate less rapidly; and the stem or branch will consequently bend towards the light. But when the light is equally diffused around a plant, the decomposition of carbonic acid will take place uniformly on all sides, and the perpendicular direction naturally be maintained. The insufficiency of this explanation is shown by the fact, that, when a stem so curved is split through, the concave side curves more than before, while the convex side springs back into the upright position. Moreover, the decomposition of carbonic acid is effected chiefly under the influence of the yellow rays, and the curvature of green stems by the blue (652[1]). The same relation of upward-growing organs to light regulates the disposition of branches and branchlets, which are invariably so arranged as to have the greatest possible exposure to the light; the uppermost branches of a tree growing nearly erect, those beneath extending more horizontally until they reach beyond their shade, when they curve upwards (unless too slender to support their own weight, as in the Weeping Willow), and the lower being still more divergent, or even turned downwards, when the foliage is dense. Certain drooping branches, however, are exceptions to this rule, such as those of the Weeping Ash, which have a constitutional tendency to turn downwards. And the widely different positions assumed by the branches of different species under the very same external influences, show that the directions are owing to differences in their own specific organization. The direct action of light is confined to the green parts of plants. Where the surface has lost its green color, branches are no longer affected by the light; and those which creep under ground beyond its influence (173), and have the white color and much the external appearance of roots, show little upward tendency so long as they remain in this situation; but whenever their extremities are exposed to the light, they first acquire a green hue by the formation of chlorophyll, and then tend to assume a vertical direction.

652. The principal exception to the rule that green parts turn to the light is that of certain tendrils, such as those of the Vine and the Virginia Creeper, which avoid it, although of a green color. This has been said to be the case in tendrils generally; but, ac-

cording to Mohl, many turn towards the light, while others appear indifferent to its influence.

652[1]. It has been shown both by Payer and Macaire, that curvature towards the light is produced in unequal degrees by the different rays of the spectrum, and this independently of their illuminating power; the blue and violet rays being most efficient, the yellow producing little effect, and the red none at all. There is no real connection, therefore, between this phenomenon and the evolution of oxygen or fixation of carbon, which does not take place under blue light.

653. In leaves it is the denser and deeper green upper surface (262) that is presented to the light, while the paler lower surface, of looser tissue, avoids it, like a rootlet. The recovery of the natural position when the leaf is artificially reversed, is the more promptly effected in proportion to the difference in structure and hue between the two strata. This movement in leaves is so prompt, that those of many plants follow the daily course of the sun. The leaf is more capable of executing such movements, on account of its extended surface, and its pliancy, and also on account of its usual attachment by an articulation. Herethe slender vascular bundles oppose little resistance to lateral motion, while the soft and usually cellular enlargement favors it. Indeed, the efficient cause of the movement appears to be exerted here, and to be connected with the unequal tension or turgescence of the cells on the two sides. But how the light acts in producing the movement, we are wholly ignorant. One of the most striking and general of these changes of position was termed by Linnæus

654. **The Sleep of Plants**, namely, the peculiar position which the leaves of many plants assume, either apparently drooping, or folding together their leaflets, as if in repose, when the stimulus of light is removed. This is well seen in the foliage of the Locust and of most Leguminous plants, and in those of Oxalis, or Wood-Sorrel. It is most striking in the leaflets of compound leaves. Their nocturnal position is various in different species, but uniform in the same species, showing that the phenomenon is not mechanical. Nor is it a passive state, for, instead of drooping as if by their own weight, the leaflets are more commonly turned upwards or forwards, contrary to the position into which they would fall from their own weight. De Candolle found that most plants could be made to acknowledge an artificial day and night,

by keeping them in darkness during the day, and by illuminating them with candles at night. The sensibility to light appears to reside in the petiole, and not in the blade of the leaf or leaflet: for these movements are similarly executed, when nearly the whole surface of the latter is cut away.

655. The leaves of the blossom also assume various positions, according to the intensity and duration of the light. Many expand their blossoms in the morning and close them towards evening, never to be opened again, as those of Cistus and of many Portulacaceous plants; while others, like the Crocus, close when the sun is withdrawn, but expand again the following morning. On the other hand, the Evening Primrose, Silene noctiflora, &c. unfold their petals at twilight, and close at sunrise. The White Water-Lily (Nymphæa) expands in the full light of day, but uniformly closes near the middle of the afternoon, and is then usually withdrawn beneath the surface of the water. The Morning-Glory opens at the dawn; the Lettuce and most Cichoraceous plants, a few hours later; the Mirabilis, or Four-o'clock plant, nearly at that hour in the afternoon, &c. Berthelot mentions an Acacia at Teneriffe, whose leaflets regularly close at sunset and unfold at sunrise, while its flowers close at sunrise and unfold at sunset.*

655[1]. Although these movements, both in leaves and blossoms, are undoubtedly dependent on the light, they are by no means directly governed by it. The sleep of the common Sensitive Plant, for instance, begins just before sunset, but its waking frequently precedes the dawn of day. No mere physical action of the light or other external agent, therefore, can be made really to account for such movements, any more than in the analogous cases of

656. **Movements from Irritation.** The leaflets of numerous Leguminous plants, especially of the Mimosa tribe, when roughly touched, assume their peculiar nocturnal position, or one like it, by a visible and sometimes a rapid movement. The Sensitive Plant

* The odors of flowers, also, are sometimes given off continually, as in the Orange and the Violet, or else they nearly lose their fragrance during the heat of midday, as in most cases; while others, such as Pelargonium triste, Hesperis tristis, and most dingy flowers, which are almost scentless during the day, exhale a powerful fragrance at night. The night-flowering Cereus grandiflorus emits its powerful fragrance at intervals; sudden emanations of odor being given off about every quarter of an hour, during the brief period of the expansion of the flower.

of the gardens (Mimosa pudica) is a familiar instance of the kind, suddenly depressing its leaflets upon the secondary petioles on being touched or jarred, and applying them one over the other upon the secondary petiole ; if more strongly irritated, the secondary petioles also bend forward and approach each other, and the general petiole itself sinks by a bending at the articulation with the stem. Similar but less vivid irritability is shown by the Mimosa strigillosa and the Schrankia of the Southern States, where the leaflets promptly fold up when brushed with the hand. The most remarkable instance of the kind, however, is presented by another native plant of the United States, the Dionæa muscipula, or Venus's Fly-trap (Fig. 228) ; in which the touch even of an insect, alighting upon the upper surface of the outspread lamina, causes its sides to close suddenly, the strong bristles of the marginal fringe crossing each other like the teeth of a steel-trap, and the two surfaces pressing together with considerable force, so as to retain, if not to destroy, the intruder, whose struggles only increase the pressure which this animated trap exerts. This most extraordinary plant abounds in the damp, sandy savannas in the neighborhood of Cape Fear River, from Wilmington to Fayetteville, North Carolina, where it is exceedingly abundant; but it is not elsewhere found.

657. A familiar, although less striking, instance of the same kind is seen in the stamens of the common Barberry, which are so excitable, that the filament approaches the pistil with a sudden jerk, when touched with a point, or brushed by an insect, near the base on the inner side. The object of this motion seems plainly to be the dislodgement of the pollen from the cells of the anther, and its projection upon the stigma. But in the Dionæa it is difficult to conceive what end is subserved by the capture of insects. In a species of Stylidium of New Holland, not uncommon in conservatories, the column, consisting of the united stamens and styles, is bent over to one side of the corolla ; but if slightly irritated, it instantly springs over to the opposite side of the flower.

657[1]. Anatomical investigation brings to view no peculiar structure of parts in these cases; except that the abundant parenchyma at the articulation, where the movement is effected, is found to be in a state of unusual turgescence, and its cells are compressed in the direction of the longitudinal axis of the joint: a small piece cut out suddenly expands about one fifth in size, as if the vascular bundle were too short for the parenchyma. Consequently, if a portion of the parenchyma be cut away down to the vascular bundle on one side of the joint, in the Sensitive Plant, the leaflet or stalk is immediately pushed to that side by the expansion of the tissue of the opposite side, which has now lost its antagonist. The irritability in the Sensitive Plant principally resides in the under side of the joints, which become concave in the motion; but the irritation of other parts is promptly propagated to the seat of motion. How the irritation brings about the movement is unknown.

658. Some other movements, which have been likened to these, are entirely mechanical in their nature; as that of Kalmia, or Sheep Laurel, where the ten anthers are in the bud received into as many pouches of the monopetalous corolla, and, being retained by a glutinous exudation, are carried outwards and downwards when the corolla expands. In this way the slender filaments are strongly recurved, like so many springs, until the anthers open, and the pollen absorbs the glutinous matter that confines them, when they fly upwards elastically, throwing the pollen in the direction of the stigma. The elastic dehiscence of the Balsam, or Touch-me-not (Impatiens), and the bursting of similar fruits, are also due to the great turgescence of the cells of the outer layer, and are therefore wholly mechanical.

658[1]. The twining of stems round a support and the coiling of tendrils are attributed by Mohl to a dull irritability; and this is the most plausible explanation that has been offered. The inner side, which becomes concave and has smaller cells, is in this, as in other cases, the irritable portion. When a foreign body, such as a stem, is reached, a contraction of this side causes the tendril partially to embrace the support: this brings the portion just above into contact with it, which is in like manner stimulated to the curving movement, and so the hold is secured; or in a twining stem the growing organ continues to wind around the support. In tendrils this irritability, propagated downward along the concave side, causes its contraction (or more probably the increased turgescence of the opposite side), which throws the whole into a spiral coil, and brings the growing stem nearer to the supporting body.

659. **Automatic Movements.** A few plants are known which execute brisk and repeated movements irrespective of extraneous excitation, and which, indeed, are arrested by the touch. An instance of such spontaneous and continued motion, of the most remarkable kind, is furnished by the trifoliate leaves of Desmodium

gyrans, an East Indian Leguminous plant. The terminal leaflet does not move, except to change from the diurnal to the nocturnal position, and the contrary; but the lateral ones are continually rising and falling, both day and night, by a succession of little jerks, like the second-hand of a time-keeper; the one rising while the other falls. Exposure to cold, or cold water poured upon the plant, stops the motion, which is immediately renewed by warmth. The late Dr. Baldwin of Georgia is said by Nuttall to have witnessed the same thing in Desmodium cuspidatum! In several tropical Orchideous plants, and especially in a species of Megaclinium, one of the petals executes similar and perfectly spontaneous automatic movements.

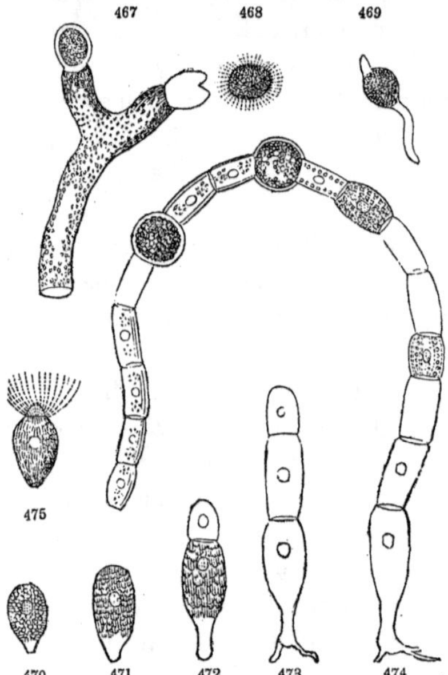

660. Free Movements of the Spores of Algæ. The spores of many of the lower Algæ are now known to exhibit a peculiar active state at the time of their discharge from the parent cell, when, for some moments, or usually for several hours, they behave like infusory animals, executing free, and to all appearance spontaneous, movements in the water, until they are about to germinate. This singular movement was first detected, many years ago, in Vaucheria. In Fig. 71 (p. 67) we see the manner in which the spore is formed; and in Fig. 72, the mode in which it is discharged: also, on a larger scale, in Fig. 467. It at once begins to move freely in the water, and continues to do so for some hours, when it fixes it-

FIG. 467. Fruiting end of a plant of Vaucheria geminata (after Thuret); one of the branches still containing its spore. 468. Moving spore just escaped from the apex of the other branch; the ciliary apparatus seen over the whole surface. 469. Spore in germination.

FIG. 470–473. Successive steps in the germination of Conferva vesicata. 474. The plant developed into a series of cells, four of which display the successive steps in the formation of a spore. 475. The locomotive spore with its vibratile cilia (copied from Thuret). When the movement ceases, and it begins to germinate, it appears as in 470.

self and begins to grow (Fig. 469). Its movements, moreover, like those of the antheridial filaments or corpuscules (646^2), may be enfeebled or arrested by the application of a weak solution of opium or of chloroform. Through these means it has been ascertained that they are caused by the vibrations of minute cilia which cover the surface, which are rendered visible by thus enfeebling their movement, and which exhibit the closest resemblance to the vibratile cilia of animals, especially those of the polygastric animalcules! In Conferva vesicata the vibratile cilia occupy one end of the spore (Fig. 475). In other species they are likewise restricted to some part of the surface, and are only two or three in number.

661. In Oscillaria (Fig. 66, p. 66) the fully developed plant exhibits occasional writhing movements, so well marked that the vegetable character of the genus was long questioned. The Closteria (Fig. 77) and other minute Desmidiaceous plants exhibit occasional well-marked spontaneous movements of translation: and the nearly allied Diatomaceæ — the lowest and most ambiguous of plants — were long referred to the animal kingdom, on account of the motions they exhibit. The lowest tribe of plants, in this as in other respects, makes the closest approach to the lowest tribes of animals.

662. Not only, therefore, do many, if not all, plants manifest impressibility or *sensitiveness* to external agents, and more or less decided, though slow, movements; but many species of the higher grades exhibit certain vivid motions, either spontaneous or automatic, or in consequence of extraneous irritation; while the lowest tribes of aquatic plants, as they diminish in size and in complexity of organization, habitually exhibit, at some period of their lives at least, varied spontaneous movements, that we are unable to distinguish in character from those of the lowest animals (16) which are likewise made by cilia.

663. When we consider that the excitability of sensitive plants is often transmitted, as if by a sort of sympathy, from one part to another; that it is soon exhausted by repeated excitation (as is certainly the case in Dionæa, the Sensitive Plant, &c.), and is only renewed after a period of repose; that all plants require a season of repose; that they evolve heat under special circumstances (372 – 374); that, as if by a kind of instinct, the various organs of the vegetable assume the positions or the directions most favorable to the proper exercise of their functions and the supply of their wants,

to this end surmounting intervening obstacles; — when we consider in this connection the still more striking cases of spontaneous motion that the lower Algæ exhibit; and that all these motions are arrested by narcotics, or other poisons, — the narcotic and acrid poisons even producing effects upon vegetables respectively analogous to their different effects upon the animal economy; — we can hardly avoid attributing to plants a *sensibility* and a power of "making movements tending to a determinate end," not different in nature, perhaps, from those of the lowest animals. Probably the *vitality* is essentially the same in the two kingdoms; and to this, faculties are superadded in the lower animals, some of which are here and there not indistinctly foreshadowed in plants.

664. Finally, if called upon to define a plant, or draw the line between the animal and the vegetable kingdoms, we can only say, — 1. That plants alone, under the solar influence, create organic matter from inorganic materials, and alone live, or are capable of living, by direct aggression upon the mineral world. Consequently, they alone decompose carbonic acid, and render free oxygen gas to the atmosphere (Chap. VI.): the action of animals upon the air is uniformly and continually the reverse. 2. In its structure, a plant may be reduced to a single simple vesicle of cellular tissue (94), containing chlorophyll, or its equivalent. But a developed animal of the lowest grade has a more complex structure: from the necessity of the case it possesses a mouth and a stomach. Indeed, we have reason to believe that the polygastric animacules are considerably complicated in structure. 3. As to chemical composition, the tissue of plants, or the material of which the fabric is constructed, is a neutral ternary product (27, 347), composed of carbon, hydrogen, and oxygen. Although the plant necessarily contains and produces the quaternary organic products, these do not enter into the composition of its permanent fabric. The animal tissue, on the contrary, is directly composed of neutral quaternary products, consisting of carbon, hydrogen, oxygen, and nitrogen.* Although such distinctions as these are, in all probability, absolute, yet it is often difficult, and frequently, perhaps, impossible, to apply them to the actual discrimination of the lower plants from the lower animals.

* To this, indeed, an exception has been announced, in the case of certain Mollusca, whose tissue is said to be identical with cellulose!

PART II.

SYSTEMATIC BOTANY.

---

665. We have now to contemplate the vegetable creation from a different point of view. In studying the structure and physiology of plants, we have been struck with the countless varieties which they present, — the almost infinite number of particular modes or forms in which the general plan of vegetation has been worked out, as it were, in detail. The vegetable kingdom, that is, vegetation taken as a great whole, presents to our view an immense number of different kinds or sorts of plants, more or less resembling each other, more or less nearly related to each other. It is the object of Systematic Botany to consider them in respect to these resemblances and differences, — to contemplate the relations which the individual members of the great whole sustain to each other (5, 6). In this view, the botanist classifies them, so as to exhibit their relationships, or points of resemblance, arranges them in an orderly manner, designates them by proper names, and distinguishes them by clear and precise descriptions; so that the name and place in the system, the known properties, and the whole history of any given plant, may be readily and surely obtained by the learner.

CHAPTER I.

OF CLASSIFICATION AND ITS PRINCIPLES.

666. **Individuals.** The vegetables with which the earth is adorned are presented to our view as Individuals only, more or less

resembling, or differing from, each other. Among these, some are so essentially alike, that we involuntarily apply to them the same name. A field of Wheat is filled with similar individuals, which we can *separate*, but cannot *distinguish*. Or, although it be possible to distinguish separate individuals, from any peculiarity of size, &c., we still inevitably associate them, as being much more like each other than like any surrounding forms, — so like, that we view the difference as an accidental circumstance. Furthermore, the Wheat *tillers*, that is, branches from the ground, and shoots forth a number of stalks from the same root, — stalks which are separable, or separate spontaneously, from the primary one. So, also, the branches of trees, which may grow indefinitely as a part of an original tree (148), become, when detached and planted by themselves in the soil, independent, but perfectly similar individuals (167, 229). Probably all the Weeping Willows, or Lombardy Poplars, of this country have sprung in this way from a single shoot. The grain of wheat, also, will reproduce similar individuals, and none other. Now, upon such universal and inevitable conceptions as these rests the idea of

667. **Species.** We mentally assemble, under this name, those individuals which we observe or judge to have arisen from one parent stock, or which, although met with widely dissociated, resemble each other so closely that we infer them to have had a common parentage. A SPECIES we have already defined (14) to be, abstractly, the *type* or *original* of each sort of plant, or animal, thus represented in time by a perennial succession of like individuals, or, concretely, the sum of such individuals. It embraces all those individuals which, slightly differing, perhaps, in size, color, or such unimportant respects, resemble each other more nearly than they resemble any other plants, so that we infer them to have sprung from a common original stock, and which preserve their characters unchanged when propagated by seed. All classification and system in natural history rests upon the fundamental idea of the original creation of certain forms, which have naturally been perpetuated unchanged, or with such changes only as we may conceive or prove to have arisen from varying physical influences, accidental circumstances, or from cultivation. Whether the original stock consisted of one individual or pair, or of numerous individuals, is not material to the view. (On the latter supposition, however, we can readily perceive that certain *varieties* or *races* may have been aboriginal.)

668. **Varieties.** This fraternal resemblance, or specific identity, however, is not incompatible with individual peculiarity. If two seeds from the same pod are sown in different soils, and submitted to different conditions as respects heat, light, and moisture, the plants that spring from them will show marks of this different treatment in their appearance. Such differences are continually arising in the natural course of things. To produce and increase, and by artificial management to perpetuate, differences of this sort, forms an important part of the art of cultivation. These minor deviations, not incompatible with the idea of a common origin, constitute VARIETIES. Whenever the conditions that give rise to varieties are carried to excess, these individuals fail to fructify, or perish. When the conditions vary less widely from those most propitious to the constitution of the particular species, a few years or a few generations may suffice to bring the variety back to the original form. In either case, the variation is transient. It must either return to the common character of the species, or perish. A certain flexibility is allowable; but accidental and individual variations tend to disappear with the causes which originate them, or are destroyed by the continued operation of those causes.

669. To this there is one class of exceptions, which is exceedingly common in domesticated plants; where the habit, once established, outlasts the cause, and continues throughout the life of the individual. The new buds and branches partake of the peculiarity, and the variety may consequently be perpetuated by cuttings, grafts, &c.; as is the case with our Apples, Pears, &c. But this tendency does not inhere in the seed.

670. **Races.** There is still another and more strongly marked kind of variety, — though unknown, perhaps, in a perfectly wild state, — in which the characteristics are transmissible by seed. Particular varieties of Peas, Radishes, Lettuce, &c., are thus perpetuated in our gardens; and in agriculture, various sorts of grain have thus been preserved from time immemorial. They have received the name of *Races.* It is not known how they originate. They start up, as it were, accidentally, from time to time, in cultivated plants. The cultivator selects the most promising sorts, or "sports," for preservation, leaving the others to their fate. By peculiar care he developes and strengthens the tendency to become hereditary, and renders it paramount (under the circumstances and conditions of cultivation) to that stronger natural tendency to re-

version to the primitive type, and so secures his particular end. The races of Corn, Wheat, &c., which now preserve their character unchanged, have become fixed by centuries of domestication. Even these, at times, manifest an unequivocal disposition to return to their aboriginal state. Were cultivation to cease, they would all speedily disappear; the greater part, perhaps, would perish outright; the remainder would revert, in a few generations of spontaneous growth, to the character of the primitive stock.

671. **Hybrids or Cross-Breeds.** Variations of a still different class are artificially, and sometimes spontaneously, produced, by fertilizing the ovary of one plant with the pollen of a nearly allied species; from which arise what are called *Cross-breeds*, or *Hybrids*. Crosses between different species, however, are almost always incapable of producing fertile seed, and therefore are not perpetuated in nature: those between distinct varieties of the same species are usually fertile, and give rise to new sets of varieties (also termed *Races*), in which the particular qualities of their immediate parents are variously modified or blended; but which, by a continuation of the same influences, revert to one or the other parent stock.

672. **Genera.** If but a moderate number of species were known, no system of generalizing, or arranging them in groups, would be necessary for ordinary purposes; though a consideration of the various degrees of resemblance between different species could not fail to suggest some form of generalization, like that which the great number of species early rendered necessary. The first step in proper classification, the bringing together of species into kinds, according as they are seen to resemble each other, is almost as natural and inevitable an operation of the mind, as is the idea of species involuntarily deduced from the assemblage of like individuals. The generic association, however, implies only resemblance, or similarity of kind, not identity of origin. A GENUS, therefore, is an assemblage of nearly related species, formed after the same pattern, and therefore agreeing with one another in general structure and appearance. Thus, the wild Swamp Rose, the Sweetbrier, the Dog Rose, French Rose, Cinnamon Rose, and others, constitute the universally recognized genus Rosa; the various species of Raspberry and Blackberry compose the genus Rubus; the Apple, Pear, &c. form the genus called by botanists Pyrus: so the different Oaks, Willows, Poplars, Birches, &c. form as many

separate genera. The languages of the most barbarous people show that they have formed such associations. Naturalists merely give to these generalizations a greater degree of precision, and endeavor to indicate what the points of common agreement are. A single species, also, may be deemed to constitute a genus, when its peculiarities are equivalent in degree to those which characterize other genera, — a case which often occurs. If only one species of Oak were known, the Oak genus would have been as explicitly recognized as it is now that the species amount to two hundred; it would have been equally distinguished by its acorn and cup from the Chestnut, Beech, Hazel, &c. A genus, then, is a group of species which present the same particular plan, and whose mutual resemblance is greater than that of any one of them to any other plant.

673. When two or more species of a genus resemble each other in particular points more nearly than they do the other species, intermediate sections are often recognized; which, when marked by characters of considerable importance, receive the title of SUBGENERA.

674. **Orders or Families.** If the genera were few, there would be little necessity for higher generalizations; although one could not but remark that the Oaks, Chestnuts, Beeches, and Hazels have a strong common resemblance, or family likeness; and that they are more unlike Birches and Alders, or Walnuts and Hickories; that they are still more unlike Maples or Ashes, and have yet fewer points in common with Pines and Firs. But, since the 100,000 species of known plants are distributed among nearly 8,000 genera, it is necessary to consider these family resemblances, for the purpose of grouping the genera into still higher, and therefore fewer, groups; just as genera are formed by the reunion of related species. The groups thus established are termed FAMILIES, or ORDERS (names which are for the most part used interchangeably in botany). Thus, the Rose, the Raspberry and Blackberry, with the Strawberry, the Apple, the Thorn, the Plum and Cherry, &c., all agreeing in their general plan of structure, are brought together into one order or family, and termed *Rosaceæ;* that is, Rosaceous or Rose-like plants.

675. But, viewed subordinately, the Plum and Cherry are evidently more nearly akin than the Cherry and Apple, &c.; and so the Raspberry, Blackberry, and Strawberry on one hand, and the

Apple and Thorn on the other, exhibit a closer relationship than that which connects them all in one common group. Hence they are respectively distinguished into groups of a rank intermediate between genera and orders, which are variously termed SUBORDERS, or TRIBES.*

676. **Classes** are groups of orders, associated in a similar manner from some higher point of view. SUBCLASSES bear the same relation to classes that suborders do to orders.

677. By this regular subordination of groups, the various degrees of relationship among plants may be expressed; and upon this Systematic Botany essentially depends. Only four of these divisions are universally employed, namely, Classes, Orders, Genera, and Species: these are common to all methods of classification, both in the animal and vegetable kingdoms, and are always arranged in the same sequence. But a more elaborate analysis is often requisite on account of the large number of objects to be arranged, and the various degrees of affinity to be expressed; when the additional members, and if need be several others, are introduced; as in the following descending series, beginning with the primary division of natural objects into kingdoms, and indicating by small capitals those of fundamental importance and universal use.

KINGDOMS,
Series,
CLASSES,
Subclasses,
ORDERS, or FAMILIES,
Suborders,
Tribes,
Subtribes,
GENERA,
Subgenera,
SPECIES,
Varieties.

678. **Characters.** An enumeration of the distinguishing marks,

* When the groups which an order embraces are distinguished by characters of nearly equal value with those commonly employed for orders themselves, they are termed SUBORDERS. Thus, the Plum, Cherry, Apricot, Peach, &c. form one suborder of Rosaceæ; the Raspberry, Blackberry, Strawberry, Cinquefoil, with the Rose and other genera, constitute another suborder; and the Apple, the Quince, Thorn, &c., a third. The name of

or points of difference between one class or order, &c. and the others, is termed its *character*. The characters of the classes, and other primary divisions, embrace only those important points of structure upon which they are constituted: the *ordinal character* describes the general structure of the included plants, especially of their flowers and fruit: the *generic character* points out the particular modifications of the ordinal structure in a given genus; and the *specific character*, those less important modifications of form, relative size of parts, color, &c., which serve to distinguish kindred species. A complete system of Botany will therefore comprise a methodical distribution of plants according to their organization, with their characters arranged in proper subordination; so that the investigation of any one particular species will bring to view, not only its name (which separately considered is of little importance), but also its floral structure, affinities, and whole natural history.

679. Such a system must of course be *natural;* that is, the groups, of whatever rank, must be composed of plants more closely related to each other than to any different groups, and so arranged that each shall stand, as far as practicable, next to those which it most nearly resembles in structure. These conditions are so far fulfilled by the Natural System (which, sketched by the master-hand of Jussieu, and augmented by succeeding botanists, is now generally adopted), as to render it on the whole far the readiest, as well as the only philosophical and satisfactory, method of acquiring any considerable amount of botanical knowledge.

680. But the relationships of plants, even when appreciated by botanists, could not be made available for the purpose of classification, until just views prevailed in vegetable organography and physiology, which constitute the very foundation of Systematic Botany, but which have only recently been placed upon a philosophical basis. Hence the immortal Linnæus, finding it impossible in his day to characterize the natural groups which his practised eye detected, proposed, as a temporary substitute, the elegant

---

TRIBE is applied to groups comprised in a suborder (thus the Rose constitutes a separate tribe from the Raspberry, Strawberry, &c.), or to the primary divisions of an order, when they are not founded on characters of high importance. In a loose and popular sense, the name of tribe is sometimes used as if synonymous with that of Order or Family.

artificial scheme which bears his name. As this system is identified with the history of the science, which in its time it so greatly promoted, and as most systematic works have until recently been arranged upon its plan, it is still necessary for the student to understand it. Fortunately, its principles are so simple that a brief space will amply suffice for its explanation.

CHAPTER II.

OF THE ARTIFICIAL SYSTEM OF LINNÆUS.

681. It must be kept in mind, that an artificial scheme does not attempt to fulfil all the conditions of natural history classification. Its principal object is to furnish an easy mode of ascertaining the names of plants; their relationships being only so far expressed as the plan of the scheme admits. All higher considerations are of course sacrificed to facility. In the Linnæan scheme, the species of a genus are always kept together, whether or not they all accord with the class or order under which they are placed. Its lower divisions, therefore, namely, the genera and species, are the same as in a natural system. But the genera are arranged in artificial classes and orders, founded on some single technical character, and have no necessary agreement in any other respect; just as words are alphabetically arranged in a dictionary, for the sake of convenience, although those which stand next each other have, it may be, nothing in common beyond the initial letter.

682. The classes and orders Linnæus founded entirely upon the number, situation, and connection of the stamens and pistils; the office and importance of which he had just set in a clear light.

683. The classes, twenty-four in number, were founded upon modifications of the stamens, and have names of Greek derivation expressive of their character. The first eleven comprise all plants with perfect flowers, and a definite number of equal and unconnected stamens; they are distinguished by the absolute number of these organs, and are designated by names compounded of Greek numerals and the word *andria* (from ἀνήρ), which is used metaphorically for stamen; as follows: —

Class 1. MONANDRIA includes all such plants with one stamen to the flower; as in Hippuris (Fig. 703).

2. DIANDRIA, those with two stamens, as in the Lilac.

3. TRIANDRIA, with three stamens, as in the Valerian, &c. (Fig. 764, 767).

4. TETRANDRIA, with four stamens, as in the Plantain (Fig. 831).

5. PENTANDRIA, with five stamens, the most frequent case (Fig. 256, 335).

6. HEXANDRIA, with six stamens, as in the Lily Family (Fig. 1108), &c.

7. HEPTANDRIA, with seven stamens, as in the Horsechestnut (Fig. 657).

8. OCTANDRIA, with eight stamens, as in the Dirca (Fig. 1009).

9. ENNEANDRIA, with nine stamens, as in the Rhubarb.

10. DECANDRIA, with ten stamens, as in Fig. 285, 288.

11. DODECANDRIA, with twelve stamens, as in Asarum (Fig. 968) and the Mignonette; extended also to include those with from thirteen to nineteen stamens.

The two succeeding classes include plants with perfect flowers, having twenty or more unconnected stamens, which, in

12. ICOSANDRIA, are inserted on the calyx (perigynous, 466), as in the Rose Family; and in

13. POLYANDRIA, on the receptacle (hypogynous), as in the Buttercup, Anemone (Fig. 325), &c.

Their essential characters are not designated by their names; the former merely denoting that the stamens are twenty in number; the latter, that they are numerous. The two following depend upon the relative length of the stamens, namely,

14. DIDYNAMIA, including those with two long and two short stamens (481, Fig. 855); and

15. TETRADYNAMIA, those with four long and two short stamens, as in Cruciferous flowers (Fig. 526).

Their names are Greek derivatives, signifying in the former that two stamens, and in the latter that four stamens, are most powerful. The four succeeding are founded on the connection of the stamens:—

16. MONADELPHIA (meaning a single fraternity), with the filaments united in a single set, tube, or column, as in Fig. 307, and in all the Mallow Family, Fig. 617.

Class 17. Diadelphia (two fraternities), with the filaments united in two sets or parcels (Fig. 296, 308, 320).

18. Polyadelphia (many fraternities), with the filaments united in more than two sets or parcels (Fig. 300, 306).

19. Syngenesia (from Greek words signifying to grow together), with the anthers united in a ring or tube (Fig. 309, 310), as in all Composite flowers.

The next class, as its name denotes, is founded on the union of the stamens to the style : —

20. Gynandria, with the stamens and styles consolidated, as in the Orchis Family (Fig. 1097).

In the three following, the stamens and pistils are separated (306) : thus,

21. Monœcia (one household) includes plants where the stamens and pistils are in separate flowers on the same individual; as in the Oak (Fig. 1042), &c.

22. Diœcia (two households), where they occupy separate flowers on different individuals; as in the Willow (Fig. 326 – 328), Prickly Ash (Fig. 639 – 644), &c.

23. Polygamia, where the stamens and pistils are separate in some flowers and united in others, either on the same, or two or three different plants; as in most Maples (Fig. 647 – 649).

The remaining class,

24. Cryptogamia, is said to have concealed stamens and pistils (as the name imports), and includes the Ferns, Mosses, Lichens, &c., which are now commonly termed Cryptogamous or Flowerless plants (459).

The characters of the classes may be presented at a single view, as in the subjoined analysis : —

Synoptical View of the Linnæan Artificial Classes.

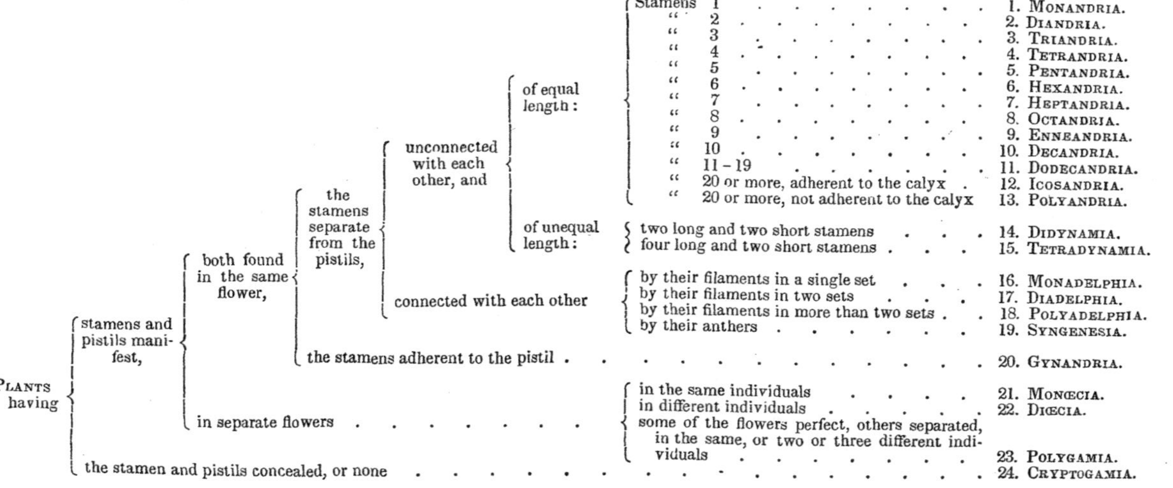

Plants having	stamens and pistils manifest,	both found in the same flower,	the stamens separate from the pistils,	unconnected with each other, and	of equal length:	Stamens 1	1. Monandria.
						" 2	2. Diandria.
						" 3	3. Triandria.
						" 4	4. Tetrandria.
						" 5	5. Pentandria.
						" 6	6. Hexandria.
						" 7	7. Heptandria.
						" 8	8. Octandria.
						" 9	9. Enneandria.
						" 10	10. Decandria.
						" 11 – 19	11. Dodecandria.
						" 20 or more, adherent to the calyx	12. Icosandria.
						" 20 or more, not adherent to the calyx	13. Polyandria.
					of unequal length:	two long and two short stamens	14. Didynamia.
						four long and two short stamens	15. Tetradynamia.
				connected with each other		by their filaments in a single set	16. Monadelphia.
						by their filaments in two sets	17. Diadelphia.
						by their filaments in more than two sets	18. Polyadelphia.
						by their anthers	19. Syngenesia.
			the stamens adherent to the pistil				20. Gynandria.
		in separate flowers				in the same individuals	21. Monœcia.
						in different individuals	22. Diœcia.
						some of the flowers perfect, others separated, in the same, or two or three different individuals	23. Polygamia.
	the stamen and pistils concealed, or none						24. Cryptogamia.

684. The orders, in the first thirteen classes of the Linnæan artificial system, depend on the number of styles, or of the stigmas when the styles are wanting; and are named by Greek numerals prefixed to the word *gynia*, used metaphorically for pistil, as follows:—

Order 1. MONOGYNIA embraces all plants of any of the first thirteen classes, with one style to each flower.
2. DIGYNIA embraces those with two styles.
3. TRIGYNIA, those with three styles.
4. TETRAGYNIA, those with four styles.
5. PENTAGYNIA, those with five styles.
6. HEXAGYNIA, those with six styles.
7. HEPTAGYNIA, those with seven styles.
8. OCTOGYNIA, those with eight styles.
9. ENNEAGYNIA, those with nine styles.
10. DECAGYNIA, those with ten styles.
11. DODECAGYNIA, those with eleven or twelve styles.
12. POLYGYNIA, those with more than twelve styles.

The orders of class 14, Didynamia, are only two; namely,

1. GYMNOSPERMIA, meaning seeds naked, the achenia-like fruits having been taken for naked seeds.
2. ANGIOSPERMIA, with the seeds evidently in a seed-vessel or pericarp.

The 15th class, Tetradynamia, is also divided into two orders, which are distinguished by the mere form of the pod:—

1. SILICULOSA; the fruit a silicle (615), or short pod.
2. SILIQUOSA; fruit a silique (615), or more or less elongated pod.

The orders of the 16th, 17th, 18th, 20th, 21st, and 22d classes depend merely on the number of stamens; that is, on the characters of the first thirteen classes, whose names they likewise bear: thus,

Order 1. MONANDRIA; 2. DIANDRIA; and so on.

The orders of the 19th class, Syngenesia, are six; namely,

1. POLYGAMIA ÆQUALIS, where the flowers are in heads (compound, 394), and all perfect.
2. POLYGAMIA SUPERFLUA, the same as the last, except that the rays, or marginal flowers of the head, are pistillate only (473).

3. Polygamia frustranea, those with the marginal flowers neutral (473, note), the others perfect.
4. Polygamia necessaria, where the marginal flowers are pistillate and fertile, and the central (those of the disc) staminate and sterile.
5. Polygamia segregata, where each flower of the head has its own proper involucre.
6. Monogamia, where solitary flowers (that is, not united into a head) have united anthers, as in Lobelia. This order was abolished by succeeding Linnæan botanists.

The 23d class, Polygamia, has three orders, founded on the characters of the two preceding classes; namely,

1. Monœcia, where both separated and perfect flowers are found in the same individual.
2. Diœcia, where they occupy different individuals.
3. Triœcia, where one individual bears the perfect, another the staminate, and a third the pistillate flowers.

The orders of the 24th class, Cryptogamia, are natural, and therefore indefinable by a single character. They are,

1. Filices, the Ferns.
2. Musci, the Mosses.
3. Algæ, which, as left by Linnæus, comprised the Hepaticæ, Lichens, &c., as well as the Seaweeds.
4. Fungi, Mushrooms, &c.

CHAPTER III.

OF THE NATURAL SYSTEM.

685. The object proposed by the Natural System of Botany is to bring together into groups those plants which most nearly resemble each other, not in a single and perhaps unimportant point (as in an artificial classification), but in all essential particulars; and to combine the subordinate groups into larger natural assemblages, and these into still more comprehensive divisions, so as to embrace the whole vegetable kingdom in a methodical arrangement. All the characters which plants present, that is, all the

points of agreement or difference, are employed in their classification; those which are common to the greatest number of plants being used for the primary grand divisions; those less comprehensive for subordinate groups, &c.; so that the *character* (678), or description of each group, when fully given, actually expresses all the known particulars in which the plants it embraces agree among themselves, and differ from other groups of the same rank. This complete analysis being carried through the system, from the primary divisions down to the species, it is evident that the study of a single plant of each group will give a correct general idea of the structure, habits, and even the sensible properties, of the whole.

686. What we call a natural method, it may here be remarked, is so termed merely because it expresses the natural relationship of plants as far as practicable; for every form yet contrived, or likely to be devised, is, to a considerable extent, artificial:— 1st. Because the affinities of a particular group cannot be fully estimated until all its members are known; and thus the progress of discovery leads to changes, or modifies our views, as in every other department of knowledge. 2d. Because the boundaries of groups are not so arbitrarily circumscribed in nature, as they necessarily are in our classifications; but individuals depart from the assigned limits in various directions (like rays from a centre); the "edge of difference being, as it were, softened down by an easy transition." 3d. Because, even supposing the groups to be perfectly natural, and their affinities completely understood, it is impossible to arrange them in a single continuous series, in such a manner that each shall be preceded and followed by its nearest allies; since the same family, for instance, may be about equally related to three or four others, only two of which points, at best, can be indicated in the lineal series, which must be adopted in books. And 4th. We are still obliged to use avowedly artificial characters, for the sake of convenience; as in the arrangement of the numerous orders of Exogenous plants into the Polypetalous, Monopetalous, and Apetalous divisions of the series, although different genera of the same order, or different species of the same genus, may present these very diversities.

687. In explaining the general principles of classification, we proceeded from the individual to the class; showing how groups of successive rank arise from the consideration of points of agreement. In applying them to the actual distribution of plants ac-

cording to the received mode of classification, it will be more convenient to pursue the analytical course, and to show how the vegetable kingdom, taken as a whole, is divided and subdivided, by regarding the points of difference.

688. The general plan upon which the vegetable creation is constituted, it has been the object of the whole former part of this treatise to illustrate: the fundamental principles of natural history classification have also been cursorily expounded in a preceding chapter. In applying the one to the other, we have to consider, in the first place, how the long series, reaching from the highest Flowering plants to the lowest and minutest Fungi and Algæ, can be primarily divided. As already intimated, the most decided break in the series occurs between the flower-bearing and the flowerless plants; the first producing proper flowers (with stamens and pistils) and seeds containing a ready formed embryo; while in the second, these are replaced by a more or less analogous, but simpler and more recondite apparatus. We need only refer to those paragraphs in which the difference is brought to view (109, &c.). The vegetable kingdom, viewed under this aspect, is therefore primarily divided into two series, a higher and a lower, the FLOWERING and the FLOWERLESS, or (under other and older names) the PHÆNOGAMOUS or (Phanerogamous) and the CRYPTOGAMOUS plants.

689. Let us next consider how the higher series, embracing far the larger part as well as the most complex forms of the vegetable kingdom, may itself be divided, in view of the most general and important points of difference which the plants it comprises exhibit. Whenever they rise to arborescent forms, a difference in port and aspect at once arrests attention; that which distinguishes our common trees and shrubs from Palms and the like (Fig. 220). On examination, this difference is found to be connected with an important difference in the structure of the stem or wood, and in its mode of growth. The former present the exogenous, the latter the endogenous structure or growth (184–187). This difference is manifest, although not so striking, in the annual or herbaceous stems of these two sorts of Phænogamous plants. A difference is also apparent in their foliage; the former generally have *reticulated*, or *netted-veined*, the latter *paralled-veined* leaves (276). The leaves of the former usually fall off by an articulation; those of the latter decay on the stem (309, 310). The Phænogamous series, therefore, divides into two great classes, namely, into EXOGE-

NOUS and ENDOGENOUS plants, more briefly named EXOGENS and ENDOGENS. The difference between the two not only pervades their whole port and aspect, but is manifest from the earliest stage. The embryo of Exogens, as already shown, is provided with a pair of cotyledons, that of Endogens with only one; whence the former are also termed DICOTYLEDONOUS, and the latter MONOCOTYLEDONOUS plants: names introduced by Jussieu, the father of this branch of botany. We employ sometimes the one and sometimes the other of the two sorts of names for these two great classes.

690. In contemplating the Exogenous or Dicotyledonous class, we find that two sets of the plants it comprises are specially distinguished by a great simplicity in their organs of fructification, approximating not indistinctly to that still greater simplicity which characterizes the highest Cryptogamous plants (108). These are the Coniferous trees, such as Pines, Firs, &c., and that small and singular tribe of Endogenous port but essentially Exogenous structure, which comprises the Cycas and Zamia (Fig. 403): in these cases, not only are the sterile or staminate flowers reduced to the last degree of simplicity, but the fertile consist of naked ovules merely, borne on the margins or surface of a sort of open leaf, instead of being inclosed in an ovary (560, 111). They are therefore named GYMNOSPERMOUS (that is, naked-seeded) plants; and form a subordinate group, or subclass, of Exogens. When it is needful to contradistinguish the great mass of Exogens from which these are thus separated, we call them ANGIOSPERMOUS Exogenous plants; a name denoting that their seeds are inclosed in a pericarp. No such reduction occurs in the Endogenous class.

691. We must next consider the systematic divison of the Flowerless, or Cryptogamous series. This is most readily accomplished by conceiving them to present a series of reductions or degradations of a higher type. In their general mode of growth, and in their anatomical structure, the higher Flowerless plants, such as Equisetums, Club-Mosses, and Ferns, do not essentially differ from Flowering plants. All the various kinds of elementary tissue, proper woody fibre, vessels, &c., enter into their composition (108, 109). If we had chosen to take anatomical structure as the basis of our primary division of the whole vegetable kingdom, we might have divided the whole into *Vascular* and *Cellular* plants (107), as was done by De Candolle; the former comprising the whole series from Ferns upward, the latter embracing the Mosses and all below

them. Having effected the primary division, however, upon other grounds, we turn this difference to subordinate account; and therefore consider the higher Flowerless plants, which agree with the series above them in so many respects, and which in their composition have woody tissue and vessels, to constitute the distinct class of VASCULAR FLOWERLESS plants. For reasons already explained (108), they have also been termed ACROGENS. All the kinds below these, being composed of cellular tissue exclusively, (though the cells are often drawn into filaments, which may even have a spiral fibre generated upon their walls,) are CELLULAR plants.

692. But the higher Cellular plants, such as Mosses, still display the proper type of vegetation; they agree with those of higher grades in having an opposite growth, forming a distinct axis or stem, which grows upward by buds and is for the most part symmetrically clothed with distinct leaves (105); while the Lichens, Seaweeds, and Fungi, the most imperfect of vegetables, present no distinction into stem, root, and leaves, no polarity, or growth in two opposite directions, no buds, and no organs which are clearly analogous to flowers. Their homogeneous tissue often tends to the formation of flat, more or less definite expansions (the *thallus*), which is the nearest approach to any thing like leaves; in which their simple spores are imbedded. Hence they are termed *Thallophytes*. If the line of primary division be drawn in view of these important distinctions, as proposed by Unger and Endlicher, the vegetable kingdom would be separated into two great, but unequal series; namely, 1st, the *Cormophytes*, or *Stem-growing plants*, — those with a distinct axis of growth, elongating downward into roots, and upward into stems, provided with leaves, and with flowers or their analogues; and 2d, the *Thallophytes*, which are stemless, rootless, leafless, and in every sense flowerless (106).

693. Following the plan we have adopted, however, we have only to distinguish this higher grade of Flowerless Cellular plants, exhibiting a distinct stem, &c., as a separate class, the ANOPHYTES, represented by the Mosses, which although of the simplest anatomical structure, still emulate the higher or typical forms (105). The remainder (94 – 103), embracing the Lichens, Fungi, and Algæ, form the last and lowest class, the THALLOPHYTES. To consider their subordinate arrangement would quite surpass our limits.

694. The general plan may be analytically expressed by the following schedule.

SYNOPTICAL VIEW OF THE CLASSES IN THE NATURAL SYSTEM.

Ser. I. PHÆNOGAMOUS PLANTS, with	Exogenous growth and a dicotyledonous embryo.		Class I. EXOGENS, or DICOTYLEDONS.
		Seeds in a pericarp.	Subclass 1. ANGIOSPERMS.
		Seeds naked.	" 2. GYMNOSPERMS.
	Endogenous growth and a monocotyledonous embryo.		" II. ENDOGENS, or MONOCOTYLEDONS.
Ser. II. CRYPTOGAMOUS PLANTS, with	a distinct axis, or stem and foliage, containing	woody and vascular tissue.	Class III. ACROGENS.
		cellular tissue only.	" IV. ANOPHYTES.
	no distinction of stem and foliage, but all confounded in a thallus.		" V. THALLOPHYTES.

695. These five classes are very unequal, in respect to the number of plants they embrace; the Exogenous class containing much the largest number of species as well as orders; the Endogens also comprising numerous types; but the others very few in comparison. Convenience of analysis therefore requires that the larger classes should be broken up into divisions, alliances, cohorts, or by whatever name groups intermediate between the classes and orders may be termed: and the accomplishment of this object, so as to form natural groups, is at present the great desideratum in Systematic Botany. But until this be well done, we are obliged to use artificial analyses of the classes, or to throw the orders into groups, which, in proportion as they are rendered natural, it becomes impossible strictly to circumscribe. In this view, the great class of Exogenous plants is usually broken up into three very convenient, but nearly artificial portions, founded on the presence, absence, or union of the petals; namely:

1. Polypetalæ, the polypetalous Exogens; where the calyx and corolla are both present, and the latter composed of distinct petals.
2. Monopetalæ or Gamopetalæ, the Monopetalous Exogens; where the petals are united.
3. Apetalæ, the Apetalous Exogens; where the petals are wanting, and the floral envelopes, if present at all, consist of the calyx alone.

696. These divisions, as well as the other classes, are subdivided by different authors in various ways, which need not be specified; since it is only the classes and the orders that are considered to rest upon a stable basis.

697. The orders, or families, are to be viewed rather as natural groups of genera, than as subdivisions of the classes. The kind of characters employed in distinguishing them will best be learned from the succeeding illustrations.

698. **Nomenclature.** Their names, which are always plural, sometimes express a characteristic feature of the group; as, for instance, *Leguminosæ*, or the Leguminous plants, such as the Pea, Bean, &c., whose fruit is a legume (603); *Umbelliferæ*, or Umbelliferous plants, so named from having the flowers in umbels; *Compositæ*, an order having what were termed compound flowers by the earlier botanists (394); *Labiatæ*, so called from the labiate

or two-lipped corolla (511), which nearly all the species exhibit; *Cruciferæ*, which have their four petals disposed somewhat in the form of a cross, &c. But more frequently, and indeed as a general rule, the name is formed from that of some leading or well-known genus, which is prolonged into the adjective termination *aceæ*. Thus, the plants of the order which comprises the Mallow (*Malva*) are called *Malvaceæ;* that is, *Plantæ Malvaceæ*, or, in English, Malvaceous plants: those of which the Rose (*Rosa*) is the well-known representative are *Rosaceæ*, or Rosaceous plants, &c. This termination in *aceæ*, being reserved for orders, should not be applied to suborders or tribes, which usually bear the name of their principal or best-known genus in an adjective form, without such prolongation. Thus the genus *Rosa* gives name to a particular tribe, *Roseæ*, of the order *Rosaceæ;* the genus *Malva* to the tribe *Malveæ* of the order *Malvaceæ*, &c.

699. The number of genera in an order is quite as indefinite as that of the orders in a class, or other great division. While some orders are constituted of a single genus, as Equisetaceæ, Grossulaceæ, &c. (just as many genera contain but a single known species), others comprise a large number of genera; nearly nine hundred being embraced in the last general enumeration of the Compositæ.

700. The names of genera are Latin substantives, in the singular number, and mostly of Greek or Latin derivation. Those which were known to the ancients generally preserve their classical appellations (Ex. *Fagus*, *Prunus*, *Myrtus*, *Viola*, &c.); and even the barbarous or vulgar names of plants are often adopted, when susceptible of a Latin termination, and not too uncouth; for example, *Thæa* and *Coffæa*, for the Tea and Coffee plants, *Bambusa* for the Bamboo, *Yucca*, *Negundo*, &c. But, more commonly, generic names are formed to express some botanical character, habit, or obvious peculiarity of the plants they designate; such as *Arenaria*, for a plant which grows in sandy places; *Dentaria*, for a plant with toothed roots; *Lunaria*, for one with moon-shaped pods; *Sanguinaria*, for the Bloodroot; *Crassula*, for some plants with remarkably thick leaves. These are instances of Latin derivatives; but recourse is more commonly had to the Greek language, especially for generic names composed of two words; such as *Menispermum*, or Moonseed; *Lithospermum*, for a plant with stony seeds; *Melanthium*, for a genus whose flowers turn of a black or

dusky color; *Epidendrum*, for Orchideous plants which grow upon trees; *Liriodendron*, for a tree which bears lily-shaped flowers, &c. Genera are also dedicated to distinguished persons, a practice commenced by the ancients; as in the case of *Pæonia*, which bears the name of Pæon, who is said to have employed the plant in medicine; and *Euphorbia*, *Artemisia*, and *Asclepias* are also examples of the kind. Modern names of this kind are given in commemoration of botanists, or of persons who have contributed to the advancement of natural history. *Magnolia*, *Bignonia*, *Lobelia*, and *Lonicera*, dedicated to Magnol, Bignon, Lobel, and Lonicer, are early instances of the practice; Linnæa, Tournefortia, Jussiæa, Gronovia, &c. bear the names of more celebrated botanists; and at the present day almost every devotee or patron of the science is thus commemorated.

701. The names of species, as a general rule, are adjectives, written after those of the genera, and established on similar principles; as, *Magnolia grandiflora*, the Large-flowered Magnolia; *M. macrophylla*, the Large-leaved Magnolia; *Bignonia radicans*, the Rooting Bignonia, &c. The generic and specific names, taken together, constitute the proper scientific appellation of the plant. Specific names sometimes distinguish the country which a plant inhabits (Ex. *Viola Canadensis*, the Canadian Violet); or the station where it naturally grows (as *V. palustris*, which grows in swamps, *V. arvensis*, in fields, &c.); or they express some obvious character of the species (as *V. rostrata*, where the corolla bears a remarkably long spur; *V. tricolor*, which has tricolored flowers; *V. rotundifolia*, with rounded leaves; *V. lanceolata*, with lanceolate leaves; *V. pedata*, with pedately parted leaves; *V. primulæfolia*, where the leaves are compared to those of the Primrose; *V. asarifolia*, where they are likened to those of Asarum; *V. pubescens*, which is hairy throughout, &c.). Frequently the species bears the name of its discoverer or describer, when it takes the genitive form, as *Viola Muhlenbergii*, *V. Nuttallii*, &c. When such commemorative names are merely given in compliment to a botanist unconnected with the discovery or history of the plant, the adjective form is preferred; as *Carex Torreyana*, *C. Hookeriana*, &c.: but this rule is not universally followed. Specific names are sometimes substantive; as *Ranunculus Flammula*, *Hypericum Sarothra*, *Linaria Cymbalaria*, &c.; when they do not necessarily accord with the genus in gender. These, as well as all specific

32 *

names derived from those of persons or countries, should always be written with a capital initial letter.

702. In an exposition of the natural system, some authors (such as Jussieu and Endlicher) commence with the lower extremity of the series, and end with the higher; while others (as De Candolle) pursue the opposite course, beginning with the most perfect Flowering plants, and concluding with the lowest grade of flowerless plants. The first mode possesses the advantage of ascending by successive steps from the simplest to the most complex structure; the second, that of passing from the most complete and best understood to the most reduced and least known forms; or, in other words, from the easiest to the most difficult; and is therefore the best plan for the student.

703. The arrangement of De Candolle, being most in use, has been followed as nearly as practicable in the following illustrations, so far as relates to the sequence of the orders. In the conspectus, these have been thrown into small, and more or less natural groups, the characters of which, imperfect as they must be, will serve as a kind of key to the orders of each class or subclass, and facilitate in some degree the student's investigation.* It is by no means necessary, or desirable, to introduce into our elementary illustrations the little known and unimportant orders, especially those which have no indigenous, naturalized, or commonly cultivated representatives in the United States. Those more important exotic families, however, which would otherwise be omitted, are mentioned in the form of notes, placed at the bottom of the page, under the indigenous orders to which they are respectively related. Full descriptions of the orders have not been attempted, but the easier distinguishing characters are given, to the exclusion of the non-essential. An explanation of the technical terms, which, for obvious reasons, are freely employed, (and which will serve to initiate the student into the language of descriptive botany,) may be sought in the combined Glossary and Index at the end of the volume.

---

* In a Flora, or other systematic work based on the natural system, artificial analyses, contrived in various ways, are necessary to the unpractised student, and afford him great assistance in disentangling the more or less complicated characters of the orders. But they are hardly necessary in our sketch, which is intended to give a cursory general view of the principal natural orders, rather than a particular and systematic analysis.

CHAPTER IV.

ILLUSTRATIONS OF THE NATURAL ORDERS OR FAMILIES.

Series I. FLOWERING, OR PHÆNOGAMOUS PLANTS.

PLANTS furnished with flowers (essentially consisting of stamens and pistils), and producing proper seeds (110, 414).

Class I. EXOGENOUS OR DICOTYLEDONOUS PLANTS.

Stem consisting of a distinct bark and pith, which are separated by an interposed layer of woody fibre and vessels, forming *wood* in all perennial stems: increase in diameter effected by the annual deposition of new layers between the old wood and the bark, which are arranged in concentric zones (189-205), and traversed by medullary rays. Leaves commonly articulated with the stem (310), their veins branching and reticulated (276). Sepals and petals, when present, more commonly in fives or fours, and very rarely in threes. Embryo with two or more cotyledons (633, 640).

Subclass 1. ANGIOSPERMOUS EXOGENOUS PLANTS.

Ovules produced in a closed ovary, and fertilized by the action of pollen through the medium of a stigma. Embryo with a pair of opposite cotyledons (633).

Division I. POLYPETALOUS EXOGENOUS PLANTS.

Floral envelopes consisting of both calyx and corolla; the petals distinct.*

CONSPECTUS OF THE ORDERS.

Group 1. Ovaries several or numerous (in a few cases solitary), distinct, when in several rows sometimes cohering in a mass, but not united into a compound pistil. Petals and stamens hypogynous. Seeds albuminous.

* Stamens or pistils (one or both) numerous or indefinite.

Herbs without stipules. RANUNCULACEÆ, p. 384.

---

* Some cases of polypetalous flowers also occur in the orders Ericaceæ, Aquifoliaceæ, and Plumbaginaceæ, which are placed in the Monopetalous part of the series; and some genera of several orders placed here are apetalous, such as Anemone, &c.

Shrubs or trees, with stipules; æstivation imbricative. MAGNOLIACEÆ, p. 385.
Without stipules; æstivation valvular. ANONACEÆ, p. 386.

* * Stamens few or definite. Pistils few or solitary.

Climbing plants. Mono-diœcious. MENISPERMACÆ, p. 387.
Shrubs or herbs. Flowers perfect. BERBERIDACEÆ, p. 388.

Group 2. Ovaries several, either distinct, or perfectly united into a compound pistil of several cells. Stamens definite or indefinite, inserted on the receptacle or torus. Embryo inclosed in a sac at the end of the albumen, or in Nelumbium without albumen. Aquatic herbs.

Carpels distinct and free. Stamens 6-18. CABOMBACEÆ, p. 389.
Carpels distinct, immersed in a dilated torus. NELUMBIACEÆ, p. 390.
Carpels united in a several-celled many-ovuled ovary. NYMPHÆACEÆ, p. 391.

Group 3. Ovary compound, 5-celled, with the placentæ in the axis. Stamens hypogynous, indefinite. Seeds numerous, anatropous, albuminous, with a small embryo. Marsh herbs, with pitcher-shaped or tubular leaves. SARRACENIACEÆ, p. 391.

Group 4. Ovary compound, with parietal placentæ. Calyx and corolla 2-4-merous, deciduous. Stamens hypogynous. Flower unsymmetrical. Embryo small in copious albumen, or coiled when there is no albumen.

Seeds albuminous: embryo small or minute.
Polyandrous: flower regular. PAPAVERACEÆ, p. 391.
Diadelphous or hexandrous: flower irregular. FUMARIACEÆ, p. 393.
Seeds without albumen: styles and stigmas united.
Pod two-celled. Radicle folded on the cotyledons. CRUCIFERÆ, p. 393.
Pod one-celled. Embryo rolled up. CAPPARIDACEÆ, p. 394.
Seeds without albumen. Styles or stigmas several. RESEDACEÆ, p. 395.

Group 5. Ovary compound, with parietal placentæ. Floral envelopes mostly 5-merous; the calyx persistent. Stamens hypogynous. Seeds albuminous.

Anthers (5) adnate, introrse, connate. Corolla irregular. VIOLACEÆ, p. 395.
Anthers extrorse, or innate, distinct. Corolla regular.
Vernation cercinate. Petals marcescent. DROSERACEÆ, p. 396.
Vernation straight. Petals usually caducous. CISTACEÆ, p. 397.

Group 6. Ovary compound with the placentæ parietal, or 2-5-celled from their meeting in the axis. Stamens hypogynous. Seeds with a straight embryo and little or no albumen.

Sterile filaments or a lobed appendage before each petal. PARNASSIEÆ, p. 397.
Sterile filaments none: leaves opposite.
Stipules none; leaves dotted. Stam. unsymmetrical. HYPERICACEÆ, p. 398.
Stipules present: leaves dotless. Stam. symmetrical. ELATINACEÆ, p. 399.

Group 7. Ovary compound, one-celled with a free central placenta, or 2-several-celled with the placenta in the axis. Calyx free or nearly so. Stamens hypogynous or perigynous. Embryo peripheric, coiled more or less around the outside of mealy albumen.

Petals numerous. Ovary many-celled. MESEMBRYANTHEMACEÆ, p. 402.
Floral envelopes symmetrical. Stamens no more than 10. CARYOPHYLLACEÆ, p. 399.
Floral envelopes unsymmetrical, or polyandrous. PORTULACACEÆ, p. 401.
Petals 3 – 5 or 6, sometimes wanting.

Group 8. Ovary compound and several-celled, with the placentæ in the axis; or the numerous carpels more or less coherent with each other or with a central axis. Calyx free from the ovary, with a valvate æstivation. Stamens mostly indefinite, monadelphous, or polyadelphous, inserted with the petals into the receptacle or base of the petals.

Anthers 1-celled. Stamens monadelphous. MALVACEÆ, p. 402.
Anthers 2-celled. Fertile stam. few, monadelphous. BYTTNERIACEÆ, p. 403.
Anthers 2-celled. Stamens polyandrous or 5-adelphous. TILIACEÆ, p. 403.

Group 9. Ovary compound, with two or more cells, and the placentæ in the axis, free from the calyx, which is imbricated in æstivation. Stamens indefinite, or twice as many as the petals, usually monadelphous, hypogynous. — Trees or shrubs.

Leaves simple, not dotted. Stamens indefinite. TERNSTRŒMIACEÆ, p. 405.
Leaves pellucid-punctate, mostly compound. AURANTIACEÆ, p. 405.
Leaves compound, dotless. Stamens 10 or less, monadelphous.
Seeds single in each cell, wingless. MELIACEÆ, p. 405.
Seeds several in each cell, winged. CEDRELACEÆ, p. 406.

Group 10. Ovary compound, or of several carpels adhering to a central axis, free from the calyx, which is mostly imbricated in æstivation. Stamens as many or twice as many as the petals, inserted on the receptacle, often monadelphous at the base. Embryo large. Albumen little or none. Flowers perfect.

* Flower irregular and unsymmetrical. Albumen none.

Stamens connate. Ovules several in each cell. BALSAMINACEÆ, p. 408.
Stamens distinct. Ovules single in each cell. TROPÆOLACEÆ, p. 408.

* * Flower regular and symmetrical throughout.

Leaves not glandular-dotted.
Calyx valvate. Albumen none. LIMNANTHACEÆ, p. 409.
Calyx imbricated in æstivation.
Embryo conduplicate: cotyledons convolute. GERANIACEÆ, p. 407.
Embryo straight or nearly so.
Leaves entire. Fertile stamens 5. LINACEÆ, p. 406.
Leaves compound. Stamens 10.
Styles separate. Leaves alternate. OXALIDACEÆ, p. 408.
Styles united. Leaves opposite. ZYGOPHYLLACEÆ, p. 408.
Leaves glandular-dotted. RUTACEÆ, p. 409.

Group 11. Ovary compound, with 2 – several cells; or one-celled by suppression; or carpels distinct and barely connected by their styles. Calyx free. Petals as many as the sepals, or rarely wanting. Stamens once or twice as many as the sepals, distinct, inserted into the receptacle or base of the calyx. — Embryo large: albumen little or none. Flowers mostly diœcious or polygamous.

Leaves dotted. Ovaries or cells 2-ovuled. ZANTHOXYLACEÆ, p. 409.
Leaves dotless. Ovule solitary.
Ovaries 4 or 5, distinct in fruit. OCHNACEÆ, p. 410.
Ovary one: ovule on a long ascending funiculus. ANACARDIACEÆ, p. 411.

Group 12. Ovary compound, 2-3-lobed, 2-3-celled, free from the calyx, which is imbricated in æstivation. Petals often irregular, or one fewer than the sepals, or sometimes wanting. Stamens definite, distinct, inserted on or around a hypogynous disc. Ovules 1 or 2 in each cell. Embryo curved or coiled. Albumen none. — Flowers often polygamous.

Leaves opposite.
Entire. Gynæcium trimerous. MALPIGHIACEÆ, p. 412.
Lobed, or compound. Gynæcium dimerous. ACERACEÆ, p. 412.
Leaves chiefly alternate. Gynæcium trimerous. SAPINDACEÆ, p. 413.

Group 13. Ovary compound, 2-5-celled. Calyx free from, or adherent to the base of, the ovary. Petals and stamens equal in number to the lobes of the calyx, and inserted either into its base or throat, or upon the disc that covers it. Seeds solitary or few in each cell, albuminous. Embryo mostly large. — Shrubs or trees. Flowers regular.

* Stamens alternate with the petals.

Ovaries partly separated. Leaves compound. STAPHYLEACEÆ, p. 415.
Ovaries wholly united. Seed arillate. Leaves simple. CELASTRACEÆ, p. 414.

* * Stamens opposite the petals!

Sepals valvate in æstivation. Cells 1-ovuled. RHAMNACEÆ, p. 414.
Petals valvate, caducous. Cells 2-ovuled. VITACEÆ, p. 415.

Group 14. Ovary compound, 2-celled, free from the calyx. Sepals and petals very irregular. Stamens monadelphous; the tube of filaments split on one side, and more or less united with the claws of the hypogynous petals: the anthers one-celled, and opening by a pore at the apex! Seeds albuminous. Embryo large. POLYGALACEÆ, p. 416.

Group 15. Ovary simple and solitary, free from the calyx; the fruit a pod. Flower 5-merous, the odd sepal anterior. Corolla papilionaceous, irregular, or sometimes regular. Stamens monadelphous, diadelphous, or distinct, mostly perigynous. Seeds destitute of albumen.

Stamens hypogynous. Stipules none. KRAMERIACEÆ, p. 417.
Stamens mostly perigynous. Fruit a legume. LEGUMINOSÆ, p. 417.

Group 16. Ovaries one or several, simple and distinct, or combined into a compound ovary with two or more cells and the placentæ in the axis. Petals and the distinct stamens perigynous. Seeds destitute of albumen.

* Calyx free, although often inclosing the ovaries in its tube, except when the latter are united, when it is adnate to the compound ovary, and the stamens are indefinite.

Leaves alternate, stipulate. Cotyledons plane. ROSACEÆ, p. 419.
Leaves opposite, exstipulate, not dotted. CALYCANTHACEÆ, p. 422.
Leaves opposite, exstipulate, pellucid-punctate. MYRTACEÆ, p. 423.

* * Calyx free from the comp. ovary. Stam. definite. LYTHRACEÆ, p. 424.

* * * Calyx-tube adnate to the compound ovary. Stamens definite.

Anthers opening by a pore at the apex. MELASTOMACEÆ, p. 424.
Anthers opening longitudinally.
Stipules interpetiolar. Leaves opposite. RHIZOPHORACEÆ, p. 424.
Stipules none. Calyx valvate.
Cotyledons convolute. COMBRETACEÆ, p. 424.
Cotyledons plane. ONAGRACEÆ, p. 424.

Group 17. Ovary compound, one-celled, with parietal placentæ. Petals and (with one exception) stamens inserted on the throat of the calyx. Flowers perfect, except in Papayaceæ.

* Calyx adherent to the ovary.

Albumen none or very little. Petals and stam. indefinite. CACTACEÆ, p. 426.
Albumen very copious. Embryo minute. Stam. 5. GROSSULACEÆ, p. 426.
Albumen present: embryo rather large. Stam. indefinite. LOASACEÆ, p. 427.

* * Calyx free from the ovary.

Flowers perfect. Stamens 5.
Stamens distinct, perigynous. TURNERACEÆ, p. 427.
Stamens monadelphous, adnate to the gynophore. PASSIFLORACEÆ, p. 427.
Flowers diœcious. Stamens 10, on the corolla. PAPAYACEÆ, p. 428.

Group 18. Ovary compound, 2-several-celled (or one-celled by obliteration); the placentæ parietal, arising from the axis, but carried outwards to the walls of the pericarp. Calyx adnate. Corolla frequently monopetalous. Stamens united either by their filaments or anthers. Flowers diœcious or monœcious. Albumen none. CUCURBITACEÆ, p. 428.

Group 19. Ovaries two or more, many-ovuled, distinct, or partly, sometimes completely, united, when the compound ovary is one-celled with parietal placentæ, or 2-many-celled with the placentæ in the axis. Calyx either free from the ovary or adherent. Petals and stamens inserted on the calyx; the latter mostly definite. Seeds albuminous, numerous.

Pistils as many as the sepals. CRASSULACEÆ, p. 429.
Pistils fewer than the sepals, more or less united. SAXIFRAGACEÆ, p. 430.

Group 20. Ovary compound, 2- (rarely 3-5-) celled, with a single ovule suspended from the apex of each cell. Stamens usually as many as the petals, or the lobes of the adherent calyx. Embryo small, in hard albumen.

* Summit of the (often 2-lobed) ovary free from the calyx; the petals and stamens inserted on the throat of the calyx. HAMAMELACEÆ, p. 431.

* * Calyx-tube entirely adherent to the ovary. Stamens and petals epigynous. Flowers umbellate.

Fruit separable into two dry carpels. UMBELLIFERÆ, p. 431.
Fruit drupaceous, usually of more than two carpels. ARALIACEÆ, p. 433.
Flowers cymose or capitate. Drupe 2-celled. CORNACEÆ, p. 433.

704. **Ord. Ranunculaceæ** (*the Crowfoot Family*). Herbaceous, occasionally climbing plants, with an acrid watery juice, and usually palmately or ternately lobed or divided leaves, without stipules. Calyx of three to six, usually five, distinct sepals, deciduous, except in Pæonia and Helleborus. Petals five to fifteen, or sometimes none. Stamens indefinite, distinct. Ovaries numerous, rarely few or solitary, distinct. Embryo minute, at the base of firm albumen (Fig. 455, 456). — *Ex.* Ranunculus, the Buttercup, which has regular flowers with petals. Clematis (Virgin's Bower), Anemone, and Hepatica (Liver-leaf), which have no petals, but the calyx is petaloid: the latter has an involucre entirely resembling a calyx, and the leaves of the former are opposite. In all these examples the ovaries are one-seeded, and the flowers regular. In others, the ovaries contain several seeds, and the flowers are irregular, or with the petals in the form of spurs or differently shaped bodes. Actæa (Cohosh, Baneberry) and one Larkspur have a solitary ovary: in the latter the petals are consolidated.

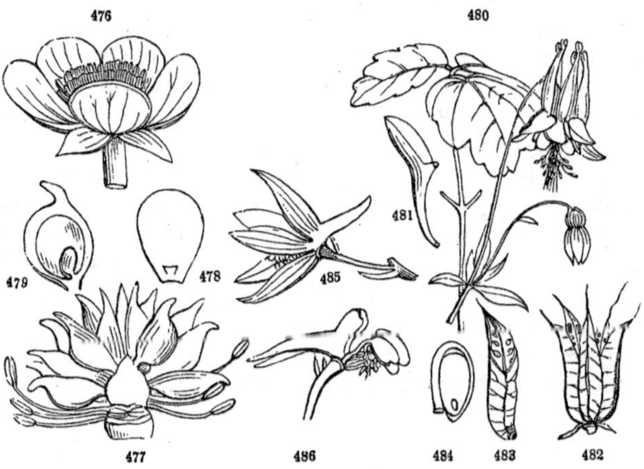

Zanthorhiza (Yellow-root) has only five or ten stamens. — The

FIG. 476. Flower of a Ranunculus. 477. Vertical section through the receptacle; the sepals, petals, and most of the stamens taken away. 478. A petal, with the nectariferous scale at its base. 479. Section through an ovary, showing the solitary ovule attached to the base of the cell. 480. Flower and part of a leaf of Aquilegia Canadensis (Wild Columbine). 481. A detached petal. 482. The five carpels of the fruit. 483. A separate follicle. 484. Vertical section of the seed, showing the minute embryo. 485. Flower of Delphinium, or Larkspur, with its spurred calyx; which is removed in 486, to show the four irregular petals and the stamens.

juice of all Ranunculaceous plants is acrid, or even caustic: some are virulent narcotico-acrid poisons.

705. **Ord. Magnoliaceæ** (*the Magnolia Family*). Trees or shrubs; with ample and coriaceous, alternate, entire or lobed leaves, usually punctate with minute transparent dots: stipules membranaceous, enveloping the bud, falling off when the leaves expand. Flowers solitary, large and showy, mostly odorous. Calyx of three to six deciduous sepals, colored like the petals; the latter three or several, often in several rows. Stamens numerous, mostly with short filaments, and adnate anthers. Carpels either several in a single row, or numerous and spicate on the prolonged receptacle; in the latter case usually more or less cohering with each other, and forming a fruit like a cone or strobile. Seeds mostly one or two in each carpel, often with a pulpy exterior coat, and suspended, when the carpels open, by an extensile funiculus, composed of unrolled spiral vessels. Embryo minute, at the base of homogeneous fleshy albumen. There are three well-marked suborders; viz.: —

706. **Subord. Magnolieæ** (*the true Magnolia Family*), characterized

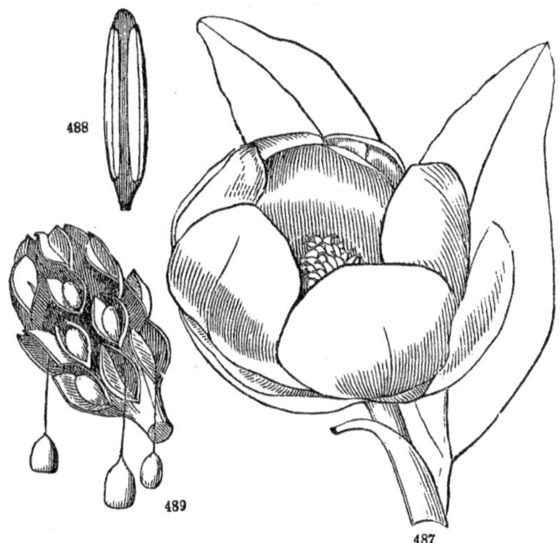

principally as above, especially by the stipules and the imbricated

FIG. 487. Magnolia glauca. 488. A stamen, seen from the inside, showing the two lobes of the adnate anther. 489. The carpels in fruit, persistent on the receptacle, and opening by the dorsal suture; the seeds suspended by their extensile cord of spiral vessels.

spiked carpels. — *Ex.* Magnolia, in which the hard or woody carpels are persistent, and accordingly open by the dorsal suture; Liriodendron (the White-wood or Tulip-tree), in which the winged carpels fall away from the receptacle, but are themselves indehiscent. Bitter, and somewhat acrid-aromatic.

707. **Subord. Wintereæ** (*the Winter's-Bark Family*) has no stipules, and the carpels occupy only a single verticil. These have pungent aromatic properties, as in Illicium, the Star-Anise, the seeds and pods of which furnish the aromatic oil of this name.

708. **Subord. Schizandreæ** is monœcious or diœcious, with the pistils spicate or capitate on a prolonged receptacle; the stamens often monadelphous. Leaves sometimes toothed, destitute of stipules. — *Ex.* Schizandra. Mucilaginous, with little aroma.

709. **Ord. Anonaceæ** (*the Custard-Apple Family*). Trees or

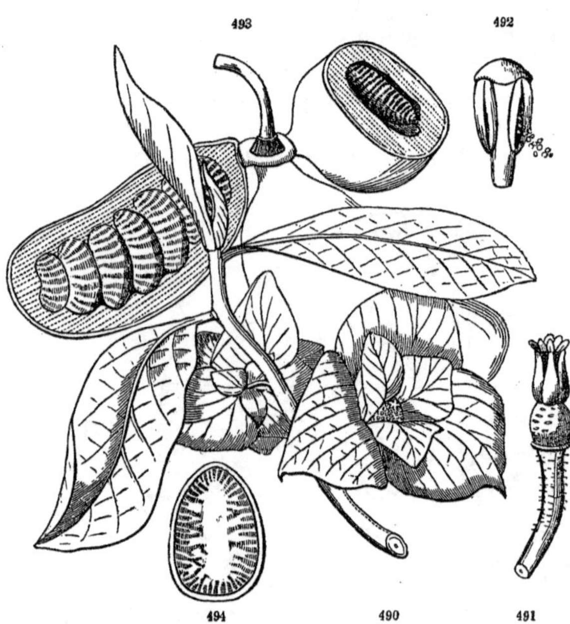

shrubs, with alternate entire leaves, destitute of stipules. Flowers

FIG. 490. Flowering branch of the Papaw (Uvaria triloba) of the natural size. 491. The receptacle, with all but the pistils removed. 492. A stamen, magnified. 493. View of three baccate pods from the same receptacle (much reduced in size); one cut across, another lengthwise, to show the large bony seeds. 494. Section of the seed, to show the ruminated albumen.

large, but dull-colored. Sepals 3. Petals 6, in two rows, with a valvate æstivation. Stamens numerous, in many rows, with extrorse anthers. Carpels few, or mostly numerous and closely packed together, sometimes cohering and forming a fleshy or pulpy mass in the mature fruit. Seeds one or more in each carpel, with a brittle testa: embryo minute, at the base of hard, ruminated albumen. *Ex.* The four species of Papaw (Asimina) are our only representatives of this chiefly tropical order, which furnishes the luscious Custard-apples of the East and West Indies, &c. Aromatic, and sometimes rather acrid, properties prevail in the order.*

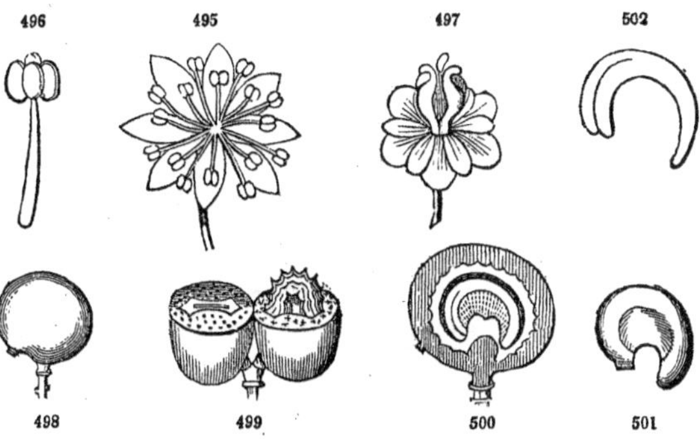

710. Ord. Menispermaceæ (*the Moonseed Family*). Climbing or twining shrubby plants; with alternate and simple palmately-veined leaves, destitute of stipules; and small flowers in racemes or panicles, diœcious, monœcious, or polygamous. Calyx of three to twelve sepals, in one to three rows, deciduous. Petals as many as the sepals or fewer, small, or sometimes wanting in the pistillate

---

* ORD. MYRISTICACEÆ, consisting of a few tropical trees (which bear nutmegs), differs from Anonaceæ in having monœcious or diœcious and apetalous flowers. The aril and the albumen of the seeds are fine aromatics. The common *nutmeg* is the seed of Myristica moschata (a native of the Moluccas) deprived of the testa: *mace* is the aril of the same species. The *ruminated* albumen (627) is nearly peculiar to this family and the Anonaceæ.

FIG. 495. Staminate flower of Menispermum Canadense. 496. A stamen, with its four-lobed anther. 497. A pistillate flower of the same. 498. A solitary fruit. 499. Two drupes on the same receptacle, cut across; one through the pulpy exocarp only, the other through the bony endocarp and seed. 500. A drupe divided vertically (the embryo here is turned the wrong way). 501. The seed, and 502, the coiled embryo detached.

flowers. Stamens as many as the petals, and opposite them, or two to four times as many: anthers often four-celled. Carpels usually several, but only one or two of them commonly fructify, at first straight, but during their growth often curved into a ring; in fruit becoming berries or drupes. Seeds solitary, filling the cavity of the bony endocarp; embryo large, inclosed in the thin, fleshy albumen. — *Ex.* Menispermum, or Moonseed (Fig. 495 – 502), Cocculus. The roots are bitter and tonic (e. g. *Colombo Root* of the materia medica); but the fruit is often narcotic and acrid; as, for instance, the *Cocculus Indicus* of the shops, once used for rendering malt liquors more intoxicating, and for stupefying fishes.

711. **Ord. Berberidaceæ** (*the Barberry Family*). Herbs or shrubs,

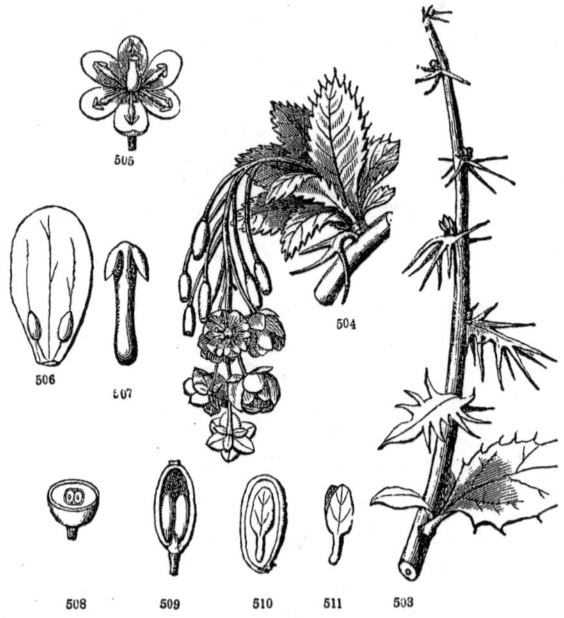

with a watery juice; the leaves alternate, compound or divided,

FIG. 503. A shoot of Berberis vulgaris, the common Barberry. 504. A flowering branch from the axil of one of its leaves or spines the following year. 505. An expanded flower. 506. A petal, nectariferous near the base. 507. A stamen; the anther opening by uplifted valves. 508. Cross-section of a young fruit. 509. Vertical section; the seeds attached at the base. 510. Vertical section of a seed enlarged, showing the large embryo with foliaceous cotyledons and a taper radicle, surrounded by albumen. 511. The embryo separate.

usually without stipules. Flowers perfect. Calyx of three to nine sepals, imbricated in one to several rows, often colored. Petals as many as the sepals and in two sets, or twice as many, with a pore, spur, or glandular appendage at the base. Stamens equal in number to the petals and opposite them, or rarely more numerous; anthers extrorse, the cells commonly opening by an uplifted valve. Carpel solitary, often gibbous or oblique, forming a one-celled pod or berry in fruit. Seeds sometimes with an aril: embryo (often minute) surrounded with fleshy or horny albumen. — *Ex.* The Barberry (Fig. 503 – 511), the sharp spines of which are transformed leaves (296); the Mahonias are Barberries with pinnated leaves. Leontice (Caulophyllum) thalictroides, the Blue Cohosh, is remarkable for its evanescent pericarp (559), and the consequent naked seeds, which resemble drupes. Podophyllum (the Mandrake) presents an exception to the ordinal character, having somewhat numerous stamens, with anthers which do not open by valves; but the latter anomaly is also found in Nandina. The order is remarkable for this valvular dehiscence of the anthers, and for the situation of both the stamens and petals opposite the sepals. But this latter peculiarity is doubtless owing to the production of two or three whorls both of the petals and the stamens, which does away with the anomaly. The æstivation in Berberis shows this to be the case. The fruit is innocent or eatable; the roots and also the herbage sometimes poisonous.

712. **Ord. Cabombaceæ** (*the Water-shield Family*). Aquatic herbs, with the floating leaves entire and centrally peltate, involute in vernation; the submersed foliage sometimes dissected. Flowers solitary, rather small. Calyx of three or four sepals, colored inside, persistent. Corolla of as many persistent petals. Stamens six to thirty-six, with slender filaments and innate anthers. Carpels two to eighteen, indehiscent, with two or few (anatropous) ovules in each, inserted on the dorsal suture! Seeds pendulous, with a minute embryo inclosed in a membranous bag (the persistent embryo-sac, 575), which is half immersed in the albumen at the end next the hilum. — *Ex.* Brasenia, the Water-shield (Fig. 512), and Cabomba, compose this very small order; the apparently single species of the former grows both in the United States and in New Holland. They are only reduced forms of Nymphæaceæ.

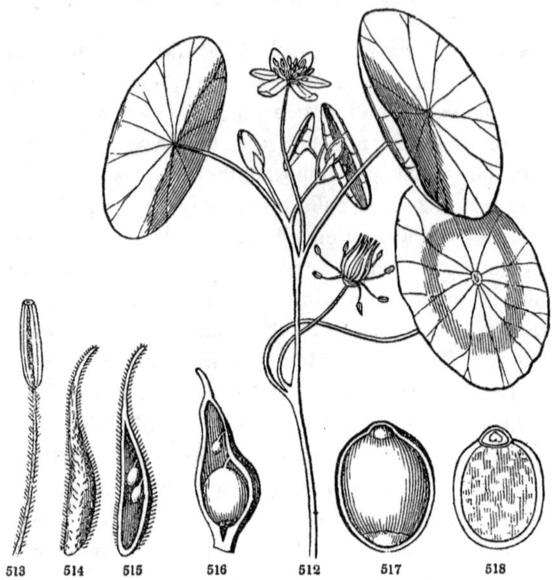

713. Ord. Nelumbiaceæ (*the Nelumbo Family*). Aquatic herbs, with very large leaves and flowers, on long stalks arising from a prostrate trunk or rhizoma, which has a somewhat milky juice: the leaves orbicular and centrally peltate. Calyx of four or five sepals, deciduous. Petals numerous, inserted in several rows into the base of a large and fleshy obconical torus, deciduous. Stamens inserted into the torus in several rows: the filaments petaloid; the anthers adnate and introrse. Carpels several, separately immersed in hollows of the enlarged flat-topped torus or receptacle (Fig. 351), each containing a single anatropous ovule; in fruit forming hard, round nuts. Seed without albumen: embryo very large, with two fleshy cotyledons, and a highly developed plumule. — *Ex.* The order consists of the single genus Nelumbium, embracing two species; one a native of Asia, the other of the United

FIG. 512. Brasenia peltata (Water-shield); the lower flower with the floral envelopes and a part of the stamens removed. 513. A magnified stamen. 514. A magnified carpel. 515. The same divided lengthwise, showing the ovules attached to the outer or dorsal suture! 516. Section of a carpel, in fruit. 517. A magnified seed, with half the outer integument removed, displaying at the upper extremity the bag which contains the embryo. 518. A magnified section through the middle of the albumen, &c.; bringing to view the minute embryo inclosed in its sac, lying outside of the albumen, which forms the principal bulk of the seed.

States. They are chiefly remarkable for their very large and showy leaves and flowers. The nuts are eatable.

714. **Ord. Nymphæaceæ** (*the Water-Lily Family*). Aquatic herbs, with showy flowers, and cordate or peltate leaves arising from a prostrate trunk or rhizoma, and raised on long stalks above the water, or floating on its surface. Calyx and corolla of several or numerous imbricated sepals and petals, which gradually pass into each other; persistent; the latter inserted on the fleshy torus which surrounds or partly incloses and adheres to the pistil; the inner series gradually changing into stamens. Stamens numerous, in several rows, inserted into the torus with or above the petals; many of the filaments petaloid, the adnate anthers introrse. Fruit indehiscent, pulpy when ripe, many-celled, crowned with the radiate stigmas; the anatropous seeds covering the spongy dissepiments. Embryo small, inclosed in a membranous bag, which is situated next the hilum, and half immersed in the mealy albumen. — *Ex.* Nymphæa, the White Water-Lily (Fig. 265 – 268); Nuphar, the Yellow Pond-Lily. Here belongs the magnificent Victoria of tropical South America, the most gigantic and showy of aquatics, both as to its flowers and its leaves.

715. **Ord. Sarraceniaceæ** (*the Water-Pitcher Family*). Perennial herbs, growing in bogs; the (purplish or yellowish-green) leaves all radical and hollow, pitcher-shaped (Fig. 223, 224), or trumpet-shaped. Flower solitary on a long scape. Calyx of five persistent sepals, with three small bracts at its base. Corolla of five petals. Stamens numerous. Summit of the combined styles very large and petaloid, five-angled, covering the five-celled ovary, persistent. Fruit five-celled, five-valved, with a large placenta projecting from the axis into the cells. Seeds numerous, albuminous, with a small embryo. — *Ex.* Sarracenia, from which the above character is taken, was the only known genus of the order, until the recent discovery of Heliamphora in Guiana. The scape of the latter bears several flowers without petals, &c. A third genus, Darlingtonia, *Torr.*, has recently been discovered in California, with calyx and corolla not very unlike those of Sarracenia, but without the umbrella-like style. The species of Sarracenia are all North American, and, excepting S. purpurea, are confined to the Southern States east of the Alleghanies.

716. **Ord. Papaveraceæ** (*the Poppy Family*). Herbs, with a milky or colored juice, and alternate leaves without stipules. Calyx of

two (rarely three) caducous sepals. Corolla of four to six regular petals. Stamens eight to twenty-four, or numerous. Fruit one-celled, either pod-shaped with two to five, or capsular with numerous parietal placentæ, from which the valves often separate in dehiscence. Seeds numerous, with a minute embryo, and copious fleshy and oily albumen. — *Ex.* The Poppy (Papaver), the leading representative of this small but important family, is remarkable for the extension of the placentæ so as nearly to divide the cavity of the ovary into several cells, and for the dehiscence of the capsule

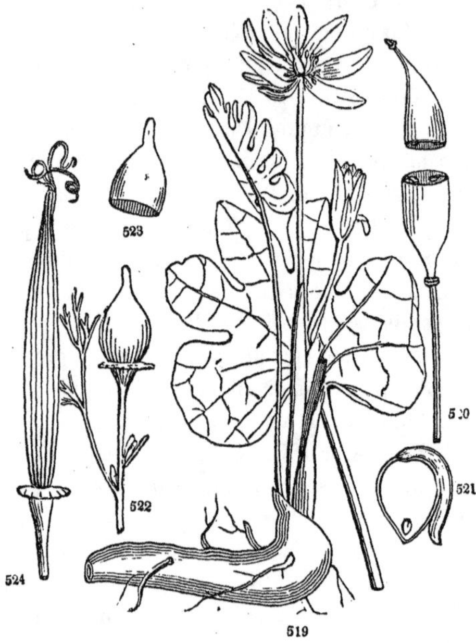

by mere chinks or pores under the edge of the crown formed by the radiate stigmas. The Eschscholtzia, now common in gardens, is remarkable for the expanded apex of the peduncle, and for the union of the two sepals into a *calyptra*, like a candle-extinguisher, which, separating at the base, is thrown off by the expansion of the petals (Fig. 522, 523). The colored juice is narcotic and

FIG. 519. Sanguinaria Canadensis (the Bloodroot). 520. The pod, divided transversely, showing the parietal attachment of the seeds. 521. Longitudinal section of a magnified seed with its large rhaphe, showing the minute embryo, near the extremity of the albumen. 522. Flower-bud of Eschscholtzia. 523. The calyptriform calyx detached from the base. 524. Pod of the same.

stimulant. That of the Poppy yields *opium.* That of the Celandine, and of the Bloodroot (Sanguinaria, Fig. 519), is acrid.

717. **Ord. Fumariaceæ** (*the Fumitory Family*). Smooth herbs, with brittle stems, and a watery juice, alternate dissected leaves, and no stipules. Flowers irregular. Calyx of two sepals. Corolla of four petals, in pairs; the two outer, or one of them, spurred or sac-like at the base; the two inner callous and cohering at the apex, including the anthers and stigma. Stamens six, in two parcels opposite the outer petals; the filaments of each set usually more or less united; the middle one bearing a two-celled anther; the lateral with one-celled anthers. Fruit a one-celled and two-valved pod, or round and indehiscent. Seeds with fleshy albumen and a small embryo. — *Ex.* Fumaria, Dicentra, Corydalis. A small and unimportant tribe of plants, chiefly remarkable for their singular irregular flowers; by which alone they are distinguished, and that not very definitely, from the preceding family. Its floral structure has already been explained (455, Fig. 294 - 299).

718. **Ord. Cruciferæ** (*the Mustard Family*). Herbs, with a pungent or acrid watery juice, and alternate leaves without stipules; the flowers in racemes or corymbs, with no bracts to the pedicels. Calyx of four sepals, deciduous. Corolla of four regular petals, with claws, their spreading limbs forming a cross. Stamens six, two of them shorter (*tetradynamous*, 519). Fruit a pod (called a *silique* when much longer than broad, or a *silicle* when short, 615), which is two-celled by a membranous partition that unites the two marginal placentæ, from which the two valves usually fall away. Seeds with no albumen: embryo with the cotyledons folded on the radicle. — *Ex.* The Water-Cress, Radish, Mustard, Cabbage, &c. A very natural order, found in every part of the world, perfectly distinguished by having six tetradynamous stamens along with four petals and four sepals, and by the peculiar pod. The peculiarity of the stamens is explained, and the symmetry of the flower shown, on p. 250. These plants have a peculiar volatile acridity (and often an ethereal oil, which abounds in sulphur) dispersed through every part, from which they derive their peculiar odor and sharp taste, and their stimulant, rubefacient, and antiscorbutic properties. The roots of some perennial species, such as the Horseradish, or the seeds of annual species, as the Mustard, are used as condiments. In some cultivated plants, the acrid principle is dispersed

among abundance of saccharine and mucilaginous matter, affording wholesome food; as the root of the Turnip and Radish; the leaves, &c. of the Cabbage and Cauliflower. None are really poisonous plants, although some are very acrid. Several species are in cultivation, for their beauty or fragrance; such as the Wallflower, Stock, &c.

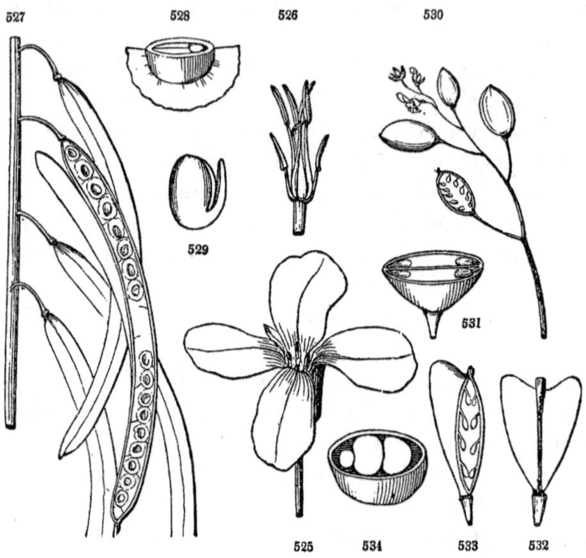

719. **Ord. Capparidaceæ** (*the Caper Family*). Herbs, or in the tropics often shrubs or trees; differing from Cruciferæ in the one-celled pod (which is often stalked) being destitute of any false partition; in the kidney-shaped seeds; and in the stamens, which, when six, are scarcely tetradynamous, and are often more numerous. — *Ex.* Cleome, and Polanisia (Fig. 525–534); chiefly tropi-

FIG. 525. A Cruciferous flower. 526. The same, with the calyx and corolla removed, showing the tetradynamous stamens. 527. *Siliques* of Arabis Canadensis; one of them with one of the valves detached, showing the seeds lying on the false partition; the other valve also falling away. 528. A magnified cross-section of one of the winged seeds, showing the embryo with the radicle applied to the edge of the cotyledons (cotyledons *accumbent*). 529. The embryo detached. 530. The raceme of Draba verna, in fruit. 531. A cross-section of one of the *silicles*, magnified, exhibiting the parietal insertion of the seeds, and the false partition. 532. A silicle of Shepherd's Purse (Capsella Bursa Pastoris). 533. The same, with one of the boat-shaped valves removed, presenting a longitudinal view of the narrow partition, &c. 534. A magnified cross-section of one of the seeds, showing the embryo with the radicle applied to the side of the cotyledon (cotyledons *incumbent*).

cal or subtropical. Many have have the pungency of Cruciferæ, but are more acrid. *Capers* are the pickled flower-buds of Capparis spinosa of the Levant, &c. The roots and herbage or bark are bitter, nauseous, and sometimes poisonous.

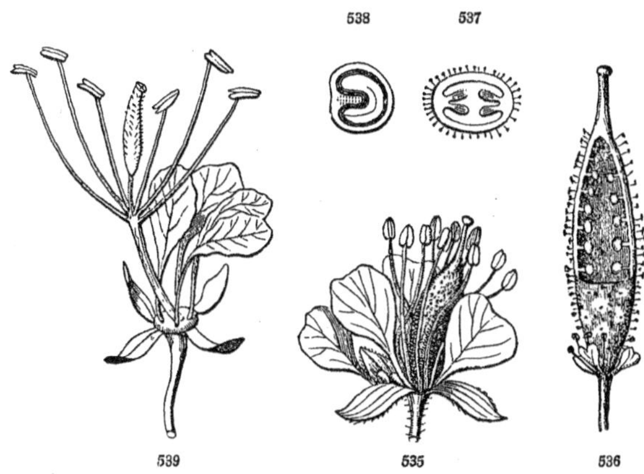

720. **Ord. Resedaceæ** (*the Mignonette Family*). Herbs, with a watery juice, and alternate leaves without stipules, except a pair of glands be so considered: the flowers in terminal racemes, small, and often fragrant. — Calyx persistent, of four to seven sepals, somewhat united at the base. Corolla of two to seven usually unequal and lacerated petals, with broad or thickened nectariferous claws. A fleshy disc is commonly present, enlarged posteriorly between the petals and the stamens, and bearing the latter, which vary from three to forty in number, and are not covered by the petals in the bud. Fruit a one-celled pod, with three to six parietal placentæ, three to six-lobed at the apex, where it opens along the inner sutures, usually long before the seeds are ripe. Seeds several or many, curved or kidney-shaped, with no albumen; the embryo incurved. — *Ex.* The common representative of this order is the Mignonette (Reseda odorata), prized for its fragrant flowers.

721. **Ord. Violaceæ** (*the Violet Family*). Herbs (in tropical countries sometimes shrubby plants), with mostly alternate simple leaves, on petioles, furnished with stipules; and irregular flowers.

FIG. 535. Flower of Polanisia graveolens. 536. Fructified ovary of the same, a portion cut away by a vertical and horizontal section, to show the single cell and two parietal placentæ. 537. Cross-section of the ovary. 538. Section of the seed and embryo. 539. Flower of Gynandropsis.

Calyx of five persistent sepals, often auricled at the base. Corolla of five unequal petals, one of them larger than the others and commonly bearing a spur or a sac at the base: æstivation imbricative. Stamens five, with short and broad filaments, which are usually elongated beyond the (adnate introrse) anthers; two of them commonly bearing a gland or a slender appendage which is concealed in the spur of the corolla: the anthers approaching each other, or united in a ring or tube. Style usually turned to one side and thickened at the apex. Fruit a one-celled capsule, opening by three valves, each bearing a parietal placenta on its middle. Seeds several or numerous, anatropous, with a crustaceous integument. Embryo straight, nearly the length of the fleshy albumen. — *Ex.* The Violet (Viola) is the principal genus of this

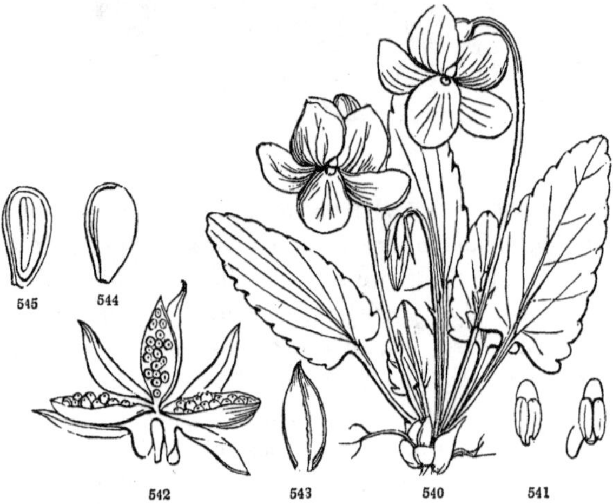

order; some species, like the Pansy, are cultivated for the beauty of their flowers; others for their delicate fragrance. The roots of all are acrid.

722. **Ord. Droseraceæ** (*the Sundew Family*). Small herbs, growing in swamps, usually covered with gland-bearing hairs; with the leaves alternate, or clustered at the base of a scape, tapering into a petiole, rolled up from the apex to the base in vernation (circin-

FIG. 540. Viola sagittata. 541. One of the stamens without appendage, seen from within; and one furnished with a spur-like appendage on the back. 542. A capsule which has opened and separated into three valves; the calyx still persistent. 543. A valve of the same, from which the seeds have fallen. 544. A magnified seed. 545. The same divided vertically, showing the large embryo in the midst of the albumen.

nate): stipules none, except a fringe of hairs or bristles at the base of the petioles. Calyx of five equal sepals, persistent. Corolla of five regular petals, withering and persistent, convolute in æstivation. Stamens as many as the petals and alternate with them, or sometimes two or three times as many, distinct, withering; anthers extrorse. Styles three to five, distinct or nearly so, and each two-parted (so as to be taken for ten styles, Fig. 390), with the divisions sometimes two-lobed or many-cleft at the apex; sometimes all united into one. Fruit a one-celled capsule, opening loculicidally by three to five valves, with three to five parietal placentæ; in Dionæa membranaceous and bursting irregularly, with a thick placenta at the base. Seeds usually numerous. Embryo small, at the base of cartilaginous or fleshy albumen. — *Ex.* Drosera, the Sundew; and Dionæa (Fig. 228), so remarkable for its sensitive leaves, which suddenly close when touched.

723. **Subord. Parnassieæ** consists of the genus Parnassia (belonging to the northern temperate and frigid zones, and to the high mountains of tropical Asia); which differs from Droseraceæ in the want of glandular hairs, in the introrse anthers, exalbuminous seeds, imbricated æstivation of the petals, and curious appendages before each petal. These are explained, and the plan of the flower shown, on p. 253 (Fig. 304, 305). In the ovary, also, the four short stigmas are situated opposite the four parietal placentæ. The genus has been placed, probably rightly, in Saxifragaceæ, on account of its slightly perigynous stamens, &c.

724. **Ord. Cistaceæ** (*the Rock-Rose Family*). Low shrubby plants or herbs, with simple and entire leaves (at least the lower opposite). Calyx of five persistent sepals, the three inner with a convolute æstivation; the two outer small or sometimes wanting. Corolla of five, or rarely three, regular petals, convolute in æstivation in the direction contrary to that of the sepals, often crumpled, usually ephemeral, sometimes wanting, at least in a portion of the flowers. Stamens few or numerous, distinct, with short innate anthers. Fruit a one-celled capsule with parietal placentæ, or imperfectly three to five-celled by dissepiments arising from the middle of the valves (dehiscence therefore loculicidal), and bearing the placentæ at or near the axis. Seeds few or numerous, orthotropous (with few exceptions), with mealy albumen. Embryo curved, or variously coiled or bent. — *Ex.* Cistus, Helianthemum (Fig. 546): a small family; the flowers often showy. No im-

portant properties. Several exude a balsamic resin, such as *Ladanum* from a Cistus of the Levant.*

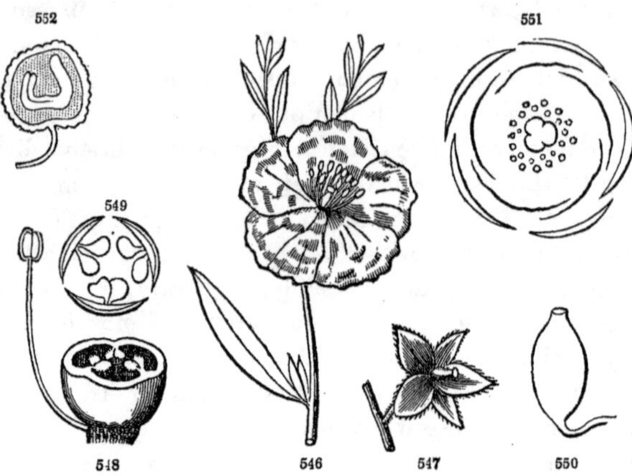

725. **Ord. Hypericaceæ** (*the St. John's-wort Family*). Shrubs or herbs, with a resinous or limpid juice, and opposite entire leaves, destitute of stipules, and punctate with pellucid or blackish dots. Flowers regular. Calyx of four or five persistent sepals, the two exterior often smaller. Petals four or five, twisted in æstivation, often with black dots. Stamens commonly polyadelphous and numerous. Capsule with septicidal dehiscence, many-seeded. — *Ex.* Hypericum (St. John's-wort, Fig. 553) is the type of this small family. The plants yield a resinous acrid juice, and a bitter, balsamic extractive matter.† Embryo straight; albumen little or

* Ord. BIXACEÆ consists of tropical trees or shrubs, not resembling any of the other orders with parietal placentæ, and is here mentioned because Bixa Orellana, of tropical America, yields the *Arnotto* of commerce; which is the waxy, orange-red pulp that surrounds the seeds, and is separated from them by washing. It is chiefly used for staining cheese, and in the preparation of chocolate.

† Ord. GUTTIFERÆ, or CLUSIACEÆ, consisting of tropical trees, with a yellow, resinous juice, large flowers, and thick and shining entire leaves, is nearly allied to Hypericaceæ, and exhibits the acrid properties of the latter family in a much higher degree. — *Gamboge* is the hardened resinous

FIG. 546. The Rock-Rose, Helianthemum Canadense. 547. Flower from which the petals and stamens have fallen. 548. Magnified cross-section of the ovary; with a single stamen, showing its hypogynous insertion. 549. Cross-section of a capsule, loculicidally dehiscent; the seeds therefore borne on the middle of each valve. 550. An ovule. 551. Plan of the flower. 552. Section of a seed, showing the curved embryo.

none. The peculiarity of the stamens is explained, and a diagram of the flower of Elodea is given, on p. 252.

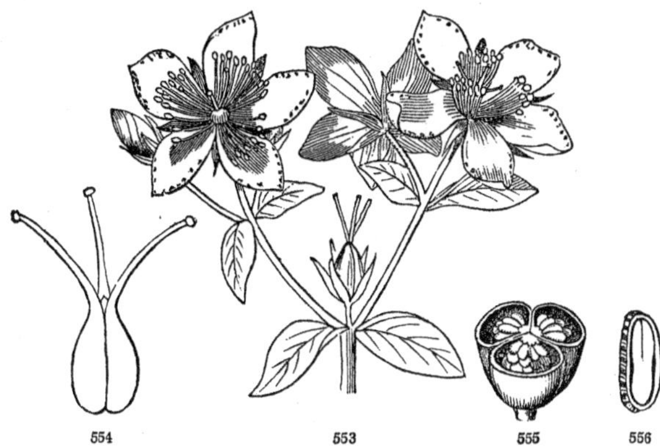

726. **Ord. Elatinaceæ** (*the Waterwort Family*). Small annual weeds, with opposite leaves, membranaceous stipules, and minute axillary flowers. Sepals and petals three to five. Stamens as many or twice as many as the petals, distinct. Capsule 2 – 5-celled, septicidal or septifragal; the numerous seeds attached to a persistent central axis. Albumen none. — *Ex.* Elatine is the type of this order, containing a few insignificant weeds.

727. **Ord. Caryophyllaceæ** (*the Pink Family*). Herbs, with opposite entire leaves; the stems tumid at the nodes. Flowers regular. Calyx of four or five sepals. Corolla of four or five petals, or sometimes wanting. Stamens as many, or commonly twice as

---

juice of the Hebradendron cambogioides of Ceylon; but the tree is supposed to have been imported from Siam by the Buddhists, to whom it is sacred, on account of the yellow color it yields. The gamboge from Siam forms the best pigment. Clusia flava yields the *Hog-gum* of Jamaica. The hot aromatic *Canella bark*, or *False Winter's-bark*, is derived from the Canella alba of the West Indies. Notwithstanding the acrid properties of this order, Garcinia Mangostana of Malacca yields one of the most delicious of fruits, the *Mangosteen*.

ORD. TAMARISCINEÆ consists of Tamarix and one or two other genera of sea-side plants, natives of Europe and Asia: they are ornamental, shrubby plants, with small scale-like and somewhat fleshy leaves, and an astringent bark.

FIG. 553. Hypericum perforatum (St. John's-wort). 554. Its tricarpellary pistil. 555. Cross-section of the capsule. 556. Vertical section of a seed and its embryo.

many, as the petals, sometimes reduced to two or three. Styles two to five, stigmatose down the inside. Ovary mostly one-celled, with a central or basilar placenta. Capsule two to five-valved, or opening only at the apex with twice as many valves as stigmas. Embryo peripheric, curved or coiled around the outside of mealy albumen. — There are five principal suborders, viz. : —

728. **Subord. Sileneæ** (*the proper Pink Family*) ; in which the sepals are united into a tube, and the petals (mostly convolute in æstivation) and stamens are inserted on the stipe of the ovary, the former with long claws ; and there are no stipules. — *Ex.* Silene, Dianthus (the Pink, Carnation).

729. **Subord. Alsineæ** (*the Chickweed Family*) ; in which there are no stipules, the ovary is sessile, the sepals and petals (imbricated in æstivation) are nearly or quite distinct ; the petals destitute of claws ; and the stamens are inserted into the margin of a small hypogynous disc, which, however, occasionally coheres with the base of the calyx, and becomes perigynous. — *Ex.* Stellaria, Arenaria, &c. (Chickweeds). Some are ornamental ; others, such as the common Chickweed, are insignificant weeds.

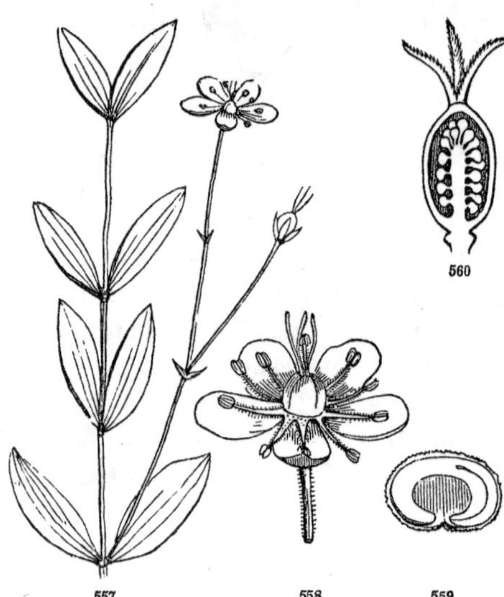

730. **Subord. Illecebreæ** (*the Knotwort Family*) ; differing from the last in having scarious stipules ; the sepals often united below ; the petals often wanting or rudimentary ; the stamens manifestly perigynous, and the fruit more commonly a one-seeded utricle. — *Ex.* Paronychia and Anychia. Spergula has conspicuous petals, and many-seeded capsules ; and so differs from Alsineæ only in its stipules. Insignificant weeds of no economical account.

FIG. 557. Arenaria lateriflora. 558. A magnified flower. 559. Magnified section of a seed, showing the embryo coiled into a ring around the albumen. 560. Vertical section of a pistil of Spergularia.

731. Subord. Scleranthеæ (*the Knawel Family*) is like the last, only there are no stipules, and the calyx-tube is urceolate in fruit, inclosing the utricle. — *Ex.* Scleranthus.

732. Subord. Mollugineæ (*the Carpet-weed Family*) is apetalous, exstipulate, and has the stamens alternate with the sepals when of the same number; thus effecting a transition to the next order. — *Ex.* Mollugo.

733. Ord. Portulacaceæ (*the Purslane Family*). Succulent or fleshy herbs, with alternate or opposite entire leaves, destitute of proper stipules, and usually ephemeral flowers. Calyx mostly of two or three sepals, cohering with the base of the ovary. Petals five, or rarely more numerous. Stamens variable in number, but when equal to the petals situated opposite them. Styles two to eight, united below. Capsule with few or numerous seeds, attached to a central basilar placenta, often by slender funiculi. Seed and embryo as in Caryophyllaceæ. — *Ex.* Portulaca (Purslane),

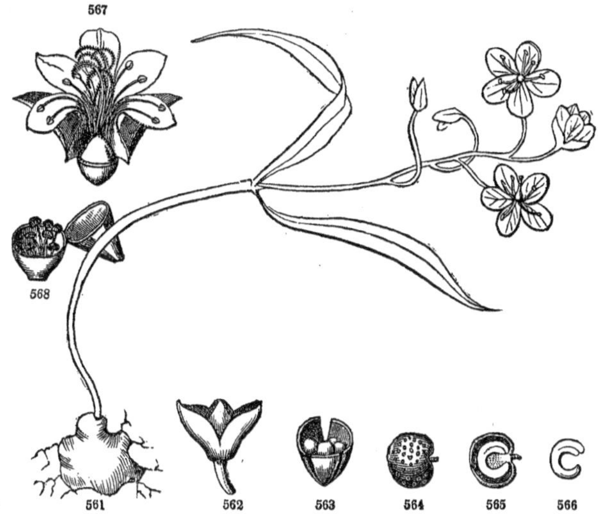

Claytonia (Fig. 561). Chiefly natives of dry and arid places in

FIG. 561. Claytonia Virginica (Spring Beauty). 562. Young fruit and the persistent two-leaved calyx. 563. Section of the dehiscing capsule. 564. A seed. 565. The same, vertically divided. 566. The embryo, detached.

FIG. 567. Flower of the Purslane; the calyx cut away at the point where it adheres to the ovary, and laid open. 568. A capsule (pyxis, 616) of the same, transversely dehiscent.

the warmer parts of the world; except Claytonia. Insipid or slightly bitter: several are used as pot-herbs, as the Purslane. Some are ornamental. The farinaceous root of Lewisia rediviva, a native of dry plains in the interior of Oregon, is an important article of food with the natives.

739. **Ord. Mesembryanthemaceæ** (*the Fig-Marigold Family*) consists of succulent plants, with showy flowers opening only under bright sunshine, containing an indefinite number of petals and stamens, and a many-celled and many-seeded capsule; otherwise much as in Caryophyllaceæ.—The thickened leaves are often oddly shaped.—*Ex.* Mesembryanthemum (Fig-Marigold, Ice-plant); the numerous species are chiefly natives of the Cape of Good Hope, flourishing in the most arid situations.

740. **Ord. Malvaceæ** (*the Mallow Family*). Herbs, shrubs, or rarely trees. Leaves alternate, palmately veined, furnished with stipules. Flowers regular, generally showy, often with an involucel, forming a double calyx. Calyx mostly of five sepals, more or less united at the base, valvate in æstivation. Petals as many as the sepals, convolute in æstivation, hypogynous. Stamens indefinite, monadelphous; inserted with the petals, united with their claws: anthers reniform, one-celled. Pollen hispid. Ovary several-celled, with the placentæ in the axis; or ovaries several, separate or separable at maturity. Styles as many as the carpels, distinct or united below. Fruit capsular, or the carpels separate or separable. Seeds with little mucilaginous or fleshy albumen.

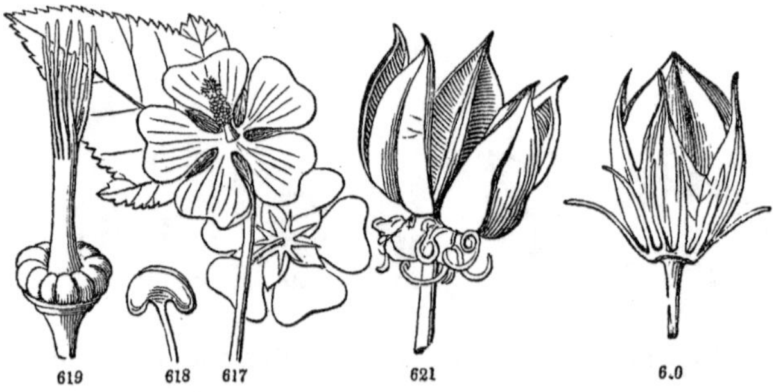

Embryo large, with foliaceous cotyledons, variously incurved or

FIG. 617. The Marsh Mallow (Althæa officinalis). 618. One of the kidney-shaped one-celled anthers, magnified. 619. The pistils, magnified. 620. Capsule of Hibiscus Moscheutos, with the persistent calyx and involucel. 621. The same, loculicidally dehiscent.

folded. — *Ex.* Malva (Mallows), Althæa (Hollyhock), Gossypium (Cotton), &c.: a pretty large and important family. Malvaceous plants commonly abound in mucilage, and are entirely destitute of unwholesome qualities. The unripe fruit of Hibiscus esculentus (Okra) is used as an ingredient in soups. Althæa officinalis is the Marsh Mallow of Europe, the Guimauve of the French. The tenacious inner bark of many species is employed for cordage. Cotton is the hairy covering of the seeds of Gossypium: the long and slender tubes, or attenuated cells, collapse and twist as the seed ripens, which renders the substance capable of being spun. Cotton-seed yields a fixed oil in large quantity, which may be used for lamps, &c. Numerous species are cultivated for ornament.

741. **Ord. Byttneriaceæ** is distinguished from the foregoing by its usually definite stamens, and the two-celled anthers (the cells parallel), with smooth pollen. The carpels are few and consolidated. — A Melochia and a Hermannia are found in Texas. The rest of the order is tropical or subtropical. *Chocolate* is made of the roasted and comminuted seeds of Theobroma Cacao (a South American tree), mixed with sugar, arnotto, vanilla, and other ingredients, and pressed into cakes. The roasted integuments of the seeds, also, are used as a substitute for coffee.*

742. **Ord. Tiliaceæ** (*the Linden Family*). Trees or shrubby plants, with alternate leaves, furnished with deciduous stipules, and small flowers. Calyx deciduous. Petals sometimes imbricated in æstivation. Disc glandular. Stamens indefinite, often in three to

---

* ORD. STERCULIACEÆ, very closely allied to Malvaceæ and Byttneriaceæ, and consisting of tropical trees, possesses the same mucilaginous properties (as well as oily seeds), with which bitter and astringent qualities are often combined. The seeds of Bombax, the Silk-cotton tree, are enveloped in a kind of cotton, which belongs to the endocarp and not to the seed; and the hairs, being perfectly smooth and even, cannot be spun. Canoes are made from the trunk of Bombax, in the West Indies. To this order belongs the famous Baobab, or Monkey's-bread of Senegal (Adansonia digitata); some trunks of which are from sixty to eighty feet in circumference! The fruit resembles a gourd, and serves for vessels; it contains a subacid and refrigerant, somewhat astringent pulp; the mucilaginous young leaves are also used for food in time of scarcity; the dried and powdered leaves (*Lalo*) are ordinarily mixed with food, and the bark furnishes a coarse thread, which is made into cordage or woven into cloth. Cheirostemon platanoides is the remarkable Hand-flower tree of Mexico. A plant of the family (Fremontia, *Torr.*) nearly allied to Cheirostemon has recently been found in California, by Col. Fremont.

five clusters, distinct or somewhat united, one of each parcel often transformed into a petaloid scale; anthers two-celled. Styles united into one. Fruit two to five-celled, or, by obliteration, one-celled when ripe. In other respects nearly as in Malvaceæ. — *Ex.* Tilia, the Linden, or Lime-Tree (Fig. 622), represents the order in northern temperate regions; the other genera are tropical. All are mucilaginous, with a tough, fibrous inner bark. From this *bast* or *bass* of the Linden, the Russian mats, &c. are made, whence the name of Basswood. Gunny-bags and fishing-nets are made in India from the bark of Corchorus capsularis; the fibre of which is called *Jute*, and is spun and woven. The light wood of the Linden is excellent for wainscoting and carving: its charcoal is used for the manufacture of gunpowder. It is said that a little sugar may be obtained from the sap: and the honey made from the odorous flowers is thought to be the finest in the world. The acid berries of Grewia sapida are employed in the East in the manufacture of sherbet.*

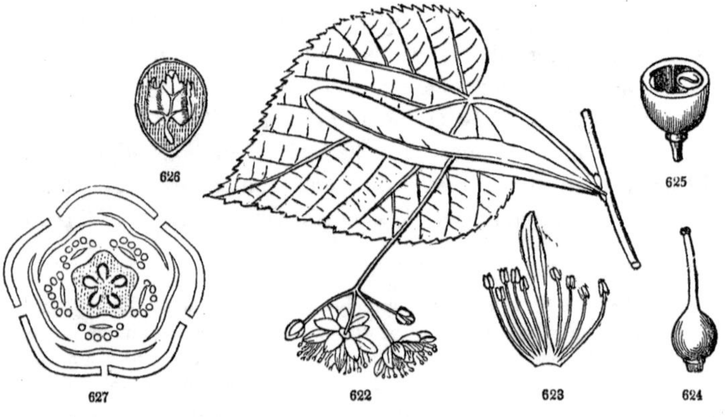

* Ord. DIPTEROCARPEÆ, intermediate in some respects between Tiliaceæ and Ternstrœmiaceæ, consists of a few tropical Indian trees, with a resinous or balsamic juice. Dryobalanops aromatica, a large tree of Sumatra and Borneo, yields in great abundance both a *camphor oil* and solid *camphor:*

FIG. 622. Flowering branch of Tilia Americana, the common American Linden; the flower-stalk cohering with the bract. 623. One of the clusters of stamens adhering to the *staminodium*, or petaloid scale. 624. The pistil. 625. Cross-section of the fruit, which has become one-celled by the obliteration of the partitions, and one-seeded. 626. Vertical section of the seed, magnified, to show the large embryo with its taper radicle and foliaceous crumpled cotyledons. (A better section of the seed, cut in the direction across the cotyledons, is shown in Fig. 451.) 627. Diagram of the flower.

743. **Ord. Ternstrœmiaceæ** (*the Tea Family*). Trees or shrubs, with a watery juice, alternate simple leaves without stipules, and large and showy flowers. Calyx of three to seven coriaceous and concave imbricated sepals. Petals five or more, imbricated in æstivation. Stamens hypogynous, indefinite, monadelphous or polyadelphous at the base. Capsule several-celled, usually with a central column. Seeds few in each cell, large, with or without albumen. — *Ex.* Gordonia (Loblolly Bay), Stuartia, Thea (Tea), Camellia. Ornamental plants, natives of tropical America, except two genera in the Southern United States, and of Eastern Asia. The leaves of *Tea* contain a peculiar extractive matter, and a somewhat stimulant ethereal oil.

744. **Ord. Aurantiaceæ** (*the Orange Family*). Trees or shrubs, with alternate leaves (compound, or with jointed petioles), destitute of stipules, dotted with pellucid glands full of volatile oil. Flowers fragrant. Calyx short, urceolate or campanulate. Petals three to five. Stamens inserted in a single row upon a hypogynous disc, often somewhat monadelphous or polyadelphous. Style cylindrical: stigma thickish. Fruit a many-celled berry, with a leathery rind, filled with pulp. Seeds without albumen. — *Ex.* Citrus, the Orange and Lemon. Nearly all natives of tropical Asia; now dispersed throughout the warmer regions of the world, and cultivated for their beauty and fragrance, and for their grateful fruit. The acid of the Lemon, &c. is the citric and malic. The rind abounds in a volatile oil (such as the *Oil of Bergamot* from the Lime), and an aromatic, bitter principle.

745. **Ord. Meliaceæ.** Trees or shrubs, with alternate, usually compound leaves, destitute of stipules. Calyx of three to five sepals. Petals three to five. Stamens twice as many as the petals, monadelphous, inserted with the petals on the outside of a hypogynous disc; the anthers included in the tube of filaments. Ovary

---

both are found deposited in cavities of the trunk, the latter sometimes in pieces as long as a man's arm, weighing ten or twelve pounds. It is more solid than common camphor, and is not volatile at ordinary temperatures. It bears a high price, and is seldom found in Europe or this country, but is chiefly carried to China and Japan. A thin balsam, called *wood-oil* in India, and used for painting ships and houses, is yielded by some species of Dipterocarpus and Shorea. Shorea robusta yields the *Dammer-pitch.* Vateria Indica exudes a kind of copal, the *Gum Animi* of commerce; and a somewhat aromatic fatty matter, called *Piney Tallow*, is derived from the seeds.

several-celled, with one or two ovules in each cell: styles and stigmas united into one. Fruit a drupe, berry, or capsule; the cells one-seeded. Seeds without albumen, wingless. — *Ex.* Melia Azedarach (Pride of India), naturalized, as an ornamental tree, in the Southern States. An acrid and bitter principle pervades this tropical order.

746. **Ord. Cedrelaceæ** (*the Mahogany Family*). Trees (tropical or Australian), with hard and durable, usually fragrant and beautifully veined wood; differing botanically from Meliaceæ chiefly by their capsular fruit, with several winged seeds in each cell. — *Ex.* The *Mahogany* (Swietenia Mahagoni) of tropical America, reaching to Southern Florida. The *Red-wood* of Coromandel is the timber of Soymida febrifuga; the *Satin-wood*, of Chloroxylon Swietenia of India; *Yellow-wood*, of the Australian Oxleya xanthoxyla, &c. All the species are bitter, astringent, tonic, often aromatic and febrifugal.

747. **Ord. Linaceæ** (*the Flax Family*). Herbs, with entire and sessile leaves, either alternate, opposite, or verticillate, and no stipules, except minute glands occasionally. Flowers regular and symmetrical. Calyx of three or five persistent sepals, strongly imbricated. Petals as many as the sepals, convolute in æstivation. Stamens as many as the petals, and usually with as many intermediate teeth representing an abortive series, all united at the base into a ring, hypogynous. Ovary with as many styles and cells as there are sepals, each cell with two suspended ovules; the cells in the capsule each more or less perfectly divided into two, by a false partition which grows from the back (dorsal suture); the spurious cells one-seeded. Embryo straight: cotyledons flat, fleshy and oily, surrounded by a thin albumen. — *Ex.* Linum, the

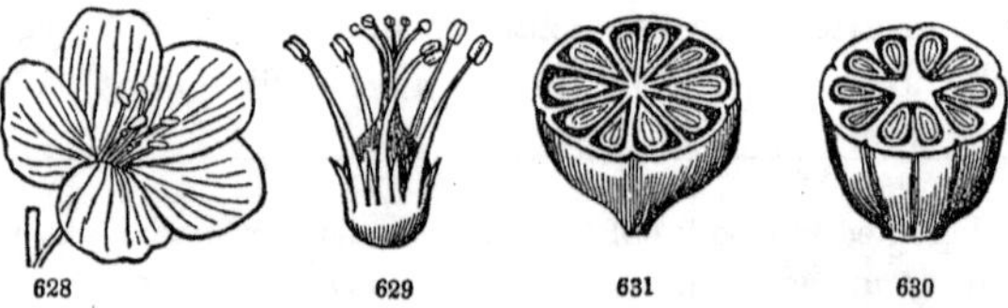

Flax (Fig. 628–631), is the principal representative of this small

FIG. 628. Flower of Linum perenne. 629. Its stamens and pistils. 630. Cross-section of its capsule, showing the incomplete false partition from the back of each cell. 631. Section of the fruit of the common Flax, where the false partitions completely divide each proper cell into two.

family. The tough woody fibre of the bark (*flax*) is of the highest importance: the seeds yield a copious mucilage, and the fixed oil expressed from them is applied to various uses in the arts. The flowers are commonly handsome. The flowers of the succeeding families are formed on the same general plan.

748. **Ord. Geraniaceæ** (*the Cranesbill Family*). Herbs or shrubby plants, commonly strong-scented; with palmately veined and usually lobed leaves, mostly with stipules; the lower opposite. Flowers regular, or somewhat irregular. — Calyx of five persistent sepals, imbricated in æstivation. Petals five, with claws, mostly

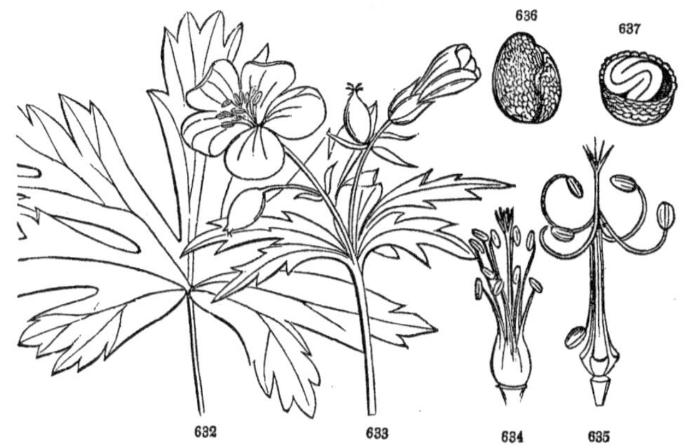

convolute in æstivation. Stamens 10, the five exterior hypogynous, occasionally sterile; the filaments all broad and united at the base. Ovary of five two-ovuled carpels, attached to the base of an elongated axis (*gynobase*) to which the styles cohere: in fruit the distinct one-seeded carpels separate from the axis, by the twisting or curling back of the persistent indurated styles from the base upwards. Seeds with no albumen: cotyledons convolute and plaited together, bent on the short radicle. — *Ex.* Geranium (Fig. 632–638), or Cranesbill.

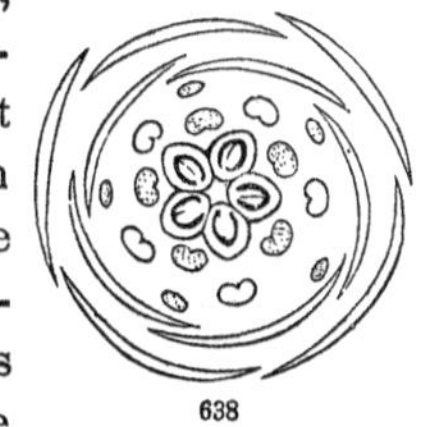

FIG. 632. Radical leaf of Geranium maculatum (Cranesbill). 633. A flowering branch. 634. A flower with the calyx and corolla removed, showing the stamens, &c. 635. The pistil in fruit; the indurated styles separating below from the prolonged axis, and curving back elastically, carrying with them the membranous carpels. 636. A magnified seed. 637. A cross-section of the same, showing the folded and convolute cotyledons.

FIG. 638. Diagram of the flower of a Geranium.

Our cultivated Geraniums, so called, from the Cape of Good Hope, are species of Pelargonium. The roots are simply and strongly astringent. The foliage abounds with aromatic resinous matter and an ethereal oil. — The proper symmetry of the flower is explained on p. 267.

749. **Ord. Oxalidaceæ** (*the Wood-Sorrel Family*). Low herbs, with an acid juice, and alternate compound leaves; the leaflets usually obcordate. Flowers regular, of the same general structure as in the preceding family, except the gynæcium. Carpels five, united into a compound ovary, with the styles distinct; in fruit forming a membranaceous five-lobed and five-celled capsule. Seeds with a fleshy outer coat, which bursts elastically when ripe, with a large and straight embryo in thin albumen. — *Ex.* Oxalis, the Wood-Sorrel. The herbage is sour, as the name denotes, and contains oxalic acid. The foliage is remarkably sensitive in some species. The tubers of some South American species (called Arracacha), filled with starch, have been substituted for potatoes.

750. **Ord. Zygophyllaceæ** differs from the last in the opposite, mostly abruptly pinnate leaves, distinct stamens (the filaments commonly furnished with an internal scale, Fig. 303), and the styles united into one. — *Ex.* Tribulus and Kallstrœmia (introduced into the Southern States) are exalbuminous; the latter is 10-coccous, just as Linum is, by a false partition. Guaiacum and Larrea, both in Texas, and the rest of the family, have a corneous albumen. The wood of Guaiacum (*Lignum-vitæ*) is extremely hard and heavy, and yields a gum-resinous, bitter, and acrid principle (*Gum Guaiacum*), well known in medicine.

751. **Ord. Balsaminaceæ** (*the Balsam Family*). Annual herbs, with succulent stems filled with a watery juice. Leaves simple, without stipules. Flowers irregular, and one of the colored sepals spurred or saccate. Stamens five, cohering by an internal appendage. Compound ovary five-celled; stigmas sessile. Capsule bursting elastically by five valves. Seeds several, without albumen, and with a straight embryo. — *Ex.* Impatiens, the Balsam, or Touch-me-not. Remarkable for the elastic force with which the capsule bursts in pieces, and expels the seeds. Somewhat differently irregular blossoms are presented by the

752. **Ord. Tropæolaceæ** (*the Indian-Cress or Nasturtium Family*). Straggling or twining herbs, with a pungent watery juice, and peltate or palmate leaves. Flowers irregular. Calyx of five colored,

united sepals, the lower one spurred. Petals five; the two upper arising from the throat of the calyx, remote from the three lower, which are stalked. Stamens, eight, unequal, distinct. Ovary three-lobed, composed of three united carpels; which separate from the common axis when ripe, are indehiscent, and one-seeded. Seed filling the cell, without albumen: cotyledons large, thick, and consolidated. — *Ex.* Tropæolum, the Garden Nasturtium, from South America, where there are a few other species, one of which bears edible tubers. They possess the same acrid principle and antiscorbutic properties as the Cruciferæ. The unripe fruit of Tropæolum majus is pickled, and used as a substitute for capers.

753. **Ord. Limnanthaceæ** differs from the last only in its regular and symmetrical blossoms, and the erect instead of suspended seeds; the calyx valvate in æstivation. — *Ex.* Limnanthes of California, and Flœrkea of the Northern United States.

754. **Ord. Rutaceæ** (*the Rue Family*). Herbs, shrubs, or trees; the leaves dotted and without stipules. Flowers perfect. Calyx of four or five sepals. Petals four or five. Stamens as many or two or three times as many as the petals, inserted on the outside of a hypogynous disc. Ovary three- to five-lobed, three- to five-celled, with the styles united, or distinct only at the base, during ripening usually separating into its component carpels, which are dehiscent by one or both sutures. Seeds few, mostly with albumen; and a curved embryo. — *Ex.* Ruta (the Rue), Dictamnus (Fraxinella), of Europe, &c., and Rutosma of Texas. Diosma and its allies, of the Cape of Good Hope, New Holland, &c., form a suborder. Remarkable for their strong and usually unpleasant odor, and their bitterness (as in the common Rue of the gardens), owing to a volatile oil and a resinous matter; the former is so abundantly exhaled by the Fraxinella in a hot, dry day, that it is said the air which surrounds it may be set on fire. Many plants of the Diosma tribe, especially those of Equinoctial America, contain a bitter alkaloid principle, and possess valuable febrifugal properties. The most important is the Galipea which furnishes the *Angostura bark*.

755. **Ord. Zanthoxylaceæ** (*the Prickly-Ash Family*) ought properly to be received only as a suborder of Rutaceæ, differing from the Diosmeæ merely in their more or less diclinous flowers. Trees or shrubs; the leaves without stipules, and punctate with pellucid dots. Flowers polygamous or diœcious. Calyx of three to nine sepals. Petals as many as the sepals, or wanting. Stamens

as many or twice as many as the petals. Carpels two or more, borne on the convex or elevated receptacle, either united or separate; in the latter case the styles usually cohere when young. Seeds one or two in each cell or carpel, with a smooth and shining crustaceous testa, albuminous, embryo rather large, straight.—*Ex.* Zanthoxylum (Prickly Ash) is the type of this order, of chiefly

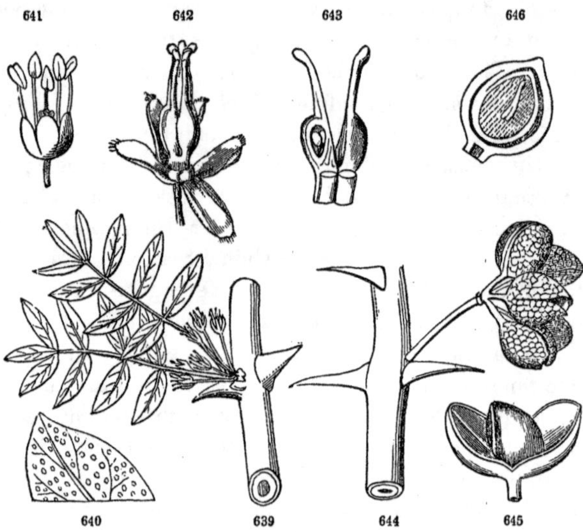

American, and nearly all tropical, plants. They are aromatic, pungent, stimulant, and bitter; these properties chiefly resident in the bark.

756. **Ord. Ochnaceæ** is a small group, nearly allied to the last, but with simple dotless leaves, not aromatic, and endowed with purely bitter qualities. Some plants of the family have a scale on the inner side of each filament, as in Zygophyllaceæ, and make a near approach to Simarubaceæ * (to which Castela has been justly transferred).

---

* Ord. SIMARUBACEÆ, composed of a few tropical, and chiefly American, trees and shrubs, is of some importance in medicine. The wood abounds

FIG. 639. A flowering branch of Zanthoxylum Americanum (the Northern Prickly Ash). 640. A piece of a leaf, to show the pellucid dots. 641. Staminate flower. 642. A pistillate flower, the sepals spread open. 643. Two of the pistils; one of them divided vertically to show the ovules. 644. A branch in fruit. 645. One of the dehiscent pods, and the seed. 646. Vertical section of an unripe pod and seed; the latter pendent from a descending funiculus, showing a slender embryo in copious albumen.

757. **Ord. Anacardiaceæ** (*the Cashew Family*). Trees or shrubs, with a resinous or milky, often acrid juice, which turns blackish in drying: the leaves alternate, without stipules, and not dotted. Flowers small, often polygamous or diœcious. Calyx of three to five sepals, united at the base. Petals, and usually the stamens, as many as the sepals, inserted into the base of the calyx or into a hypogynous disc. Ovary one-celled, but with three styles or stigmas, and a single ovule. Fruit a berry or drupe. Seed without albumen. Embryo curved or bent. — *Ex.* Rhus, Anacardium (the Cashew), Pistacia. Chiefly tropical; but several species of Rhus are indigenous to the United States. The acrid resinous juice is used in varnishes; but it often contains a caustic poison. Even the exhalations from Rhus Toxicodendron (Poison Oak, Poison Ivy), and R. venenata (Poison Sumach, Poison Elder), as is well known, severely affect many persons, producing erysipelatous swellings, &c. Their juice is a good indelible ink for marking linen. But the common Sumachs (R. typhina and R. glabra) are innocuous; their astringent bark is used for tanning; and their sour berries (which contain bimalate of lime) for acidulated drinks. The oily seeds of Pistacia vera (the Pistachio-nut) are edible. The drupe of Mangifera Indica (Mango) is one of the most grateful of tropical fruits. The kernel of the Cashew-nut (Anacardium occidentale) is eatable; and so is the acid enlarged and fleshy peduncle on which the nut rests: but the coats of the latter are filled with a caustic oil, which blisters the skin; while from the bark of the tree a bland gum exudes.*

---

in an excessively bitter extractive principle, called *Quassine*. The Quassia-wood of the shops is derived from the Quassia amara of Surinam and Guiana, or more commonly, at least of late years, from Picræna excelsa of Jamaica. It has been used as a substitute for hops in the manufacture of beer.

* Ord. BURSERACEÆ, including a great part of what were formerly called Terebinthaceæ, consists of tropical trees, with a copious resinous juice, compound leaves usually marked with pellucid dots, and small, commonly perfect flowers; with valvate petals, a two- to five-celled ovary, and drupaceous fruit. Their balsamic juice, which flows copiously when the trunk is wounded, usually hardens into a resin. The *Olibanum*, used as a fragrant incense, the *Balm of Gilead*, or *Balsam of Mecca*, *Myrrh*, and the *Bdellium*, are derived from Arabian species of the order; the East Indian *Gum Elemi*, from Canarium commune; *Balsam of Acouchi*, and similar substances, from various American trees of this family.

Ord. AMYRIDACEÆ consists of a few West Indian plants, intermediate as it were between Burseraceæ and Leguminosæ, and distinguished from the

758. **Ord. Malpighiaceæ** is a large tropical family (with one or two representatives in Texas), which differs from Aceraceæ in its more symmetrical flowers, trimerous gynæcium, solitary ovules, the want of a disc, and in the entire leaves, &c.

759. **Ord. Aceraceæ** (*the Maple Family*). Trees or shrubs, with opposite leaves and no stipules. Flowers small, polygamous, regular, sometimes perfect, in racemes, corymbs, or fascicles, often preceding the leaves. Calyx mostly of five sepals, more or less united. Petals as many as the sepals, or none. Stamens three to twelve, seldom agreeing in number with the sepals, inserted on or around a hypogynous disc. Ovary of two more or less united carpels; each carpel forming a samara in fruit. Ovules two in each cell. Seeds solitary, destitute of albumen. Embryo coiled.

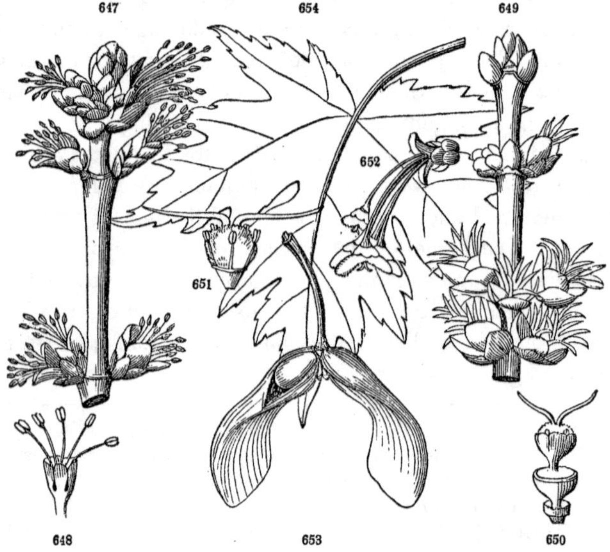

— *Ex.* Acer, the Maple; useful timber-trees of northern temper-

---

former chiefly by their simple and solitary ovary. One species of Amyris grows in Florida. Their properties are the same as the preceding; the trunks abounding in a fragrant resinous juice.

FIG. 647. A branch of Acer dasycarpum (the White Soft Maple) with staminate flowers. 648. A separate, enlarged, staminate flower. 649. Branch with pistillate flowers. 650. A separate fertile flower; the bracts, &c. of the cluster cut away. 651. The same, enlarged, with the calyx cut away. 652. A cluster showing the fruiting ovaries expanding into wings (reduced in size). 653. Ripe fruit; one of the samaras cut open to show the seed. 654. A leaf.

ate regions. Sugar is yielded by the vernal sap of Acer saccharinum, and in less quantity by A. dasycarpum and other species.

760. **Ord. Sapindaceæ** (*the Soapberry Family*). Trees, shrubs, or climbers, with tendrils, rarely herbs (nearly all tropical and American); with alternate and mostly compound leaves. Flowers small, unsymmetrical, usually irregular and polygamous. Calyx of four or five sepals. Petals irregular and often one fewer than the sepals, sometimes wanting. Stamens eight to ten. Ovary two- or three-celled; the styles or stigmas more or less united. Seeds usually with an aril, destitute of albumen. Embryo coiled; the cotyledons usually thick and fleshy. — *Ex.* Sapindus (Soapberry, one species of which is indigenous to the southern borders of the United States); and Cardiospermum, which is a climbing herb, with a bladdery capsule, often met with in gardens. They are astringent and bitter. The fruit of Sapindus is used for soap. The leaves of true Sapindaceæ are alternate. Inseparably connected with this order is the

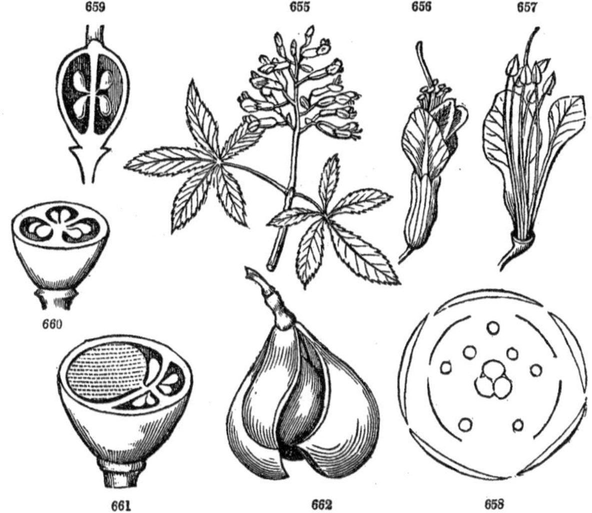

761. **Subord. Hippocastanaceæ.** Trees or shrubs; with opposite

FIG. 655. Flowering branch of Æsculus Pavia, a species of Buckeye. 656. A flower. 657. Flower with the calyx and two of the petals removed. 658. A ground-plan of the flower, showing that its parts are unsymmetrical. 659. Vertical section of an ovary, showing two of the cells with a pair of ovules in each, one ascending, one descending. 660. Cross-section of an ovary. 661. Cross-section of the immature fruit; only one fertile seed; the others abortive. 662. The dehiscent fruit.

digitate leaves, without stipules. Fruit roundish, coriaceous, dehiscent, with one to three very large seeds, resembling chestnuts. Embryo very large and fleshy, showing a two-leaved plumule: the cotyledons united. — *Ex.* Æsculus, the Horsechestnut, and Buckeye: fine ornamental trees. The large, starchy seeds are nutritious, but they contain a bitter principle which is more or less noxious. Those of Æ. Pavia are used to stupefy fish. The root, according to Elliott, is employed as a substitute for soap.

762. **Ord. Celastraceæ** (*the Spindle-tree Family*). Shrubs or trees, with alternate or opposite simple leaves. Calyx of four or five sepals, imbricated in æstivation. Petals as many as the sepals, inserted under the flat expanded disc which closely surrounds the ovary, imbricated in æstivation. Stamens as many as the petals, and alternate with them, inserted on the margin or upper surface of the disc. Ovary free from the calyx. Fruit a capsule or berry, with one or few seeds in each cell. Seeds usually arilled, albuminous, with a large and straight embryo. — *Ex.* Celastrus (False Bittersweet), Euonymus (Burning Bush, Spindle-tree): they are all somewhat bitter and acrid; but of little economical importance. The crimson capsules and bright scarlet arils of Euonymus atropurpureus and E. Americanus (sometimes called Strawberry-tree), present a striking appearance when the fruit is ripe.

763. **Ord. Rhamnaceæ** (*the Buckthorn Family*). Shrubs or trees, often with spinose branches; the leaves mostly alternate, simple. Flowers small. Calyx of four or five sepals, united at the base, valvate in æstivation. Petals four or five, cucullate or convolute, inserted on the throat of the calyx, sometimes wanting. Stamens as many as the petals, inserted with and opposite them! Ovary sometimes coherent with the tube of the calyx, and more or less immersed in a fleshy disc, with a single erect ovule in each cell. Fruit a capsule, berry, or drupe. Seeds not arilled. Embryo straight, large, in sparing albumen. — *Ex.* Rhamnus (Buckthorn) is the type of the order. Ceanothus is peculiar to North America; just as some genera are to the Cape, and others to New Holland. The berries of most species of Rhamnus are somewhat nauseous; but those of Zizyphus are edible. The genuine *Jujube paste* is prepared from those of Z. Jujuba and Z. vulgaris of Asia. *Syrup of Buckthorn* and the pigment called *Sap-green* are prepared from the fruit of Rhamnus catharticus. The herbage and bark in this order are more or less astringent and bitter. An infusion of the

leaves of Ceanothus Americanus (thence called New Jersey Tea) has been used as a substitute for tea.

764. **Ord. Staphyleaceæ** (*the Bladder-nut Family*), consisting chiefly of Staphylea, is intermediate between the order Sapindaceæ, from which it differs in its more symmetrical flowers and straight embryo in fleshy albumen, and Celastraceæ, from which the compound leaves, partly separate pistils, and bony seeds distinguish it.

765. **Ord. Vitaceæ** (*the Vine Family*). Shrubby plants climbing by tendrils, with simple or compound leaves, the upper alternate. Flowers small, often polygamous or diœcious. Calyx very small,

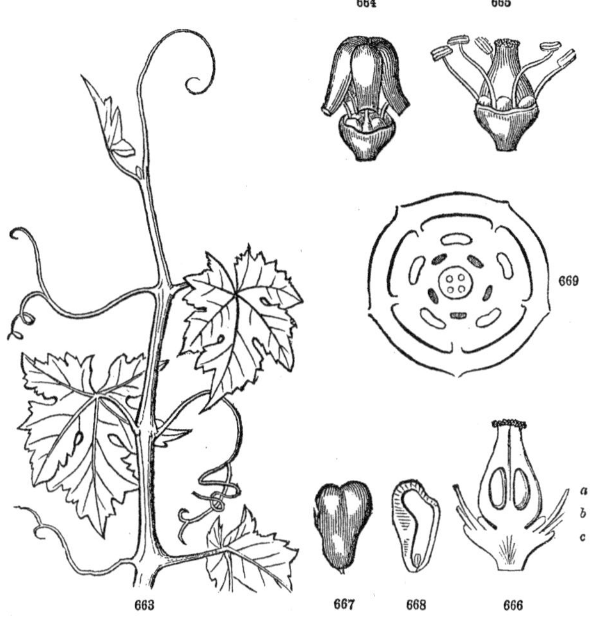

entire or four- or five-toothed, lined with a disc. Petals four or

FIG. 663. A branch of the Grape; showing the nature of the tendrils. 664. A flower; the petals separating from the base, and falling off together without expanding. 665. A flower from which the petals have fallen; the lobes of the disc seen alternate with the stamens. 666. Vertical section through the ovary and the base of the flower: *a*, calyx, the limb of which is a mere rim: *b*, petal; having the stamen, *c*, directly before it; and the lobes of the disc are shown between this and the ovary. 667. A seed. 668. Section of the seed; showing the thick crustaceous testa, and the albumen, at the base of which is the minute embryo. 669. A horizontal plan of the flower.

five, inserted upon the outside of the disc, valvate in æstivation, sometimes cohering by their tips, and caducous. Stamens as many as the petals, and opposite them! Ovary two-celled, with two erect ovules in each cell. Fruit a berry. Seeds with a bony testa, and a small embryo in hard albumen. — *Ex.* Vitis (the Vine), Ampelopsis (the Virginia Creeper). The fruit of the Vine is the only important product of the order. The acid of the grape, which also pervades the young shoots and leaves, is chiefly the tartaric. Grape-sugar is very distinct from cane-sugar, and the only kind that can long exist in connection with acids. — The symmetry of the flower is explained on p. 269.

766. **Ord. Polygalaceæ** (*the Milkwort Family*). Herbs or shrubby plants, with simple entire leaves, destitute of stipules; the roots sometimes with a milky juice. Pedicles with three bracts. Flowers perfect, unsymmetrical, and irregular, falsely papilionaceous. Calyx of five irregular sepals; the odd one superior, the two inner (*wings*) larger, and usually petaloid. Petals usually three, inserted

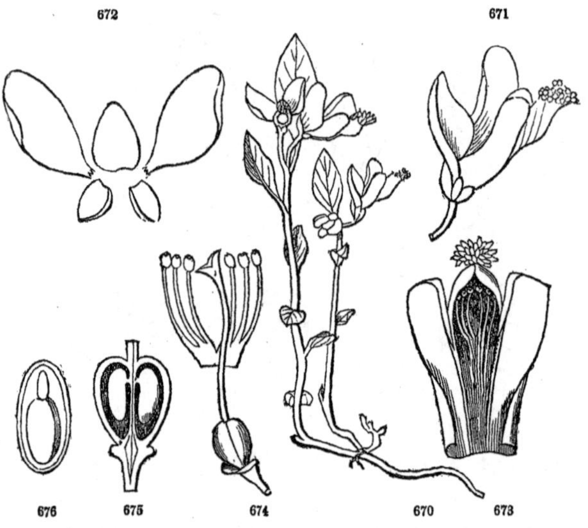

on the receptacle, more or less united; the anterior (*keel*) larger

FIG. 670. Polygala paucifolia. 671. A flower, enlarged. 672. The calyx displayed. 673. The corolla and stamineal tube laid open. 674. The pistil and the free portion of the stamens. 675. Vertical section of the ovary. 676. Vertical section of the seed, showing the large embryo and scanty albumen.

than the rest. Stamens six to eight, combined in a tube, which is split on the upper side, and united below with the claws of the petals: anthers innate, mostly one-celled, opening by a pore at the apex. Ovary compound, two-celled, with a single suspended ovule in each cell: style curved and often hooded. Capsule flattened. Seeds usually with a caruncle. Embryo straight, large, in fleshy, thin albumen. — *Ex.* Polygala, the type of the order, is dispersed nearly throughout the world. A bitter principle pervades the order; and many species also yield an acrid extractive matter. The Polygala Senega (Seneca Snakeroot) is the most important medicinal plant of the family. Other species are employed medicinally in Brazil, Peru, Nepaul, &c.; where, like our own, they are reputed antidotes to the bites of venomous reptiles.

767. **Ord. Krameriaceæ** (*the Rhatany Family*) consists of the genus Krameria only, which has ordinarily been annexed to the Polygalaceæ; but it is much nearer the Leguminosæ, having the odd sepal inferior, a simple unilocular pistil, and an exalbuminous seed. In fact it is distinguishable from the latter only by the hypogynous stamens and the want of stipules. The roots contain a red coloring matter, and are astringent without bitterness. Rhatany-root, used to adulterate port-wine, and as an ingredient in tooth-powders, &c., is the produce of K. triandra of Peru. That of our own Southern K. lanceolata possesses the same properties.

768. **Ord. Leguminosæ** (*the Pulse Family*). Herbs, shrubs, or trees, with alternate and usually compound leaves, furnished with stipules. Calyx mostly of five sepals, more or less united; the odd sepal inferior (Fig. 382). Corolla of five petals, either papilionaceous or regular. Stamens perigynous, or sometimes hypogynous. Ovary single and simple. Fruit a legume, various forms of which are shown in Fig. 438-446. Seeds destitute of albumen. — This immense family is divided into three principal suborders; viz.: —

769. **Subord. Papilionaceæ** (*the Proper Pulse Family*); which has the papilionaceous flower, already illustrated (468, Fig. 317-321), ten stamens (or rarely fewer), which are diadelphous (Fig. 308), sometimes monadelphous (Fig. 307, 324), or rarely distinct (Fig. 322), inserted into the base of the calyx. Radicle bent on the large cotyledons. Leaves only once compound, the leaflets entire.

(*Ex.*, the Pea, Bean, Locust, Clover, &c.) The vexillum is the largest petal, and external in æstivation, in all true papilionaceous corollas, as in the diagram, Fig. 382. But in the

770. **Subord. Cæsalpineæ** (to which Cassia, Cercis, and the Honey-Locust belong), the corolla gradually loses its papilionaceous character, and always has the vexillum, or superior petal, covered by the lateral ones in æstivation; the stamens are distinct, and the embryo straight. The leaves are often bipinnate.

771. **Subord. Mimoseæ** (a large group to which the Acacia and the Sensitive Plant belong) has a perfectly regular calyx and corolla, the latter mostly valvate in æstivation and hypogynous, as well as the stamens, which are sometimes definite, but often very numerous; and the embryo is straight. The leaves are frequently tri-pinnate.

772. Papilionaceæ are found in every part of the world, from the tropics to the frigid zones: Cæsalpineæ and Mimoseæ are confined to the tropical and warmer temperate regions. — A full account of the useful plants and products of this large order would require a separate volume. Many, such as Clover, Lucerne (Medicago sativa), &c., are extensively cultivated for fodder; Peas and Beans, for pulse. The roots of the Licorice (Glycirrhiza glabra of Southern Europe) abound in a sweet mucilaginous juice, from which the pectoral extract of this name is prepared. The sweet pulp of the pods of Ceratonia Siliqua (Carob-tree of the South of Europe, &c.), of the Honey-Locust (Gleditschia), &c., is likewise eaten. The laxative pulp of Cathartocarpus Fistula, and of the Tamarind, is well known; the latter is acidulated with malic, and a little tartaric and citric acid. — A peculiar volatile principle (called *Coumarin*) gives its vanilla-like fragrance to the well-known *Tonka-bean*, and to the Melilotus, or Sweet Clover. The flowers and seeds of the latter and of Trigonella cærulea give the peculiar odor to *Scheipzeiger* cheese. — Astringents and tonics are also yielded by this order: such as the African Pterocarpus erinaceus, the hardened red juice of which is *Gum Kino*; that of P. Draco, of Carthagena, &c., is *Dragon's Blood*. The bark of most Acacias and Mimosas contains a very large quantity of tannin, and is likely to prove of great importance in tanning. The valuable astringent called *Catechu* is obtained by boiling and evaporating the heart-wood of the Indian Acacia Catechu. — Leguminosæ yield the most important coloring matters; such as the *Brazil-wood*, the

Logwood of Campeachy (the peculiar coloring principle of which is called *Hæmatin*), and the *Red Sandal-wood* of Ceylon. Most important of all is *Indigo*, which is prepared from the fermented juice of the Indigofera tinctoria (a native of India), and also from I. cærulea, and other species of the genus. This substance is highly azotized, and is a violent poison. — To the same order we are indebted for valuable resins and balsams; such as the Mexican *Copal*, *Balsam of Copaiva* of the West Indies, Para, and Brazil, the bitter and fragrant *Balsam of Peru*, and the sweet, fragrant, and stimulant *Balsam of Tolu.* — It also furnishes the most useful gums; of which we need only mention *Gum Tragacanth*, derived from Astragalus verus of Persia, &c.; and *Gum Arabic*, the produce of numerous African species of Acacia. The best is said to be obtained from Acacia vera, which extends from Senegal to Egypt; while *Gum Senegal* is yielded by A. Verek, and some other species of the River Gambia. The *Senna* of commerce consists of the leaves of several species of Cassia, of Egypt and Arabia. C. Marilandica of this country is a succedaneum for the officinal article. — More acrid, or even poisonous properties, are often met with in the order. The roots of Baptisia tinctoria (called Wild Indigo, because it is said to yield a little of that substance), of the Broom, and of the Dyers' Weed (Genista tinctoria, used for dying yellow), possess such qualities; while the seeds of Laburnum, &c. are even narcotico-acrid poisons. The branches and leaves of Tephrosia, and the bark of the root of Piscidia Erythrina (Jamaica Dogwood, which is also found in Southern Florida), are commonly used in the West Indies for stupefying fish. *Cowitch* is the stinging hairs of the pods of Mucuna pruriens of the West, and M. prurita of the East, Indies. — Among the numerous valuable timber-trees, our own Locust (Robinia Pseudacacia) must be mentioned; and also the Rosewood of commerce, the produce of a Brazilian species of Mimosa. Few orders furnish so many plants cultivated for ornament.

773. **Ord. Rosaceæ** (*the Rose Family*). Trees, shrubs, or herbs, with alternate leaves, usually furnished with stipules. Flowers regular. — Calyx of five (rarely three or four) more or less united sepals, and often with as many bracts. Petals as many as the sepals (rarely none), mostly imbricated in æstivation, inserted on the edge of a thin disc that lines the tube of the calyx. Stamens perigynous, indefinite, or sometimes few, distinct. Ovaries with soli-

tary or few ovules: styles often lateral. Albumen none. Embryo straight, with broad and flat or plano-convex cotyledons (Fig. 457). — This important order is divided into four suborders; viz.: —

774. **Subord. Chrysobalaneæ** (*the Cocoa-plum Family*). Ovary solitary, free from the calyx, or else cohering with it at the base on one side only, containing two erect ovules: style arising from the apparent base. Fruit a drupe. Trees or shrubs. — *Ex.* Chrysobalanus.

775. **Subord. Amygdaleæ** (*the Almond or Plum Family*). Ovary solitary, free from the deciduous calyx, with two suspended ovules, and a terminal style. Fruit a drupe (Fig. 447, 448). Trees or shrubs. — *Ex.* Amygdalus (the Almond, Peach, &c.), Prunus (the Plum), Cerasus (the Cherry).

776. **Subord. Rosaceæ proper.** Ovaries several, numerous, or rare-

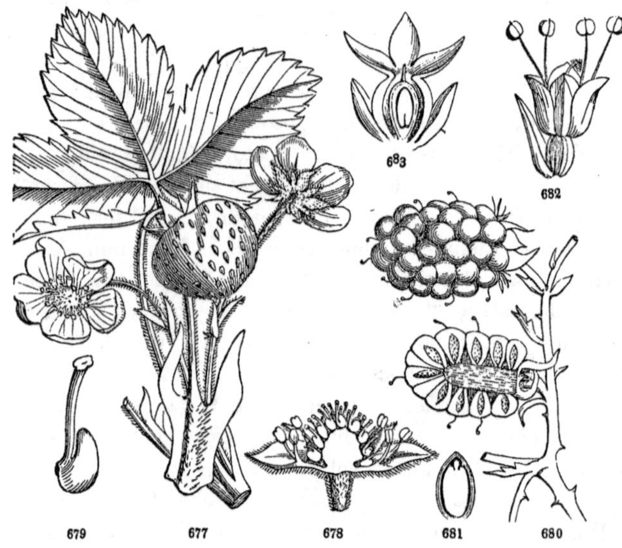

FIG. 677. The Strawberry (Fragaria). 678. Half of a flower, divided vertically, from which the petals are removed; showing the perigynous insertion of the stamens, and the enlarged receptacle, which, increasing in size, forms the pulpy, edible fruit, bearing the achenia, or real fruits, on its surface. 679. One of the carpels magnified, showing the lateral style. 680. Fruit of the Blackberry (Rubus villosus), with a longitudinal section: here the elongated receptacle does not enlarge, but the ovaries become drupes. 681. Section of the endocarp; the cavity of which is filled by the seed, and that by the embryo, with its large cotyledons. 682. A flower of Sanguisorba Canadensis, enlarged. 683. Vertical section of the same in fruit; the solitary ovary inclosed by, but not coherent with, the persistent calyx-tube; the single seed with its large embryo filling the achenium.

ly solitary, free from the calyx (which is often bracteolate, as i double), but sometimes inclosed in its persistent tube, in fruit becoming either follicles or achenia. Styles terminal or lateral. Herbs or shrubs. — The three tribes of this suborder are Tribe 1. SPIREÆ, where the fruit is a follicle. *Ex.* Spiræa and Gillenia. Tribe 2. DRYADEÆ, where the fruits are achenia, or sometimes little drupes, and when numerous crowded on a conical or hemispherical torus. *Ex.* Dryas, Agrimonia, Potentilla, Fragaria (Strawberry), Rubus (Raspberry aud Blackberry). Tribe 3. ROSEÆ, where numerous achenia cover the hollow torus which lines the urn-shaped calyx-tube; and the latter, being contracted at the mouth, and becoming fleshy or berry-like, forms a kind of false pericarp; as in the Rose.

777. **Subord. Pomeæ** (*the Pear Family*). Ovaries two to five, or rarely solitary, cohering with each other and with the thickened and fleshy or pulpy calyx-tube; each with one or few ascending seeds. Trees or shrubs. — *Ex.* Cratægus (the Thorn), Cydonia (the Quince), Pyrus (the Apple, Pear, &c.).

778. This important order is diffused through almost every part

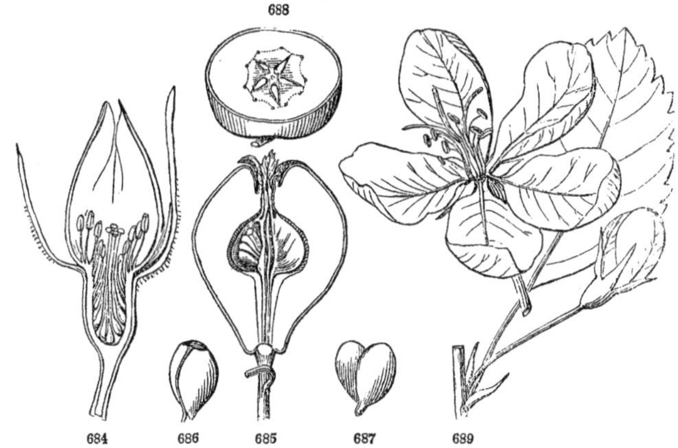

of the world; but chiefly abounds in temperate climates, where it

FIG. 684. Vertical section of an unexpanded Rose, showing the attachment of the carpels to the lining of the calyx-tube, and of the stamens and petals to its summit or edge. 685. Vertical section of the fruit of the Quince, exhibiting the carpels invested by the thickened calyx which forms the edible part of the fruit; one of the ovaries laid open to show the seeds. 686. A magnified seed; the rhaphe and chalaza conspicuous. 687. The embryo. 688. Cross-section of an apple. 689. Flower, &c. of the American Crab-apple (Pyrus coronaria).

furnishes the most important fruits. It is destitute of unwholesome qualities, with one or two exceptions; viz.: — 1st. The bark, leaves, and kernel of Amygdaleæ contain prussic acid, as is indicated by their peculiar odor, — a trace of which is perceived insome species of Spiræa, and in the Mountain Ash, &c., among Pomeæ; and 2d. The root of Gillenia (Bowman's Root, Indian Physic) is emetic in large doses, but in small doses it acts as a tonic. The bark and root in all are astringent. The bark of Amygdaleæ also exudes gum. That of the Wild Black Cherry is febrifugal; and the timber is useful in cabinet-work. The leaves of Cerasus Caroliniana contain so much prussic acid as to destroy cattle that feed upon them. It takes the place in this country of the Cerasus Lauro-cerasus (Cherry-Laurel) of the Old World, from which the poisonous *Laurel-water* and the virulent *Oil of Laurel* are obtained. Sweet and bitter almonds are the seeds of varieties of Amygdalas communis (indigenous to the East), differing in the quantity of the prussic acid they contain: the oil of the former resembles olive-oil; that of the latter is a deadly poison. Of the Peach, Apricot, Nectarine, Plum, and Cherry, it is unnecessary to speak. The kernels, as well as the flowers, of the former, especially, abound in prussic acid. — The strawberry, raspberry, and blackberry are the principal fruits of the proper Rosaceæ. The leaves of Rosa centifolia are more commonly distilled for *Rose-water*: and *Attar of Roses* is obtained from R. Damascena, &c. — Pomaceous fruits, such as the apple, pear, quince, services, medlar, &c., yield to none in importance: their acid is usually the malic.

779. **Ord. Calycanthaceæ.** Shrubs, with quadrangular stems (which when old exhibit four axes of growth exterior to the old wood), opposite entire leaves without stipules, and solitary, axillary and terminal, lurid flowers. Calyx of numerous somewhat thickened colored sepals, in several rows, confounded with the petals, all united below into a fleshy tube or cup, bearing numerous stamens upon its rim. Outer stamens with adnate extrorse anthers: the inner sterile. Ovaries indefinite, two-ovuled, becoming hard achenia in fruit, inserted on the whole inner surface of the disc which lines the calyx-tube, in which they are inclosed, as in the Rose. Albumen none. Cotyledons convolute. — Consists of two genera; namely, Calycanthus (Carolina Allspice, Sweet-scented Shrub, &c.), and Chimonanthus, of Japan. They are cultivated for their fragrant flowers. The bark and foliage of Caly-

canthus exhale a camphoric odor; and the flowers a fragrance not unlike that of strawberries.

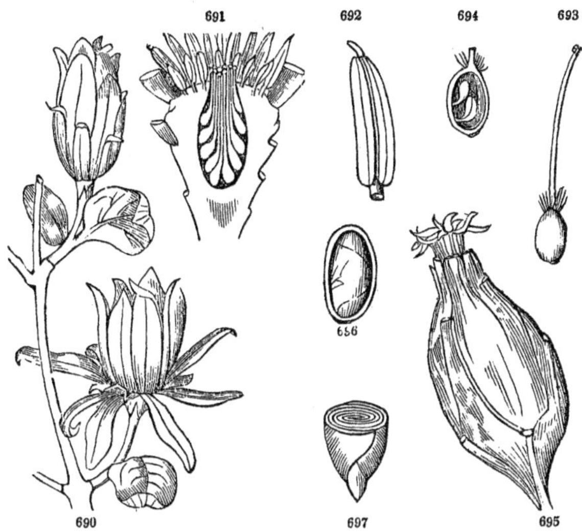

780. **Ord. Myrtaceæ** (*the Myrtle Family*). Trees or shrubs, with opposite and simple entire leaves, which are punctate with pellucid dots, and usually furnished with a vein running parallel with and close to the margin; without stipules. Calyx-tube adherent to the compound ovary; the limb four- or five-cleft, valvate in æstivation. Petals four or five, or sometimes wanting. Stamens indefinite, usually with long filaments and small round (introrse) anthers. Style one. Seeds usually numerous, destitute of albumen. — *Ex.* Myrtus, the Myrtle, is the most familiar representative of this beautiful tropical and subtropical order; which is well distinguished from its allies by its opposite dotted leaves and aromatic properties. The species abound in a pungent and aromatic volatile oil, and an astringent principle. *Cloves* are the dried flower-buds of Caryophyllus aromaticus. *Pimento* (Allspice) is the dried fruit of Eugenia Pimenta. *Cajeput oil*, a powerful sudorific, is distilled

FIG. 690. Flowers of Calycanthus floridus. 691. Vertical section of a flower, showing the hollow receptacle, &c.; the floral envelopes cut away. 692. A stamen, seen from without. 693. A pistil. 694. Section of the ovary, showing the two ascending ovules. 695. The closed pod-shaped receptacle in fruit. 696. A vertical section of an achenium, showing the embryo of the seed. 697. Cross-section of an embryo, showing the finely convolute cotyledons.

from the leaves and fruit of a Melaleuca of the Moluccas. Numerous Australian species of Eucalyptus, which compose a great part of the forests of that country, yield a large quantity of tannin. The aromatic fruits of many species, filled with sugar and mucilage, and acidulated with a free acid, are highly prized; such, for instance, as the Pomegranate, the Guava, Rose-Apple, &c.

781. **Ord. Melastomaceæ.** Trees, shrubs, or herbs, with opposite ribbed leaves, and showy flowers, with as many or twice as many stamens as petals; the anthers mostly appendaged and opening by pores, inflexed in æstivation: further distinguished from Myrtaceæ by the leaves not being dotted; and from Lythraceæ by the adnation of the calyx-tube (at its nerves at least) with the ovary.— *Ex.* The beautiful species of Rhexia represent this remarkable order in the United States: all the rest are tropical. The berries of Melastoma are eatable, and tinge the lips black, like whortleberries; whence the generic name.

782. **Ord. Lythraceæ** (*the Loosestrife Family*) is distinguished among these perigynous orders, with exalbuminous seeds, by its tubular calyx inclosing the 2–4-celled ovary, but entirely free from it. The styles are perfectly united into one: the fruit is a thin capsule. The stamens are inserted on the tube of the calyx below the petals. — *Ex.* Lythrum. Chiefly tropical, of little consequence.

783. **Ord. Rhizophoraceæ** (*the Mangrove Family*) consists of a few tropical trees (extending into Florida and Louisiana), growing in maritime swamps; with the ovary often partly free from the calyx, two-celled, with two pendulous ovules in each cell; they are remarkable for their opposite leaves, with interpetiolar stipules, and for the germination of the embryo while within the pericarp (645). — *Ex.* Rhizophora, the Mangrove (Fig. 118). The astringent bark has been used as a febrifuge, and for tanning.

784. **Ord. Combretaceæ** consists of tropical trees or shrubs (which have one or two representatives in Southern Florida), often apetalous, but with slender colored stamens; distinguishable from any of the preceding orders of this group by their one-celled ovary, with several suspended ovules, but only a solitary seed, and convolute cotyledons. — *Ex.* Combretum. Some species cultivated for ornament; some are used by tanners. The seeds of Terminalia Catappa (which extends into Florida) are eaten like almonds.

785. **Ord. Onagraceæ** (*the Evening-Primrose Family*). Herbs, or rarely shrubby plants, with alternate or opposite leaves, not dotted

nor furnished with stipules. Flowers usually showy, tetramerous. Calyx adherent to the ovary, and usually produced beyond it into a tube. Petals usually four (rarely three or six, occasionally absent), and the stamens as many, or twice as many, inserted into the throat of the calyx. Ovary commonly four-celled: styles united; the stigmas four, or united into one. Fruit mostly capsular. — *Ex.* Chiefly an American order; many are ornamental in cultivation. Fuchsia, remarkable for its colored calyx and berried fruit; Œnothera (Evening Primrose); Epilobium, where the seeds bear a coma; Gaura, where the petals are often irregular; Ludwigia, which is sometimes apetalous; and Circæa, where the lobes of the calyx, petals, stamens, cells of the ovary, and the seeds, are reduced to two; showing a connection with the appended

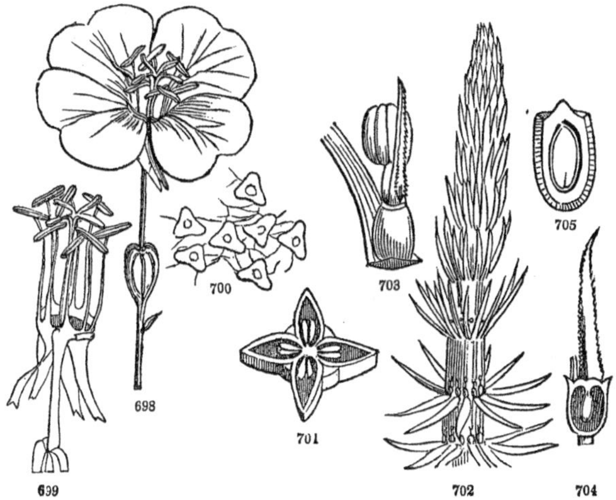

786. **Subord. Halorageæ**, which are a sort of reduced aquatic Onagraceæ, often apetalous: the solitary seeds furnished with a little albumen, as in Myriophyllum (Water-Milfoil) and Hippuris (Horse-tail), where the limb of the calyx is almost wanting; the petals none; the stamens reduced to a single one, and the ovary to a single cell, with a solitary seed.

FIG. 698. Flower of Œnothera fruticosa. 699. The same, about the natural size, with the petals removed. 700. Magnified grains of pollen, with some of the intermixed cellular threads. 701. Cross-section of the four-lobed and four-celled capsule.

FIG. 702. Hippuris vulgaris (suborder Halorageæ). 703. Magnified flower, with the subtending leaf. 704. Vertical section of the ovary. 705. Vertical section of the fruit and seed.

787. **Ord. Cactaceæ** (*the Cactus Family*). Succulent shrubby plants, peculiar in habit, with spinous buds, usually leafless; the stems either subglobose and many-angled, columnar with several angles, or flattened and jointed. Flowers usually large and showy. Calyx of numerous sepals, imbricated, coherent with and crowning the one-celled ovary, or covering its whole surface; the inner usually confounded with the indefinite petals. Stamens indefinite, with long filaments, cohering with the base of the petals. Styles united: stigmas and parietal placentæ several. Fruit a berry. Seeds numerous, with little or no albumen. — All American, the greater part Mexican or on the borders of Mexico. The common Opuntia (Prickly Pear) extends north to New England. The mucilaginous fruit is eatable.

788. **Ord. Grossulaceæ** (*the Gooseberry Family*). Small shrubs, either spiny or prickly, or unarmed; with alternate, palmately-lobed and veined leaves, usually in fascicles, often sprinkled with

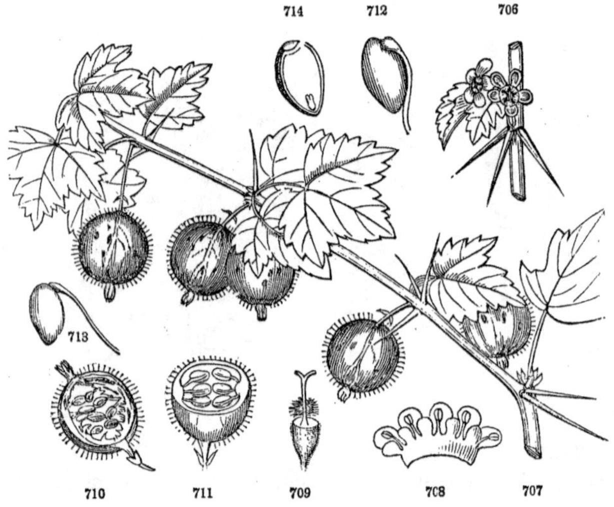

resinous dots. Flowers in racemes or small clusters. Calyx-tube

FIG. 706. The Gooseberry (Ribes Uva-crispa); a branch in flower. 707. Branch in fruit. 708. The calyx, bearing the petals and stamens, cut away from the summit of the ovary (709), and laid open. 710, 711. Sections of the unripe fruit. 712. Magnified seed (anatropous). 713. The same from the ripe fruit, where the rhaphe separates from the side of the seed, and forms a part of the funiculus. 714. Longitudinal section of the same, showing the minute embryo at the extremity of the albumen.

adherent to the one-celled ovary, and more or less produced beyond it, five-lobed, sometimes colored. Petals (small) and stamens five, inserted on the calyx. Ovary with two parietal placentæ: styles more or less united. Fruit a many-seeded berry, crowned with the shrivelled remains of the flower. Embryo minute, in hard albumen. — *Ex.* Ribes (Gooseberry and Currant). Natives of temperate and colder regions, chiefly of the northern hemisphere. Never unwholesome: the fruit usually esculent, containing mucilaginous and saccharine pulp, with more or less malic or citric acid. Several Oregon and Californian species are showy in cultivation.

789. **Ord. Loasaceæ.** Herbs usually clothed with rigid or stinging hairs; the leaves opposite or alternate, without stipules; the flowers showy. Calyx-tube adherent to the one-celled ovary; the limb mostly five-parted. Petals as many, or twice as many, as the lobes of the calyx. Stamens perigynous, indefinite, and in several parcels, or sometimes definite. Style single. Ovary with three to five parietal placentæ. Seeds few or numerous, albuminous. — *Ex.* Loasa, Mentzelia, Cevallia; the latter with solitary seeds and no albumen. All American, and in the United States nearly confined to the regions beyond the Mississippi. The bristles of Loasa sting like nettles.

790. **Ord. Turneraceæ.** Herbs, with the habit of Cistus or Helianthemum; the alternate leaves without stipules. Flowers solitary, showy. Calyx five-lobed; the five petals and five stamens inserted on its throat. Ovary free from the calyx, one-celled, with three parietal placentæ. Styles distinct, commonly branched or many-cleft at the summit. Fruit a three-valved capsule. Seeds numerous (anatropous), with a crustaceous and reticulated testa, and a membranaceous aril on one side. Embryo in fleshy albumen. — *Ex.* Turnera, of which there is one species in Georgia.

791. **Ord. Passifloraceæ** (*the Passion-flower Family*). Herbs, or somewhat shrubby plants, climbing by tendrils; with alternate, entire, or palmately lobed leaves, mostly furnished with stipules. Flowers often showy, sometimes involucrate. Calyx mostly of five sepals, united below, free from the one-celled ovary; the throat bearing five petals and a filamentous crown. Stamens as many as the sepals, monadelphous, and adhering to the stalk of the ovary, which has usually three club-shaped styles or stigmas, and as many parietal placentæ. Fruit mostly fleshy or berry-like.

Seeds numerous, with a brittle sculptured testa, inclosed in pulp. Embryo inclosed in thin, fleshy albumen. — *Ex.* Passiflora (the Passion-flower, Granadilla): nearly all natives of tropical America. Two species are found as far north as Virginia and Ohio. Many are cultivated for their singular and showy flowers. The acidulous refrigerant pulp of Passiflora quadrangularis (the Granadilla), P. edulis, and others, is eaten in the West Indies, &c. But the roots are emetic, narcotic, and poisonous. They contain a principle resembling morphine, which, in some species, extends even to the flowers and fruit.

792. **Ord. Papayaceæ** comprises merely a small genus of tropical diœcious trees, of peculiar character: the principal one is the Papaw-tree (Carica Papaya) of tropical America, which has been introduced into East Florida. The fruit, when cooked, is eatable; but the juice of the unripe fruit, as well as of other parts of the plant, is a powerful vermifuge. The juice contains so much fibrine that it has an extraordinary resemblance to animal matter: meat washed in water impregnated with this juice is rendered tender; even the exhalations from the tree produce the same effect upon meat suspended among the leaves.

793. **Ord. Cucurbitaceæ** (*the Gourd Family*). Juicy herbs, climbing by tendrils; with alternate, palmately veined or lobed, rough leaves, and monœcious or diœcious flowers. Calyx of four or five (rarely six) sepals, united into a tube, and in the fertile flowers adherent to the ovary. Petals as many as the sepals, commonly more or less united into a monopetalous corolla, which coheres with the calyx. Stamens five or three, inserted into the base of the corolla or calyx, either distinct or variously united by their filaments, and long, sinuous or contorted anthers. Ovary two- to five-celled (rarely one-celled by obliteration, and even one-ovuled); the thick and fleshy placentæ often filling the cells, or diverging before or after reaching the axis and carried back so as to reach the walls of the pericarp, sometimes manifestly parietal; the dissepiments often disappearing during its growth: stigmas thick, dilated or fringed. Fruit (pepo, 613) usually fleshy, with a hard rind, sometimes membranous. Seeds mostly flat, with no albumen. Embryo straight. Cotyledons foliaceous. — *Ex.* The Pumpkin and Squash (Cucurbita), Gourd, Cucumber, and Melon. When the acrid principle which prevails throughout the order is greatly diffused, the fruits are eatable and sometimes delicious: when con-

centrated, as in the Bottle Gourd, Bryony, &c., they are dangerous or actively poisonous. The officinal *Colocynth*, a resinoid, bitter extract from the pulp of Cucumis Colocynthis (of the Levant, India, &c.), is very acrid and poisonous; and *Elaterium*, obtained from the juice of the Squirting Cucumber (Momordica Elaterium of the South of Europe), is still more violent in its effects. Momordica Balsamina (the cultivated Balsam-Apple) contains the same principle in smaller quantity. The seeds of all are harmless.

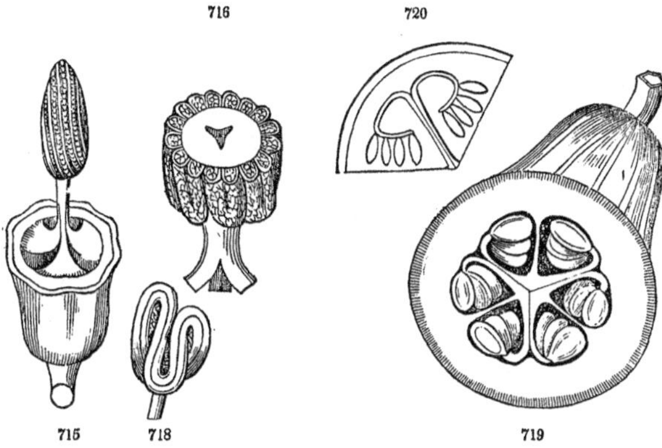

794. **Ord. Crassulaceæ** (*the Orpine Family*). Herbs, or slightly shrubby plants, mostly fleshy or succulent; with scattered leaves, and flowers usually in cymes or racemes. Calyx of three to twenty sepals, more or less united at the base, free from the ovaries, persistent. Petals as many as the sepals, rarely combined into a monopetalous corolla. Stamens as many or twice as many as the sepals, inserted with the petals on the base of the calyx. Pistils always as many as the sepals, distinct, or rarely (in Penthorum and Diamorpha) partly united: ovaries becoming follicles in fruit, several-seeded. Embryo straight, in thin albumen. — *Ex.* Sedum (Stone-crop, Orpine, Live-for-ever), Crassula, Sempervivum, or Houseleek, &c. Distinguished by their completely symmetrical flowers, on which account they have already been illustrated (449, 450). They mostly grow in arid places: of no economical importance.

FIG. 715. Staminate flower of the Gourd; the calyx and corolla cut away. 716. Cross-section of the united anthers. 718. Separate stamen of the Melon. 719. Section of the ovary of the Gourd. 720. Plan of one of the three constituent carpels.

795. **Ord. Saxifragaceæ** (*the Saxifrage Family*). Herbs or shrubs, with alternate or opposite leaves. Calyx of four or five more or less united sepals, either free from or more or less adherent to the ovary, persistent. Petals as many as the sepals, rarely wanting. Stamens as many, commonly twice as many, or rarely three or four times as many, as the sepals, perigynous. Ovaries mostly two (sometimes three or four), usually united below and distinct at the summit. Seeds numerous, with a straight embryo in fleshy albumen. There are three principal divisions, or suborders; viz.: —

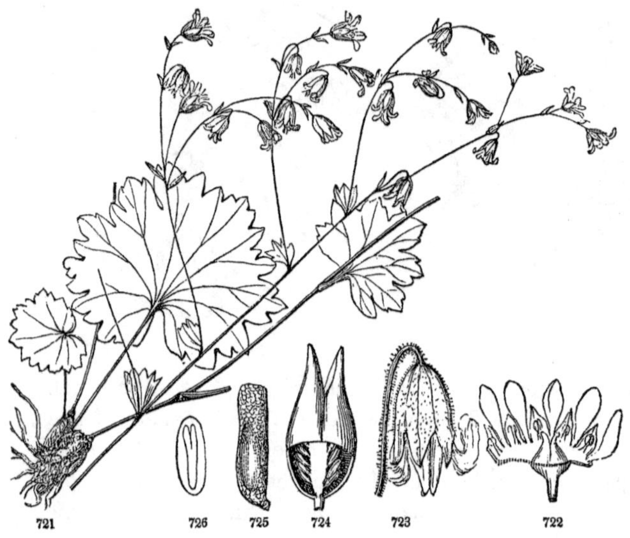

796. **Subord. Saxifrageæ** (*the true Saxifrage Family*). Herbs. Petals imbricate in æstivation. Capsule (when the carpels are united) either two-celled with the placentæ in the axis, or one-celled with parietal placentæ. — *Ex.* Saxifraga, Sullivantia (Fig. 721), Heuchera. Of little consequence, except as ornamental plants. The roots are generally astringent; powerfully so in Heuchera, especially in the common H. Americana (sometimes called Alum-root).

FIG. 721. Sullivantia Ohionis. 722. Flower with the calyx laid open, somewhat enlarged. 723. Fruit surrounded by the persistent calyx and withered petals, enlarged. 724. Section of the lower part of the capsule, magnified; showing the central placenta covered with the ascending seeds. 725. A magnified seed, with its cellular, wing-like testa. 726. Section of the nucleus, showing the embryo in the midst of albumen.

797. **Subord. Hydrangeæ** (*the Hydrangea Family*). Shrubs: Petals valvate in æstivation. Capsule two- (rarely five- to ten-) celled: the styles or stigmas distinct or united. Stamens sometimes numerous. — *Ex.* Hydrangea, Decumaria.

798. **Subord. Philadelpheæ** (*the Mock Orange Family*). Shrubs. Petals convolute in æstivation. Capsule three- or four-celled: styles more or less united. Stamens mostly numerous. — *Ex.* Philadelphus, the Mock Orange.

799. **Ord. Hamamelaceæ** (*the Witch-Hazel Family*). Shrubs or small trees, with alternate simple leaves, without stipules. Flowers often polygamous. Petals valvate in æstivation. Stamens twice as many as the petals, half of them sterile; or numerous, and the petals none. Summit of the ovary free from the calyx, a single ovule suspended from the summit of each cell: styles two, distinct. Capsules cartilaginous or bony. Seeds bony, with a small embryo in hard albumen. — *Ex.* Hamamelis (Witch-Hazel), Fothergilla. A small order, of little importance. Hamamelis is remarkable for flowering late in autumn, just as its leaves are falling, and perfecting its fruit the following spring.

800. **Ord. Umbelliferæ** (*the Parsley Family*). Herbs, with hollow stems, and alternate, dissected leaves, with the petioles sheathing or dilated at the base. Flowers in simple or mostly compound umbels, which are occasionally contracted into a kind of head. Calyx entirely coherent with the surface of the dicarpellary ovary; its limb reduced to a mere border, or to five small teeth. Petals five, valvate in æstivation, inserted, with the five stamens, on a disc which crowns the ovary; their points inflexed. Styles two; their bases often united and thickened, forming a stylopodium. Fruit dry, separating from each other, and often from a slender axis (*carpophore*), into two indehiscent carpels (called *mericarps*): the face by which these cohere receives the technical name of *commissure*: they are marked with a definite number of *ribs* (*juga*), which are sometimes produced into wings: the intervening spaces (*intervals*), as well as the commissure, sometimes contain canals or receptacles of volatile oil, called *vittæ*: these are the principal terms peculiarly employed in describing the plants of this difficult family. Embryo minute. Albumen hard or corneous. — *Ex.* The Carrot, Parsnip, Celery, Caraway, Anise, Coriander, Poison Hemlock, &c. are common representatives of this well-known family. Nearly all Umbelliferous plants are furnished with a volatile oil or

balsam, chiefly accumulated in the roots and in the reservoirs of the fruit, upon which their aromatic and carminative properties depend: sometimes it is small in quantity, so as merely to flavor the saccharine roots which are used for food; as in the Carrot and Parsnip. But in many an alkaloid principle exists, pervading the foliage, stems, and roots, especially the latter, which renders them

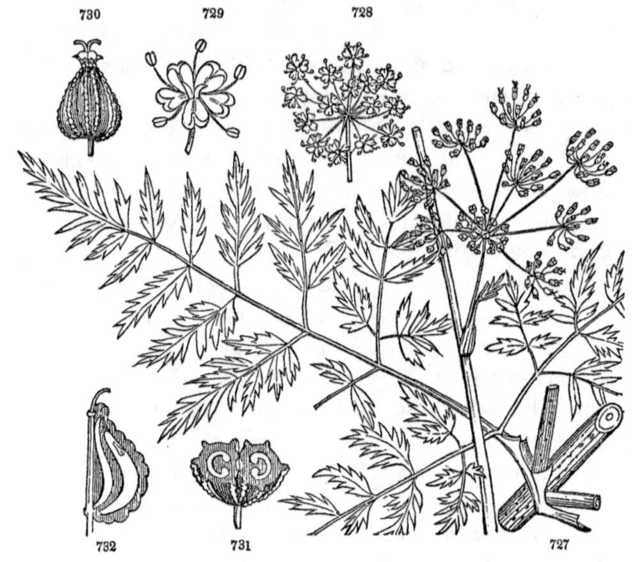

acrid-narcotic poisons. And finally, many species of warm regions yield odorous gum-resins (such as Galbanum, Assafœtida, &c.), which have active stimulant properties. The stems of Celery (Apium graveolens), which are acrid and poisonous when the plant grows wild in marshes, &c., are rendered innocent by cultivation in dry ground, and by blanching. Among the virulent acrid-narcotic species, the most famous are the Hemlock (Conium maculatum, naturalized in this country), and Cicuta maculata (Cowbane, Water-Hemlock) indigenous to this country, the root of which (like that of the C. virosa of Europe) is a deadly poi-

FIG. 727. Conium maculatum (Poison Hemlock), a portion of the spotted stem, with a leaf; and an umbel with young fruit. 728. A flower umbellet. 729. A flower, enlarged. 730. The fruit. 731. Cross-section of the same, showing the involute (*cœlospermous*) albumen of the two seeds. 732. Longitudinal section of one mericarp, exhibiting the minute embryo near the apex of the albumen.

son. A drachm of the fresh root has killed a boy in less than two hours.

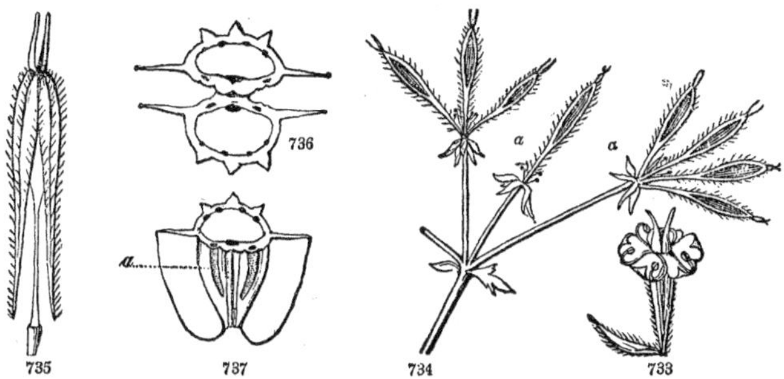

801. **Ord. Araliaceæ** (*the Spikenard Family*). A small family, scarcely differing from Umbelliferæ in botanical character, except that the ovary is mostly composed of more than two carpels, which do not separate when ripe, but become drupes or berries; and the albumen is not hard like horn, but only fleshy. — *Ex.* Aralia (the Spikenard, the Wild Sarsaparilla, and the Angelica-tree), Panax (Ginseng), and Hedera (the Ivy). Their properties are aromatic, stimulant, somewhat tonic, and alterative.

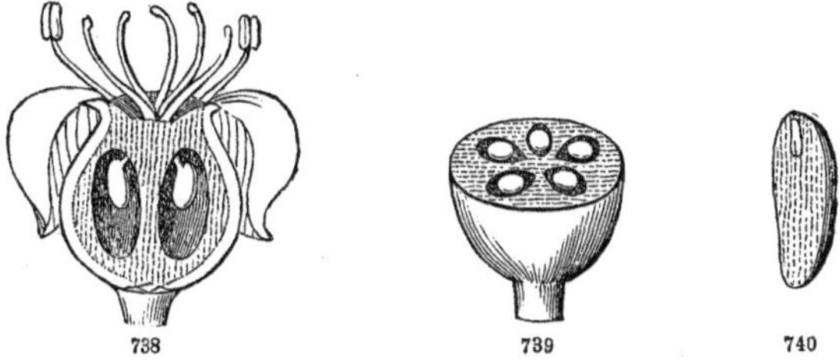

802. **Ord. Cornaceæ** (*the Cornel Family*). Chiefly trees or shrubs; with the leaves almost always opposite, destitute of stipules. Flowers in cymes, sometimes in heads surrounded by colored involucres. Calyx coherent with the two-celled ovary; the very small

FIG. 733. Flower of Osmorrhiza longistylis. 734. Umbel of the same in fruit: *a*, the involucels. 735. The ripe mericarps separating from the axis or carpophore. 736. Cross-section of the fruit of Angelica, where the lateral ribs are produced into wings: the black dots represent the vittæ, as they appear in a cross-section. 737. One of the mericarps of the same, showing the inner face, or commissure, as well as the transverse section, with two of the vittæ, *a*.

FIG. 738. Flower of Aralia nudicaulis (Wild Sarsaparilla); a vertical section, displaying two of the cells of the ovary. 739. Cross-section of the ovary. 740. Longitudinal section of a seed, magnified, showing the small embryo at the upper end.

limb four-toothed. Petals four, valvate in æstivation. Stamens four, alternate with the petals. Styles united into one. Fruit a two-celled drupe. — *Ex.* Cornus, the Dogwood. Chiefly remarkable for their bitter and astringent bark, which in this country has been substituted for Cinchona. The peculiar principle they contain is named *Cornine.* Cornus Canadensis (Fig. 240) is a low and herbaceous species.

Division II. — Monopetalous or Gamopetalous Exogenous Plants.*

Floral envelopes consisting of both calyx and corolla: the petals more or less united (corolla gamopetalous).

Conspectus of the Orders.

Group 1. Ovary coherent with the calyx, two- to several-celled, with one or many ovules in each cell. Seeds albuminous, with a small embryo. Stamens inserted on the corolla. Leaves opposite.

Stipules wanting. Caprifoliaceæ, p. 436.
Stipules interpetiolar (or leaves whorled). Rubiaceæ, p. 437.

Group 2. Ovary coherent with the calyx, one-celled and one-ovuled; rarely three-celled with two of the cells empty. Seeds with little or no albumen. Stamens inserted on the corolla. Calyx a mere ring, crown, or pappus, or obsolete. Fruit like an achenium.

Stamens distinct. Seeds suspended.
Stamens 3 or fewer. Valerianaceæ, p. 439.
Stamens 4. Heads involucrate. Dipsaceæ, p. 440.
Stamens syngenesious. Seed erect. Compositæ, p. 441.

Group 3. Ovary coherent with the calyx, with two or more cells and numerous ovules. Seeds albuminous. Satmens inserted with the corolla (epigynous): anthers not opening by pores.

Corolla irregular. Stamens united in a tube. Lobeliaceæ, p. 443.
Corolla regular. Stamens distinct. Campanulaceæ, p. 444.

Group 4. Ovary free from the calyx, or sometimes coherent with it, with two or more cells and numerous ovules. Seeds albuminous. Stamens inserted with the corolla, or rarely coherent with its base, as many, or twice as many, as its lobes: anthers mostly opening by pores or chinks.

---

* Cucurbitaceæ, placed in the Polypetalous series, are commonly somewhat gamopetalous: so are some exotic Crassulaceæ, &c.

Anthers two-celled. ERICACEÆ, p. 444.
Anthers one-celled. EPACRIDACEÆ, p. 447.

Group 5. Ovary free, or rarely coherent with the calyx, several-celled, with a single ovule (or at least a single seed) in each cell. Seeds mostly albuminous. Stamens definite, as many as the lobes of the (sometimes almost polypetalous) corolla and alternate with them, or two to four times as many: anthers not opening by pores. — Trees or shrubs.

Stamens as many as the lobes of the corolla and alternate with them. AQUIFOLIACEÆ, p. 447.
Stamens more numerous and all fertile.
Flowers polygamous: calyx free. EBENACEÆ, p. 447.
Flowers perfect: calyx more or less adnate. STYRACACEÆ, p. 448.
Stamens as many fertile as there are lobes of the corolla and opposite them; and with a sterile series alternate between them. SAPOTACEÆ, p. 448.

Group 6. Ovary free, or with the base merely coherent with the tube of the calyx, one-celled, with a free central placenta. Stamens inserted into the regular corolla opposite its lobes! which they equal in number. Seeds albuminous.

Shrubs or trees: fruit drupaceous. MYRSINACEÆ, p. 448.
Herbs: fruit capsular. PRIMULACEÆ, p. 448.

Group 7. Ovary free, one-celled, with a single ovule; or two-celled with several ovules attached to a thick central placenta. Stamens as many as the lobes of the regular corolla or the nearly distinct petals. Seeds albuminous.

Ovary two-celled: style single: stamens 4. PLANTAGINACEÆ, p. 449.
Ovary one-celled: styles and stamens 5. PLUMBAGINACEÆ, p. 450.

Group 8. Ovary free, one- or two- (or spuriously four-) celled, with numerous ovules. Corolla bilabiate or irregular; the stamens inserted upon its tube, and mostly fewer than its lobes.

Ovary one-celled with a central placenta. Stam. 2. LENTIBULACEÆ, p. 451.
Ovary one-celled with parietal placentæ. OROBANCHACEÆ, p. 451.
Ovary spuriously 4–5-celled: seeds exalbuminous. Subord. SESAMEÆ, p. 452.
Ovary two-celled: placentæ in the axis.
Seeds indefinite, winged: albumen none. BIGNONIACEÆ, p. 452.
Seeds, few, wingless: albumen none. Corolla convolute in æstivation. ACANTHACEÆ, p. 452.
Seeds mostly indefinite: albumen copious. Corolla imbricative in æstivation. SCROPHULARIACEÆ, p. 453.

Group 9. Ovary free, two- to four-lobed, and separating or splitting into as many one-seeded nuts or achenia, or drupaceous. Corolla regular or irregular; the stamens inserted on its tube, equal in number or fewer than its lobes. Albumen little or none.

Stamens 4, didynamous, or 2. Corolla more or less irregular.
Ovary not 4-lobed. VERBENACEÆ, p. 454.

Ovary 4-lobed, forming 4 achenia. LABIATÆ, p. 455.
Stamens 5. Flower regular. Leaves alternate. BORAGINACEÆ, p. 456.

Group 10. Ovary free, compound, or the carpels two or more and distinct: the ovules usually several or numerous. Corolla regular; the stamens inserted upon its tube, as many as the lobes and alternate with them. Seeds albuminous.

* Ovary compound (of two or more united carpels).

Placentæ 2, parietal (sometimes expanded). Embryo minute.
Corolla not valvate in æstivation.
Leaves lobed, mostly alternate. Seeds few. HYDROPHYLLACEÆ, p. 457.
Leaves entire, opposite. Seeds indefinite. GENTIANACEÆ, p. 462.
Corolla valvate-induplicate in æstivation. Subord. MENYANTHIDEÆ, p. 462.
Placentæ in the axis: ovary 2 – 3-celled.
Embryo large, bent or coiled, with little albumen. Seeds one or two in each cell. CONVOLVULACEÆ, p. 459.
Embryo straight or arcuate, in copious albumen.
Styles 2, distinct. Seeds indefinite. HYDROLEACEÆ, p. 458.
Styles united nearly or quite to the summit.
Ovary 3-celled. Cor. convolute in æstivation. POLEMONIACEÆ, p. 458.
Ovary 3-celled. Cor. imbricated in æstivation. DIAPENSIACEÆ, p. 458.
Ovary 2-celled. Corolla plaited or valvate in æstivation. SOLANACEÆ, p. 461.

* * Ovaries mostly two and distinct, at least in fruit.

Anthers introrse: pollen granular. APOCYNACEÆ, p. 463.
Anthers extrorse: pollen in waxy masses. ASCLEPIADACEÆ, p. 463.

Group 11. Ovary free, two-celled, few-ovuled; the cells of the fruit one-seeded. Corolla regular (sometimes nearly polypetalous or wanting); the stamens (two) fewer than its lobes. — Shrubs or trees.

Seeds erect. Cor. imbricated or contorted in æstivation. JASMINACEÆ, p. 464.
Seeds suspended. Corolla valvate in æstivation. OLEACEÆ, p. 465.

803. Ord. Caprifoliaceæ (*the Honeysuckle Family*). Mostly shrubs, often twining, with opposite leaves, and no stipules. Calyx-tube adnate to the 2 – 5-celled ovary; the limb 4 – 5-cleft. Corolla regular or irregular. Stamens inserted on the corolla, as many as the petals of which it is composed, and alternate with them, or rarely one fewer. Fruit mostly a berry or drupe. Seeds pendulous, albuminous. — *Ex.* The Honeysuckles (Lonicera), which have usually a peculiar bilabiate corolla (470, Fig. 743), the Snowberry (Symphoricarpus), Diervilla, which has a capsular fruit, &c., compose the tribe LONICEREÆ, characterized by their tubular flowers and filiform style: while the Elder (Sambucus) and Viburnum, which have a rotate or urn-shaped corolla, form the tribe SAMBUCEÆ. These plants chiefly belong to temperate regions. Several are widely cultivated for ornament. They are generally bitter, and rather active or nauseous in their properties: but the fruit of some few is edible.

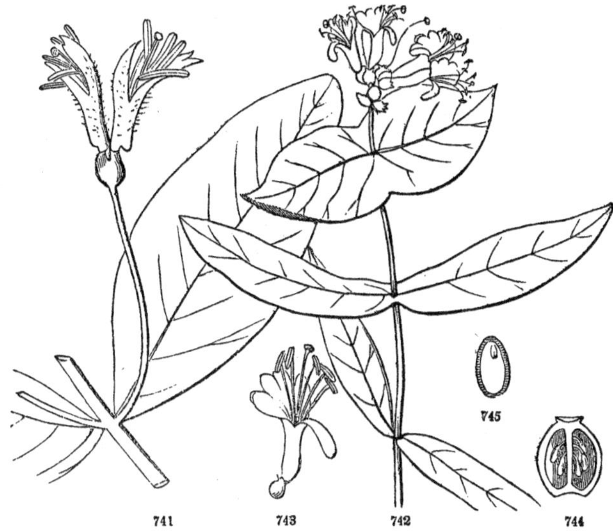

804. **Ord. Rubiaceæ** (*the Madder Family*). Shrubs or trees, or often herbs, with the entire leaves either in whorls, or opposite and furnished with stipules. Calyx-tube completely, or rarely incompletely, adnate to the 2 – 5-celled ovary; the limb four- or five-cleft or toothed, or occasionally obsolete. Stamens as many as the lobes of the regular corolla, and alternate with them, inserted on the tube. Fruit various. Seeds albuminous. — This extensive family divides into two suborders, to which a third may be appended, which differs in the free ovary, and is by most botanists deemed a distinct order.

805. **Subord. Stellateæ** (*the true Madder Family*). Herbs, with the leaves in whorls; but all except a single pair are generally supposed to take the place of stipules. — *Ex.* Galium, Rubia (the Madder), &c., nearly all belonging to the colder parts of the world.

806. **Subord Cinchoneæ** (*the Peruvian-Bark Family*). Shrubs, trees, or herbs; the leaves opposite and furnished with stipules,

FIG. 741. Branch of Lonicera (Xylosteon) oblongifolia: the two ovaries united! 742. Lonicera (Caprifolium) parviflora. 743. A flower about the natural size. 744. Longitudinal section of the ovary. 745. Longitudinal section of a magnified seed, showing the albumen and minute embryo.

which are very various in form and appearance. — *Ex.* Cephalanthus (Button-bush), Hedyotis, and an immense number of tropical genera. Their stipules distinguish them from Caprifoliaceæ.

807. **Subord. Loganieæ**, or **Spigelieæ**, have opposite stipulate leaves, and the ovary nearly or entirely free from the persistent calyx. — *Ex.* Mitreola, Spigelia (the Pink-root), and other genera intermediate between Rubiaceæ and Apocynaceæ.

808. Very active, and generally febrifugal properties prevail in this large order. The roots of Madder yield a most important dye: and many Galiums have a similar red coloring matter. —

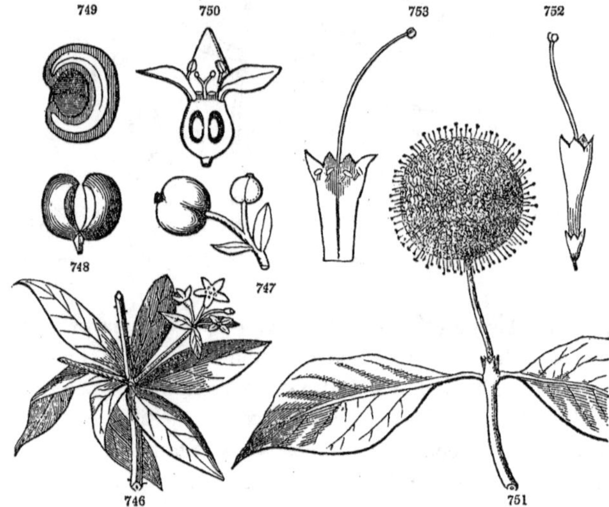

The division Cinchoneæ furnishes two of the most valuable known remedial agents, namely, *Peruvian bark*, or *Cinchona*, and *Ipecacuanha*. The febrifugal properties of the former depend on the presence of two alkalis, *Cinchonia* and *Quinia*, both combined with *Kinic acid*. The *Quinquina barks*, which are derived from some species of Exostemma and other West Indian, Mexican, and Brazilian genera, contain neither cinchonia nor quinia. The bark of Pinckneya pubens, of the Southern United States, has been substituted for Cinchona. — The true *Ipecacuanha* is furnished by the

FIG. 746. Piece of Rubia tinctoria (the Madder) in flower. 747. The fruit. 748. The two constituent portions of the fruit separating. 749. Vertical section of one carpel, showing the curved embryo. 750. Section of a flower of Galium.

FIG. 751. Cephalanthus occidentalis, the Button-Bush. 752. A flower, taken from the head. 753. The corolla laid open.

roots of Cephaælis Ipecacuanha of Brazil and the mountains of New Granada. Its emetic principle (called *Emetine*) also exists in Psychotria emetica of New Granada, which furnishes the *striated*, *black*, or *Peruvian Ipecacuanha*. *Coffee* is the horny seed (albumen) of Coffæa Arabica. According to Blume, the leaves of the Coffee-plant are used as a substitute for tea in Java. — The roots and leaves of Spigelia Marilandica (Carolina Pink-root) form a well-known vermifuge.

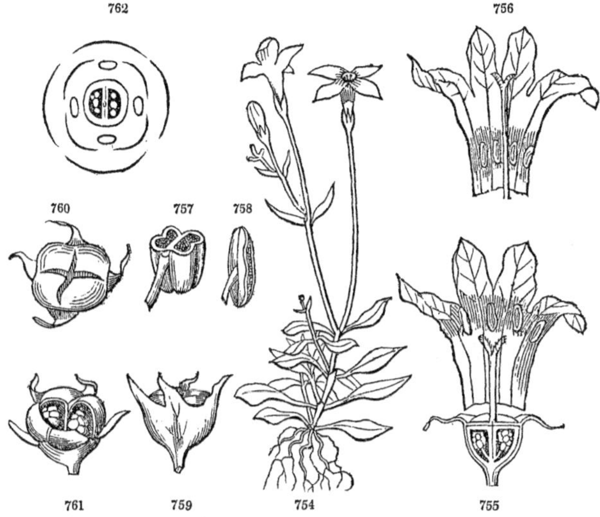

809. **Ord. Valerianaceæ** (*the Valerian Family*). Herbs with opposite leaves, and no stipules. Flowers often in cymes, panicles, or heads. Limb of the adnate calyx two- to four-toothed, obsolete, or else forming a kind of pappus. Corolla tubular or funnel-form, sometimes with a spur at the base, four- or five-lobed. Stamens distinct, inserted on the corolla, usually fewer than its lobes. Ovary one-ovuled, with one perfect cell and two abortive ones. Fruit a kind of achenium. Seed suspended, exalbuminous. Embryo straight. Radicle superior. — *Ex.* Valeriana, the Valerian; Fedia, the Lamb-lettuce: the latter is eaten as a salad. The roots, &c. of the perennial species exhale a heavy and peculiar

FIG. 754. Hedyotis (Houstonia) cærulea. 755, 756. The two sorts of flowers that different individuals bear, with the corolla laid open; one with the stamens at the base, the other at the summit of the tube: the lower figure shows also a section of the ovary. 757. Cross-section of an anther, magnified. 758. Anther less enlarged, opening longitudinally. 759. Capsule with the calyx. 760, 761. Views of the capsule in dehiscence. 762. Diagram of a cross-section of the unexpanded flower.

odor, have a somewhat bitter, acrid taste, and are antispasmodic and vermifugal. The *Valerian* of the shops is chiefly derived from Valeriana officinalis of the South of Europe. It produces a peculiar intoxication in cats. The roots of V. edulis are used for food by the aborigines of Oregon. The *Spikenard* of the ancients, esteemed as a stimulant medicine as well as a perfume, is the root of Nardostachys Jatamansi of the mountains of the North of India.

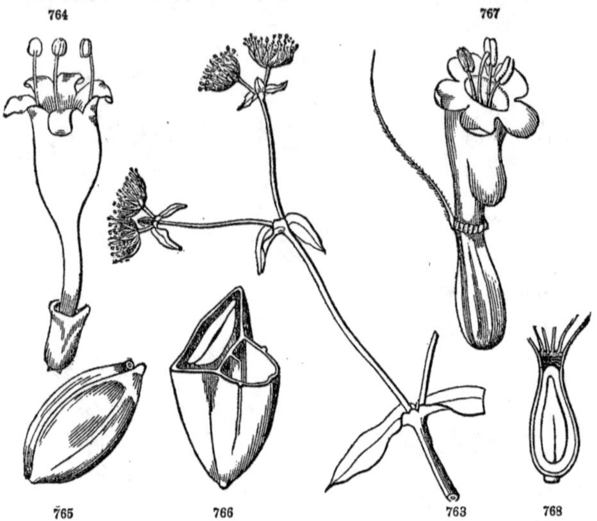

810. **Ord. Dipsaceæ** (*the Teasel Family*). Herbs, with opposite or whorled sessile leaves, destitute of stipules. Flowers in dense heads, which are surrounded by an involucre. Limb of the adnate calyx cup-shaped and entire or toothed, or forming a bristly or plumose pappus. Corolla tubular; the limb four- or five-lobed, somewhat irregular. Stamens four, distinct, or rarely united in pairs, often unequal, inserted on the coralla. Ovary one-celled, one-ovuled. Seed suspended, albuminous. — *Ex.* Dipsacus, the Teasel, and Scabiosa, or Scabious. All natives of the Old World. Some are cultivated for ornament. *Teasels* are the dried heads of Dipsacus Fullonum, covered with stiff and spiny bracts, with recurved points.

FIG. 763. Branch of Fedia Fagopyrum. 764. A magnified flower. 765. A fruit. 766. An enlarged cross-section of the same, and the cotyledons of the seed in the single fertile cell: the two empty cells are confluent into one.

FIG. 767. Flower of a Valerian, with one of the pappus-like bristles of the calyx unrolled. 768. Section through the ovary and embryo; the bristles of the calyx broken away.

811. **Ord. Compositæ** (*the Composite or Sunflower Family*). Herbs or shrubs; with the flowers in heads (compound flowers of the older botanists), crowded on a receptacle, and surrounded by a set of bracts (*scales*) forming an involucre; the separate flowers often furnished with bractlets (*chaff*, *paleæ*). Limb of the adnate calyx obsolete, or a *pappus* (305), consisting of hairs, bristles, scales, &c. Corolla regular or irregular. Stamens five, as many as the lobes or teeth of the regular corolla, inserted on its tube: anthers united into a tube (*syngenesious*, Fig. 769). Style two-cleft. Fruit an achenium, with a single erect exalbuminous seed, either naked or crowned with a pappus. Embryo straight. — This vast but very natural family is divided into three sets or suborders; viz.: —

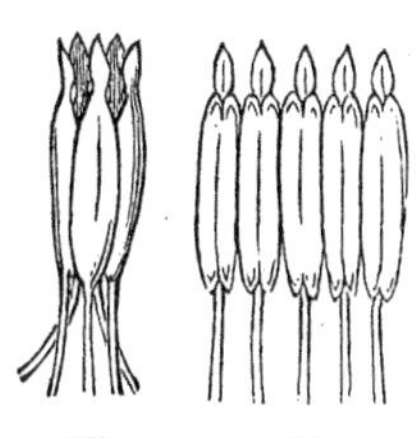

812. **Subord. Tubulifloræ.** Corolla tubular and regularly four- or five-lobed, either in all the flowers (when the head is *discoid*), or in the central ones (those of the *disc*) only, the marginal or *ray-flowers* presenting a *ligulate* or strap-shaped corolla. — *Ex.* Liatris, Eupatorium, &c.; where the heads are *homogamous*, that is, the flowers all tubular, similar, and perfect: Helianthus (Sunflower), Helenium, Aster, &c.; where the heads are *heterogamous;* the *disc* flowers being tubular and perfect, while those of the *ray* are *ligulate*, and either pistillate only, or *neutral* (473, note), that is, destitute of both stamens and pistils.

813. **Subord. Labiatifloræ.** Corolla of the disc-flowers bilabiate. — *Ex.* Chaptalia, of the United States, Mutisia, Chætanthera, &c., of South America.

814. **Subord. Liguliforæ.** Corolla of all the flowers (both disc and ray) ligulate; all perfect. — *Ex.* The Dandelion, Lettuce, Cichory, &c.

815. This vast family comprises about a tenth part of all Phænogamous plants. A bitter and astringent principle pervades the whole order; which in some is tonic (as in the Camomile, Anthemis nobilis, the Boneset, or Thoroughwort, Eupatorium perfoliatum, &c.); in others, combined with mucilage, so that they are demulcent as well as tonic (as in Elecampane and Coltsfoot); in many, aromatic and extremely bitter (such as Wormwood and all the spe-

FIG. 769. Syngenesious stamens of a Composita. 770. The anthers laid open.

cies of Artemisia); sometimes accompanied by acrid qualities, as in the Tansy (Tanacetum vulgare), and the Mayweed (Maruta Cotula), the bruised fresh herbage of which blisters the skin. The species of Liatris, which abound in terebinthine juice, are among the reputed remedies for the bites of serpents. The juice of Silphium and of some Sunflowers is resinous. The leaves of Solidago odora, which owe their pleasant anisate fragrance to a peculiar volatile oil, are infused as a substitute for tea. From the seeds of Sunflower, and several other plants of the order, a bland

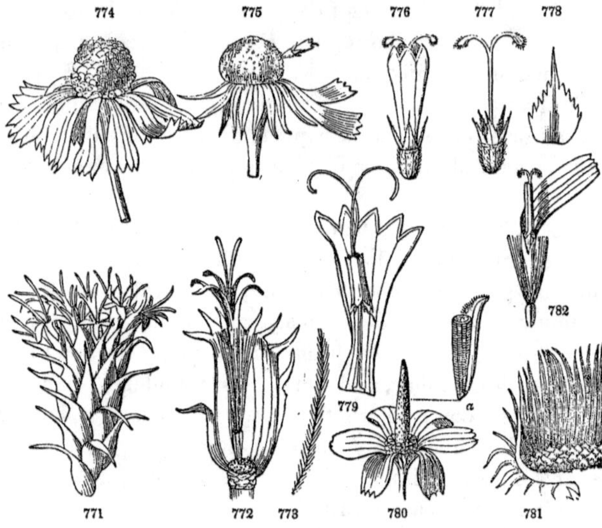

oil is expressed. The tubers of Helianthus tuberosus are eaten

FIG. 771. Head of Liatris squarrosa (discoid; the flowers all tubular and perfect). 772. The same, with the scales of one side of the imbricated involucre removed; and also all the flowers but one, showing the naked flat receptacle. 773. Portion of one of the plumose bristles of the capillary pappus. 774. Head of Helenium autumnale (heterogamous); the rays neutral, consisting merely of a ligulate corolla. 775. The same, with the flowers all removed from the roundish receptacle, except a single disc-flower and one or two rays: the reflexed scales of the involucre in a single series. 776. Magnified disc-flower of the same: the corolla exhibiting the peculiar venation of the family; namely, the veins corresponding to the sinuses, and sending a branch along the margins of the lobes. 777. The same with the corolla removed; the achenium crowned with the limb of the calyx in the form of a chaffy pappus, of about five scales. 778. A chaff of the pappus more magnified. 779. A tubular corolla of this family laid open, showing the venation; and also the five syngenesious anthers united in a tube, through which the two-cleft style passes. 780. Head of Dracopis amplexicaulis, with the flowers removed from the elongated spike-like receptacle, except a few at the base: *a*, achenium of one of the disc-flowers, magnified, partly inclosed by its bracelet (chaff or palea); the pappus obsolete. 781. Part of the involucre and alveolate (honeycomb-like) receptacle of Onopordon or Cotton-Thistle. 782. A perfect and ligulate flower of the Dandelion, with its hair-like or capillary pappus.

under the name of *Jerusalem artichokes*. True *artichokes* are the fleshy receptacle of Cynara Scolymus. The flowers of Carthamus tinctorius, often called Saffron, yield a yellow dye. — The Ligulifloræ, or Cichoraceæ, all have a milky juice, which is narcotic, and has been employed as a substitute for *opium*. The bland young leaves of the Garden Lettuce are a common salad. The roasted roots of the Wild Succory (Cichorium Intybus) are extensively used to adulterate coffee: and the roots of some species of Tragopogon (Salsify, Oyster-plant) and Scorzonera are well-known esculents.

816. **Ord. Lobeliaceæ** (*the Lobelia Family*). Herbs or somewhat shrubby plants, often yielding a milky juice, with alternate leaves and usually showy flowers. Limb of the adnate calyx five-cleft. Corolla irregularly five-lobed, usually appearing bilabiate, cleft on one side nearly or quite to the base. Stamens 5, epigynous, co-

herent into a tube. Stigma fringed. Fruit capsular, two- or three-

FIG. 783. Campanula rotundifolia, much reduced in size. 784. Lobelia inflata, reduced in size. 785. A flower, enlarged. 786. The united filaments and anthers inclosing the style; the corolla and limb of the calyx cut away. 787. The stigma surrounded by a fringe. 788. Transverse section of a capsule. 789. Section of a magnified seed, showing the embryo.

(rarely one-) celled, many-seeded. Seeds albuminous. — *Ex.* Lobelia. All narcotico-acrid poisons. The well-known Lobelia inflata (Indian Tobacco) is one of the most powerful articles of the materia medica, and the most dangerous in the hands of the reckless quacks who use it. Less than a teaspoonful of the seeds or powdered leaves will destroy life in a few hours.

817. **Ord. Campanulaceæ** (*the Campanula Family*). Herbs, with a milky (slightly acrid) juice, alternate leaves, and usually showy flowers. Tube of the calyx adnate, the limb commonly five-cleft, persistent. Corolla regular, campanulate, usually five-lobed, withering. Stamens five, distinct. Style furnished with collecting hairs. Capsule two- to several-celled, many-seeded. Seeds albuminous. — *Ex.* Campanula (Bell-flower, Harebell). Of little importance, except for ornament.

818. **Ord. Ericaceæ** (*the Heath Family*). Shrubs or sometimes herbs. Flowers regular or nearly so, 4 – 5-merous, the petals sometimes distinct. Stamens mostly distinct, free from the corolla, as many or twice as many as its lobes, and inserted with it (either hypogynous or epigynous), anthers two-celled, often appendaged, commonly opening by terminal pores. Styles and stigmas united into one. Ovary with two or more cells and usually numerous ovules, free, or in Vaccineæ coherent with the calyx-tube. Seeds usually indefinite, albuminous. — Some botanists give the rank of orders to the following suborders.

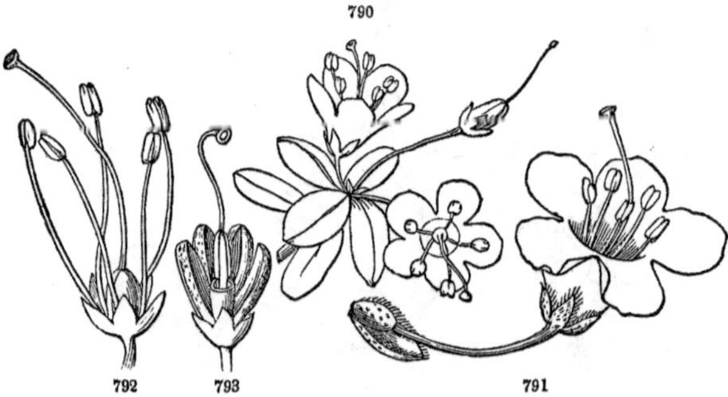

819. **Subord. Vaccinieæ** (*the Whortleberry Family*). Ovary adnate

FIG. 790. Branch of Rhododendron Lapponicum. 791. Enlarged flower, with its pedicel and bracts. 792. Flower with the corolla removed, more enlarged. 793. Capsule of R. maximum, opening by septicidal dehiscence; the valves breaking away from the persistent axis, or columella.

to the tube of the calyx, becoming a berry or a drupe-like fruit. Shrubs, with scattered leaves, often evergreen. — *Ex.* Vaccinium (Whortleberry), Oxycoccus (the Cranberry).

820. **Subord. Ericineæ** (*the proper Heath Family*). Ovary free from the calyx. Fruit capsular, sometimes baccate or drupaceous. Testa conformed to the nucleus of the seed. Mostly shrubs. Leaves various, often evergreen. Petals rarely almost or entirely distinct. — *Ex.* Erica (Heath), Kalmia, Rhododendron, Gaultheria, Andromeda, &c.

821. **Subord. Pyroleæ** (*the Pyrola Family*). Ovary free from the calyx. Petals distinct or nearly so. Fruit a capsule. Seeds with

FIG. 794. Gaultheria procumbens (Wintergreen, &c.). 795. The enlarging calyx in the immature fruit. 796. Vertical section of the pulpy or berry-like calyx and the included capsule (the seeds removed from the placenta in one cell). 797. Horizontal section of the same, showing the five-celled capsule, with a placenta proceeding from the inner angle of each cell. 798. Section of a seed, magnified. 799. Flower of a Vaccinium (Whortleberry). 800. Vertical section of the ovary and adherent calyx. 801. Anther of Vaccinium Vitis-Idæa; each cell prolonged into a tube, and opening by a terminal pore. 802. Anther of Vaccinium Myrtillus; the connectivum furnished with two appendages. 803. Stamen of an Andromeda (Cassiope), showing the appendages of the connectivum. 804. Stamen of Arctostaphylos Uva-Ursi, showing the separate anther-cells, opening by a terminal pore, the appendages of the connectivum, and the filament, which is swollen at the base.

a loose cellular testa, not conformed to the nucleus. Mostly herbs. Leaves flat and broad. — *Ex.* Pyrola, Chimaphila.

822. **Subord. Monotropeæ** (*the Indian-Pipe Family*). Ovary free from the calyx. Petals distinct or united. Anthers opening longitudinally or by transverse chinks. Fruit a capsule. Seeds with a loose or winged testa. Parasitic herbs, destitute of green color, and with scales instead of leaves. — *Ex.* Monotropa, the Indian Pipe. — In this widely diffused order the bark and foliage are generally astringent, often stimulant or aromatic from a volatile oil or a resinous matter, and not seldom narcotic. Thus, the leaves of Rhododendron, Kalmia, and all the related plants, are deleterious (being stimulant narcotics), or suspicious. The honey made from their flowers is sometimes poisonous. The Uva-Ursi and the Chi-

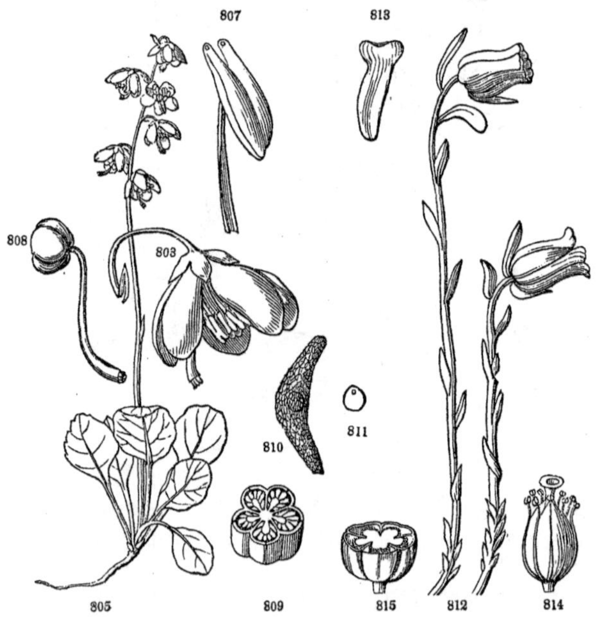

maphila (Pipsissewa) are the chief medicinal plants of the order.

FIG. 805. Pyrola chlorantha, reduced in size. 806. Enlarged flower. 807. Magnified stamen. 808. Pistil. 809. Cross-section of the capsule. 810. A highly magnified seed. 811. The nucleus removed from the loose cellular testa, and divided, showing the very minute embryo.

FIG. 812. Monotropa uniflora. 813. A petal. 814. Capsule, with the stamens. 815. Transverse section of the same; the thick and lobed placenta covered with very minute seeds.

The berries are generally edible (Whortleberries, Wintergreen, &c.). Many are very ornamental plants.

823. **Ord. Epacridaceæ,** which takes the place of Heaths in Australia, essentially differs from them only in the one-celled anthers.

824. **Ord. Aquifoliaceæ** (*the Holly Family*). Trees or shrubs, commonly with coriaceous leaves, and small axillary flowers. Calyx of four to six sepals. Corolla four- to six-parted or cleft: the stamens as many as its segments and alternate with them, inserted on the base of the corolla. Anthers opening longitudinally. Ovary two- to six-celled; the cells with a single suspended ovule. Fruit drupaceous, with two to six stones or nucules. Embryo minute, in hard albumen. — *Ex.* Ilex (the Holly) and Prinos. The bark and leaves contain a tonic, bitter, extractive matter. The leaves of a species of Ilex are used for tea in Paraguay: and the famous *black drink* of the Creek Indians is prepared from the leaves of Ilex vomitoria (Cassena); which are still used as a substitute for tea in some parts of the Southern States.

825. **Ord. Ebenaceæ** (*the Ebony Family*). Trees or shrubs, destitute of milky juice, with alternate, mostly entire leaves, and polygamous flowers. Calyx three- to six-cleft, free from the ovary. Corolla three- to six-cleft, often pubescent. Stamens twice to four times as many as the lobes of the corolla, inserted on them. Ovary three- to several-celled; the style with as many divisions. Fruit a kind of berry, with large and bony seeds.

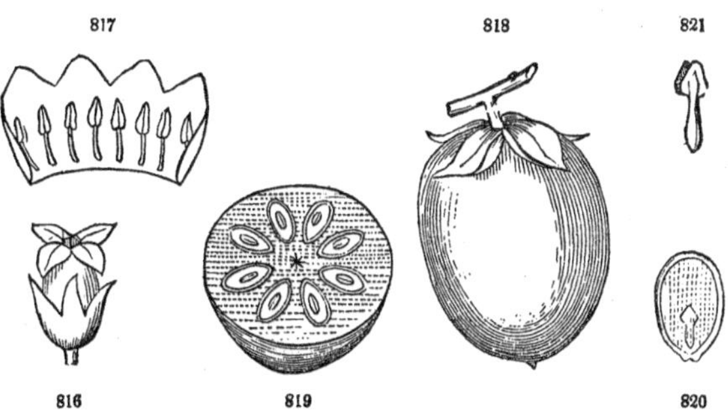

Embryo shorter than the hard albumen. — *Ex.* Diospyros, the

FIG. 816. Perfect flower of Diospyros Virginiana, the Persimmon. 817. The corolla, laid open, and stamens. 818. The fruit. 819. Section through the fruit and bony seeds. 820. Vertical section of a seed. 821. The detached embryo.

Persimmon. The fruit, which is extremely austere and astringent when green, is sweet and eatable when fully ripe. The bark is powerfully astringent. *Ebony* is the wood of D. Ebenus and other African and Asiatic species.

826. **Ord. Styracaceæ** (*the Storax Family*). Shrubs or trees with perfect flowers. Calyx-tube coherent either with the base of the ovary, or with its whole surface. Styles and stigmas perfectly united into one. Stamens more or less united. Cells of the ovary opposite the calyx-lobes. Otherwise much as in the last family. — *Ex.* Styrax, Halesia, Symplocos. Some yield a fragrant, balsamic resinous substance; such as Storax and Benzoin, containing *Benzoic acid.* The sweet leaves of our Symplocos tinctoria afford a yellow dye.

827. **Ord. Sapotaceæ** (*the Sapodilla Family*). Trees or shrubs, usually with a milky juice; the leaves alternate, entire, coriaceous, the upper surface commonly shining. Flowers perfect, regular, axillary, usually in clusters. Calyx four- to eight-parted. Corolla four- to eight- (or many-) cleft. Stamens distinct, inserted on the tube of the corolla, commonly twice as many as its lobes, half of them fertile and opposite the lobes, the others petaloid scales or filaments and alternate with them: anthers extrorse. Ovary 4 – 12-celled, with a single ovule in each cell. Styles united into one. Fruit a berry. Seeds with a bony testa, with or without albumen. — *Ex.* Bumelia of the Southern United States. The fruit of many species is sweet and eatable; such as the Sapodilla Plum, the Marmalade, the Star-Apple, and other West Indian species. The large seeds, particularly of some kinds of Bassia, yield a bland fixed oil, which is sometimes thick and like butter, as in the Chee of India (B. butyracea), and the African Butter-tree, or Shea, described by Mungo Park.

828. **Ord. Myrsinaceæ.** Trees or shrubs, mostly with alternate coriaceous leaves, which are often dotted with glands, and with all the characters of Primulaceæ, except the drupaceous fruit and arborescent habit. — Nearly all tropical (Ardisia, Myrsine).

829. **Ord. Primulaceæ.** Herbs, with opposite, whorled, or alternate leaves, often with naked scapes and the leaves crowded at the base. Calyx four- or five-cleft or toothed, usually persistent Corolla rotate, hypocrateriform, or campanulate. Stamens inserted on the tube of the corolla, as many as its lobes and opposite them! Ovary free, one-celled with a free central placenta!

Ovules mostly indefinite and amphitropous. Style and stigma single. Fruit capsular: the fleshy central placenta attached to the base of the cell. Seeds albuminous. Embryo transverse. — *Ex.* Primula (Primrose), Cyclamen, Anagallis. In Samolus, the calyx coheres with the base of the ovary, and there is a row of sterile filaments occupying the normal position of the first set of stamens, namely, alternate with the lobes of the corolla. Of little consequence, except for their beauty.

830. **Ord. Plantaginaceæ** (*the Plantain Family*). Chiefly low herbs, with small spiked flowers on scapes, and ribbed radical leaves. — Calyx four-cleft, persistent. Corolla tubular or urn-shaped, scarious and persistent; the limb four-cleft. Stamens four, inserted on the tube of the corolla alternate with its seg-

FIG. 822. Primula pusilla. 823. The corolla removed; its tube laid open. 824. The calyx divided vertically, showing the pistil. 825. Vertical section of the ovary and of the free central placenta, covered with ovules, which nearly fills the cell. 826. Capsule of Primula veris, dehiscent at the summit by numerous teeth. 827. A magnified seed. 828. Section of the same, exhibiting the transverse embryo.

FIG. 829. Branch of Anagallis arvensis (Pimpernel), with a capsule showing the line of circumscissile dehiscence. 830. The capsule (pyxis, 616), with the lid falling away.

ments; the persistent filaments long and flaccid. Ovary two-celled: style single. Capsule (pyxis) membranaceous, opening by circumscissile dehiscence; the cells one- to several-seeded. Embryo large, straight, in fleshy albumen. — *Ex.* Plantago, the Plantain, or Ribgrass, is the principal genus of the order. It is destitute of any important economical qualities.

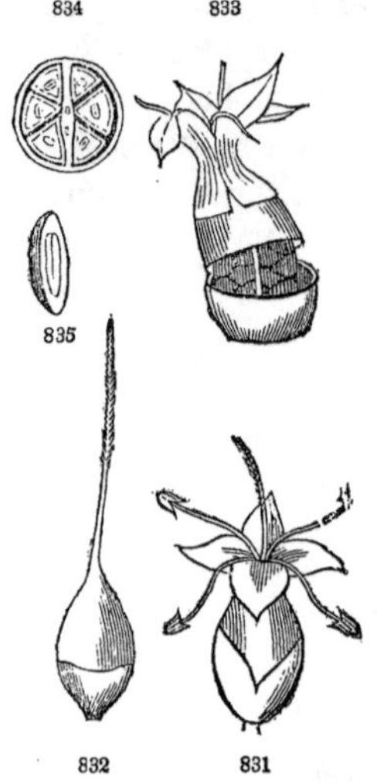

831. Ord. Plumbaginaceæ (*the Leadwort Family*). Perennial herbs, or somewhat shrubby plants; with the flowers often on simple or branching scapes; and the leaves crowded at the base, entire, mostly sheathing or clasping. — Calyx tubular, plaited, five-toothed, persistent. Corolla hypocrateriform, with a five-parted limb, the five stamens inserted on the receptacle opposite its lobes (Plumbago); or else of five almost distinct unguiculate (scarious or coriaceous) petals, with the stamens inserted on their claws! (Statice, &c.) In the former case the five styles are united nearly to the top; but in the latter they are separate! Ovary one-celled, with a single ovule pendulous from a strap-shaped funiculus which rises from the base of the cell. Fruit a utricle, or opening by five valves. Embryo large, in thin albumen. — *Ex.* Statice (Marsh Rosemary, Sea Lavender), and Armeria (Thrift); sea-side or saline plants. The Statices have astringent roots: none more so than those of our own Marsh Rosemary or Sea Lavender (S. Caroliniana), one of the best and most intense astringents of the materia medica.

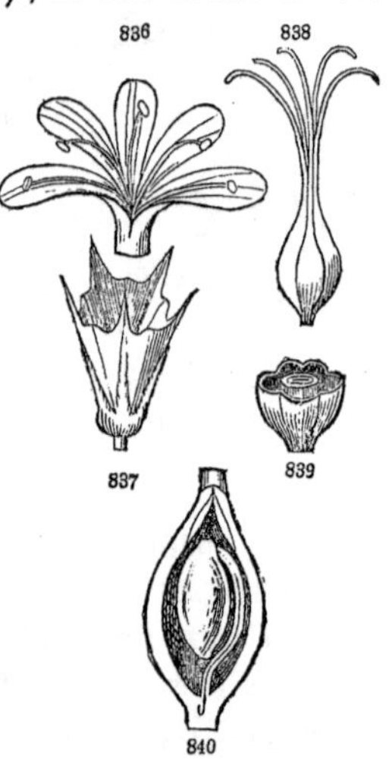

FIG. 831. A flower enlarged. 832. Pistil. 833. Capsule (pyxis, 616) with the marcescent corolla. 834. Cross-section of the capsule and seeds. 835. Vertical section of a seed.

FIG. 836. Corolla, and 837, calyx of Thrift (Armeria vulgaris). 838. Pistil with distinct styles. 839. Cross-section of the pod and seed. 840. Vertical section of the ovary, magnified, to show the ovule.

832. **Ord. Lentibulaceæ** (*the Bladderwort Family*). Herbs, growing in water, or wet places, with the flowers on scapes; the leaves either submersed and dissected into filiform segments resembling rootlets, and commonly furnished with air-bladders to render them buoyant; or, when produced in the air, entire and somewhat fleshy, clustered at the base of the scape. Flowers showy, very irregular. Calyx of two sepals, or unequally five-parted. Corolla bilabiate, personate; the very short tube spurred. Stamens two, inserted on the upper lip of the corolla: anthers one-celled. Ovary free, one-celled, with a free central placenta! bearing numerous ovules. Fruit a capsule. Seeds destitute of albumen. Embryo straight. — *Ex.* Utricularia (Bladderwort), Pinguicula. Unimportant plants.

833. **Ord. Orobanchaceæ** (*the Broom-Rape Family*). Herbs, des-

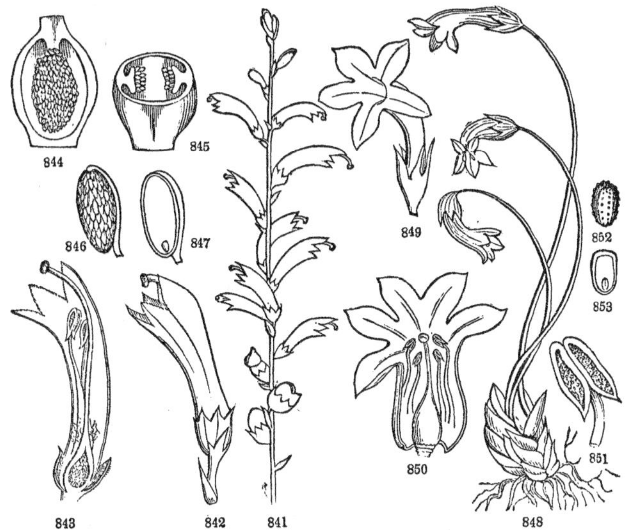

titute of green foliage, and with scales in place of leaves, parasitic

FIG. 841. Branch of Epiphegus Virginiana (Beech-drops), nearly of the natural size: the lower flowers, with short imperfect corollas, alone producing ripe seeds. 842. A flower enlarged. 843. Longitudinal section of the same. 844. Longitudinal section of the ovary, more magnified, showing one of the parietal placentæ covered with minute ovules. 845. Cross-section of the same, showing the two parietal placentæ. 846. A highly magnified seed. 847. Section of the same, exhibiting the minute embryo next the hilum.

FIG. 848. Orobanche uniflora, reduced in size. 849. A flower about the size of nature. 850. The same laid open, showing the didynamous stamens and the pistil. 851. A magnified anther. 852. A magnified seed. 853. Section of the same.

on the roots of other plants; the flowers solitary or spicate. Calyx persistent, four- or five-toothed or bilabiate. Corolla withering or persistent, with a bilabiate or more or less irregular limb. Stamens four, didynamous, inserted on the corolla. Ovary free, one-celled, with two parietal placentæ! which are often two-lobed, or divided. Capsule inclosed in the persistent corolla. Seeds very numerous, minute. Embryo minute at the extremity of the albumen. — *Ex.* Orobanche, Epiphegus (Beech-drops), &c. Astringent, bitter, and escharotic. The pulverized root of Epiphegus (thence called Cancer-root) is applied to open Cancers.*

834. **Ord. Bignoniaceæ** (*the Bignonia Family*). Mostly trees, or climbing or twining shrubby plants, with large and showy flowers, and opposite, simple, or mostly pinnately-compound leaves. Calyx five-parted, two-parted, or bilabiate, often spathaceous. Corolla with an ample throat, and a more or less irregular five-lobed or bilabiate limb. Stamens five, inserted on the corolla, of which one, and often three, are reduced to sterile filaments or rudiments: when four are fertile, they are didynamous. Ovary two-celled, with the placentæ in the axis; the base surrounded by a fleshy ring or disc. Capsule woody or coriaceous, pod-shaped, two-valved, many-seeded. Seeds winged, destitute of albumen. Cotyledons foliaceous, flat, heart-shaped, also notched at the apex. — *Ex.* Bignonia (Trumpet-Creeper), Catalpa, and other tropical genera. Of little importance, except as ornamental plants.

835. **Subord. Sesameæ** (*the Sesamum Family*) has few and wingless seeds; the fruit indurated or drupaceous, often two- to four-horned, sometimes perforated in the centre from the dissepiments not reaching the axis before they diverge and become placentiferous, and spuriously four- to eight-celled by the various cohesion of parts of the placentæ with the walls of the pericarp. — *Ex.* Sesamum, Martynia (Unicorn-plant), and some other tropical plants.

836. **Ord. Acanthaceæ** (*the Acanthus Family*). Herbs or shrubby plants, with bracteate, often showy flowers, and opposite, simple leaves, without stipules. Calyx of five sepals united at the base, or combined into a tube, persistent. Coralla bilabiate, or some-

---

* Ord. GESNERIACEÆ, consisting of tropical herbs, with green foliage and showy flowers, the calyx often partly adherent to the ovary, agrees with Orobanchaceæ in the parietal placentation, by which both are distinguished from all other orders of this group.

times nearly equally five-lobed: æstivation convolute! Stamens four and didynamous, or only two, the anterior pair being abortive or obsolete, inserted on the corolla. Ovary two-celled, with the placentæ in the axis, often few-ovuled. Seeds (sometimes only one or two in each cell) usually supported by hooked processes of the placenta, destitute of albumen. — *Ex.* Acanthus, Dianthera. A large family in the tropics. Many are ornamental.

837. **Ord. Scrophulariaceæ** (*the Figwort Family*). Herbs, or sometimes shrubby plants; with opposite, verticillate, or alternate leaves. Calyx of four or five more or less united sepals, persistent. Corolla bilabiate, personate, or more or less irregular; the lobes imbricated in æstivation. Stamens four and didynamous, the fifth stamen sometimes appearing in the form of a sterile filament,

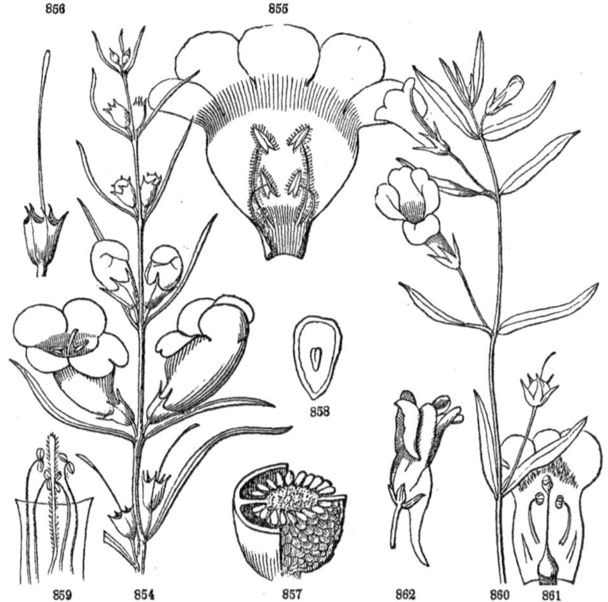

or very rarely antheriferous; or often only two, one pair being

FIG. 854. Branch of Gerardia purpurea. 855. Corolla, of the natural size, laid open. 856. Calyx and style of the same. 857. Magnified transverse section of the capsule, with one of the valves removed. 858. Magnified section of a seed.

FIG. 860. Gratiola aurea, natural size. 861. Corolla laid open, showing the two perfect stamens and two rudimentary filaments (481) as well as the pistil. 859. The perfect stamens and sterile filament of Chelone. 862. Flower of a Linaria (Toadflax, or Snapdragon), with a personate corolla (511).

either suppressed or reduced to sterile filaments, inserted on the corolla. Ovary free, two-celled, with the placentæ united in the axis. Capsule two-valved. Seeds indefinite, albuminous. Embryo small. — *Ex.* Scrophularia, Verbascum (Mullein, which is remarkable for the nearly regular corolla, with five perfect stamens), Linaria, Antirrhinum (Snapdragon), &c. — The plants of this large and important order are generally to be suspected of deleterious (bitter, acrid, or drastic) properties. The most important medicinal plant is the Foxglove (Digitalis purpurea), so remarkable for its power of lowering the pulse. Numerous species are cultivated for ornament.

838. **Ord. Verbenaceæ** (*the Vervain Family*). Herbs, shrubs, or even trees in the tropics, mostly with opposite leaves. Calyx tubular, four- or five-toothed, persistent. Corolla bilabiate, or the four- or five-lobed limb more or less irregular. Stamens mostly four and didynamous, occasionally only two, inserted on the corolla. Ovary free, entire, two- to four-celled. Fruit drupaceous, baccate, or dry, and splitting into two to four indehiscent one-seeded portions. Seeds with little or no albumen. Embryo straight, inferior. — *Ex.* Verbena (Vervain, Fig. 863 – 871) is the

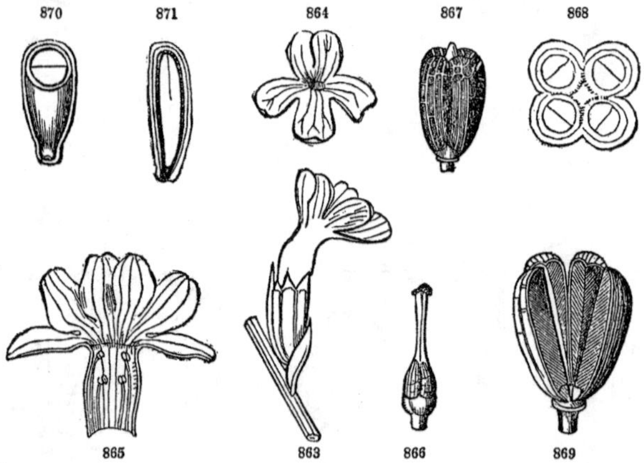

principal representative in cooler regions. There are many others

FIG. 863 and 864. Flower of a Verbena enlarged. 865. The corolla laid open. 866. Pistil. 867. The fruit. 868. Cross-section of the young fruit and the contained seeds. 869. Fruit separating into its four nucules. 870. Cross-section of one nucule or pericarp, and a vertical section of the lower part, showing the surface of the contained seed. 871. Vertical section through the nucule, seed, and embryo.

in the tropics, mostly trees; one of which is the gigantic Indian Teak (Tectona grandis), remarkable for its very heavy and durable wood, which abounds in silex. The leaves of the Aloysia citriodora of the gardens yield an agreeable perfume. Others are bitter and aromatic.

839. **Subord. ? Phrymàceæ** (Phryma) is separated on account of its simple pistil, uniovulate ovary, spirally convolute cotyledons, and superior radicle.

840. **Ord. Labiatæ** (*the Labiate or Mint Family*). Herbs, or somewhat shrubby plants, with quadrangular stems, and opposite or sometimes whorled leaves, replete with receptacles of volatile oil. Flowers in axillary or terminal cymules (412), rarely solitary. Calyx tubular, persistent, five-toothed or cleft, or bilabiate. Corolla bilabiate. Stamens inserted on the corolla, four, didynamous, or only two, one of the pairs being abortive or wanting. Ovary free, deeply four-lobed; the central style proceeding from the base of the lobes. Fruit consisting of four (or fewer) little nuts or achenia, included in the persistent calyx. Seeds with little or no albumen.—*Ex.* The Sage, Rosemary,

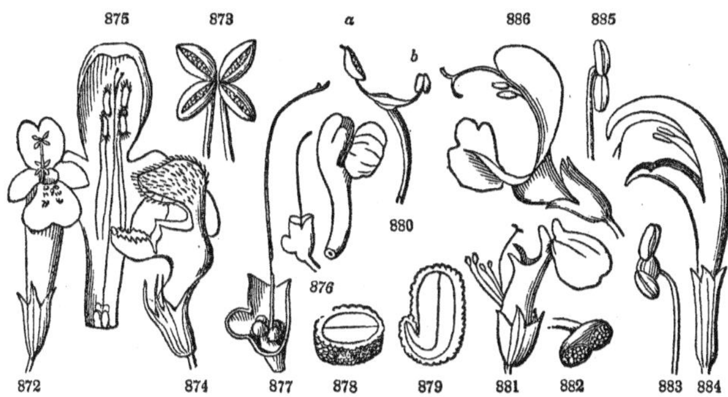

Lavender, Thyme, Mint, &c., are familiar representatives of this

FIG. 872. Flower of Glechoma hederacea, or Ground Ivy. 873. Approximate anthers of one pair of stamens, magnified 874. Flower of a Lamium. 875. Corolla of L. amplexicaule (Dead Nettle), laid open, showing the didynamous stamens, &c. 876. Calyx and corolla of Scutellaria galericulata (Skull-cap). 877. Section of the enlarged calyx of the same, bringing to view the deeply four-lobed ovary, raised on a short gynobase. 878. Cross-section of a magnified achenium. 879. Vertical section of the same, showing the embryo. 881. Flower of Teucrium Canadense. 882. Magnified anther of the same. 883. Stamen of the Thyme. 884. Flower of Monarda. 885. Magnified anther of the same. 886. Flower of a Salvia; the calyx as well as the corolla bilabiate. 880. Magnified stamen of the same, with widely separated anther-cells, one of which (*a*) is polliniferous, the other (*b*) imperfect.

universally recognized order. Their well-known cordial, aromatic, and stomachic qualities depend upon a volatile oil, contained in glandular receptacles which abound in the leaves and other herbaceous parts, with which a bitter principle is variously mixed. None are deleterious.

841. **Ord. Boraginaceæ** (*the Borage Family*). Herbs, or sometimes shrubby plants; with round stems, and alternate, rough leaves; the flowers often in one-sided clusters (406), which are spiral before expansion. Calyx of five leafy and persistent sepals, more or less united at the base, regular. Corolla regular; the limb five-lobed, often with a row of scales in the throat. Stamens inserted on the corolla, as many as its lobes and alternate with them. Ovary deeply four-lobed, the style proceeding from the base of the lobes, which in fruit become little nuts or hard achenia. Seeds with little or no albumen. — *Ex.* Borago (Borage), Lithospermum, Myosotis, Cynoglossum (Hound's-tongue), Heliotropium, &c. In Echium, the limb of the corolla is somewhat irregular, and the stamens unequal. Innocent mucilaginous plants, with a

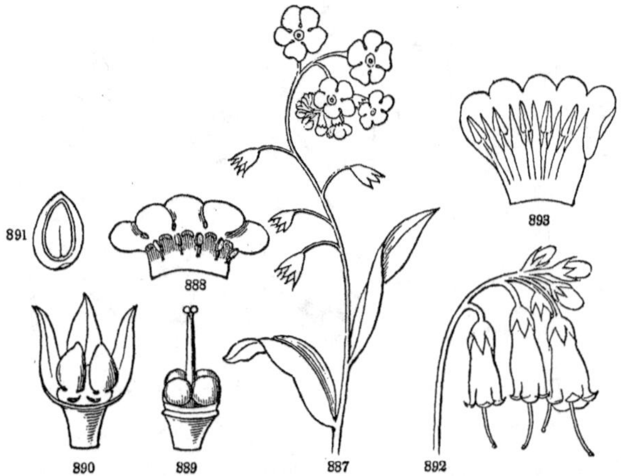

slight astringency: hence demulcent and pectoral; as the roots of

FIG. 887. Myosotis, or Forget-me-not. 888. The rotate corolla laid open, showing the scales of the throat, and the short stamens. 889. The pistil, with its four-lobed ovary. 890. The calyx in fruit; two of the little nuts having fallen away from the receptacle. 891. Section of a nut, or rather achenium, showing the embryo. 892. Raceme of Symphytum officinale (Comfrey). 893. A corolla laid open; exhibiting the lanceolate and pointed scales of the throat, alternate with the stamens.

the Comfrey. The roots of Anchusa tinctoria (Alkanet) and Batschia canescens (used by the aborigines under the name of Puccoon) yield a red dye.

842. **Ord. Hydrophyllaceæ** (*the Water-leaf Family*). Herbs, usually with alternate and lobed or pinnatifid leaves; the flowers mostly in cymose clusters or unilateral racemes. Calyx five-cleft, with the sinuses often appendaged, persistent. Corolla regular, imbricated or convolute in æstivation, usually furnished with scales or honey-bearing grooves inside; the five stamens inserted into its base, alternate with the lobes. Ovary free, with two parietal placentæ, which sometimes dilate in the cell and appear like a kind of inner pericarp in the capsular fruit. Styles partly united. Seeds few, crustaceous. Embryo small, in hard albumen. — *Ex.* Hydrophyllum, Nemophila, and Phacelia; nearly all North American plants, some of them handsome in cultivation.

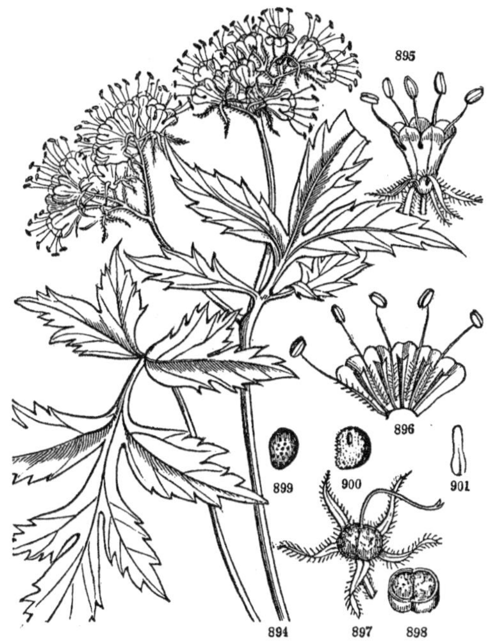

FIG. 894. Hydrophyllum Virginicum. 895. A flower, nearly of the natural size. 896. Corolla laid open. 897. Capsule, with the persistent calyx and style. 898. Cross-section of the same, the cavity filled by two seeds. 899. Magnified seed. 900. Section of the same. 901. Highly magnified embryo.

843. **Ord. Hydroleaceæ** differs (not sufficiently) from the last by the simple and entire leaves, the two-celled ovary, the two distinct styles, and the numerous seeds. — *Ex.* Hydrolea, Nama: chiefly natives of warm regions.

844. **Ord. Polemoniaceæ** (*the Polemonium Family*). Herbs, with alternate or opposite leaves, and panicled, corymbose, or clustered flowers. Calyx five-cleft. Corolla regular, with a five-lobed limb, convolute in æstivation. Stamens five, inserted on the corolla alternate with its lobes, often unequal. Ovary free, three-celled, with a thick axis, bearing few or numerous ovules: styles united into one: stigmas three. Capsule three-valved, loculicidal; the valves also usually breaking away from a thick central column which bears the seeds. Embryo straight, in fleshy or horny albumen. — *Ex.* Polemonium (Greek Valerian), Phlox, Gilia. Chiefly North American; many are very common ornamental plants in cultivation.

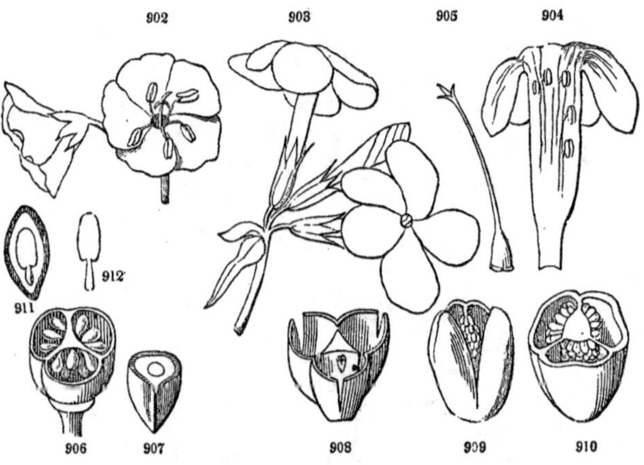

845. **Ord. Diapensiaceæ.** Low, prostrate, and tufted suffruticose plants; with crowded and evergreen heath-like leaves, and solitary terminal flowers. Differing from the last family chiefly in the transversely two-valved anthers, and amphitropous seeds; and doubtless to be united to it. Consists of two plants only, viz. the Alpine

FIG. 902. Flowers of Polemonium. 903. Flowers of Phlox. 904. Corolla of the same, laid open, showing the stamens unequally inserted on its tube. 905. Pistil of the same. 906. Cross-section of the capsule of Polemonium. 907. Cross-section of a magnified seed. 911. Perpendicular section of the same. 912. Magnified embryo. 908. Cross-section of the dehiscent capsule of Collomia. 909, 910. Capsule of Leptodactylon.

Diapensia, and Pyxidanthera, of the Pine-barrens of New Jersey, &c.

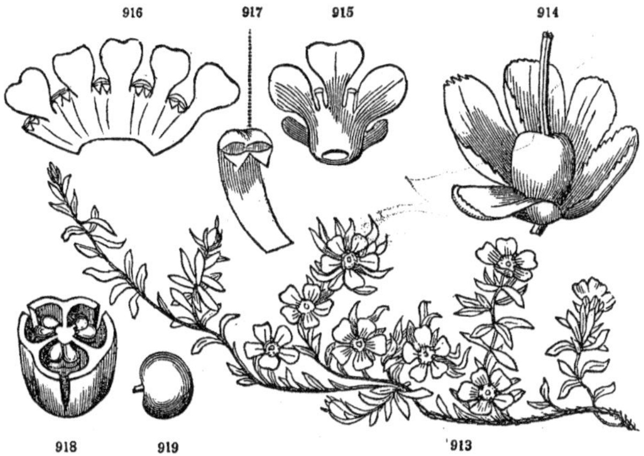

846. **Ord. Convolvulaceæ** (*the Convolvulus Family*). Twining or trailing herbs or shrubs, often with milky juice; the leaves alternate, and the flowers regular. Calyx of five sepals, imbricated, or usually more or less united, persistent. Corolla supervolute in æstivation (Fig. 363); the limb often entire. Stamens five, inserted on the tube of the corolla near the base. Ovary free, two- to four-celled, with one or two erect ovules in each cell: styles united, or more or less distinct. Capsule two- to four- (or by obliteration one-) celled; the valves falling away from the persistent dissepiments (septifragal). Seeds large, with a little mucilaginous albumen: embryo curved, and the foliaceous cotyledons usually crumpled. — *Ex.* Convolvulus (Morning-Glory, Bindweed). They all contain a peculiar strongly purgative resinous matter, which is chiefly found in the acrid, milky juice of their thickened or tuberous roots. Convolvulus Jalapa, and other Mexican species, furnish the *Jalap* of the shops. The more drastic *Scammony* is derived from the roots of C. Scammonia of the Levant. There is much less of this in those of Convolvulus panduratus (Mechameck, Man-of-the-Earth, Wild Potato-vine): while those of C. macrorhizus of the Southern States, which sometimes weigh 40 or 50 pounds, are farinaceous, with so slight an admixture of the peculiar resin

FIG. 913. Pyxidanthera barbulata, natural size. 914. Pistil, in fruit, and the persistent calyx, enlarged. 915. Corolla and stamens. 916. Same, laid open. 917. A separate stamen, magnified. 918. Section of the dehiscent capsule. 919. A seed.

as to be quite inert; as is also the case with the Babatas, or Sweet Potato, an important article of food. — To this family are appended

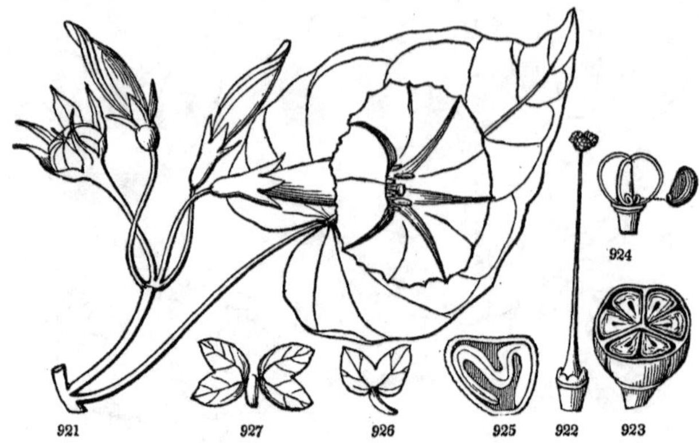

847. **Subord. Dichondreæ.** Ovaries two to four, either entirely distinct or with their basilar styles united in pairs. Creeping

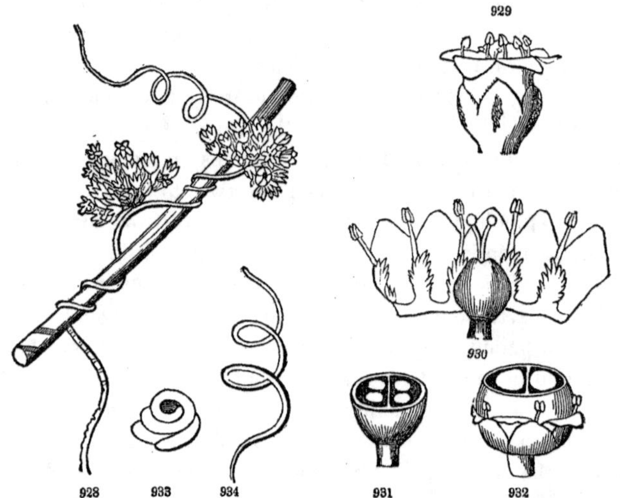

FIG. 921. Convolvulus purpureus. 922. The pistil. 923. Section of the capsule, and of the two seeds in each cell. 924. Capsule (reduced in size), when the valves have fallen away from the dissepiments; and one of the seeds. 925. Magnified cross-section of a seed. 926. Embryo, with the leaf-like two-lobed cotyledons spread out. 927. Same, with the two cotyledons separated and laid open.

FIG. 928. A piece of Cuscuta Gronovii, the common Dodder of the Northern United States, of the natural size. 929. A flower, enlarged. 930. The same, laid open. 931. Section of the ovary. 932. Section of the capsule and seeds. 933. The spiral embryo detached. 934. The same in germination.

plants, with axillary and scape-like one-flowered peduncles. — *Ex.* Dichondra.

848. **Subord. Cuscutineæ.** Ovary two-celled; the capsule opening by circumscissile dehiscence, or bursting irregularly. Embryo filiform, and spirally coiled in fleshy albumen, destitute of cotyledons! Parasitic, leafless, twining herbs, destitute of green color (135). Stamens usually furnished with fringed scales within. — *Ex.* Cuscuta (Dodder).

849. **Ord. Solanaceæ** (*the Nightshade Family*) differs from Scrophulariaceæ chiefly in the regular (rarely somewhat irregular) flowers, with as many fertile stamens as there are lobes to the corolla (four or five), and the plaited or valvate æstivation of the corolla. Fruit either capsular or baccate. Embryo small, mostly curved, in fleshy albumen. — *Ex.* Solanum (Potato), Nicotiana. The fruit of Datura is spuriously four-celled. — Stimulant narcotic properties pervade the order, the herbage and fruits of which are mostly deleterious, often violently poisonous, and furnishing some of the most active medicines; such as the Tobacco, the Henbane (Hyoscyamus niger), the Belladonna (Atropa

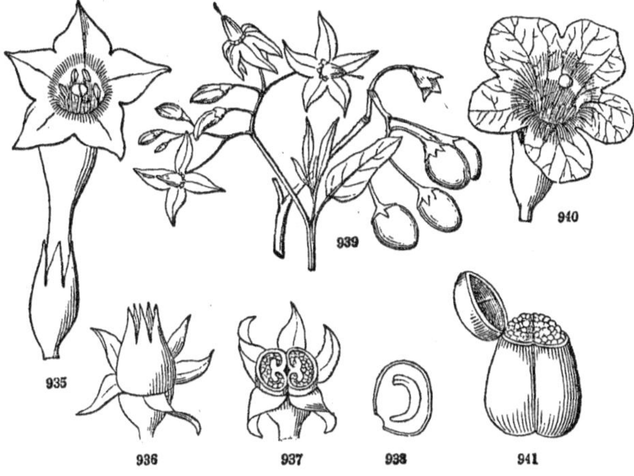

Belladonna), the Thorn-apple or Jamestown Weed (Datura Stramonium), and the Bittersweet (Solanum Dulcamara); the last

FIG. 935. Flower of Tobacco (Nicotiana Tabacum). 936. The capsule, dehiscent at the apex, with the persistent calyx. 937. Cross-section of the same. 938. Magnified section of the seed of Solanum. 939. Flowers and berries of Solanum Dulcamara. 940. Flower of Hyoscyamus niger. 941. Fruit (pyxis, 616) of the same.

only slightly narcotic. Yet the berries of some Solanums are eatable when cooked (as Tomatoes, the Egg-Plant, &c.), and the starchy tubers of the Potato are a most important article of food. But the fruit and seeds of Capsicum (*Cayenne pepper*) are stimulant.

850. **Ord. Gentianaceæ** (*the Gentian Family*). Herbs, with a watery juice; the leaves opposite and entire. Flowers regular, often showy. Calyx of usually four or five persistent, more or less united sepals. Corolla mostly convolute in æstivation; the stamens inserted on its tube. Ovary one-celled, with two parietal, but often introflexed, placentæ; styles united or none. Capsule many-seeded. Seeds with fleshy albumen and a minute embryo. — *Ex.* Gentiana, Frasera (the American Columbo). A pure bitter and tonic principle (*Gentianine*) pervades the whole order. Gentiana lutea of Middle Europe furnishes the officinal *Gentian*, for which almost any of our species may be substituted.

851. **Subord. Menyanthideæ** (*the Buckbean Family*) has alternate, sometimes trifoliolate or toothed leaves, and a valvate-induplicate æstivation of the corolla. — *Ex.* Menyanthes, Limnanthemum (this bears the peduncles on the petiole, Fig. 949).

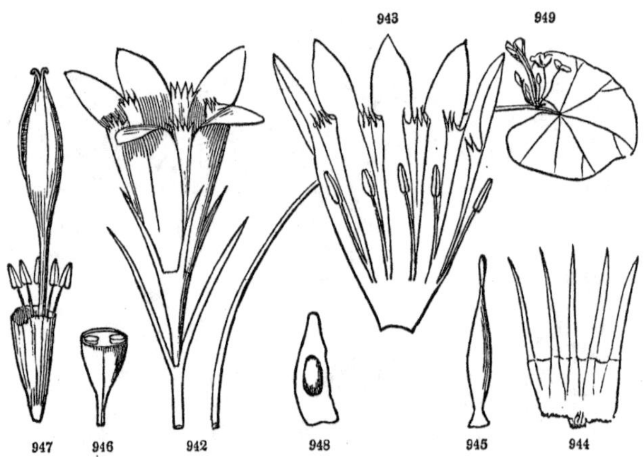

852. **Subord. Obolarieæ** has an imbricative æstivation of the co-

FIG. 942. Flower of Gentiana angustifolia. 943. Corolla, and 944, the calyx, laid open. 945. The pistil. 946. Cross-section of the pistil, showing the parietal attachment of the ovules. 947. Ripe capsule of G. Saponaria, raised on a stype: the persistent withering corolla, &c. torn away. 948. A magnified seed, with its large and loose testa. 949. Leaf of Limnanthemum (Villarsia), bearing the flowers on its petiole.

rolla, opposite leaves, and the whole internal surface of the ovary ovuliferous! — *Ex.* Obolaria.

853. **Ord. Apocynaceæ** (*the Dogbane Family*). Trees, shrubs, or herbs, with milky juice, and opposite entire leaves, without stipules. Flowers regular. Calyx five-cleft, persistent. Corolla five-lobed, twisted in æstivation. Filaments distinct; the anthers sometimes slightly connected: pollen granular. Ovaries two, distinct, or rarely united, but their styles or stigmas combined into one: in fruit usually forming two follicles. Seeds often with a coma. Embryo large and straight, in sparing albumen. — *Ex.* Apocynum (Dog's-bane, Fig. 950), Vinca (Periwinkle); and a great number of tropical shrubs and trees. In all, the juice is drastic or poisonous, and often yields *caoutchouc;* which in Sumatra is obtained from Urceola elastica. The well-known *Nux vomica* is the seed of Strychnos Nux-vomica of India. S. toxifera yields the famous *Woorari* poison of Guiana. One kind of *Upas* is obtained from the bark of the root of S. Tieute in Java. The poisonous principle in these plants is an alkaloid, called *Strychnia.*

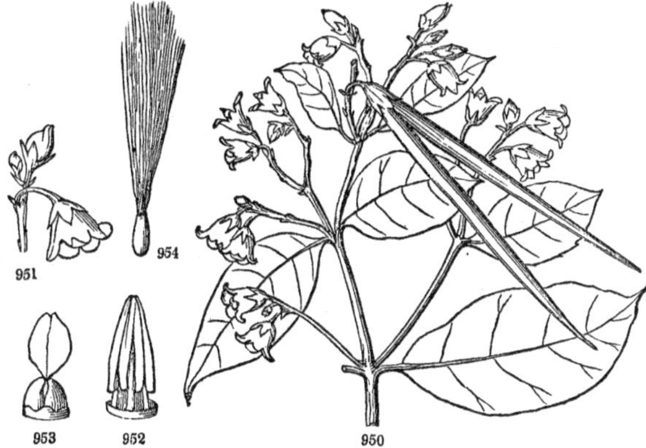

854. **Ord. Asclepiadaceæ** (*the Milkweed Family*). Herbs or shrubs, with milky juice, and opposite entire leaves; mainly differing from the preceding order (as they do from all other Exogenous plants) by the peculiar connection of the stamens with the stigma, and the cohesion of the pollen into wax-like masses, which are attached in

FIG. 950. Apocynum androsæmifolium. 951. Flower, of the natural size. 952. Stamens with the anthers connivent around the pistils. 953. The pistils with their large common stigma. 954. Seed with its coma, or tuft of silky hairs.

pairs to five glands of the stigma, and removed from the anther-cells usually by the agency of insects. Fruit consisting of two follicles. Seeds usually with a silky coma. — *Ex.* Asclepias (Milkweed, Wild Cotton). The juice of A. tuberosa (Pleurisy-root, Butterfly-weed) is not milky. In all, it is bitter and acrid, and contains *caoutchouc*.

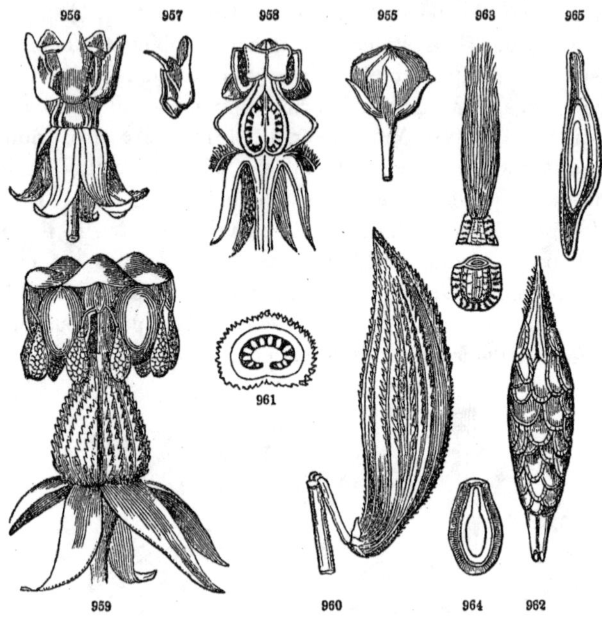

855. **Ord. Jasminaceæ** (*the Jessamine Family*) consists of a few

FIG. 955. Flower-bud of the Common Milkweed (Asclepias Cornuti). 956. Expanded flower; the calyx and corolla reflexed; showing the stamineal crown. 957. One of the hooded appendages of the latter removed and seen sidewise, with its included process or horn. 958. A vertical section of a flower (the hooded appendages removed) through the tube of stamens, the thick stigma, ovaries, &c. 959. Flower with the calyx, and the fertilized enlarging ovaries, crowned with the large stigma common to the two, from the angles of the peltate summit of which the pairs of pollen-masses, detached from the anther cells, hang by their stalks or caudicle from a gland. [See page 315; Fig. 420. An anther, from which the hooded appendage is cut away. 421. One more magnified: its two pollen-masses still in the open cells, but attached by their stalks each to one of the glands, to which a pollen-mass of an adjacent stamen on each side is already similarly attached. 422. One of these pairs of pollen-masses separate. 423. Pollen-masses of Asclepias incarnata, connected by their emitted pollen-tubes (much magnified) with the stigma. 424. Section through the stigma and into one of the styles, showing the course of the pollen-tubes.] 960 Fruit (follicle) of the Common Milkweed. 961. Cross-section of the last, in an early state. 962. Detached placenta in fruit, covered with seeds. 963. Seed (cut across), with its coma. 964. Section of the seed, as it lies in 963, parallel with the cotyledons. 965. Vertical section of the seed perpendicular to the face of the cotyledons.

chiefly Asiatic shrubs, with compound leaves and fragrant flowers; differing from Oleaceæ by the imbricated or twisted æstivation of the hypocrateriform corolla, erect seeds, &c. — *Ex.* Jasminum, the Jessamine. Cultivated for ornament, and for their very fragrant blossoms.

856. **Subord. Bolivarieæ** consists of a few American (three or four of them Texan) plants, and one from the Cape of Good Hope, sometimes with simple leaves, and scarcely differing from the true Jasminaceæ; though most of them have four ovules in each cell.

857. **Ord. Oleaceæ** (*the Olive Family*). Trees or shrubs, with opposite leaves, either simple or pinnate. Calyx persistent. Corolla four-cleft, or of four separate petals, valvate in æstivation, sometimes none. Stamens mostly two, adnate to the base of the corolla. Ovary free, two-celled, with two pendulous ovules in each cell. Fruit by suppression usually one-celled and one- or two-seeded. Seed albuminous. Embryo straight. — *Ex.* Olea (the Olive), and Chionanthus (Fringe-tree), where the fruit is a drupe. Syringa, the Lilac, which has a capsular fruit. Fraxinus, the Ash; where the fruit is a samara, the flowers are polygamous, and often destitute of petals. *Olive oil* is expressed from the esculent drupes of Olea Europæa. The bark, like that of the Ash, is bitter, astringent, and febrifugal. *Manna* exudes from the trunk of Fraxinus Ornus of Southern Europe, &c. — Forestiera, of doubtful affinity, is perhaps to follow this order, although entirely apetalous.

Division III. — Apetalous Exogenous Plants.*

Corolla none; the floral envelopes consisting of a single series (calyx), or sometimes entirely wanting.

Conspectus of the Orders.

Group 1. Flowers perfect, with a conspicuous or colored mostly adnate calyx. Ovary several-celled and many-ovuled. Capsule or berry many-seeded. — Herbs or climbing shrubs. ARISTOLOCHIACEÆ, p. 467.

---

* Numerous plants of the Polypetalous orders are apetalous, such as Clematis, Anemone, and other Ranunculaceæ, some Rhamnaceæ, Caryophyllaceæ, Onagraceæ, Portulacaceæ, Crassulaceæ, Rosaceæ, Aceraceæ, &c. Also some Oleaceæ and Primulaceæ of the Gamopetalous series are apetalous.

Group 2. Flowers perfect, or rarely polygamous, with a regular and often petaloid calyx. Ovary free. Ovules solitary in each ovary or cell. Embryo curved or coiled around mealy albumen, rarely in the axis or exalbuminous. — Chiefly herbs.

Ovary several-celled, consisting of a whorl of several one-ovuled carpels. PHYTOLACCACEÆ, p. 468.
Ovary one-celled, with a single ovule.
  Stipules none. Ovule campylotropous or amphitropous.
    Calyx herbaceous. CHENOPODIACEÆ, p. 469.
    Calyx and bracts scarious. AMARANTACEÆ, p. 470.
    Calyx corolline, the persistent base indurated. NYCTAGINACEÆ, p. 470.
  Stipules sheathing (ochreæ). Calyx corolline. Ovule orthotropous. POLYGONACEÆ, p. 470.

Group 3. Flowers perfect, polygamous or diœcious, not disposed in aments, with a regular, and often petaloid calyx. Ovary one-celled, or rarely two-celled, with one or few ovules in each cell: but the fruit one-celled and one-seeded. Embryo not coiled around albumen. — Trees or shrubs.

* Style or stigma one.

Calyx free from the ovary, and not enveloping the fruit.
  Flowers polygamo-diœcious. Anth. opening by valves. LAURACEÆ, p. 471.
  Flowers perfect. Anthers opening longitudinally. THYMELACEÆ, p. 472.
Calyx free, but baccate in fruit and inclosing the achenium. ELEAGNACEÆ, p. 472.
Calyx adnate to the ovary.
  Ovules several, pendulous from a stipe-like placenta. SANTALACEÆ, p. 473.
  Ovule solitary, suspended.
    Parasitic shrubs. Ovule without integuments. LORANTHACEÆ, p. 474.
    Trees. Fruit a drupe. NYSSACEÆ, p. 473.
* * Styles or stigmas two, divergent. ULMACEÆ, p. 474.

Group 4. Flowers perfect, entirely destitute of calyx as well as corolla. Embryo minute, inclosed in the persistent embryo-sac at the apex of the albumen. — Herbs or suffrutescent plants. SAURURACEÆ, p. 475.

Group 5. Flowers perfect or diclinous, frequently destitute of both calyx and corolla. — Submersed or floating aquatic herbs.

Flowers monœcious. Fruit one-celled and one-seeded. CERATOPHYLLACEÆ, p. 476.
Flowers mostly perfect. Fruit four-celled and four-seeded. CALLITRICHACEÆ, p. 476.
Flowers mostly perfect. Capsule several-celled, several-seeded. PODOSTEMACEÆ, p. 477.

Group 6. Flowers monœcious or diœcious, not amentaceous. Fruit capsular or drupaceous, with two or more cells, and one (or rarely two) seeds in each. — Herbs, shrubs, or trees.

Fruit mostly dry. Juice milky. Pollen simple. EUPHORBIACEÆ, p. 477.
Fruit drupaceous. Pollen-grains quaternary. EMPETRACEÆ, p. 478.

Group 7. Flowers monœcious or diœcious; the sterile, and frequently the fertile also, in aments, or in heads or spikes. Ovary often two- to several-celled, but the fruit always one-celled.— Trees, shrubs, or (only in Urticaceæ) herbs.

* Fruit drupaceous. Calyx adherent. JUGLANDACEÆ, p. 479.

* * Fruit a nut, involucrate. Calyx adherent. CUPULIFERÆ, p. 479.

* * * Fruit one-seeded, indehiscent. Fertile and sterile flowers both in aments, and entirely destitute of calyx.

Ovary one-celled; ovule solitary, erect. MYRICACEÆ, p. 480.
Ovary two-celled, two-ovuled: ovule pendulous. BETULACEÆ, p. 480.

* * * * Fruit dehiscent, many-seeded. Seeds with a coma. Fertile and sterile flowers both in aments, and destitute of calyx. SALICACEÆ, p. 481.

* * * * * Fruit a nut or a two-celled and few-seeded capsule. Fertile and sterile flowers both in aments or heads, and destitute of calyx.

Capsule two-beaked, many-seeded. BALSAMIFLUÆ, p. 482.
Nut club-shaped, one-seeded, bristly-downy. PLATANACEÆ, p. 482.

* * * * * * Fruit an achenium, often inclosed in a baccate calyx. Flowers variously disposed, sometimes collected in fleshy heads. — Juice milky, when trees or shrubs. URTICACEÆ, p. 482.

858. **Ord. Aristolochiaceæ** (*the Birthwort Family*). Herbaceous

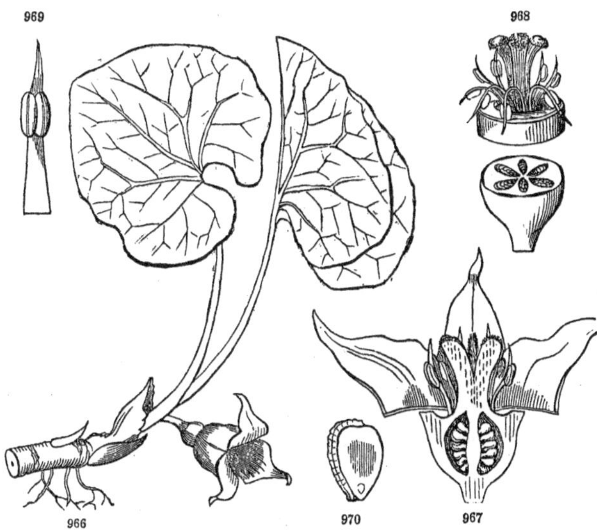

FIG. 966. Asarum Canadense. 967. Calyx displayed, and a vertical section through the rest of the flower. 968. Cross-section of the ovary; the upper portion (from which the limb of the calyx is cut away) showing the stamens, the united styles, &c. 969. A separate stamen, enlarged. 970. Vertical section of a seed.

or climbing shrubby plants, with alternate leaves. Flowers brown or greenish, usually solitary. Calyx-tube more or less united with the ovary; the limb valvate. Stamens six to twelve, epigynous, or adherent to the base of the short and thick style: anthers adnate, extrorse. Stigmas radiate. Ovary 3–6-celled. Capsule or berry three- to six-celled, many-seeded. Embryo minute, in fleshy albumen. — *Ex.* Asarum (Wild Ginger, Canada Snake-root), Aristolochia (Virginia Snake-root). Pungent, aromatic, or stimulant tonics; generally termed Snake-roots, being reputed antidotes for the bites of venomous snakes.*

859. **Ord. Phytolaccaceæ** (*the Poke-weed Family*). Chiefly repre-

sented by the common Poke (Phytolacca decandra), which has a

* The Ord. RAFFLESIACEÆ, and perhaps other RHIZANTHEÆ, consisting of most remarkable fungus-like parasites (136, and Fig. 125) are to be placed somewhere in this vicinity.

FIG. 988, 989. Phytolacca decandra (Poke). 990. A flower. 991. Unripe fruit. 992. Cross-section of the same, a little enlarged. 993. Magnified seed. 994. Section of the same across the embryo. 995. Vertical section, showing the embryo coiled around the albumen into a ring. 996. Magnified detached embryo.

compound ovary of ten confluent (one-seeded) carpels, the short styles or stigmas distinct; the fruit a flattened berry. The root is acrid and emetic: yet the young shoots in the spring are used as a substitute for Asparagus. The berries yield a copious deep-crimson juice. Other genera connect the order with the next; but are distinguished, when the stamens are of the same number as the sepals, by their position alternate with them, as in Portulacaceæ.

860. **Ord. Chenopodiaceæ** (*the Goosefoot Family*). Chiefly weedy herbs, with alternate and more or less succulent leaves, and small herbaceous flowers. Calyx sometimes tubular at the base, persistent; the stamens as many as its lobes, or fewer, and inserted at their base. Ovary free, one-celled, with a single ovule arising from its base. Fruit a utricle or achenium. Embryo curved or coiled around the outside of mealy albumen, or spiral, without any albumen (in Salsola, &c.). — *Ex.* Chenopodium, Atriplex, Beta (the Beet), &c. Sea-side plants, or common weeds: some are pot-herbs, such as Spinach: a few are cultivated for their esculent roots; as the Beet, which contains sugar. Soda is largely extracted from the maritime species, especially from those of Salsola and Salicornia (Samphire, Glass-wort). Chenopodium anthelminticum yields the *Worm-seed oil.*

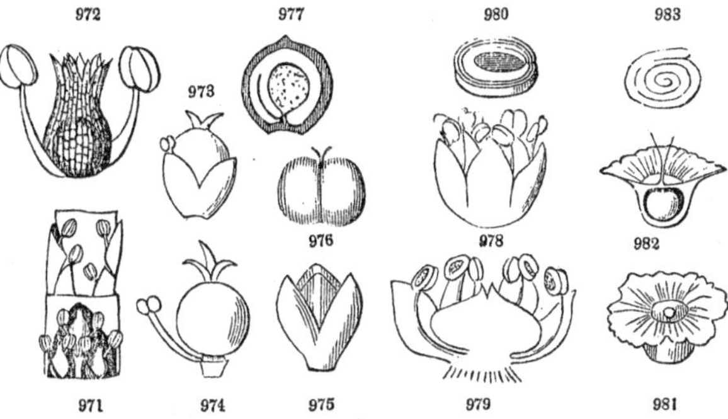

FIG. 971. Part of the spike of Salicornia herbacea: the flowers placed three together in excavations of the stem, protected by a fleshy scale. 972. Separate flower. 973. A flower of Blitum, with its fleshy calyx and single stamen. 974. Same, more enlarged, with the thickened juicy calyx (975) removed. 976. The ripe fruit. 977. Same, divided vertically, showing the embryo coiled around the central albumen. 978. Flower of Chenopodium album (common Goosefoot). 979. Section of the same, more enlarged. 980. Section of the utricle and seed, showing the embryo. 981. Calyx of Salsola kali (Saltwort), in fruit, with its wing-like border. 982. Section of the same, bringing the ovary into view. 983. The spirally coiled embryo of Chenopodina maritima.

861. **Ord. Amarantaceæ** (*the Amaranth Family*). Herbs, with opposite or alternate leaves; the flowers in heads, spikes, or dense clusters, imbricated with dry and scarious bracts which are usually colored. Calyx of three to five sepals, which are dry and scarious, like the bracts. Stamens five or more, hypogynous, distinct or monadelphous: anthers frequently one-celled. Embryo annular, always vertical. Otherwise nearly as in Chenopodiaceæ. — *Ex.* Amarantus, Gomphrena, &c. Weeds. A few Amaranths are cultivated for their dry and enduring richly-colored flowers.

862. **Ord. Nyctaginaceæ.** Herbs or shrubs, with opposite leaves; distinguished by their tubular and infundibuliform calyx, the upper part of which resembles a corolla, and at length separates from the base, which hardens and incloses the one-celled achenium-like fruit, appearing like a part of it. Stamens hypogynous, 1–20. Embryo coiled around the outside of mealy albumen. Flowers involucrate, often showy. Mirabilis (Four-o'clock) has a one-flowered involucre exactly like a calyx, while the latter resembles the corolla of a Morning-Glory. Plants of warm latitudes; many on our Southwestern frontiers.

863. **Ord. Polygonaceæ** (*the Buckwheat Family*). Herbs with al-

FIG. 984. Polygonum Pennsylvanicum. 985. Flower, laid open. 986. Section of the ovary, showing the erect ovule. 987. Section of the seed, showing the embryo, at one side of albumen.

ternate leaves; remarkable for their stipules (ochreæ, 304), which usually form sheaths around the stems above the leaves, and for their orthotropous ovules. Stamens definite, inserted on the petaloid calyx. Fruit achenium-like, compressed or triangular. Embryo curved, or nearly straight, applied to the outside (rarely in the centre) of starchy albumen. — *Ex.* Polygonum, Rumex (Dock, Sorrel), Rheum (Rhubarb). The stems and leaves of Rhubarb and Sorrel are pleasantly acid: while several Polygonums (Knotweed, Smart-weed, Water Pepper, &c.) are acrid or rubefacient. The farinaceous seeds of P. Fagopyrum (the Buckwheat) are used for food. The roots of most species of *Rhubarb* are purgative: but it is not yet known what particular species of Tartary yields the genuine officinal article. The ERIOGONEÆ (of southern and western North America) form a tribe remarkable for their exstipulate leaves and involucrate flowers.

864. **Ord. Lauraceæ** (*the Laurel Family*). Trees or shrubs, with pellucid-punctate alternate leaves, their margins entire. Flowers sometimes polygamo-diœcious. Calyx of four to six somewhat united petaloid sepals, which are imbricated in two series, free from the ovary. Stamens definite, but usually more numerous than the sepals, inserted on the base of the calyx: anthers two- to four-celled, opening by recurved valves! Fruit a berry or drupe, the pedicel often thickened. Seed with a large almond-like em-

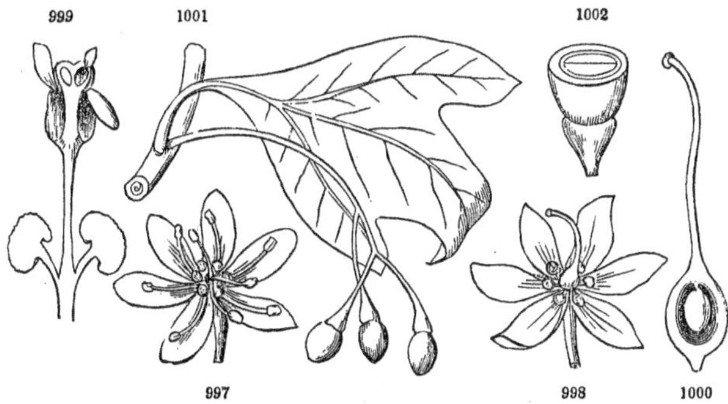

bryo, destitute of albumen. — *Ex.* Laurus, Sassafras, Benzoin. All aromatic plants, almost every part abounding in warm and

FIG. 997. A staminate, and 998, a pistillate flower of Sassafras. 999. A stamen with its glands at the base: the anthers opening by two sets of valves. 1000. Pistil; the ovary divided. 1001. Branch in fruit. 1002. Section of the drupe and seed.

stimulant volatile oil, to which their qualities are due. *Camphor* is obtained from Camphora officinarum of Japan, China, &c. *Cinnamon* is the bark of Cinnamomum Zeylanicum; *Cassia bark*, of Cinnamomum aromaticum of China. The aromatic bark and wood and the very mucilaginous leaves of our own Sassafras are well known. Our Benzoin odoriferum is the Spice-wood, or Fever-bush. Laurus nobilis is the true Laurel, or Sweet Bay. Persea gratissima, of the West Indies, bears the edible Avocado pear.

865. **Ord. Thymelaceæ** (*the Mezereum Family*). Shrubby plants, with perfect flowers, and a very tough bark; the tube of the petaloid calyx being free from the (one-ovuled) ovary; its lobes imbricated in æstivation; the pendulous seed destitute of albumen. Stamens often twice as many as the lobes of the calyx, inserted upon its tube or throat. — *Ex.* Daphne, &c., of Europe and Middle Asia; and Dirca (Leather-wood, Moose-wood, Wickopy), which is the only North American genus. The tough bark is acrid, or even blistering, and is also useful for cordage. The reticulated fibres may be separated into a kind of lace in the Lagetta or Lace-bark of Jamaica. The fruit of all the species is deleterious.

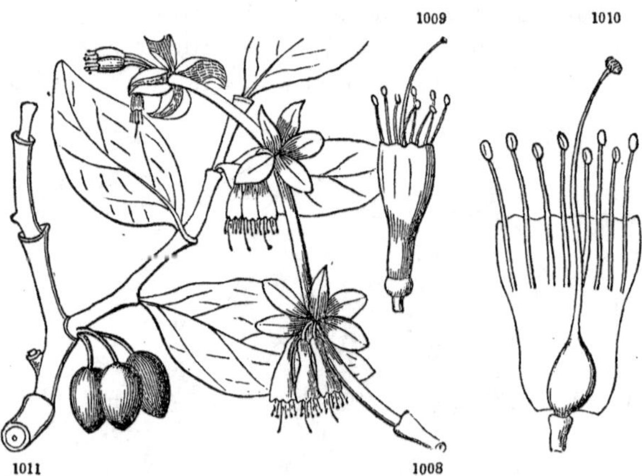

866. **Ord. Eleagnaceæ** (*the Oleaster Family*). Shrubs or small trees, with the flowers more commonly diœcious, the leaves either opposite or alternate; readily distinguished from the preceding by

FIG. 1008. Flowering branch of Dirca palustris. 1009. A flower. 1010. The same, laid open and enlarged. 1011. Branch in fruit.

having the foliage and shoots covered with scurf, by the ascending albuminous seed, and the persistent tube of the calyx, which, although free from the ovary, becomes succulent, like a berry in fruit, and constricted at the throat, inclosing the crustaceous achenium! — *Ex.* Eleagnus, Shepherdia; cultivated for their silvery foliage. The fruit is sometimes eaten.

867. **Ord. Santalaceæ** (*the Sandal-wood Family*). Trees, shrubs, or sometimes herbs; with alternate entire leaves, and small (very rarely diœcious) flowers. Calyx-tube adherent to the ovary; the limb four- or five-cleft, valvate in æstivation; its base lined with a fleshy disc, the edge of which is often lobed. Stamens as many as the lobes of the calyx, and opposite them, inserted on the edge of the disc. Ovules several, destitute of proper integuments, pendulous from the apex of a stipe-like basilar placenta. Style one. Fruit indehiscent, crowned with the limb of the calyx. Seed albuminous. Embryo small. — *Ex.* Comandra, Pyrularia, &c. The fragrant *Sandal-wood* is obtained from several Indian and Polynesian species of Santalum. The large seeds of Pyrularia oleifera (Buffalo-tree, Oil-nut) would yield a copious fixed oil.

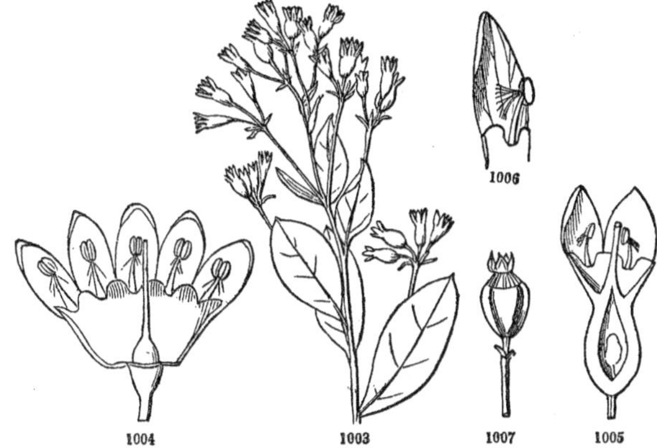

868. **Ord. Nyssaceæ** (*the Tupelo Family*). Trees, with diœcio-polygamous flowers, differing from the last in the solitary ovule suspended from the summit of the cell, and furnished with integuments in the ordinary manner. Style one, stigmatose down one

FIG. 1003. Branch of Comandra umbellata. 1004. Enlarged flower, laid open. 1005. Vertical section of a flower. 1006. One of the segments of the calyx, enlarged, showing the tuft of hairs which connects its surface with the anther! 1007. The fruit, reduced in size.

side. Drupe baccate. Embryo large, in sparing albumen. — Consists only of the genus Nyssa. The Black Gum-tree, &c. is remarkable for the toughness of the interlaced fibres, so that it is very difficult to split the timber. The acid berries give the name of Sour Gum to Nyssa capitata.

869. **Ord. Loranthaceæ** (*the Mistletoe Family*) consists of shrubby plants, with articulated branches, and opposite coriaceous and dull greenish entire leaves, parasitic on trees. The floral envelopes are various. In Mistletoe (which is diœcious) the anthers are sessile and adnate to the face of the sepals, one to each. The ovary is one-celled, with a single suspended ovule, consisting of a nucleus without integuments. Fruit a one-seeded berry. Embryo small, in fleshy albumen. — *Ex.* Loranthus; Viscum, the Mistletoe, from the glutinous berries of which *birdlime* is made. The bark is astringent.

870. **Ord. Ulmaceæ** (*the Elm Family*). Trees or shrubs, with a watery juice, and alternate rough leaves, furnished with deciduous

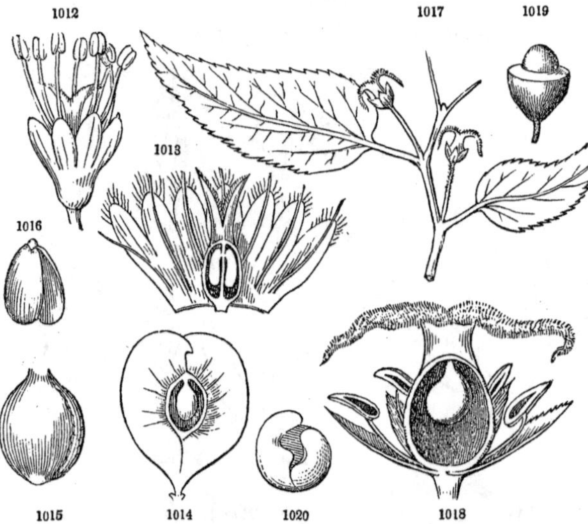

stipules. Flowers in axillary clusters or fascicles, rarely solitary,

FIG. 1012. Flower of the Slippery Elm. 1013. Calyx laid open and the ovary divided vertically. 1014. Fruit, the cell laid open to show the single seed. 1015. The latter magnified. 1016. Its embryo.

FIG. 1017. Branch of Celtis Americana, in flower. 1018. Enlarged flower, divided vertically. 1019. Drupe, the flesh divided to show the stone. 1020. The coiled embryo.

perfect or polygamous. Calyx campanulate, four- or five-cleft, free from the ovary; the lobes imbricated in æstivation. Stamens inserted on the base of the calyx, as many as its lobes and opposite them, or more numerous. Ovary one- or two-celled, with a single suspended ovule in each: styles or stigmas two. Fruit one-celled and one-seeded, either a samara with a straight embryo and no albumen, as in the Elm (Ulmus); or a drupe with a curved embryo and scanty albumen, as in Celtis (Hackberry), the type of the suborder or tribe CELTIDEÆ. Timber-trees. The inner bark of the Slippery Elm is charged with mucilage. Hackberries are edible.

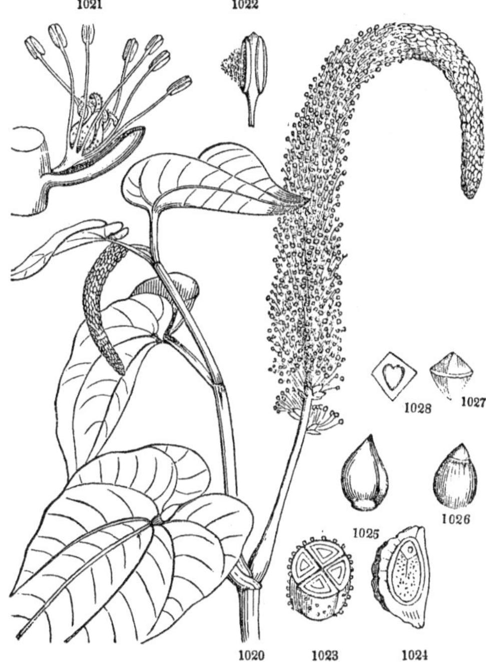

871. **Ord. Saururaceæ** (*the Lizard's-tail Family*). Herbs (growing in swampy places), with the stems jointed at the nodes; the

FIG. 1020. Saururus cernuus. 1021. A separate flower, with its bract and a part of the axis magnified. 1022. A more magnified anther, discharging its pollen from one cell. 1023. Cross-section of the ovary. 1024. Vertical section of one of the carpels in fruit, and of the contained seed, with the sac at the extremity of the albumen, containing the minute embryo. 1025. A seed. 1026. Same, with the outer integument (testa) removed, showing the sac of the amnios. 1027. The latter, highly magnified. 1028. Section of the same, showing the inclosed heart-shaped embryo.

leaves alternate, entire, with somewhat sheathing petioles; the flowers perfect, in racemes or spikes, destitute of all floral envelopes. Stamens definite. Ovary composed of three to five, more or less united, few-ovuled carpels, with distinct styles or stigmas. Capsule or berry with usually a single seed in each cell. Embryo heart-shaped, minute, inclosed in the persistent embryo-sac, at the apex of the albumen! — *Ex.* Saururus (Lizard's-tail). Slightly pungent plants. They are scarcely distinct from the Pepper Family.*

872. **Ord. Ceratophyllaceæ** (*the Hornwort Family*) consists of the single genus Ceratophyllum (growing in ponds and streams in many parts of the world); distinguished by the whorled and dissected leaves with filiform segments; the flowers monœcious, and sessile in the axil of the leaves; the stamens indefinite, with sessile anthers; and the simple one-celled ovary, which forms a beaked achenium in fruit, containing an orthotropous suspended seed, with four cotyledons! and a manifest plumule.

873. **Ord. Callitrichaceæ** (*the Water-Starwort Family*), formed of

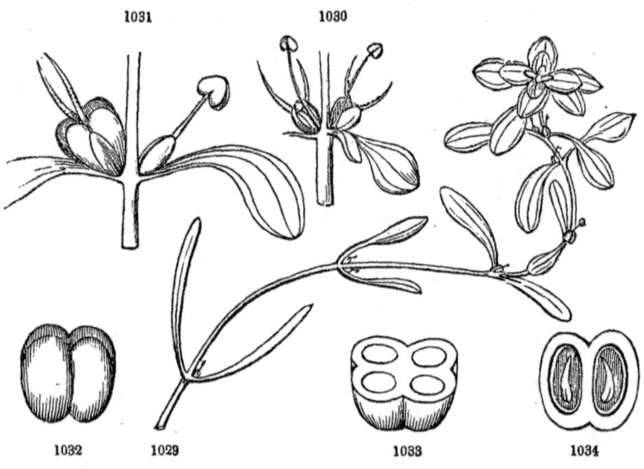

* Ord. PIPERACEÆ (*the Pepper Family*), a chiefly tropical order with the embryo inclosed in the persistent embryo-sac, differing from Saururaceæ principally in the one-celled simple ovary, with a solitary ovule (fruit a berry), and the extrorse anthers; the leaves often opposite or whorled; the jointed

FIG. 1029. Callitriche verna, about the natural size. 1030. Perfect flowers, magnified. 1031. A staminate and pistillate flower, magnified. 1032. The fruit. 1033. Cross-section of the fruit. 1034. Vertical section through the pericarp, seeds, and embryo.

the genus Callitriche; aquatic annuals, with opposite entire leaves; the axillary flowers (either perfect or monœcious) with a two-leaved involucre, but entirely destitute of calyx and corolla; stamen one (or rarely two), hypogynous, with a slender filament, and a reniform one-celled anther; the ovary four-lobed, four-celled, indehiscent in fruit; the seeds albuminous.

874. **Ord. Podostemaceæ** (*the River-weed Family*) comprises a few (American and Asiatic) aquatics, with the aspect of Mosses or Hepaticæ; their small flowers arising from a kind of spathe; the calyx often entirely wanting; the stamens frequently reduced to one, or two and monadelphous; the ovary two- or three-celled, with distinct styles; in fruit forming a ribbed capsule, containing numerous exalbuminous seeds attached to a central column. — *Ex.* Podostemum.

875. **Ord. Euphorbiaceæ** (*the Spurge Family*). Herbs, shrubs, or even trees, often with a milky juice: in northern temperate climes chiefly represented by the genus Euphorbia (Fig. 344 – 349); which is remarkable for having numerous staminate flowers, reduced to a single stamen (484), inclosed in an involucre along with one pistillate flower, reduced to a compound pistil, and also achlamydeous, or with an obsolete calyx. But other genera have a regular calyx both to the staminate and pistillate flowers; and a few are likewise provided with petals. Ovary of two to nine more or less united carpels, coherent to a central prolongation of the axis: styles distinct, often two-cleft. Fruit mostly capsular, separating into its elementary carpels, or cocci (usually leaving a persistent axis), which commonly open elastically by one or both sutures. Seed with a large embryo in fleshy albumen, suspended. — *Ex.* Euphorbia (Spurge), Croton, Buxus (the Box). Acrid and deleterious qualities pervade this large order, chiefly resident in the (usually) milky juice. But the starchy accumulations in the rhizoma, or underground portion of the stem, as in the Mandioc or Cassava (Janipha Manihot) of tropical America, are perfectly innocuous, when freed from the poisonous juice by washing

---

stems sometimes woody, but scarcely exhibiting annual layers. They all possess stimulant, aromatic, and pungent qualities, the common *Pepper* (the dried berries of the Indian Piper nigrum) representing the ordinary properties of the order. The intoxicating *Betel* of the Malays consists of the leaves of Piper Betle. The *Ava* of the Society and Sandwich Islands, from which an inebriating drink is made, is Piper methysticum.

and exposure to heat. The starch thus obtained is the *Cassava*, which, when granulated, forms the *Tapioca* of commerce. The farinaceous albumen of the seed is also innocent, and the fixed oil which it frequently contains is perfectly bland. But the oil procured by expression abounds in the juices of the embryo and integuments of the seed, and possesses more or less active properties. The seeds of Ricinus communis yield the *Castor oil:* and those of Croton Tiglium, and some other Indian species, yield the violently drastic *Croton oil* or *Oil of Tiglium.* Some plants of the family are most virulent poisons; as, for example, the Manchineal-tree of the West Indies (Hippomane Manicella), which is said even to destroy persons who sleep under its shade; and a drop of the juice falling upon the hand produces an instantaneous blister. The hairs of some species (such as Jatropha stimulosa) sting like Nettles. The hard and close-grained wood of the box is invaluable to the wood-engraver. The purple dye called *Turnsole* is derived from Crozophora tinctoria. Another most important product of this order is *caoutchouc*, which is yielded by various plants of different families; but the principal supply of the article (that of Para, Demarara, and Surinam) is furnished by the tree named Hevea Guianensis by Aublet, the Siphonia elastica of Persoon.

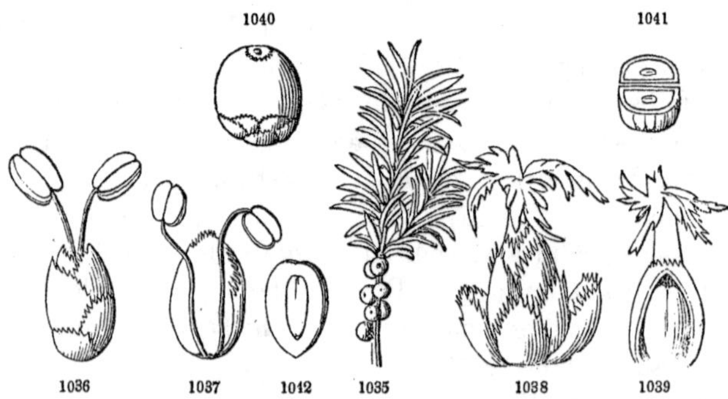

876. Ord. Empetraceæ (*the Crowberry Family*). Low, shrubby

FIG. 1035. Branch of Ceratiola ericoides in fruit. 1036. Magnified staminate flower, with its bracts. 1037. The two stamens, with an inner bract or sepal. 1038. Magnified pistillate flower, with its imbricated bracts. 1039. The pistil separate; one of the cells laid open by a vertical section, showing the erect ovule. 1040. Drupe, with the persistent scales at the base. 1041. Transverse section of its endocarp, or two nucules, with the inclosed seed and embryo. 1042. Vertical section of the seed.

evergreens, with the aspect of Heaths; the leaves crowded and acerose, with small (diœcious or polygamous) flowers produced in the axils of the uppermost. Calyx consisting of regular imbricated sepals, or represented by imbricated bracts. Stamens few: pollen of four grains coherent in one, as in Heaths. Ovary three- to nine-celled, with a single erect ovule in each cell: style short or none: stigmas lobed and often laciniated. Fruit a drupe, with from three to nine bony nucules. Seeds albuminous; the radicle inferior. — *Ex.* Empetrum, Ceratiola, Corema; unimportant plants.

877. **Ord. Juglandaceæ** (*the Walnut Family*). Trees, with alternate pinnated leaves, and no stipules. Flowers monœcious. Sterile flowers in aments, with a membranous irregular calyx, and indefinite stamens. Fertile flowers few, clustered, with the calyx adherent to the incompletely two- to four-celled but one-ovuled ovary, the limb small, three- to five-parted; sometimes with as many small petals. Ovule orthotropous. Fruit drupaceous; the epicarp fibrous-fleshy and coherent, or else coriaceous and dehiscent: endocarp bony. Seed four-lobed, without albumen. Embryo oily: cotyledons corrugate, 2-cleft. — *Ex.* Juglans (Walnut, Butternut), Carya (Hickory, Pecan, &c.). — The greater part of the order is North American. The timber is valuable; especially that of Black Walnut, for its rich dark-brown color when polished; that of Hickory, for its great elasticity and strength. The young fruit is acrid: the often edible seeds abound in a drying oil.

878. **Ord. Cupuliferæ** (*the Oak Family*). Trees or shrubs, with alternate and simple straight-veined leaves, and deciduous stipules. Flowers usually monœcious. Sterile flowers in aments, with a scale-like or regular calyx, and the stamens one to three times the number of its lobes. Fertile flowers solitary, two to three together, or in clusters, furnished with an involucre which incloses the fruit or forms a cupule at its base. Ovary adnate to the calyx, and crowned by its minute or obsolete limb, two- to six-celled with one or two pendulous ovules in each cell: but the fruit is a one-celled and one-seeded nut (585). Seed without albumen. Embryo with thick and fleshy cotyledons, which are sometimes coalescent. — *Ex.* Quercus (the Oak), Fagus (the Beech), Corylus (the Hazelnut), Castanea (the Chestnut), &c. Some of the principal forest-trees in northern temperate regions. Their valuable timber and

edible seeds are too well known to need enumeration. The astringent bark and leaves of the Oak abound in tannin, gallic acid, and a bitter extractive called *Quercine ;* they are used in tanning and dyeing. *Quercitron* is obtained from the Quercus tinctoria. *Galls* are swellings on the leafstalks, &c., when wounded by certain insects ; those of commerce are derived from Q. infectoria of Asia Minor. *Cork* is the exterior bark of the Spanish Quercus Suber.

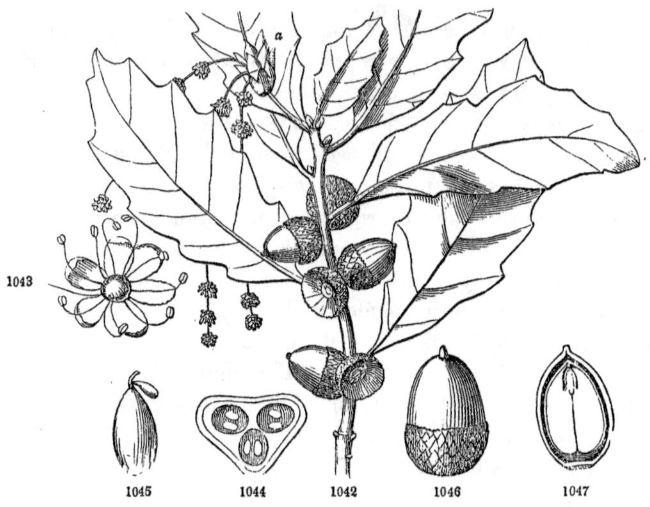

879. **Ord. Myricaceæ** (*the Sweet-Gale Family*). Shrubs, with alternate and simple aromatic leaves, dotted with resinous glands ; monœcious or diœcious. Differs from the next principally by the one-celled ovary, with a single erect orthotropous ovule, and a drupe-like nut. — *Ex.* Myrica, Comptonia, the Sweet Fern. The drupes of M. cerifera (our Candleberry) yield a natural wax.

880. **Ord. Betulaceæ** (*the Birch Family*). Trees or shrubs, with alternate and simple straight-veined leaves, and deciduous stipules. Flowers monœcious ; those of both kinds in aments and commonly achlamydeous, placed three together in the axil of each three-lobed

FIG. 1042. Quercus Chinquapin in fruit : *a*, cluster of sterile aments. 1043. A magnified staminate flower. 1044. Transverse section of an ovary, showing the three cells with two ovules in each. 1045. The immature seed, with the accompanying abortive ovule. 1046. The nut (acorn), in its scaly involucre, or cupule. 1047. Vertical section of the same, and of the included seed and embryo, showing the thick cotyledons.

bract. Stamens definite. Ovary two-celled, each cell with one suspended ovule: styles or stigmas distinct. Fruit membranaceous or samara-like, one-celled and one-seeded, forming with the three-lobed bracts a kind of strobile. Albumen none. — *Ex.* Betula (the Birch), Alnus (Alder). The bark is sometimes astringent, and that of the Birch is aromatic. The peculiar odor of Russia leather is said to be owing to a pyroligneous oil obtained from Betula alba.

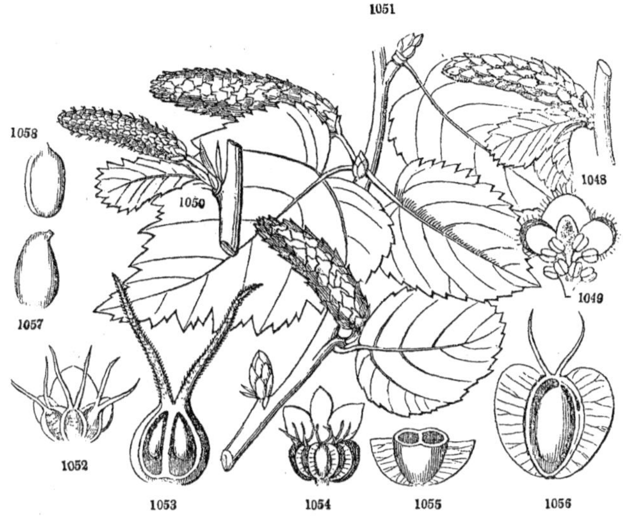

881. **Ord. Salicaceæ** (*the Willow Family*). Trees or shrubs, with alternate simple leaves, furnished with stipules. Flowers diœcious; both kinds in aments, and destitute of floral envelopes (achlamydeous), one under each bract. Stamens two to several, sometimes monadelphous. Ovary one-celled, many-ovuled! Styles or stigmas two, often two-cleft. Fruit a kind of follicle opening by two valves. Seeds numerous, ascending, furnished with a silky

FIG. 1048. Ament of staminate flowers of Betula fruticosa? 1049. One of the three-lobed scales of the same, enlarged, showing the flowers (stamens) on the inner side. 1050. Ament of pistillate flowers. 1051. Branch in fruit. 1052. One of the scales with its three flowers (pistils) seen from within. 1053. Magnified section of one of the two-celled pistils, displaying the ovule suspended from the summit of each cell. 1054. The pistils (with their subtending bract) in a more advanced state. 1055. Magnified cross-section of one of the ovaries. 1056. The mature fruit, with the cell divided vertically; the single seed occupying the cavity; a mere trace of the other cell being visible. 1057. The seed removed. 1058. The embryo.

coma! Albumen none. — *Ex.* Salix (Willow, already illustrated, 473, Fig. 326 - 329), and Populus (the Poplar). Trees with light and soft wood: the slender flexible shoots of several Willows are employed for wicker-work. The bark is bitter and tonic; containing a peculiar substance (*Salicine*), which possesses febrifugal qualities. The buds of several Poplars exude a fragrant balsamic resin.

882. **Ord. Balsamifluæ** (*the Sweet-Gum Family*) consists of a single genus of three or four species (natives of Eastern India, the Levant, and North America): which are trees, with alternate palmately-lobed leaves, and deciduous stipules; the monœcious flowers in rounded aments or heads, destitute of floral envelopes; the indurated capsules and scales forming a kind of strobile; the former two-celled, two-beaked, opening between the beaks, several-seeded: the seeds with a little albumen. It has recently been referred to the order Hamamelaceæ (799). — *Ex.* Liquidambar, or Sweet-Gum: so called from the fragrant balsam or *Storax* it exudes.

883. **Ord. Platanaceæ** (*the Plane-tree Family*) consists of the single genus Platanus (Plane-tree, Button-ball), with one Asiatic and one or more North American species: which are fine trees, with a watery juice, and alternate palmately-lobed leaves, with sheathing stipules. Flowers in globose amentaceous heads; both kinds destitute of floral envelopes. Fruit a one-seeded club-shaped little nut, the base furnished with bristly hairs. Seed albuminous.

884. **Ord. Urticaceæ** (*the Nettle Family*). Trees or shrubs with milky juice, or herbs with a watery juice. Leaves often stipulate. Flowers monœcious, diœcious, or polygamous, sometimes collected in aments or fleshy heads, furnished with a regular calyx. Stamens definite. Ovary free from the calyx, simple, with a solitary ovule. Fruit an achenium or utricle, often inclosed in a fleshy or baccate calyx. The order comprises the following principal divisions, viz.: —

885. **Subord. Artocarpeæ** (*the Bread-fruit Family*); which are trees or shrubs with a milky or yellow juice; the flowers mostly aggregated into fleshy heads, and forming a compound baccate fruit, or else inclosed in a dry or succulent involucre. Albumen none. — *Ex.* Artocarpus (the Bread-fruit), Antiaris (Upas): all tropical.

886. **Subord. Moreæ** (*the Mulberry Family*); which are shrubs or

trees, very rarely herbs, with a milky juice; the staminate and pistillate flowers either in separate aments or spikes, or often intermixed and included in the same hollow and closed fleshy receptacle (as in the Fig): the calyx, &c. becoming succulent, and forming a compound fruit. Seeds albuminous. — *Ex.* Morus (the Mulberry, Fig. 244 – 246), Maclura (the Osage Orange), Ficus (the Fig, Fig. 241 – 243): nearly all tropical.

887. **Subord. Urticeæ** (*the proper Nettle Family*); which are herbs in colder countries, but often shrubs or trees in the tropics, with a watery juice, often with stinging hairs; the flowers mostly loose, spicate, or panicled; the achenium usually surrounded by a dry and membranous calyx. Embryo straight, in fleshy albumen. — *Ex.* Urtica (the Nettle), Parietaria.

888. **Subord. Cannabineæ** (*the Hemp Family*); which are annual erect herbs, or perennial twining plants, with a watery juice; the staminate flowers racemose or panicled; the pistillate glomerate, or imbricated with bracts, and forming a kind of strobile-like ament. Embryo curved: albumen none. — *Ex.* Cannabis (the Hemp), Humulus (the Hop): natives of northern temperate regions.

889. The fruit in this large and polymorphous family is mostly innocent and edible, at least when cooked; while the milky juice is more or less acrid or deleterious. It also abounds in *caoutchouc;* much of which is obtained from some South American trees of this order, and from Ficus elastica in Java. In one instance, however, the milky juice is perfectly innocent; that of the famous Cow-tree of South America, which yields copiously a rich and wholesome milk. One of the most virulent of poisons, the *Bohon Upas*, is the concrete juice of Antiaris toxicaria of the Indian Archipelago. The Bread-fruit is the fleshy receptacle and multiple fruit of Artocarpus. *Fustic* is the wood of the South American Morus tinctoria. The resin called *Gum Lac* exudes and forms small grains on the branches of the celebrated Banyan-tree (Ficus Indica, Fig. 119). Nettles are remarkable for their stinging venomous hairs, and tough fibres of the bark, which, as in those of Hemp, are used for cordage. The leaves of the Hemp are stimulant and narcotic, and are used extensively in the East for intoxication. *Hops* are the catkins of Humulus Lupulus; the bitter and sedative principle chiefly resides in the yellow grains that cohere to the scales and cover the fruit.

Subclass 2. Gymnospermous Exogenous Plants.

890. Ovules, and consequently the seeds, naked, that is, not inclosed in an ovary (560); the carpel being represented either by an open scale, as in Pines; or by a more evident leaf, as in Cycas; or else wanting altogether, as in the Yew.

891. **Ord. Coniferæ** (*the Pine Family*). Trees or shrubs, with branching trunks, abounding in resinous juice (the wood chiefly consisting of a tissue somewhat intermediate between ordinary woody fibre and vessels, which is marked with circular discs); the leaves mostly evergreen, scattered or fascicled, usually rigid and needle-shaped or linear, entire. Flowers monœcious or diœcious, commonly amentaceous. Staminate flowers consisting of one or more (often monadelphous) stamens, destitute of calyx or corolla, arranged on a common rhachis so as to form a kind of loose ament. — The particular structure of the flowers and fruit varies in the subordinate groups chiefly as follows: —

892. **Subord. Abietineæ** (*the Fir, or proper Pine Family*). Fertile aments formed of imbricated scales; which are the flat and open carpels, and bear a pair of ovules adherent to their base, with the foramen turned downwards. Scales subtended by bracts. Fruit a strobile or cone (619). Integument of the seed coriaceous or woody, more or less firmly adherent to the scale. Embryo in the axis of fleshy albumen, with two to fifteen cotyledons. (Illustrated in Fig. 391 – 401, pp. 306, 307.)

893. **Subord. Cupressineæ** (*the Cypress Family*). Fertile aments of few scales crowded on a short axis, or more numerous and peltate (Fig. 402), not bracteate. Ovules one, two, or several, borne on the base of the scale, erect (the foramen looking towards its apex, Fig. 394). Fruit an indurated strobile, or fleshy and with the scales concreted, forming a kind of drupe. Integument of the seed membranous or bony. Cotyledons two or more. Anthers of several parallel cells, placed under a shield-like connectivum. — *Ex.* Cupressus (Cypress), Taxodium (American Cypress), Juniperus (Juniper, Red Cedar).

894. **Subord. Taxineæ** (*the Yew Family*). Fertile flowers solitary, terminal, consisting merely of an ovule, forming a drupaceous seed at maturity. There are, therefore, no strobiles and no carpellary scales. Embryo with two cotyledons. — *Ex.* Taxus (the Yew), Torreya.

895. It is unnecessary to specify the important uses of this large and characteristic family, which comprises the most important timber-trees of cold countries, and also furnishes resinous products of great importance, such as *turpentine*, *resin*, *pitch*, *tar*, *Canada balsam* (obtained from the Balsam Fir), &c. The terebinthine *Juniper-berries* are the fruit of Juniperus communis. The Larch yields *Venetian turpentine*. The powerful and rubefacient *Oil of Savin* is derived from J. Sabina of Europe: for which our J. Virginiana (Red Cedar) may be substituted. The leaves of the Yew are narcotic and deleterious. The bark of Hemlock and Larch is used for tanning.

896. **Ord. Cycadaceæ** (*the Cycas Family*). Tropical plants, with an unbranched cylindrical trunk, increasing, like Palms, by a single terminal bud; the leaves pinnate and their segments rolled up from the apex (circinate) in vernation, in the manner of true Ferns. Flowers diœcious; the staminate in a strobile or cone; the pistillate also in strobiles, or else (in Cycas) occupying contracted and partly metamorphosed leaves; the naked ovules borne on its margins. — *Ex.* Cycas, *Zamia*, the dwarf Florida species of which is illustrated in Fig. 403 - 409, p. 308. — A kind of *Arrowroot* is obtained from these thickened stems; and a sort of *Sago* from the trunk of Cycas.

Class II. Endogenous or Monocotyledonous Plants.

897. Stem not distinguishable into bark, pith, and wood; but the latter consisting of bundles of fibres and vessels irregularly imbedded in cellular tissue; the rind firmly adherent; no medullary rays, and no appearance of concentric layers: increase in diameter effected by the deposition of new fibrous bundles, which, at their commencement at least, occupy the central part of the stem. Leaves seldom falling off by an articulation, commonly sheathing at the base, usually alternate, entire, and with simple parallel veins (nerved). Floral envelopes when present mostly in threes; the calyx and corolla frequently undistinguishable in texture and appearance. Embryo with a single cotyledon; or if the second is present, it is much smaller than the other and alternate with it (634).

CONSPECTUS OF THE ORDERS.

Group 1. Flowers on a spadix, furnished with a double perianth (calyx and corolla). Ovary one- to three-celled, with a single ovule in each cell. Embryo in hard albumen. — Trees with unbranched columnar trunks.
PALMÆ, p. 487.

Group 2. Flowers on a spadix; with the perianth simple, scale-like, or commonly altogether wanting. — Chiefly herbs.

Terrestrial, mostly with a spathe. Fruit baccate. ARACEÆ, p. 488.
Terrestrial. Fruit nut-like, one-seeded. TYPHACEÆ, p. 489.
Aquatic (floating or immersed).
Flowers from the edge of the floating frond. LEMNACEÆ, p. 489.
Flowers axillary or on a spadix. NAIADACEÆ, p. 490.

Group 3. Flowers not spadiceous, furnished with a double perianth (calyx and corolla). Ovaries several, distinct, or sometimes united, free. — Aquatic herbs. ALISMACEÆ, p. 490.

Group 4. Flowers with a simple or double perianth, adherent to the ovary (ovary inferior), either completely or partially. — Herbs.

* Perianth regular. Ovary one-celled, with parietal placentæ, or rarely three- to six-celled, with the placentæ in the axis.
Diœcious or polygamous; aquatic. HYDROCHARIDACEÆ, p. 491.
Flowers perfect; terrestrial. BURMANNIACEÆ, p. 491.

* * Perianth irregular. Ovary one-celled, with parietal placentæ. Stamens one or two, adherent to the style (gynandrous). ORCHIDACEÆ, p. 491.

* * * Perianth irregular. Ovary three-celled. Perfect stamens usually one.
Fertile stamen 1, inferior. ZINGIBERACEÆ, p. 492.
Fertile stamen 1, superior. CANNACEÆ, p. 493.
Fertile stamens mostly 5, the sixth abortive. MUSACEÆ, p. 493.

* * * * Perianth regular, or sometimes a little irregular. Ovary three-celled, many-ovuled (in Tillandsia free, in Lophiola nearly so). Stamens either three or six.
Anthers introrse. Stamens mostly 6.
Bulbous. AMARYLLIDACEÆ, p. 494.
Not bulbous: root fibrous: leaves indurated or scurfy. BROMELIACEÆ, p. 493,
and HŒMODORACEÆ, p. 493.
Anthers extrorse. Stamens 3. IRIDACEÆ, p. 494.

* * * * * Perianth regular. Ovary three-celled, with one or two ovules in each. Flowers diœcious. Stamens six. DIOSCOREACEÆ, p 495.

Group 5. Flowers with a regular perianth, which is more or less petaloid (the two series when present are similar), or rarely glumaceous, and free from the ovary. Embryo inclosed in albumen.

Perianth not glumaceous.
Anthers introrse. Styles or stigmas separate. SMILACEÆ, p. 495.

Anthers introrse. Styles united into one.
Terrestrial, not spathaceous. Flower regular. LILIACEÆ, p. 495.
Aquatic, spathaceous. Flower oftener irregular. PONTEDERIACEÆ, p. 496.
Anthers extrorse (except Tofieldia). MELANTHACEÆ, p. 496.
Perianth glumaceous. JUNCACEÆ, p. 497.

Group 6. Flowers with a double or imbricated perianth : the exterior herbaceous or glumaceous ; the inner petaloid, free from the one- to three-celled ovary. Seeds orthotropous; the embryo at the extremity of the albumen farthest from the hilum.

Flowers perfect. Sepals herbaceous. COMMELYNACEÆ, p. 498.
Flowers perfect, capitate. Sepals and bracts glumaceous. XYRIDACEÆ, p. 498.
Flowers monœcious or diœcious, capitate. ERIOCAULONACEÆ, p. 498.

Group 7. Flowers imbricated with bracts (glumes) and disposed in spikelets; the proper perianth none or rudimentary. Ovary one-celled, one-ovuled. Embryo at the extremity of the albumen next the hilum.

Sheaths closed. Glume or bract single. CYPERACEÆ, p. 498.
Sheaths open. Glumes in pairs. GRAMINEÆ, p. 499.

898. **Ord. Palmæ** (*Palms*). Chiefly trees, with unbranched cylindrical trunks growing by a terminal bud. Leaves large, clustered, fan-shaped or pinnated, plaited in vernation. Flowers small, perfect or polygamous, mostly with a double (6-merous) perianth; the stamens usually as many as the petals and sepals together. Ovary 1 – 3-celled, with a single ovule in each cell. Fruit a drupe or berry. Seeds with a cartilaginous albumen, often hollow; the embryo placed in a small separate cavity. — *Ex.* Palms, the most majestic race of plants within the tropics, and of the highest value to mankind, are scarcely found beyond the limits of these favored regions. The Date-tree (Phœnix dactylifera, the leaves of which are the *Palms* of Scripture), a native of Northern Africa, endures the climate of the opposite shores of the Mediterranean : while in the New World, Chamærops Palmetto (Fig. 166), the only arborescent species of the United States, and one or two low Palms with a creeping caudex (Dwarf Palmettos), extend from Florida to North Carolina. Palms afford food and raiment, wine, oil, wax, flour, sugar, salt, thread, weapons, utensils, and habitaotins. The Cocoa-nut (Cocos nucifera) is perhaps the most important, as well as the most widely diffused species. Besides its well-known fruit, and the beverage it contains, the hard trunks are employed in the construction of huts; the terminal bud (as in our Palmetto and other Cabbage Palms) is a delicious article of food ; the leaves are used for thatching, for making hats, baskets, mats, fences, for

torches, and for writing upon; the stalk and midrib for oars; their ashes yield abundance of Potash; the juice of the flowers and stems (replete with sugar, which is sometimes separated under the name of *Jagery*) is fermented into a kind of wine, or distilled into *Arrack;* from its spathes (as from some other Palms), when wounded, flows a grateful laxative beverage, known in India by the name of *Toddy;* the rind of the fruit is used for culinary vessels; its tough, fibrous, outer portion is made into very strong cordage (*Coir rope*); and an excellent fixed oil is copiously expressed from the kernel. *Sago* is procured from the trunks of many Palms, but chiefly from species of Sagus of Eastern India. *Canes* and *Rattans* are the slender, often prostrate, stems of species of Calamus. The Phytelephas of South America yields the larger sort of nuts, the hard and white albumen of which is the *vegetable ivory*, now so largely used by the turner.

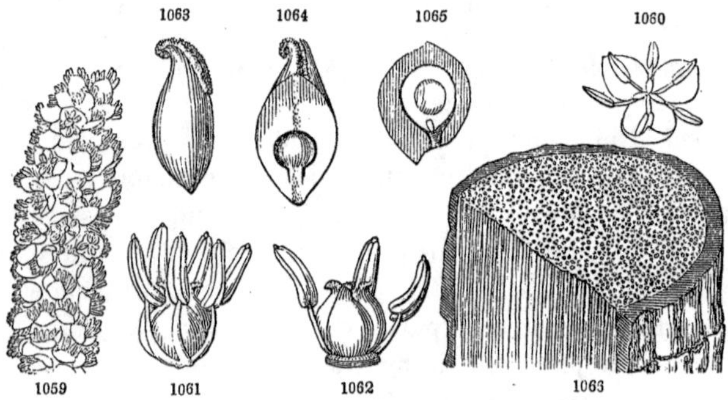

899. **Ord. Araceæ** (*the Arum Family*). Herbs, with a fleshy corm or rhizoma, occasionally shrubby or climbing plants in the tropics; the leaves sometimes compound or divided, frequently with more or less reticulated veins. Flowers mostly on a spadix (often naked at the extremity) usually surrounded by a spathe. Flowers commonly monœcious, and destitute of envelopes, or with a single perianth. Ovary one- to several-celled, with one or more

FIG. 1059. Branch of the inflorescence of Chamærops Hystrix (Blue Palmetto). 1060. A sterile flower. 1061. Perfect flower, with the calyx and corolla removed. 1062. Same, with three of the stamens removed, so as more distinctly to show the three somewhat united carpels. 1063. One of the carpels enlarged, seen laterally. 1064 Same, with a section of its inner face, showing the ovule or young seed. 1065. Vertical section of a young cocoa-nut, showing the hollow albumen; and also the small embryo in a separate little cavity. 1066. Section of a Palm-stem.

ovules. Fruit a berry. Seeds with or without albumen. — *Ex.* Arum, Calla, Symplocarpus (Skunk-Cabbage), Orontium, Acorus (Sweet Flag): the three latter bear flowers furnished with a perianth. — All are endowed with an acrid volatile principle, which is merely pungent and aromatic in *Sweet Flag* (Acorus Calamus).

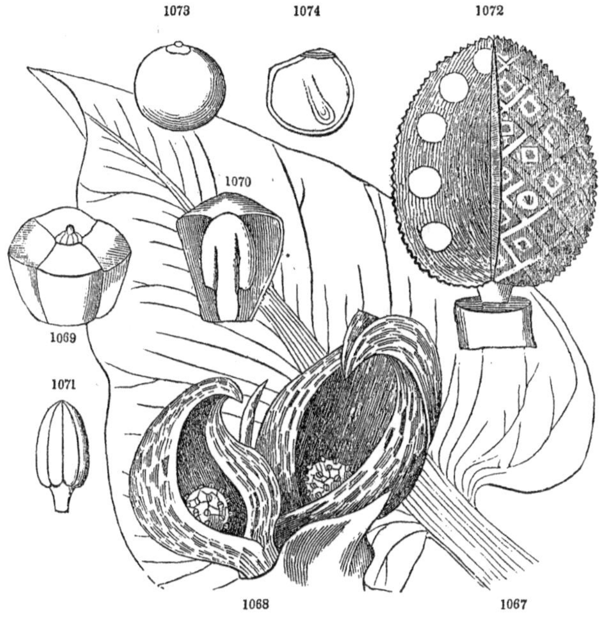

900. **Ord. Typhaceæ** (*the Cat-tail Family*) consists of two genera; namely, Typha (the Cat-tail), and Sparganium (Burr-reed), of no important use; they are somewhat intermediate between Araceæ and Cyperaceæ.

901. **Ord. Lemnaceæ** (*the Duck-weed Family*), consisting chiefly of Lemna (Duckweed, or Water Flax-seed); floating plants, with their roots arising from the bottom of a flat frond, and hanging loose in the water; their flowers produced from the margin of the frond, bursting through a membranous spathe; the sterile, of one

FIG. 1067. Young leaf, and 1068, spathes and flowers, of Symplocarpus fœtida. 1069. A separate flower. 1070. A sepal and stamen seen from within. 1071. An anther seen from the front. 1072. The spadix or collective head in fruit; a quarter-section removed, showing sections of the immersed seeds. 1073. A seed detached, of the natural size. 1074. Section of the seed, with its large globular embryo and plumule: in this plant there is no albumen.

or two stamens; the fertile, of a one-celled ovary; in fruit a utricle: they are a kind of minute and greatly reduced Araceæ.

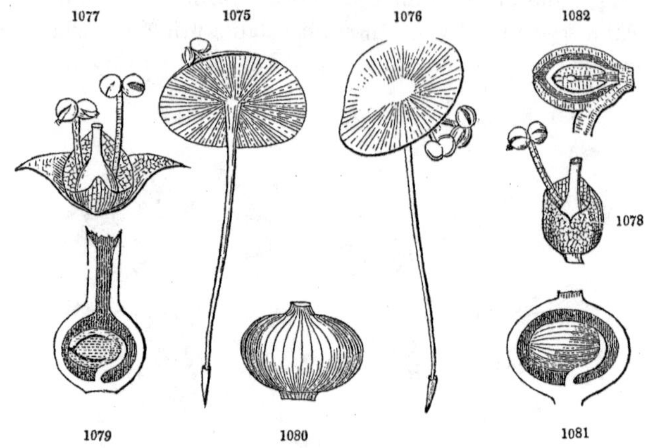

902. **Ord. Naiadaceæ** (*the Pond-weed Family*). Water-plants, with cellular leaves, and sheathing stipules or bases: the flowers inconspicuous, sometimes perfect. Perianth simple or none. Stamens definite. Ovaries solitary, or two to four and distinct, one-seeded. Albumen none. Embryo straight or curved. — *Ex.* Potamogeton (Pond-weed), Najas, Ruppia, Zostera; the two latter in salt or brackish water.

903. **Ord. Alismaceæ** (*the Water-Plantain Family*). Marsh herbs, with the leaves and scapes usually arising from a creeping rhizoma; the former either linear, or bearing a flat limb, which is ribbed or nerved, but the veinlets commonly reticulated. Flowers regular, perfect or polygamous, mostly in racemes or panicles, not on a spadix. Perianth double. Sepals three. Petals three. Seeds solitary in each carpel or cell, straight or curved, destitute of albumen. — *Ex.* Alisma (Water-Plantain), Sagittaria (Arrowhead); belonging to the proper Alisma Family, which has the seed (and consequently the embryo) curved or doubled upon itself. Triglochin and Scheuchzeria chiefly constitute the suborder JUNCAGINEÆ;

FIG. 1075. Whole plant of Lemna minor, magnified, bearing a staminate monandrous flower. 1076. An individual with a diandrous perfect flower; which at 1077 is seen separate, with its spathe, highly magnified. 1078. Flower of Lemna gibba, much magnified. 1079. Vertical highly magnified section of the pistil and the contained ovule of Lemna minor. 1080. The fruit, and 1081, its section, showing the seed. 1082. Section through the highly magnified seed and large embryo.

where the seed and embryo are straight, and the petals (if present) greenish like the calyx.

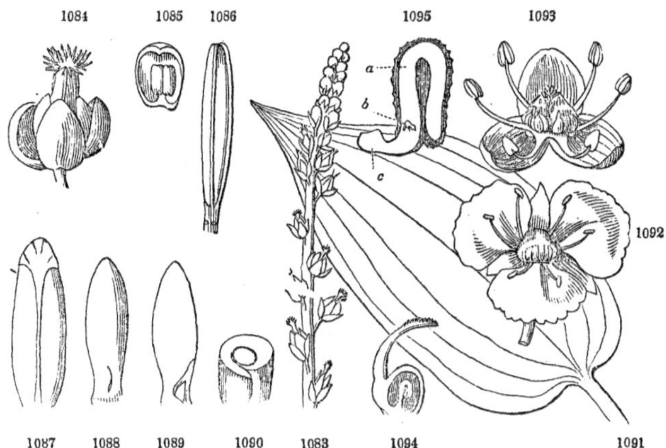

904. **Ord. Hydrocharidaceæ** (*the Frog's-bit Family*) consists of a few aquatic herbs, with diœcious or polygamous regular flowers on scape-like peduncles from a spathe, and simple or double floral envelopes, which in the fertile flowers are united in a tube, and adnate to the 1–6-celled ovary, more commonly one-celled with three parietal placentæ. Seeds numerous, without albumen. — *Ex.* Limnobium, Vallisneria, Udora.*

905. **Ord. Burmanniaceæ** consists of small, mostly tropical, annual herbs, differing from Orchidaceæ by their regular and perfect flowers with three stamens. — *Ex.* Burmannia, Apteria, of the Southern States.

906. **Ord. Orchidaceæ** (*the Orchis Family*). Herbs, of varied aspect and form; distinguished from the other orders with an adnate

* Ord. BUTOMACEÆ consists of Butomus, Hydrocleis, &c.: plants resembling the Alisma tribe, but with a milky juice, and the numerous seeds attached to the whole inner surface of the carpels!

FIG. 1083. Raceme or spike of Triglochin palustre. 1084. Enlarged flower. 1085. A petal and stamen. 1086. The club-shaped capsule. 1087. A magnified seed, exhibiting the rhaphe and chalaza. 1088. Embryo of the same, showing the lateral slit just above the radicular end (634, where this structure is explained). 1089. Vertical section of the same passing through the slit, bringing the plumule to view. 1090. Cross-section (more magnified), showing the cotyledon wrapped around the plumule.

FIG. 1091. Leaf, and 1092. flower, of Alisma Plantago. 1093. More enlarged flower, with the petals removed. 1094. Carpel, with the ovary divided, showing the double ovule. 1095. Vertical section of the germinating seed of Alisma Damasonium: *a*, the cotyledon; *b*, the plumule; *c*, the protruding radicle.

ovary, and from all other plants, by their irregular flowers, with a perianth of six parts ; their single fertile stamen (or in Cypripedium their two stamens) coherent with the style (composing the *column*) ; their pollen usually combined into two or more compact or regular masses (*pollinia*), or of the consistence and appearance of wax : the ovary one-celled, with three parietal placentæ, covered with numerous small seeds. — *Ex.* Orchis, Cypripedium (Ladies'-Slipper), Arethusa, &c. In the tropics many are Epiphytes (132, Fig. 120). Many are cultivated for their beauty and singularity. The tuberiferous roots are often filled with a very dense mucilaginous or glutinous substance (as those of our Aplectrum, thence called Putty-root). Of this nature is the Salep of commerce, the produce of some unascertained species of Middle Asia. The fragrant *Vanilla* is the fleshy fruit of the West Indian Vanilla claviculata.

907. **Ord. Zingiberaceæ** (*the Ginger Family*) consists of some tropical aromatic herbs, the nerves of their leaves diverging from

FIG. 1096. Orchis spectabilis: *a*, a separate flower. 1097. Column (somewhat magnified), from which the other parts are cut away : the two anther-cells opening and showing the pollen-masses. 1098. Magnified pollen-mass, with its stalk. 1099. Arethusa bulbosa. 1100. The column, enlarged : the anther terminal and opening by a lid. 1101. Magnified anther, with the lid removed, showing the two pollen-masses in each cell.

a midrib; the adnate perianth irregular and triple (having a corolla of two series as well as a calyx); fertile stamen one, on the anterior side of the flower, free; the fruit a three-celled capsule or berry; the seeds several: with the embryo in a little sac at one extremity of the farinaceous albumen. — There are, in fact, six stamens in the andrœcium, the three exterior petaloid and forming the so-called inner corolla, and two of the inner verticil are sterile. — *Ex.* Zingiber (Ginger), Amomum (Cardamon). Stimulant and aromatic. Some afford a coloring matter (*Turmeric*). They are all showy plants.

908. **Ord. Cannaceæ** (*the Arrowroot Family*), which are equally tropical plants, differ from the preceding chiefly in the want of aroma, and in having the single fertile stamen posterior, with a one-celled anther. — *Ex.* Maranta arundinacea (the *Arrowroot*) of the West Indies; the tubers of which are filled with pure starch.

909. **Ord. Musaceæ** (*the Banana Family*). Tropical plants, of which the Banana and Plantain are the type; distinguished by their simple perianth and five or six perfect stamens. The fruit is most important in the tropics; the gigantic leaves are used in thatching; and the fibres of Musa textilis yield Manilla hemp, as well as a finer fibre from which a delicate linen is made.

910. **Ord. Bromeliaceæ** (*the Pine-Apple Family*) consists of American and chiefly tropical plants; with rigid and dry channelled leaves, often with a scurfy surface, a mostly adnate perianth of three sepals and three petals, and six or more stamens; the seeds with mealy albumen. — *Ex.* Ananassa, the Pine-Apple; the fine fruit of which is formed by the consolidation of the imperfect flowers, bracts, and receptacle into a fleshy, succulent mass. Tillandsia, the Black Moss or Long Moss, which, like most Bromelias, grows on the trunks and branches of trees in the warmer and humid parts of America, has the ovary free from the perianth.

911. **Ord. Hæmodoraceæ** (*the Bloodwort Family*) is composed of perennial herbs, with fibrous roots, equitant or ensiform leaves; which, with the stems and flowers, are commonly densely clothed with woolly hairs or scurf. Perianth with the tube either nearly free from, or commonly adherent to, the three-celled ovary; the limb six-cleft, regular. Stamens six, or only three, with introrse anthers. Style single, the stigma standing over the dissepiments of the ovary. Embryo in cartilaginous albumen. — *Ex.* Lachnanthes (Red-Root), Lophiola.

912. **Ord. Amaryllidaceæ** (*the Amaryllis Family*). Bulbous plants (sometimes with fibrous roots), bearing showy flowers mostly on scapes. Perianth regular, or nearly so; the tube adherent to the ovary, and often produced above it, six-parted. Stamens six, distinct, with introrse anthers. Stigma undivided or three-lobed. Fruit a three-celled capsule or berry. Seeds with fleshy albumen. — *Ex.* Amaryllis, Narcissus, Crinum, &c. The bulbs acrid, emetic, &c.: those of Hæmanthus (with whose juice the Hottentots poison their arrows) are extremely venomous. The fermented juice of Agave is the intoxicating *Pulque* of the Mexicans.

913. **Ord. Iridaceæ** (*the Iris Family*). Perennial herbs; the flower-stems springing from bulbs, corms, or rhizomas, rarely with fibrous roots, mostly with equitant leaves. Flowers regular or irregular, showy, often springing from a spathe. Perianth with the tube adherent to the three-celled ovary, and usually elongated

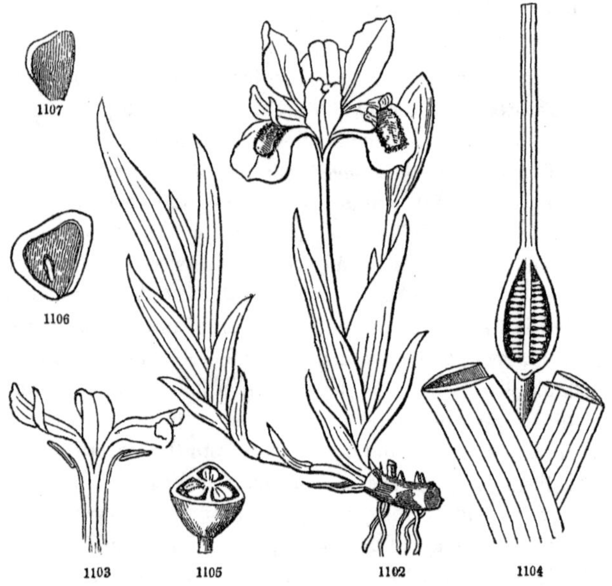

above it; the limb six-parted, in two series. Stamens three, dis-

FIG. 1102. Iris cristata. 1103. The summit of the style, petaloid stigmas, and stamens. 1104. Vertical section of the ovary (the equitant leaves cut away) and long tube of the perianth. 1105. Cross-section of the pod. 1107. Seed. 1106. Enlarged section of the same, showing the embryo, &c.

tinct or monadelphous; the anthers extrorse! Stigmas three, dilated or petaloid! Seeds with hard albumen. — *Ex.* Iris, Crocus. The rootstocks, corms, &c. contain starch, with some volatile acrid matter. *Orris-root* is the dried rhizoma of Iris florentina, of Southern Europe. *Saffron* is the dried orange stigmas of Crocus sativus.

914. **Ord. Dioscoreaceæ** (*the Yam Family*) consists of a few twining plants, with large tuberous roots or knotted rootstocks; distinguished by their ribbed and netted-veined leaves, with distinct petioles, and by their inconspicuous diœcious flowers. Perianth in the pistillate flowers adherent to the ovary; the limb six-cleft in two series. Stamens six. Ovary three-celled, with only one or two ovules in each cell: styles nearly distinct. Fruit often a three-winged capsule. Albumen cartilaginous. — *Ex.* Dioscorea. The tubers of one or more species, filled with starch and mucilage (but more or less acrid until cooked), are *Yams*, an important article of food in tropical countries.

915. **Ord. Smilaceæ** (*the Smilax Family*). Herbs or shrubby plants, often climbing, with the veins or veinlets of the leaves reticulated. Flowers perfect or diœcious. Perianth six-parted or double, the three sepals green, and the three petals colored. Stamens six: anthers introrse. Cells of the ovary and distinct styles or stigmas three. Berry few- or many-seeded. Albumen hard. — *Ex.* Smilax (Greenbrier, Catbrier, &c.). The *Sarsaparilla* of the shops consists of the roots of numerous species of Smilax, chiefly of tropical America. Trillium is the type of the suborder TRILLIACEÆ.

916. **Ord. Liliaceæ** (*the Lily Family*). Herbs, with the flower-stems springing from bulbs, tubers, or with fibrous or fascicled roots. Leaves simple, sheathing or clasping at the base. Flowers regular, perfect. Perianth colored, mostly of six parts, or six-cleft. Stamens six: anthers introrse. Ovary free, three-celled; the styles united: stigma often three-lobed. Fruit capsular or fleshy, with several or numerous seeds in each cell. Albumen fleshy. — *Ex.* This large and widely diffused order comprises a great variety of forms: the Lily and Tulip represent one division; the Polianthes (Tuberose), a second; the Aloe and Yucca, a third; the Hyacinth, the Onion, &c. (Allium), the Asphodel, Asparagus, &c., a fourth. Acrid and often bitter principles prevail in the order, and are most concentrated in the bulbs, &c., which abound in

starchy or mucilaginous matter, and are often edible when cooked. *Squills* are the bulbs of Scilla maritima of the South of Europe. *Aloes* is yielded by the succulent leaves of species of Aloe. The original *Dragon's-blood* was derived from the juice of the famous Dragon-tree (Dracæna Draco) of the East.

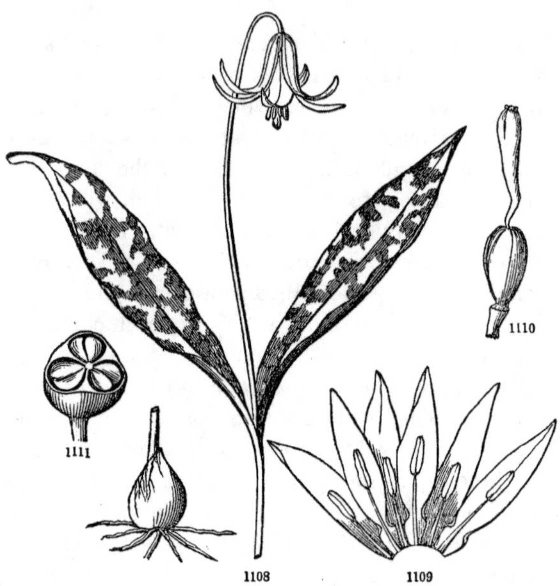

917. **Ord. Pontederiaceæ** (*the Pickerel-weed Family*) comprises a few aquatic plants, with the flowers, either solitary or spicate, arising from a spathe or from a fissure of the petiole; the six-cleft perianth persistent and withering, often adherent to the base of the three-celled ovary; the stamens three, and inserted on the throat of the perianth, or six, and unequal in situation. Ovules numerous; but the fruit often one-celled and one-seeded. — *Ex.* Pontederia (Pickerel-weed), Heteranthera, &c.

918. **Ord. Melanthaceæ** (*the Colchicum Family*). Herbs, with bulbs, corms, or fasciculated roots. Perianth regular, in a double series; the sepals and petals either distinct, or united below into a tube. Stamens six; the anthers extrorse (except in Tofieldia).

FIG. 1108. Erythronium Americanum (Dog-tooth Violet, Adder's-tongue). 1109. Perianth laid open, with the stamens. 1110. The pistil. 1111. Cross-section of the capsule.

Ovary free, three-celled, several-seeded: styles distinct. Albumen fleshy. The true Melanthaceæ, or

919. **Subord. Melanthieæ** have a mostly septicidal capsule and a marcescent or persistent perianth. — *Ex.* Colchicum has a perianth with a long tube, arising from a subterranean ovary; it is also remarkable for flowering in the autumn, when it is leafless, ripening its fruit and producing its leaves the following spring. In most of the order, the leaves of the perianth are uncombined; as in Veratrum (White Hellebore), Helonias, &c. Acrid and drastic poisonous plants, with more or less narcotic qualities; chiefly due to a peculiar alkaloid principle, named *Veratria*, which is largely extracted from the seeds of Sabadilla, or Cebadilla; the produce of Schœnocaulon officinale, &c., of the Mexican Andes.

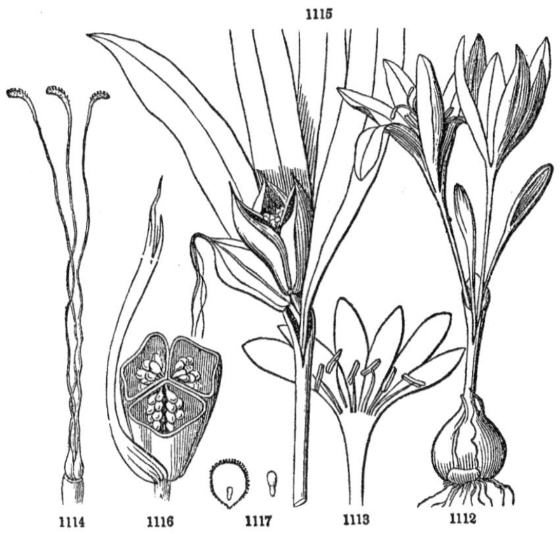

920. **Subord. Uvularieæ** (*the Bellwort Family*) has a few-seeded loculicidal capsule or berry, more or less united styles, and a deciduous perianth; the stems from rootstocks. — *Ex.* Uvularia.

921. **Ord. Juncaceæ** (*the Rush Family*). Herbaceous, mostly grass-like plants, often leafless: the small glumaceous flowers in

FIG. 1112. Colchicum autumnale; a flowering plant. 1113. Perianth laid open. 1114. Pistil, with the long distinct styles. 1115. Leafy stem and fruit (capsule opening by septicidal dehiscence). 1116. Capsule divided transversely. 1117. Section of a seed, and a separate embryo.

clusters, cymes, or heads. Perianth mostly dry, greenish or brownish, of six leaves (sepals and petals) in two series. Stamens six, or three. Ovary free, three-celled, or one-celled from the placentæ not reaching the axis; their styles united into one: stigmas three. Capsule three-valved, few- or many-seeded. Albumen fleshy. — *Ex.* Juncus (Rush).

922. **Ord. Commelynaceæ** (*the Spiderwort Family*), with usually sheathing leaves; distinguished from other Endogens (except Alismaceæ and Trillium) by the manifest distinction between the calyx and corolla; the former of three herbaceous sepals; the latter of as many delicate colored petals. Stamens six, or fewer: anthers with two separated cells: filaments often clothed with jointed hairs, hypogynous. Ovary two- or three-celled; the styles united into one. Capsule few-seeded, loculicidal. Seeds orthotropous. Embryo small, pulley-shaped, partly sunk in the apex of the albumen. — *Ex.* Commelyna, Tradescantia (Spiderwort). Mucilaginous plants.

923. **Ord. Xyridaceæ.** Swampy, rush-like plants; with ensiform, grassy or filiform radical leaves, sheathing the base of a simple scape, which bears a head of flowers at the apex, imbricated with bracts. Calyx of three glumaceous sepals, caducous. Petals three, with claws, more or less united into a monopetalous tube. Stamens six, inserted on the corolla; three of them bearing extrorse anthers, the others mere sterile filaments. Ovary one-celled, with three parietal placentæ, or three-celled: styles partly united: stigmas lobed. Capsule many-seeded. Seeds orthotropous, albuminous. — *Ex.* Xyris (Yellow-eyed Grass).

924. **Ord. Eriocaulonaceæ** (*the Pipewort Family*). Swampy or aquatic herbs, with much the aspect and structure of the preceding; their leaves cellular or fleshy; their minute flowers (monœcious or diœcious) crowded, along with scales or hairs, into a very compact head: the corolla less petaloid than in Xyridaceæ; the six stamens often all perfect; the ovules and seeds solitary in each cell. — *Ex.* Eriocaulon.

925. **Ord. Cyperaceæ** (*the Sedge Family*). Stems (*culms*) usually solid, cæspitose. Sheaths of the leaves closed. Flowers one in the axil of each glumaceous bract. Perianth none, or of a few bristles. Stamens mostly three, hypogynous. Styles two or three, more or less united. Fruit an achenium. Embryo small, at the extremity of the seed next the hilum. — *Ex.* Cyperus, Scirpus,

Carex. Sedge-Grasses. — The *papyrus* of the Egyptians was made from the stems of Cyperus Papyrus.

926. **Ord. Gramineæ** (*the Grass Family*). Stems (*culms*) cylindrical, mostly hollow, and closed at the nodes. Sheaths of the leaves split or open. Flowers in little spikelets, consisting of two-ranked imbricated bracts; of which the exterior are called *glumes*, and the two that immediately inclose each flower, *paleæ*. Perianth none, or in the form of very small and membranous hypogynous scales, from one to three in number, distinct or united (termed *squamulæ*, *squamellæ*, or *lodiculæ*). Stamens commonly three: anthers versatile. Styles or stigmas two; the latter feathery.

FIG. 1118. Scirpus triqueter, with its cluster of spikelets. 1119. A separate flower, enlarged, showing its rudimentary perianth of a few denticulate bristles, its three stamens, and pistil with a three-cleft style: *a*, section of the seed, showing the minute embryo. 1120. Carex Careyana, reduced in size (flowers monœcious, the two kinds in different spikes). 1121. Stem, with the staminate and upper pistillate spike, of the size of nature. 1122. A scale of the staminate spike, with the flower (consisting merely of three stamens) in its axil. 1123. Magnified pistillate flower, with its scale or bract: the ovary inclosed in a kind of sac (*perigynium*), formed by the union of two bractlets. 1124. Cross-section of the perigynium; with the pistil, *p*, removed. 1125. Vertical section of the achenium, showing the seed.

Fruit a caryopsis (607). Embryo situated on the outside of the farinaceous albumen, next the hilum. — *Ex.* Agrostis, Phleum, Poa, Festuca, which are the principal meadow and pasture grasses: Oryza (Rice), Zea (Maize), Milium (Millet), Avena (the Oat), Triticum (Wheat), Secale (Rye), Hordeum (Barley), are the chief cereal plants, cultivated for their farinaceous seeds. This universally diffused order, one of the largest of the vegetable kingdom, is doubtless the most important; the floury albumen of the seeds, and the nutritious herbage, constituting the chief support of man and the herbivorous animals. No unwholesome properties are known in the family, except in the seeds of Lolium temulentum, which are deleterious. The *Ergot*, or Spurred Rye, forms no real exception to this rule, as it is caused by parasitic fungus. — The

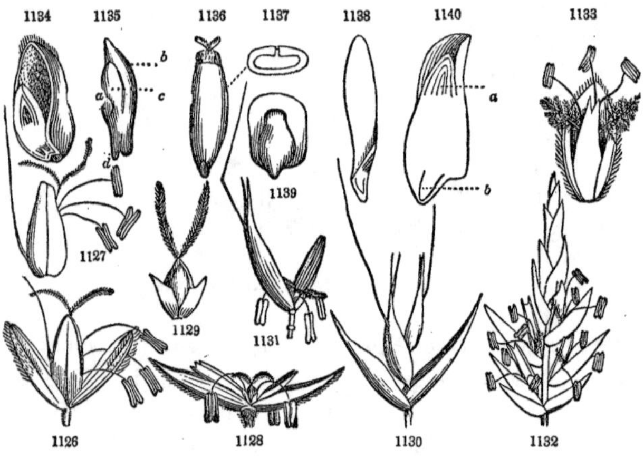

stems of grasses frequently contain sugar in considerable quantity;

FIG. 1126. One-flowered spikelet or locusta of Alopecurus, with the glumes separated. 1127. Same, with the glumes removed: an *awn* on the back of the outer palea. 1128. One-flowered spikelet of an Agrostis. 1129. Pistil of a Grass, showing the two feathery stigmas, and the two hypogynous scales or squamulæ (representing the perianth). 1130. Two-flowered spikelet of an Avena; with the glumes spreading. 1131. One of the flowers with its paleæ; the exterior pointed, with two bristles or cusps at the apex, and with a bent awn on the back. 1132. Many flowered spikelet of Glyceria fluitans. 1133. An enlarged separate flower of the same, seen from within, showing the inner palea, &c. 1134. The fruit (caryopsis) of the Wheat, with an oblique section through the integuments of the embryo, which is exterior to the albumen. 1135. Detached magnified embryo: *a*, the imperfect lower cotyledon; *b*, the large cotyledon; *c*, the plumule; *d*, the radicle. 1136. The caryopsis of Hordeum (Barley). 1137. A cross-section. 1138. A vertical section, showing the external embryo at the base. 1139. Magnified detached embryo, with its broad cotyledon and the plumule. 1140. More magnified vertical section of the same: *a*, the plumule; *b*, the radicle.

especially in the few instances where it is solid, as in the Maize, and more largely in the Sugar-Cane (Saccharum officinarum), which affords the principal supply of this article.

Series II. Cryptogamous or Flowerless Plants.

Plants destitute of proper flowers (stamens and pistils), and propagated by spores instead of seeds (101, 109).

Class III. Acrogenous Plants.

Vegetables with a distinct axis, growing from the apex; with no provision for subsequent increase in diameter (containing woody and vascular tissue), and usually with distinct foliage (108).

927. **Ord. Equisetaceæ** (*the Horse-tail Family*). Leafless plants; with striated, jointed, simple or branched stems (containing ducts and some spiral vessels), which are hollow and closed at the joints; each joint terminating in a toothed sheath, which surrounds the base of the one above it. Inflorescence consisting of peltate scales crowded in a terminal spike, or kind of strobile: each with several *thecæ* attached to its lower surface, longitudinally dehiscent. Spores numerous, with four elastic club-shaped bodies (of unknown use, called *elaters*) wrapped around them. — *Ex.* Equisetum. The epidermis of Equisetum hyemale (Scouring Rush) contains so much silex that it is used for polishing.

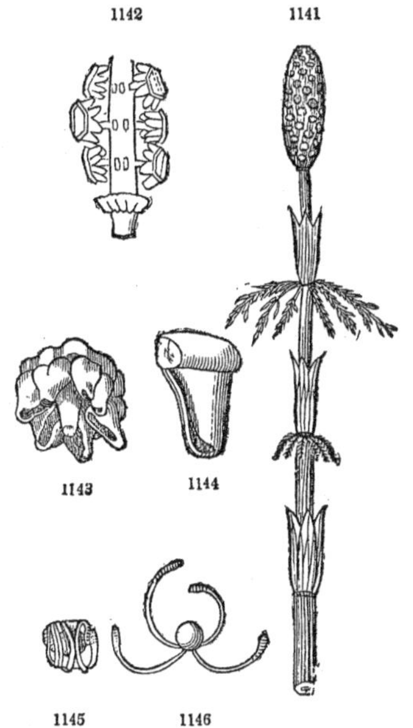

928. **Ord. Filices** (*Ferns*). Leafy plants; with the leaves (*fronds*)

FIG. 1141. Summit of the stem of Equisetum sylvaticum. 1142. Part of the axis of the cone of fructification, with some of the fruit-bearing organs, shown magnified in Fig. 1143. 1144. A separate *theca*, more magnified. 1145, 1146. Spores with elaters, still more magnified.

spirally rolled up or circinate in vernation (except in one suborder), usually rising from prostrate or subterranean rootstocks, sometimes from an erect arborescent trunk (Fig. 94), and bearing, on the veins of their lower surface, or along the margins, the simple fructification, which consists of one-celled spore-cases (*thecæ* or *sporangia*), opening in various ways, and discharging the numerous minute spores. The stalk or petiole of the frond is termed a *stipe*. — There are three principal suborders, viz.: —

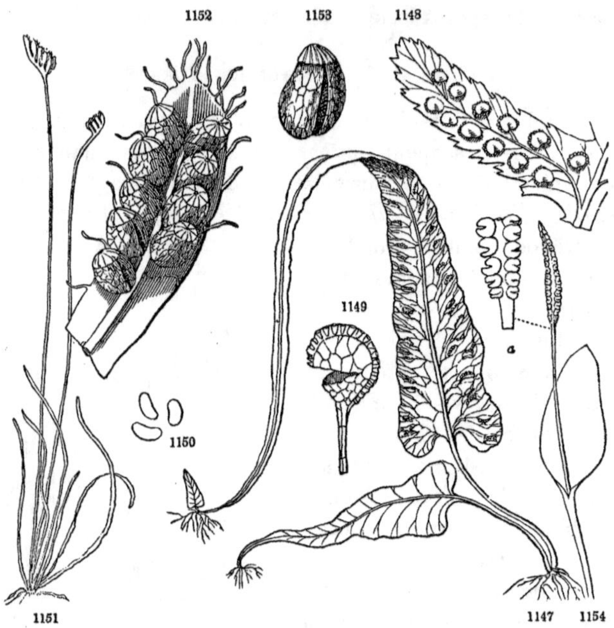

929. Subord. Polypodineæ. Sporangia collected in dots, lines, or variously shaped clusters (*sori* or *fruit-dots*) on the back or mar-

FIG. 1147. Asplenium (Camptosoros) rhizophyllum (Walking Fern); the fronds rooting, as they frequently do, at the apex; the sori occupying the reticulated veins on the back. 1148. Division (pinnula) of a frond of Aspidium (Nephrodium) Goldianum; the roundish sori attached to the simple veins, and covered with an indusium, which is fastened in the centre, and opens all around the margin. 1149. Magnified sporangium of this division of Ferns, with its stalk, and elastic ring partly surrounding it; which, tending to straighten itself when dry, tears open the sporangium, shedding the minute spores (1150). 1151. Schizæa pusilla of about the natural size, with simple and slender radical leaves; the contracted fertile frond pinnate. 1152. A division (pinna) of the fertile frond, magnified, showing the sessile sporangia occupying its lower surface. 1153. One of the sporangia more magnified; they have no proper ring, and open by a longitudinal cleft. 1154. Ophioglossum vulgatum (Adder-tongue); the sporangia forming a two-ranked spike on a transformed and contracted frond: *a*, portion of the spike enlarged, showing the coriaceous sporangia, destitute of a ring, and opening transversely.

gins of the frond or its divisions, stalked, cellular-reticulated, the stalk running into a vertical incomplete ring, which by straightening at maturity ruptures the sporangium transversely on the inner side, discharging the spores. Fruit-dots often covered, at least when young, by a membrane called the *involucre*, or *indusium*.

930. **Subord. Osmundineæ.** Sporangia variously collected, destitute of any proper ring, cellular-reticulated, opening lengthwise by a regular slit.

931. **Subord. Ophioglosseæ.** Sporangia spiked, closely sessile, naked, coriaceous and opaque, not reticulated, destitute of a ring, opening by a transverse slit into two valves, discharging the very copious spores which appear like floury dust. Fronds straight, never rolled up (circinate) in the bud!

932. **Ord. Lycopodiaceæ** (*the Club-Moss Family*). Plants with creeping or erect leafy stems, mostly branching; the crowded leaves lanceolate or subulate, one-nerved. Thecæ sessile in the axils of the leaves, sometimes all collected at the summit under leaves which are changed into bracts and crowded into a kind of ament, one-celled, or rarely two- to three-celled, dehiscent, containing either minute grains, appearing like fine powder, or a few rather large sporules; both kinds often found in the same plant. — *Ex.* Lycopodium (Club-Moss, Ground Pine, Fig. 89 – 93), Psilotum. — Appended to this family, rather than to the next, is the

933. **Subord. Isoetineæ** (*the Quillwort Family*), consisting of a few acaulescent submersed aquatics, with sporangia in the axils and immersed in the inflated base of the grassy stalk-like leaves. — *Ex.* Isoetes.

934. **Ord. Hydropterides.** Aquatic cryptogamous plants of diverse habit, with the fructification borne at the bases of the leaves, or on submerged branches, consisting of two sorts of organs, contained in indehiscent or irregularly bursting involucres (*sporocarps*): — comprising the

935. **Subord. Marsileæ** (*the Pepperwort Family*); with creeping stems; the leaves long-stalked, circinate in vernation; — of four obcordate leaflets in Marsilea, or filiform and destitute of leaflets in Pilularia (the Pillwort).

936. **Subord. Salviniеæ**; which are free floating plants, with alternate and sometimes imbricated sessile leaves; the fructification borne on the stem or branches underneath. — *Ex.* Salvinia, Azolla.

Class IV. Anophytes.

Vegetables composed of parenchyma alone, with acrogenous growth, usually with distinct foliage, sometimes the stem and foliage confluent into a frond (105, Fig. 87, 88).

937. **Ord. Musci** (*Mosses*). Low, tufted plants, always with a stem and distinct (sessile) leaves, producing spore-cases which mostly open by a terminal lid, and contain simple spores alone. Reproductive organs of two kinds: — 1. The sterile flower, consisting of numerous (4 – 20) minute cylindrical sacs (*antheridia*) which discharge from their apex a mucous fluid filled with oval particles, and then perish. 2. The fertile flower, composed of numerous (4 – 20) flask-like bodies (*pistillidia*), each having a membranous covering (*calyptra*), terminated by a long cylindrical funnel-mouthed tube (*style*). The ripened pistillidium (sel-

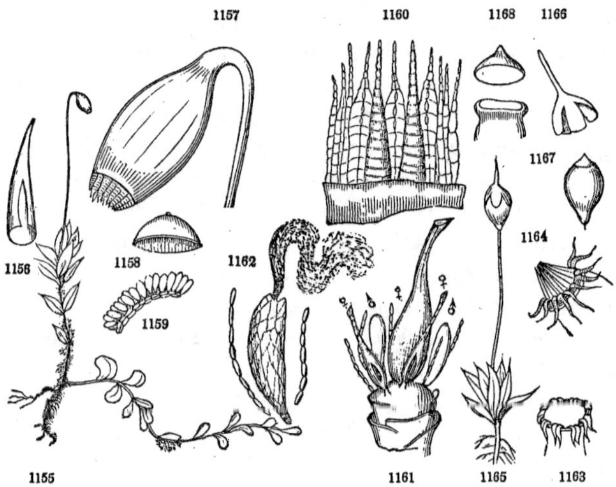

FIG. 1155. Mnium cuspidatum. 1156. The calyptra detached from the theca. 1157. Magnified theca, from which the lid or operculum, 1158, has been removed, showing the peristome. 1159. A portion of the annulus, magnified. 1160. A portion of the outer and inner peristome, highly magnified. 1161. The so-called flowers in a young state, consisting of the young thecæ ♀, and the antheridia ♂, with some cellular jointed threads intermixed; the involucral leaves cut away. 1162. One of the antheridia more magnified (with the accompanying cellular threads), opening at the apex, and discharging the fovilla. 1163. Simple peristome of Splachnum; the teeth united in pairs. 1164. Double peristome of Hypnum; the exterior spreading. 1165. Physcomitrium (Gymnostomum) pyriforme. 1166. Its calyptra, detached from 1167, the theca. 1168. The lid removed from the orifice, which is destitute of a peristome.

dom more than one in a flower maturing) becomes the *capsule*, which is rarely indehiscent or splitting by four longitudinal slits, but usually opens by a *lid* (*operculum*): beneath the lid and arising from the mouth of the capsule are commonly either one or two rows of rigid processes (collectively the *peristome*), which are always some multiple of four: those of the outer row are called *teeth*, of the inner, *cilia*. An elastic ring of cells (*annulus*) lies between the rim of the capsule and operculum. The powdery particles filling the capsule are *spores*. The thread-like stalk (*pedicel*) supporting the capsule is inserted into the elongated torus (*vaginula*) of the flower. The pedicel continued through the capsule forms the *columella*: enlarged under the capsule it sometimes forms an *apophysis*. The calyptra separating early at its base is carried up on the apex of the capsule; if it splits on one side, it is hood-shaped or *cuculliform*, if not, it is mitre-shaped or *mitriform*. Intermixed with the reproductive organs are jointed filaments (*paraphyses*). The leaves next the antheridia are called *perigonial* leaves, those around the pistillidia or pedicel the *perichætial* leaves.

938. **Ord. Hepaticæ** (*Liverworts*). Frondose or Moss-like plants, of a loose cellular texture, usually procumbent and emitting rootlets from beneath; the calyptra not separating from the base, but usually rupturing at the apex; the capsule not opening by a lid, containing spores usually mixed with elaters (which are thin, thread-like cells, containing one or two spiral fibres, uncoiling elastically at maturity). Vegetation sometimes *frondose*, i. e. the stem and leaves confluent into an expanded leaf-like mass; sometimes *foliaceous*, when the leaves are distinct from the stem, as in true Mosses, entire or cleft, two-ranked, and often with an imperfect or rudimentary row (*amphigastria*) on the under side of the stem. Reproductive organs of two kinds, viz. *antheridia* and *pistillidia*, much as in Mosses (937), variously situated. The matured pistillidium forms the *capsule*, which is either sessile or borne on a long cellular pedicel, and dehiscent by irregular openings, by teeth at its apex, or lengthwise by two or four valves. A *columella* is rarely present. The *perianth* is a tubular organ inclosing the *calyptra*, which directly includes the pistillidium. Surrounding the perianth are *involucral* leaves of particular forms. The antheridia in the foliaceous species are situated in the axils of perigonial leaves.

939. **Subord. Ricciaceæ** are chiefly floating plants, rooting from

43

beneath, with the fructification immersed in the frond, the sporangium bursting irregularly. No involucre nor elaters. — *Ex.* Riccia.

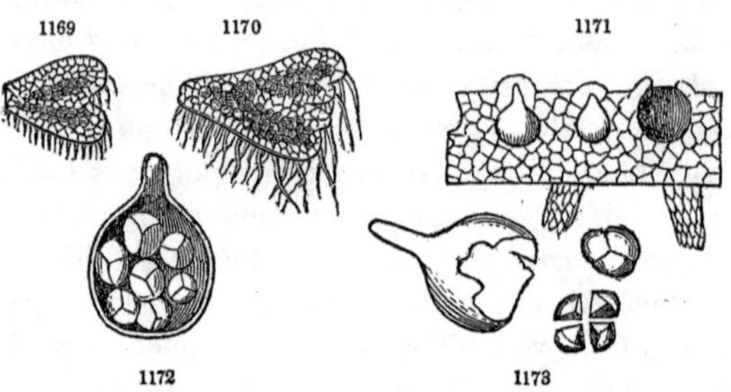

940. **Subord. Anthoceroteæ.** Terrestrial frondose annuals, with the fruit protruded from the upper surface of the frond. Perianth none. Capsule pod-like, one- to two-valved, with a free central columella. Elaters none or imperfect.

941. **Subord. Marchantiaceæ** (*true Liverworts*). Frondose and terrestrial perennials, growing in wet places, with the fertile receptacle raised on a peduncle, capitate or radiate, bearing pendent calyptrate capsules from the under side, which open variously, not four-valved. Elaters with two spiral fibres.

942. **Subord. Jungermanniaceæ.** Frondose or mostly foliaceous plants; with the sporangium dehiscent into four valves, and the spores mixed with elaters (Fig. 84 – 86).

CLASS V. THALLOPHYTES.

Vegetables composed of parenchyma alone, of congeries of cells, or even of separate cells, often vaguely combined in a *thallus*, never exhibiting a marked distinction into root, stem, and foliage, or into axis and leaves (94 – 104, 106). Fructification of the most simple kinds. (Spores often termed *sporules* or *sporidia.*)

943. **Ord. Lichenes** (*Lichens*) form the highest grade of this lower series. They consist of flat expansions, which are rather crusta-

FIG. 1169, 1170. Riccia natans, about the natural size. 1171. Magnified section through the thickness of the frond, showing the immersed sporangia; one of which has burst through and left an effete cavity. 1172. Magnified vertical section of one of the sporangia, with the contained spores. 1173. Sporangium torn away from the base, and a quaternary group of spores, united and separated.

ceous than foliaceous. Their structure is, as it were, anticipated in Riccia, above mentioned (Fig. 1170). They are by no means aquatic, however, but grow on the ground, on the bark of trees, or on the surface of exposed rocks, to which they cling by their lower surface, often with the greatest tenacity, while by the upper they draw their nourishment directly from the air (Fig. 1174). The fructification is in *cups*, or *shields* (*apothecia*, Fig. 1176), resting on the surface of the thallus, or more or less immersed in its substance (Fig. 1178), or else in pulverulent spots scattered over the surface. A magnified section through an apothecium (Fig. 1176) brings to view a stratum of elongated sacs (*asci*), with filaments

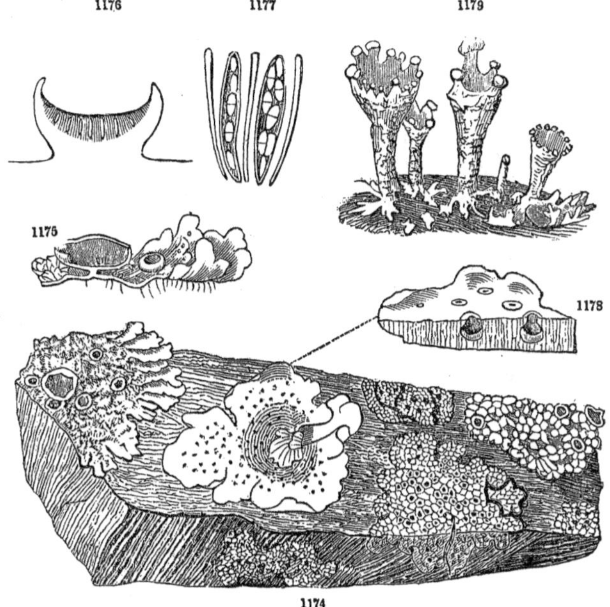

intermixed, as seen detached and highly magnified at Fig. 1177.

FIG. 1174. A stone upon which several Lichens are growing, such as (passing from left to right) Parmelia conspersa, Sticta miniata, Lecidea geographica (so called from its patches resembling the outline of islands, &c. on maps), &c., &c. 1175. Piece of the thallus of Parmelia conspersa, with a section through an apothecium. 1176. Section of a smaller apothecium, more magnified. 1177. Two asci and their contained spores, with the accompanying filaments, highly magnified. 1178. Section of a piece of the thallus of Sticta miniata, showing the immersed apothecia. 1179. Cladonia coccinea, bearing its fructification in rounded red masses on the edges of a raised cup.

Each *ascus*, or sac, contains a few spores, which divide into two, but generally remain coherent. The vegetation of some Lichens rises into a kind of axis, as in the Cladonia coccinea, which abounds on old logs (Fig. 1179); or in Cladonia rangiferina, the Reindeer Moss; also in Usnea, where it forms long, gray tufts, hanging from the boughs of old trees in our Northern forests.

944. **Ord. Fungi** (*Mushrooms*, *Moulds*, *&c.*) are parasitic (137) Flowerless plants, either in a strict sense, as living upon and drawing their nourishment from living, though more commonly languishing, plants and animals, or else as appropriating the organized matter of dead and decaying animal and vegetable bodies. Hence they fulfil an office in the economy of creation analogous to that of the infusory animalcules. Those Fungi which produce Rust, Smut, Mildew, &c. are of the first kind; those which produce Dry-rot, &c. hold a somewhat intermediate place; and Mushrooms, Puff-balls, &c. are examples of the second. Fungi are consequently not only destitute of any thing like foliage, but also of the green matter, or chlorophyll, which appears to be essential to the formation of organic out of inorganic matter (87, 135, 344). A full account of the diversified modifications of structure that Fungi display, and of the remarkable points in their economy, would require a volume. We will notice three sorts only, which may represent the highest, and nearly the lowest, forms of this vast order or class of plants. They all begin (in germination or by offsets) with the production of copious filamentous threads, or series of attenuated cells, appearing like the roots of the fungus that arises from them (Fig. 1179, 1181), and to a certain extent performing the functions of roots: this is called the *mycelium*, and is the true vegetation of Fungi. The subsequent developments properly belong to the fructification, or are analogous to tubers, rhizomas, &c. In one part of the order, the masses that arise, of various definite shapes, and often attaining a large size, contain in their interior a multitude of *asci* (Fig. 1180), inclosing simple or double sporules, just as in Lichens. The esculent Morel has this kind of fructification; as well as the less conspicuous Sphæria (Fig. 1179), which is in other respects of a lower grade. The Agarics, like the Edible Mushroom (Fig. 1181), present a different type. Rounded tubercles appear on the mycelium; some of these rapidly enlarge, burst an outer covering which is left at the base (the *volva*, or *wrapper*), and protrude a thick stalk (*stipes*),

bearing at its summit a rounded body that soon expands into the *pileus*, or *cap*. The *lamellæ*, or *gills* (*hymenium*), that occupy its

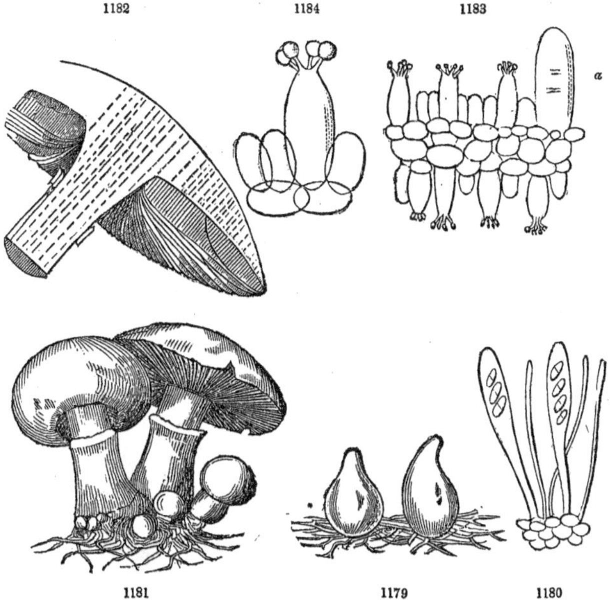

lower surface, consist of parallel platès (Fig. 1182), which bear naked sporules over their whole surface. A careful inspection with the microscope shows that these sporules are grouped in fours; and a view of a section of one of the gills shows their true origin (Fig. 1183). Certain of the cells (*basidia*), one of which is shown more magnified at Fig. 1184, produce four small cells at their free summit, apparently by gemmation and constriction: these are the *sporules*. It is maintained that the larger intermingled cells, (of which one is shown at Fig. 1183, *a*,) filled with an attenuated form of matter, are the analogues of stamens. The lowest Fungi produce from their mycelium only simple or branching series of cells (Fig. 74 – 76). The mycelium itself either ramifies through decaying organized matter, as the Moulds,

FIG. 1179. Sphæria rosella. 1180. Asci from its interior, containing sporules, highly magnified. 1181. Agaricus campestris, the Edible Mushroom, in its various stages. 1182. Section through the pileus, to display the gills. 1183. A small piece of a slice through the thickness of one of the gills, magnified; showing the spores borne on the summit of salient cells of both surfaces. 1184. One of the sporule-bearing cells, with some subjacent tissue, more magnified.

&c.; or else, like the Blight and Rust in grain, and the *Muscardine* so destructive to silkworms, it attacks and spreads throughout living tissues, often producing great havoc before its fructification is revealed at the surface. Sometimes the last cells of the stalks swell into a vesicle, in which the minute sporules are formed; as in Fig. 74. Sometimes the branching stalks bear single sporules, like a bunch of grapes (Fig. 76), or long series of cells, or sporules, in rows, like the beads of a necklace (Fig. 75), which, falling in pieces, are the rudiments of new plants.

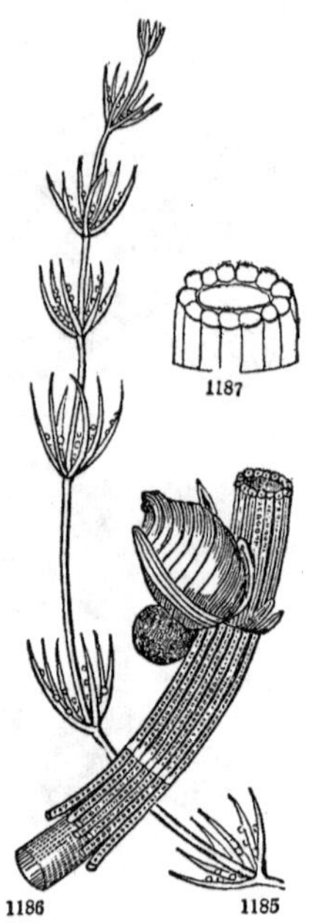

945. **Ord. Characeæ.** *The Chara Family* consists of a few aquatic plants, which have all the simplicity of the lower Algæ in their cellular structure, being composed of simple tubular cells placed end to end, and often with a set of smaller tubes applied to the surface of the main one (Fig. 1186). Hence they have been placed among Algæ. But their fructification is of a higher order. It consists of two kinds of bodies (both shown in Fig. 1186), of which the smaller (and lower) is probably a mass of antheridia of curious structure, while the upper and larger is a sporocarp, formed of a budding cluster of leaves wrapped around a nucleus, which is a spore or sporangium. The order should have been introduced between the Equisetaceæ (to which the verticillatè branches show some analogy) and the Hydropterides, which they somewhat resemble in fructification. They are, of all plants, those in which the rotary movement of the contents of the cells (36, which has been called *Cyclosis*) may be most readily observed.

946. **Ord. Algæ** (*Seaweeds*). This vast order, or rather class, consists of aquatic plants; for the most part strictly so, but some grow in humid terrestrial situations. The highest forms are the proper Seaweeds (*Wrack*, *Tang*, *Dulse*, *Tangle*, &c.); "some of which have stems exceeding in length (although not in diameter)

FIG. 1185. Branch of the common Chara, nearly the natural size. 1186. A portion magnified, showing the lateral tubes inclosing a large central one (a portion more magnified at 1187); also a spore, invested by a set of tubes twisted spirally around it; and with an antheridium borne at its base.

the trunks of the tallest forest-trees, while others have leaves (*fronds*) which rival in expansion those of the Palm." "Others again are so minute as to be wholly invisible, except in masses, to the naked eye, and require the highest powers of our microscopes to ascertain their form and structure." Some have the distinction of stems and fronds; others show simple or branching solid stems only; and others flat foliaceous expansions alone (Fig. 82), either green, olive, or rose-red in hue. From these we descend by successive gradations to simple or branching series of cells placed end to end, such as the green Confervas of our pools, and many marine forms (Fig. 81): we meet with congeries of such cells capable of spontaneous disarticulation, each joint of which becomes a new plant, so that the organs of vegetation and of fructification become at length perfectly identical, both reduced to mere cells; and finally, as the last and lowest term of possible vegetation, we have the plant reduced to a single cell, giving rise to new ones in its interior, each of which becomes an independent plant (94 – 99).

947. The fructification of Algæ exhibits four principal varieties. In the great division of olive-brown or olive-green proper Seaweeds, the MELANOSPERMEÆ of Harvey, the fructification forms tubercles immersed in the tissue of the summit of the branches of the frond (Fig. 1188 – 1191), which are filled with a mass of simple spores with filaments intermixed (1191), invested by a proper membranous coat, and finally escaping from the frond by a minute orifice. The beautiful red-colored Seaweeds, or RHODOSPERMEÆ, exhibit two kinds of spores; one large, simple, superficial, and resembling those above described, except that they have no proper integument; the others, dispersed through the interior of the frond, are formed four together in a mother cell. The bright green series, or CHLOROSPERMEÆ, have the whole green contents of certain cells, or of some part of the cell, (as in Vaucheria, Fig. 71, 72, 467, and in Conferva vesicata, Fig. 474, &c.,) condensed into a spore, in some of the ways already described (95 – 101), or else they result from the conjugation of two cells (102, Fig. 78 – 81). This conjugation occurs throughout in the

948. **Subord. Desmidieæ**, which are microscopic and infusory green Algæ of single cells (Fig. 77 – 80) often of crystal-like forms, invested with mucus, and belonging to fresh water. They *multiply* largely by division, but *propagate* only by conjugation. Many of them have long been claimed for the animal kingdom, or esteemed

of ambiguous nature, on account of the free movements they exhibit (661); but these are nearly as well marked in Oscillaria, &c. (Fig. 66). More ambiguous still, and on the lowest confines of the vegetable kingdom, are those minute vegetables, as they doubtless are, which constitute the

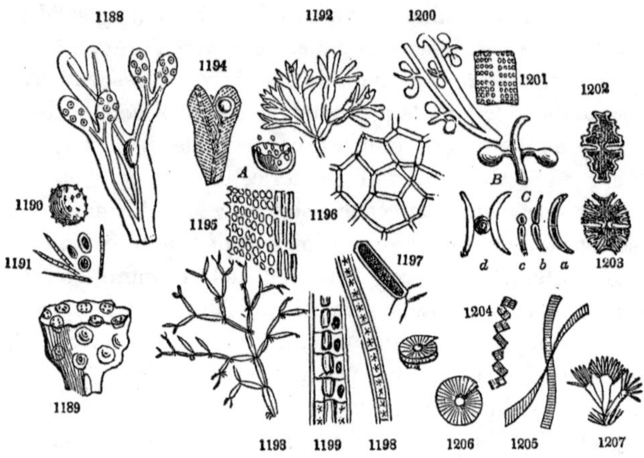

949. **Subord. Diatomaceæ.** These essentially differ from the last in the brown instead of green color of their contents, in the siliceous and durable nature of their cell-wall, and in being natives of salt instead of fresh water. Their movements, as they break up from their connections, are still more vivid and varied. Some are fixed (Fig. 1207); others are free. Some are extremely minute; others consist of clusters of cells of considerable size.

FIG. 1188. Summit of the frond of Fucus vesiculosus. 1189. Section of one of the receptacles. 1190. One of the contained globules. 1191. Spores and jointed filaments of which the globules are composed. 1192. Delesseria Le Prieurii. 1193. The sterile plant. 1194. Magnified portion of the fertile frond. 1195. Portion of the same, more magnified, showing its tissue from the midrib to the margin. *A*. Theca, opened, with the spores. 1196. Portion of the network of Hydrodyction utriculatum. 1197. A magnified joint, filled with the green matter which developes into a new plant. 1198. Single filament of Tyndaridea cruciata, showing the star-shaped bodies, enveloped in mucus. 1199. Two filaments of the same united side by side. 1200. Vaucheria geminata, in fruit. *B*. Vesicular receptacles, enlarged. — The remaining figures represent some of the ambiguous Diatomaceæ. 1201. Gonium glaucum, of Ehrenberg, who thinks it a congeries of animalcules; while Meÿen has described it as an Alga. *C*. Closterium Lunula; usually filled with floating green globules: *a*, the perfect vegetable; *b*, the same, separating into two by spontaneous division; *c*, an individual resulting from this spontaneous division, developing a second; *d*, two individuals conjugately united; the green matter all collected in the uniting globule. 1202. Euastrum Pecten, and 1203, E. Crux-Melitensis. 1204. A Diatoma, breaking up into separate individuals. 1205. A Fragillaria. 1206. Meridion circulare, front and side views. 1207. Echinella flabellata; perhaps a group of animalcules.

APPENDIX.

Of the Signs and Abbreviations employed in Botanical Writings.

Linnæus adopted the following signs for designating the duration of a plant, viz.: —

⚇ An annual plant.
♂ A biennial plant.
♃ A perennial herb.
♄ A shrub or tree.

Among the signs recently introduced, the following have come into general use: —

☉ A monocarpic plant, whether annual or biennial.
① An annual plant.
② A biennial plant.
♃ A perennial herb.
♄ A plant with a woody stem.
♂ A staminate flower, or plant.
♀ A pistillate flower, or plant.
☿ A perfect flower, or a plant bearing perfect flowers.
! The exclamation point is employed as the counterpart of the note of interrogation. When it follows the name of an author appended to the name of a plant, it imports that an authentic specimen of the plant in question, under this name, has been examined by the writer: when it is appended to a locality, it signifies that the writer has seen or collected specimens of the plant from that locality, &c.
? The note of interrogation is similarly employed in case of doubt or uncertainty; and is affixed either to a generic or specific name, or to that of an author or locality cited.
* As used by De Candolle, indicates that a good description is found at the reference to which it is appended. It is not in common use.

Those abbreviations of the names of organs which are commonly employed, such as *Cal.* for calyx, *Cor.* for corolla, *Fl.* for flower, *Fr.* for fruit, *Gen.* for genus, *Hab.* for habitat, *Herb.* for herbarium, *Hort.* for garden, *Mus.* for Museum, *Ord.* for order, *Rad.* (Radix) for root, *Syn.* for synonymy, *Sp.* or *Spec.* for species, *Var.* for variety, &c., scarcely require explanation.

V. sp. denotes, in general terms, that the writer has seen the plant under consideration.

V. s. c. (*Vidi siccam cultam*), that a dried specimen of a cultivated plant has been examined.

V. s. s. (*Vidi siccam spontaneam*), that a dried specimen of the wild plant has been examined.

V. v. c. (*Vidi vivam cultam*), that the living cultivated plant has been under examination.

V. v. s. (*Vidi vivam spontaneam*), that the wild plant has been examined in a living state.

The names of authors, when of more than one syllable, are commonly abridged by writing the first syllable, and the first letter or the first consonant of the second. Thus, *Linn.*, or *L.*, is the customary abbreviation for Linnæus; *Juss.* for Jussieu; *Willd.* for Willdenow; *Muhl.* for Muhlenberg; *Michx.* for Michaux; *Rich.* for Richard; *De Cand.*, or *DC.*, for De Candolle; *Hook.* for Hooker; *Endl.* for Endlicher; *Lindl.* for Lindley, &c.

Of Collecting and Preserving Plants.

1. The botanist's collection of specimens of plants, preserved by drying under pressure between folds of paper, is termed a *Hortus Siccus*, or commonly an *Herbarium*.

2. A complete specimen consists of one or more shoots, bearing the leaves, flowers, and fruit; and, in case of herbaceous plants, a portion of the root is also desirable.

3. Fruits and seeds which are too large to accompany the dried specimens, or which would be injured by compression with sections of wood, &c., should be separately preserved in cabinets.

4. Specimens for the herbarium should be gathered, if possible, in a dry day; and carried either in a close tin box, as is the common practice, or in a strong portfolio, containing a quire or more of firm paper, with a few loose sheets of blotting-paper to receive delicate plants. They are to be dried, under strong pressure, (but without crushing the parts,) between dryers composed of six to ten thicknesses of bibulous paper; which should be changed daily, or even more frequently, until all the moisture is extracted from the plants; — a period which varies in different species, and with the season, from two or three days to a week. All delicate speci-

mens should be laid in folded sheets of thin and smooth bibulous paper (such as tea-paper), and such sheets, filled with the freshly gathered specimens, are to be placed between the dryers, and so transferred entire, day after day, into new dryers, without being disturbed, until perfectly dry. This preserves all delicate flowers better than the ordinary mode of shifting of the papers which are in immediate contact with the specimens, and also saves much time usually lost in transferring numerous small specimens, one by one, into dry paper, often to the great injury of the delicate corolla, &c.

5. The dried specimens, properly ticketed with the name, locality, &c., and arranged under their respective genera and orders, are preserved in the herbarium, either in separate double sheets, or with each species attached by glue or otherwise to a half-sheet of strong white paper, with the name written on one corner. These are collected in folios, or else lie flat (as is the best mode) in parcels of convenient size, received into compartments of a cabinet, with close doors, and kept in a perfectly dry place.

6. The seeds of plants intended for cultivation, which are to be transported to a distance before being committed to the earth, should first be dried in the sun, wrapped in coarse paper, and preserved in a dry state. They should not be packed in close boxes, at least so long as there is danger of the retention of moisture.

7. Roots, shrubs, &c., designed for cultivation, should be taken from the ground at the close of their annual vegetation, or early in the spring before growth recommences, and packed in successive layers of slightly damp (but not wet) Peat-moss (Sphagnum). Succulent plants, however, such as Cacti, may be packed in dry sand.

8. Plants in a growing state can only be safely transported to a considerable distance, especially by sea, in the closely glazed cases invented by Mr. Ward; * where they are provided with the requisite moisture, while they are fully exposed to the light.

* On the Growth of Plants in Closely Glazed Cases, by N. B. Ward, F. L. S., London, 1842.—Ed. 2. 1853.

INDEX

AND GENERAL GLOSSARY OF BOTANICAL TERMS.

THE END.

www.ingramcontent.com/pod-product-compliance
Lightning Source LLC
LaVergne TN
LVHW021305110826
845150LV00003B/484

* 9 7 8 1 4 2 5 5 5 9 9 1 5 *